Mathematical Foundation of Computer Science

AF516806

MATHEMATICAL FOUNDATION OF COMPUTER SCIENCE

By

Dr. J. Rajendra Prasad
Professor and Head
Department of Computer Applications and Science & Humanities.
P.V.P. Siddhartha Institute of Technology, Kanuru
Vijayawada, Andhra Pradesh

T. Rama Rao
Senior Assistant Professor
Department of Mathematics,
P.V.P. Siddhartha Institute of Technology, Kanuru
Vijayawada, Andhra Pradesh

A. Madana Mohan Rao
Lecturer
V.R. Siddhartha Engg. College,
Kanuru Vijayawada,
Andhra Pradesh

UNIVERSITY SCIENCE PRESS
(An Imprint of Laxmi Publications Pvt. Ltd.)
An ISO 9001:2015 Company

BENGALURU • CHENNAI • GUWAHATI • HYDERABAD
KOCHI • KOLKATA • LUCKNOW • RANCHI
NEW DELHI

Copyright © by Laxmi Publications (P) Ltd.

All rights reserved including those of translation into other languages. In accordance with the Copyright (Amendment) Act, 2012, no part of this publication may be reproduced, stored in a retrieval system, translated into any other language or transmitted in any form or by any means, electronic, mechanical, photocopying, recording or otherwise. Any such act or scanning, uploading, and or electronic sharing of any part of this book without the permission of the publisher constitutes unlawful piracy and theft of the copyright holder's intellectual property. If you would like to use material from the book (other than for review purposes), prior written permission must be obtained from the publishers.

Printed and bound in India

First Edition : 2009; *Edition* : 2025

ISBN **:** 978-81-31805-63-3

Limits of Liability/Disclaimer of Warranty: The publisher and the author make no representation or warranties with respect to the accuracy or completeness of the contents of this work and specifically disclaim all warranties. The advice, strategies, and activities contained herein may not be suitable for every situation. In performing activities adult supervision must be sought. Likewise, common sense and care are essential to the conduct of any and all activities, whether described in this book or otherwise. Neither the publisher nor the author shall be liable or assumes any responsibility for any injuries or damages arising from here. The fact that an organization or Website if referred to in this work as a citation and/or a potential source of further information does not mean that the author or the publisher endorses the information the organization or Website may provide or recommendations it may make. Further, readers must be aware that the Internet Websites listed in this work may have changed or disappeared between when this work was written and when it is read.

All trademarks, logos or any other mark such as USP, Amanda, Golden Bells, Firewall Media, A-Cube, Triunity, Mathstick, Engpot, Laxmi appearing in this work are trademarks and intellectual property owned by or licensed to Laxmi Publications, its subsidiaries or affiliates. Notwithstanding this disclaimer, all other names and marks mentioned in this work are the trade names, trademarks or service marks of their respective owners.

Branches

✆	Bengaluru	080-26 75 69 30
✆	Chennai	044-24 34 47 26
✆	Guwahati	0361-254 36 69
✆	Hyderabad	040-27 55 53 83
✆	Kochi	0484-405 13 03
✆	Kolkata	033-40 04 77 79
✆	Lucknow	0522-430 36 13
✆	Ranchi	0651-224 24 64

Published in India by

UNIVERSITY SCIENCE PRESS

(An Imprint of Laxmi Publications Pvt. Ltd.)

An ISO 9001:2015 Company

113, GOLDEN HOUSE, GURUDWARA ROAD, DARYAGANJ,
NEW DELHI - 110002, INDIA

Corporate Office : B-33, Sector-5, Noida,
Gautambuddha Nagar (U.P.) - 201 301, INDIA

Telephone : +91-120-44 46 900

www.laxmipublications.com

info@laxmipublications.com

CONTENTS

Example. *Show that the set of transformations f_1, f_2, f_3, f_4, f_5 and f_6 on the s C is the set of complex numbers) defined by $f_1(z) = z, f_2(z) = \frac{1}{z}, f_3(z) = 1 - z,$ $\frac{1}{1-z}, f_6(z) = \frac{z-1}{z}$ forms a finite non-abelian group of order six w.r.t., comp the composition.*

Solution. If $f: A \to A$, then f is a transformation on A.

Let $$G = \{f_1, f_2, f_3, f_4, f_5, f_6\}.$$

Each f_i is a transformation since $f_i : A \to A$.

For any $z \in A, f_1(z) = z \Rightarrow f_1$ is the identity functioin.

$$\therefore \quad f_1 o f_1 = f_1, f_1 o f_2 = f_2 = o f_1, f_1 o f_3 = f_3 = f_3 o f_1$$

$$f_1 o f_4 = f_4 = f_4 o f_1, f_1 o f_5 = f_5 = f_5 o f_1, f_1 o f_6 = f_6$$

Now $$(f_2 o f_2)(z) = f_2(f_2(z)) = f_2\left(\frac{1}{z}\right) = z = f_1(z) \Rightarrow f_2 o f_2 = f$$

$$(f_2 o f_3)(z) = f_2(f_3(z)) = f_2(1 - z) = \frac{1}{1-z} = f_5(z) \Rightarrow f_2$$

$$(f_2 o f_4)(z) = f_2(f_4(z)) = f_2\left(\frac{z}{z-1}\right) = \frac{z-1}{z} = f_6(z) \Rightarrow f$$

$$(f_2 o f_5)(z) = f_2(f_5(z)) = f_2\left(\frac{1}{1-z}\right) = 1 - z = f_3(z) \Rightarrow f_2$$

$$(f_2 o f_6)(z) = f_2(f_6(z)) = f_2\left(\frac{z-1}{z}\right) = \frac{z}{z-1} = f_4(z) \Rightarrow f$$

Find remaining values, and construct composition table, which can

Table 4.11.

0	f_1	f_2	f_3	f_4	f_5
f_1	f_1	f_2	f_3	f_4	f_5
f_2	f_2	f_1	f_5	f_6	f_3
f_3	f_3	f_6	f_1	f_5	f_4
f_4	f_4	f_5	f_6	f_1	f_2
f_5	f_5	f_4	f_2	f_3	f_6
f_6	f_6	f_3	f_4	f_2	f_1

1. Since all the 36 entries in the table are in G, G is closed w.r.t. '0'
2. Clearly composition of functions is associative in G.
3. Since $f o f_1 = f \forall f \in G$, f_1 is the identity transformation in G.
4. From table, we can see that $f_1 o f_1 = f_1, f_2 o f_2 = f_1, f_3 o f_3 = f_1, f_4 o f_4$ $f_5 = f_1$.

$x \in$ A, then I $\in$ S

$\circ f)(x) = \mathrm{I}(f(x)) = f(x)$ and

$\circ \mathrm{I})(x) = f(\mathrm{I}(x)) = f(x)$.

I $o f = f = f o$ I $\Rightarrow$ I is the identity in S.

e : Let $f \in$ S $\Rightarrow f$: A $\to$ A is a bijection.

$^{-1}$: A $\to$ A is a bijection $\Rightarrow f^{-1} \in$ S.

$^{-1})(x) = f(f^{-1}(x)) = x = \mathrm{I}(x)$

$\circ f)(x) = f^{-1}(f(x)) = x = \mathrm{I}(x)$.

$f o f^{-1} = \mathrm{I} = f^{-1} o f \Rightarrow f^{-1}$ is the inverse of f in S.

v.r.t, the composition of mappings '0' as the binary operation.

the set of matrices $A_\alpha = \begin{bmatrix} \cos\alpha & -\sin\alpha \\ \sin\alpha & \cos\alpha \end{bmatrix}$

v.r.t. matrix multiplication.

$\alpha, \beta, \gamma \in$ R and G $= \{A_\alpha \mid \alpha \in R\}$.

$A_\alpha, A_\beta \in G$

$$A_\alpha A_\beta = \begin{bmatrix} \cos\alpha & -\sin\alpha \\ \sin\alpha & \cos\alpha \end{bmatrix} \begin{bmatrix} \cos\beta & -\sin\beta \\ \sin\beta & \cos\beta \end{bmatrix}$$

$$= \begin{bmatrix} \cos(\alpha+\beta) & -\sin(\alpha+\beta) \\ \sin(\alpha+\beta) & \cos(\alpha+\beta) \end{bmatrix} \in G$$

$, A_\beta, A_\gamma \in G$.

$A_\beta) A_\gamma = A_\alpha (A_\beta, A_\gamma)$, since matrix multiplication is associative.

y : Let $A_\alpha \in G$.

$$A_0 = \begin{bmatrix} \cos 0 & -\sin 0 \\ \sin 0 & \cos 0 \end{bmatrix} = \begin{bmatrix} 1 & 0 \\ 0 & 1 \end{bmatrix} = I \in G.$$

$_\alpha \cdot A_0 = A_\alpha \cdot I = A_\alpha$ and $A_0 \cdot A_\alpha = I \cdot A_\alpha = A_\alpha$

$A_0 = I$ is the identity in G.

e : Let $A_\alpha \in G$.

$$A_\alpha^{-1} = \frac{1}{|A_\alpha|}(\text{adj. } A_\alpha) = \frac{1}{1}\begin{bmatrix} \cos\alpha & \sin\alpha \\ -\sin\alpha & \cos\alpha \end{bmatrix}$$

$$= \begin{bmatrix} \cos(-\alpha) & -\sin(-\alpha) \\ \sin(-\alpha) & \cos(-\alpha) \end{bmatrix} = A_{-\alpha}.$$

t $A_\alpha \cdot A_{-\alpha} = A_0 = A_{-\alpha} \cdot A_\alpha$.

ınder matrix multiplication.

PREFACE

This textbook "Mathematical Foundation of Computer Science" has been designed to suit the needs of B. Tech. second year I semester and MCA first year students of **JNTU, Hyderabad** according to the revised syllabus . The students assigned with M. Tech., B. Sc., B.C.A. and IT related professional courses may also be benefited.

Every topic has been started with necessary introduction and developed gradually up to the standard form. The theory part in each chapter has been given in a lucid explanation and number of examples has been worked out to illustrate the application of theory. Most of the worked out examples and unsolved problems in exercise have been taken from the previous examination paper of JNTU. We are sure that this book will meet the requirements and expectations of the students and the faculty.

We are greatly indebted to Sri P.L.N. Prasad, Secretary and Sri N. Venkateswarlu, Vice President of Siddhartha Academy of General and Technical Education, Vijayawada, and Sri C. Dutt, Convener of the college for their constant encouragement. We express our sincere thanks to Dr. K. Srinivasu, Principal, Sri R. Madhava Rao, A.O., Prof. G. Sambasiva Rao, HOD, ME, Dr. A. Nageswara Rao, HOD, CSE, and Prof. M.V. Rama Krishna CSE Dept., P.V.P. Siddhartha Institute of Technology, Vijayawada and our other colleagues for their continuous support and encouragement.

We will gratefully accept for the reader any corrections, criticisms and suggestions for the improvement of the book.

We greatfully acknowledge our family members for their help, support and encouragement throughout the preparation of this book.

— Authors

CHAPTER

1

Mathematical Logic

1.1 INTRODUCTION

Many proofs in Mathematics and many algorithms in Computer Science use logical expressions such as "If P then Q" or "If P_1 and P_2, then Q_1 or Q_2". It is therefore, necessary to know the cases in which these expressions are either true or false. What we refer to as the truth values of such expressions. We discuss these issues in this unit.

We use different types of sentences say declarative interrogative, exclamatory and imperative to express our ideas. But in mathematics, sentences which are declarative and about which it is possible to judge, whether they are true or false but not both are used to draw conclusions and/or to prove theorems. Such sentences are statements. The value associated with the truthfulness or falsity of a statement is called its truth values.

1.2 STATEMENTS AND NOTATIONS

Definition. ***A Proposition (Statement) :*** *It is a well defined argument, which is either true or false, but not both. The truth or falsity of a statement is called its truth value.*

Example 1. Tirupati is in Andhra Pradesh

Example 2. $8 + 3 = 11$

Example 3. Close the door.

Example 4. Where are you going ?

Example 5. Put the home work on the black board.

Note 1. (1), (2) are statements, whereas (3), (4), (5) are not statements since neither true nor false.

Note 2. Statements are denoted by P, Q, R, ..., (or) p, q, r, ..., (or) A, B, C, ..., (or) P_1, P_2, P_3, ..., while declarative sentences are of the second type are obtained from the primitive ones by using certain symbols, called **connectives**.

Note 3. T, F will be used for true, false respectively. Sometimes use 1, 0 for T, F.

Definition. ***Logic*** *is a method concerned with all kinds of reasoning where there be legal arguments (or) mathematical proof (or) conclusions in a scientific theory based upon a set of hypothesis.*

Definition. *Any statement which do not contain any of the connectives is called* **atomic statement** *(or)* ***primary*** *or* ***primitive statement***.

1.3 CONNECTIVES

There are five (basic) logical connectives which are frequently used.

These are shown in the following table :

Table 1.1.

S. No.	*Symbol*	*Name*	*Connective Word*
1.	$\neg$	Negation	Not
2.	$\wedge$	conjunction	and
3.	$\vee$	disjunction	or
4.	$\rightarrow$	Implication (or) conditional	implies (or). If ..... then .....
5.	$\rightleftarrows$ (or) $\leftrightarrow$	Bi conditional (or) Equivalence	If and only if

If P and Q are any two statements then $\neg P$, $P \wedge Q$, $P \vee Q$, $P \rightarrow Q$, $P \rightleftarrows Q$ are also statements.

1.3.1 Negation

The **negation** of a statement is generally formed by inserting the word "not" at an appropriate place in the statement or by prefixing the statement with the phrase, "it is not the case that (or) it is not true that".

If P denotes a statement, then the negation of P is written as $\neg P$ and read as "not P".

If the truth value of P is T, then the truth value of $\neg P$ is F. And if the truth value of P is F, then the truth value of $\neg P$ is T.

The truth table of $\neg P$ as follows :

Table 1.2.

P	$\neg P$
T	F
F	T

Note. Sometimes $\neg P$ also denoted by $\sim P$ or P .

Example 1. P : There are 12 months in a year.

$\neg P$: There are not 12 months in a year.

or $\neg P$: It is not true that there are 12 months in a year.

Example 2. P : Vijayawada is the capital of Andhra Pradesh.

$\neg P$: Vijayawada is not capital of Andhra Pradesh.

1.3.2 Conjunction

The **conjunction** of two statements P, Q is a statement P ∧ Q, which is read as "P and Q". The statement P ∧ Q has the truth value T whenever both P and Q have the truth value T, otherwise it has truth value F. The conjunction is defined by the truth table as follows :

Table 1.3

P	Q	P ∧ Q
T	T	T
T	F	F
F	T	F
F	F	F

Note : ∧ is symmetric

i.e., P ∧ Q, Q ∧ P have same truth values.

Example 1. The conjunction of

P : It is raining today

Q : There are 20 tables in this room

is P ∧ Q : It is raining today and there are 20 tables in this room.

Example 2. If P : Jack went up the Hill

Q : Jill went up the Hill.

Then P ∧ Q : Jack and Jill went up the Hill.

Normally, in our everyday language the conjunction "and" is used between two statements which have some kind of relation. Thus a statement, "It is raining today and 3 + 4 = 7" sounds odd, but in logic it is a perfectly acceptable statement formed from the statements "It is raining today" and "3 + 4 = 7".

So far, we have seen that the symbol ∧ is used as a translation of the connective "and" appearing in English. However, the connective "and" is sometimes used in a different sense, and in such cases it cannot be translated by the symbol ∧ defined above. In order to see this difference, consider the statements :

1. Roses are red and violets are blue.
2. He opened the book and started to read.
3. Jack and Jill are cousins.

In statement (1) the conjunction "and" is used in the same sense as the symbol ∧. In (2) the word "and" is used in the sense of "and then", because the action described in "he" started to read" occurs after the action described in "he opened the book". In (3) the word "and" is not a conjunction. Note that our definition of conjunction is symmetric as far as P and Q are connected ; that is to say, the truth values of P ∧ Q and of Q ∧ P are the same for specific values of P and Q. Here the truth value of (1) will not change if we write it as violets are blue and roses are red.

On otherhand, we cannot write (2) as he started to read and opened the book.

These examples show that the symbol ∧ has a specific meaning which corresponds to the connective "and" in general, although "and" may also be used with some other meanings.

1.3.3 Disjunction

The disjunction of two statements P, Q is the statement P ∨ Q, we read it as "P or Q" or "P Join Q". The truth value of P ∨ Q is true whenever P is true (or) Q is true (or) both P, Q are true ; otherwise P ∨ Q is false. The disjunction is defined by the following table. Note that disjunction, ∨ is symmetric. *i.e.,* P ∨ Q, Q ∨ P have same truth values.

Table 1.4.

P	Q	P ∨ Q
T	T	T
T	F	T
F	T	T
F	F	F

Example 1. I shall go to market or a cinema.

2. There is something wrong with the bulb or wiring.

The connective ∨ is not always the same as the word "or", because of the fact that the word "or" in English is commonly used both as an "exclusive OR" and as an "inclusive OR".

For example, consider the following statements :

1. I shall watch the game on T.V. or go to the stadium.
2. There is something wrong with the bulb or wire.
3. Twenty or thirty animals were killed in the floods.

In statement (1), the connective "or" is used in the exclusive sense. *i.e.,* Any one of the possibility exists but not both.

In statement (2), the intended meaning is clearly one (or) the other (or) the both. The connective "or" is "inclusive OR".

In statement (3), the "or" used for indicating an approximate number of animals and it is not used as a connective. In practice, we use both inclusive or and exclusive or depending on the environment.

Molecular (Composite or Compound) Statements

Those statements which contain one or more atomic statements and some connectives are called **Molecular Statements.**

Example. Let P, Q be any two statements. Some of the molecular statements formed by using P and Q are ¬P, P ∧ Q, (P ∧ Q) ∨ (¬P), P ∧ ¬Q.

1.3.4 Conditional Statement

If P, Q are any two statements, then the statement P → Q which we read "If P then Q" is called a **conditional** statement. The statement P → Q has a truth value F when Q has the truth value F and P has the truth value T ; otherwise it has the truth Value T.

The conditional is defined in the following table :

Table 1.5.

P	Q	$P \to Q$
T	T	T
T	F	F
F	T	T
F	F	T

Here the statement P is called **antecedent** and Q is called **consequent** in $P \to Q$.

Other representations of $P \to Q$ are

(*i*) Q is necessary for P. (*ii*) P is sufficient for Q (*iii*) Q if P.

(*iv*) P only if Q (*v*) P implies Q.

Example 1. If I get the book then I begin to read.

Here P : I get the book

Q : I began to read.

Symbolic form is $P \to Q$.

Example 2. *Express in English the statement $P \to Q$ where*

P : The sun rises in the east

Q : $4 + 3 = 7$.

Solution. If the sun rises in the east then $4 + 3 = 7$.

Example 3. *Write the following statement in symbolic form.*

Statement : *If either John prefers tea or Jim prefers Coffee, then Rita prefers milk.*

Solution. A : John prefers tea.

B : Jim prefers coffee.

C : Rita prefers milk.

Symbolic form is $(A \vee B) \to C$.

1.3.5 Bi-Conditional Statement

If P and Q are any two statements, then the statement $P \rightleftarrows Q$, which is read as "P if and only if Q" and abbreviated as "P iff Q" is called a **bi-conditional** statement. The statement $P \rightleftarrows Q$ is also translated as "P is necessary and sufficient for Q" and it is also denoted by $P \leftrightarrow Q$.

The truth value of statement $P \rightleftarrows Q$ is T whenever both P, Q have identical truth values. The biconditional is defined in the following table.

Table 1.6.

P	Q	$P \rightleftarrows Q$
T	T	T
T	F	F
F	T	F
F	F	T

Example. *Find the truth value of each of the following statements :*

(i) *Tirupati is in A.P. if and only if* $1 + 3 = 4$

(ii) *Hyderabad is in Karnataka if and only if* $1 + 3 = 4$

(iii) *Chennai is in Tamil Nadu if and only if* $4 \div 2 = 2$.

(iv) *New Delhi is Capital of Sri Lanka if and only if* $5 - 2 = 3$.

Solution. (*i*) and (*iii*) are true

and (*ii*) and (*iv*) are false.

Example. *Determine the truth value of each of the following statements :*

(i) *Paris is in France and* $2 + 2 = 4$

(ii) *Paris is in France and* $2 + 2 = 5$

(iii) *Paris is in England and* $2 + 2 = 4$

(iv) *Paris is in England and* $2 + 2 = 5$.

Solution.

(*i*) T since $T \wedge T = T$

(*ii*) F since $T \wedge F = F$

(*iii*) F since $F \wedge T = F$

(*iv*) F since $F \wedge F = F$.

Example. *Determine the truth value of each of the following statements :*

(i) $1 + 1 = 5$ (or) $2 + 2 = 4$

(ii) $1 + 1 = 5$ (or) $3 + 3 = 4$

(iii) $7 + 1 = 5$ or $3 + 7 = 8$

(iv) $2 + 5 = 7$ or $1 + 7 = 8$

Solution.

(*i*) T since $F \vee T = T$

(*ii*) F since $F \vee F = F$

(*iii*) F since $F \vee F = F$

(*iv*) T since $T \vee T = T$

Example. *Find the negation of the propositions :*

(i) *It is cold*

(ii) *Today is Thursday*

(iii) *Ravi is rich.*

Solution. The negation of the propositions are

(*i*) It is not cold

(*ii*) Today is not Thursday

(*iii*) Ravi is not rich.

Example. *Let P : Ravi is rich, Q : Ravi is happy, write each of the following in symbolic form.*

(i) *Ravi is poor but happy*

(ii) *Ravi is neither rich nor happy*

(iii) *Ravi is rich and unhappy.*

Solution.

(*i*) $\neg P \wedge Q$ (*ii*) $\neg P \wedge \neg Q$ (*iii*) $P \wedge \neg Q$

Example. *Let P : Erik reads Newsweek,*

Q : Erik reads the New Yorker,

R : Erik reads Time,

Write each of the following in symbolic form.

(*i*) *Erik reads Newsweek or The New Yorker, but not Time.*

(*ii*) *Erik reads Newsweek and the New Yorker, or he does not read Newsweek and Time.*

(*iii*) *It is not true that Erik reads Newsweek but not Time.*

(*iv*) *It is not true that Erik reads Time or The New Yorker but not Newsweek.*

Solution.

(*i*) $(P \vee Q) \wedge \neg R$

(*ii*) $(P \wedge Q) \vee (\neg (Q \wedge R))$

(*iii*) $\neg (P \wedge \neg R)$

(*iv*) $\neg((R \vee Q) \wedge \neg P)$

Example. *Write the symbolic statement of "If Rita and Sita go to I.T. Camp and Jim and John go to P.C. Camp then the college gets the good name".*

Solution. Consider A : Rita goes to I.T. Camp

B : Sita goes to I.T. Camp

C : Jim goes to P.C. Camp

D : John goes to P.C. Camp

E : College gets good name.

Symbolic form of given statement is $((A \wedge B) \wedge (C \wedge D)) \rightarrow E$.

Example. *Write the following statements in symbolic form.*

(*i*) *If the sun is shining today, then $2 + 3 > 4$*

(*ii*) *The crop will be destroyed if there is a flood.*

Solution. Take p : The Sun is shining today

$q : 2 + 3 > 4$

Given statement can be written as in symbolic form as $p \rightarrow q$

(*ii*) Take A : The crop will be destroyed.

B : There is a flood.

Given statement can be symbolised as $B \rightarrow A$.

Example. *Write the following statement in symbolic form. If either Jerry takes calculus or Ken takes sociology, then Larry will take English.*

Solution. Consider J : Jerry takes calculus

K : Ken takes Sociology

L : Larry takes English.

Hence given statement can be symbolised as

$$(J \vee K) \rightarrow L.$$

Example. *Use Demorgan's laws to write the negation of each statement.*

(*i*) *I want a car and a worth cycle.*

(*ii*) *My cat stays out-side or it makes a mess.*

(*iii*) *I've fallen and I can't get up.*

(*iv*) *You stay or you don't get a good grade.* (November, 2006 - Set 4 JNTU)

Solution.

(*i*) I don't want a car or not a worth cycle.

(*ii*) My cat not stays outside and it doesn't make a mess.

(*iii*) I don't have fallen or I can get up.

(*iv*) You don't stay and you can get a good grade.

Example. *Let p, q and r be the propositions*

p : you have the flee.

q : you miss the final examination.

r : you pass the course.

Write the following proposition into statement form

(i) $p \to q$ (ii) $\neg p \to r$ (iii) $q \to \neg r$

(iv) $p \vee q \vee r$ (v) $(p \to \neg r) \vee (q \to \neg r)$ (vi) $(p \wedge q) \vee (\neg q \wedge r)$

[JNTU, November, 2006 - Set 1 and 2, February, 2007 (Set-3).

Solution.

(*i*) If you have flee then you miss the final examination.

(*ii*) If you have not flee, then you pass the course.

(*iii*) If you miss the final examination then you have not pass the course.

(*iv*) You have the flee or you miss the final examination or you pass the course.

(*v*) If you have the flee then you have not pass the course (or) if you miss the final examination then you have not pass the course.

(*vi*) You have the flee and you miss the final examination or you have not miss the final examination and you pass the course.

Example. *Consider the propositions :*

p : David is playing pool.

q : David is inside.

r : David is doing his homework.

s : David is listening to music.

Translate the following sentences into symbolic notation using p, q, r, s, ~, ∧, ∨ and parenthesis only.

(i) *David is playing pool and not doing his homework.*

(ii) *David is not listening to music, nor is he play pool, neither is he doing his homework.*

[JNTU - MCA - February, 2007]

Solution. (*i*) $p \wedge \neg r$

(*ii*) $\neg s \wedge \neg p \wedge \neg r.$

Example. *Write the negation of the following statements :*

(i) *Jan will take a job in industry (or) go to graduate school.*

(ii) James will bicycle or run tomorrow.

(iii) If the processor is fast, then the printer is slow.

[JNTU - November - 2006 - Set 3, February - 2007 - Set 4]

Solution.

(*i*) Jan will not take a job in industry and not go to graduate school.

(*ii*) Jame will not bicycle and not run tomorrow.

(*iii*) The processor is fast and the printer is not slow.

($\because$ $p \to q \equiv \neg p \vee q$ hence $\neg(p - q) = p \wedge \neg q$)

1.4 WELL-FORMED FORMULAS AND TRUTH TABLES

The notion of a statement formula has already been introduced. A statement formula is not a statement (although, for the sake of brevity, we have often called it a statement).

However, a statement can be obtained from it by replacing variables by statements.

Definition. *A statement formula is an expression which is a string consisting of variables, parentheses, and connective symbols. Note that every string of these symbols is a statement formula. We shall now give a recursive definition of a statement formula, often called a* ***well-formed formula (wff)****. A well-formed formula can be generated by the following rules :*

R_1 : A statement variable standing alone is a well-formed formula.

R_2 : If A is a well-formed formula, then $\neg A$ is a well-formed formula.

R_3 : If A and B are well-formed formulas, then $(A \wedge B)$, $(A \vee B)$, $(A \to B)$, and $(A \rightleftarrows B)$ are well-formed formulas.

Definition. *A string of symbols containing the statement variables, connectives, and parentheses is said to be well-formed formula if it can be obtained by finitely many applications of the rules R_1, R_2 and R_3.*

Example 1. According to this definition, the following are well-formed formulas : $\neg(P \wedge Q)$, $\neg(P \vee Q)$, $(P \to (P \vee Q))$, $(P \to (Q \to R))$, and $(((P \to Q) \wedge (Q \to R)) \rightleftarrows (P \to R))$.

Example 2. $\neg P \wedge Q$ is not well-formed formula because a wff would be either $(\neg P \wedge Q)$ or $\neg(P \wedge Q)$, Here $\neg P \wedge Q$ does not contains parentheses.

Example 3. $(P \to Q) \to (\wedge Q)$ is not well-formed formula because $\wedge Q$ is not statement variable.

Example 4. $(P \to Q$ is not well-formed formula but $(P \to Q)$is a **wff.**

Example 5. $(P \wedge Q) \to Q)$ is not **wff,** because one of the parentheses in the beginning is missing.

Example 6. $((P \wedge Q) \to Q)$is a **wff,** while $(P \wedge Q) \to Q$ is still not a **wff.**

It is possible to introduce some conventions so that the number of parentheses used can be reduced. In fact, there are conventions which, when followed, allow one to dispense with all the parentheses. We shall not discuss these conventions here. For the sake of convenience we shall omit the outer parentheses. Thus, we write $P \wedge Q$ in place of $(P \wedge Q)$, $(P \wedge Q) \to Q$ in place of $((P \wedge Q) \to Q)$, $((P \to Q) \wedge (Q \to R)) \rightleftarrows (P \to R)$ instead of $(((P \to Q) \wedge (Q \to R)) \rightleftarrows (P \to R)$. Since the only formulas we will encounter are well-formed formulas, we will refer to well-formed formulas as formulas.

Example. *From the formulas given below select those which are well-formed according to the definition.*

(*i*) $(P \to (P \vee Q))$

(*ii*) $((P \to (\neg P)) \to (\neg P))$

(*iii*) $((P \to (Q \to R)) \to ((P \to Q) \to (P \to R)))$

(*iv*) $((\neg P \to Q) \to (Q \to P)))$

(*v*) $((P \wedge Q) \rightleftarrows P)$.

Solution.

(*i*) It is well-formed formula.

(*ii*) It is well-formed formula.

(*iii*) It is not well-formed formula, because one of the parentheses in the begining is missing.

(*iv*) It is not well-formed formula because one more parenthesis in the end.

(*v*) It is well-formed formula.

Construction of Truth Tables

Truth tables have already been introduced in the definitions of the connectives. Our basic concern is to determine the truth table of a proposition for each possible combination of the truth values of the compound propositions. A truth table consists of columns and rows. The number of columns depend upon the numbers of sample propositions and connectives used to form a compound proposition. The number of rows in a truth table are found on the basis of simple propositions. For example, if there are two simple propositions, there will be $2^2 = 4$ rows. In general, for *n* simple propositions, the total number of rows will be 2^n.

Two methods of constructing truth table are shown in the following examples.

Example. *Construct the truth table for the statement formula* $\neg P \wedge Q$.

Solution. It is necessary to consider all possible values of P and Q. These values are entered in the first two columns of the table for both methods.

Method 1 : In this method, the truth values of ¬P are entered in the third column, and the truth values of ¬P → Q are entered in the fourth column.

Truth Table (Method 1)

Table 1.7 (a)

P	*Q*	¬P	¬P ∧ Q
T	T	F	F
T	F	F	F
F	T	T	T
F	F	T	F

Truth Table (Method 2)

Table 1.7 (b)

P	Q	P	∧	¬	Q
T	T	T	F	F	T
T	F	T	F	F	F
F	T	F	T	T	T
F	F	F	F	T	F
Step number		1	3	2	1

Method 2 : In this method, a column is drawn for each statement as well as for the connectives that appear. The truth values are entered step by step. The step numbers at the bottom of the table shows the sequence followed in arriving at the final step.

Example. *Construct the truth table for $P \wedge \neg P$.*

Solution.

Method 1

P	¬P	P ∧ ¬ P
T	F	F
F	T	F

Method 2

P	P	∧	¬	P
T	T	F	F	T
F	F	F	T	F
Step number	1	3	2	1

Example. *Construct the truth table for $(P \vee Q) \vee \neg P$.*

Method 1

P	Q	P ∨ Q	¬P	(P ∨ Q) ∨ ¬ P
T	T	T	F	T
T	F	T	F	T
F	T	T	T	T
F	F	F	T	T

Method 2.

P	Q	P	∨	Q	∨	¬	P
T	T	T	T	T	T	F	T
T	F	T	T	F	T	F	T
F	T	F	T	T	T	T	F
F	F	F	F	F	T	T	F
Step number		1	2	1	3	2	1

Example. *Construct the truth tables for the following formulas :*

(i) $\neg(\neg P \vee \neg Q)$ *(ii)* $\neg(\neg P \wedge \neg Q)$

Solution. (*i*) Truth Table for $\neg(\neg P \vee \neg Q)$ is

P	Q	¬P	¬Q	¬P ∨ ¬Q	¬(¬P ∨ ¬Q)
T	T	F	F	F	T
T	F	F	T	T	F
F	T	T	F	T	F
F	F	T	T	T	F

(*ii*) Truth Table for $\neg(\neg P \wedge \neg Q)$ is

P	Q	¬P	¬Q	¬P ∧ ¬Q	¬(¬P ∧ ¬Q)
T	T	F	F	F	T
T	F	F	T	F	T
F	T	T	F	F	T
F	F	T	T	T	F

Example. *Construct truth tables for the following formulas.*

(i) $P \wedge (P \vee Q)$ *(ii)* $\neg P \wedge (Q \wedge P)$ *(iii)* $\neg(P \wedge Q)$ *(iv)* $P \wedge \neg Q$

Solution.

(*i*)

P	Q	P ∨ Q	P ∧ (P ∨ Q)
T	T	T	T
T	F	T	T
F	T	T	F
F	F	F	F

(*ii*)

P	Q	¬P	Q ∧ P	¬P ∧ (Q ∧ P)
T	T	F	T	F
T	F	F	F	F
F	T	T	F	F
F	F	T	F	F

(*iii*)

P	Q	P ∧ Q	¬(P ∧ Q)
T	T	T	F
T	F	F	T
F	T	F	T
F	F	F	T

(*iv*)

P	Q	¬Q	P ∧ ¬Q
T	T	F	F
T	F	T	T
F	T	F	F
F	F	T	F

Example. *Construct the truth tables for the following formulas :*

(i) $(P \wedge Q) \vee (Q \wedge R) \vee (R \wedge P)$ *(ii)* $(P \vee Q) \wedge (R \vee S)$.

Solution. *(i)*

P	Q	R	$P \wedge Q$	$Q \wedge R$	$R \wedge P$	$(P \wedge Q) \vee (Q \wedge R)$	$(P \wedge Q) \vee (Q \wedge R) \vee (R \wedge P)$
T	T	T	T	T	T	T	T
T	T	F	T	F	F	T	T
T	F	T	F	F	T	F	T
T	F	F	F	F	F	F	F
F	T	T	F	T	F	T	T
F	T	F	F	F	F	F	F
F	F	T	F	F	F	F	F
F	F	F	F	F	F	F	F

(ii)

P	Q	R	S	$P \vee Q$	$R \vee S$	$(P \vee Q) \wedge (R \vee S)$
T	T	T	T	T	T	T
T	T	T	F	T	T	T
T	T	F	T	T	T	T
T	T	F	F	T	F	F
T	F	T	T	T	T	T
T	F	T	F	T	T	T
T	F	F	T	T	T	T
T	F	F	F	T	F	F
F	T	T	T	T	T	T
F	T	T	F	T	T	T
F	T	F	T	T	T	T
F	T	F	F	T	F	F
F	F	T	T	F	T	F
F	F	T	F	F	T	F
F	F	F	T	F	T	F
F	F	F	F	F	F	F

Example. *Construct a truth table for each of the compound statements*

(i) $(p \to q) \wedge (\neg P \to q)$ *(ii)* $p \to (\neg q \vee r)$

[JNTU November, 2006 - Set 3]

Solution.

(i)

p	q	$p \to q$	$\neg p$	$\neg p \to q$	$(p \to q) \wedge (\neg p \to q)$
T	T	T	F	T	T
T	F	F	F	T	F
F	T	T	T	T	T
F	F	T	T	F	F

(ii)

p	q	r	$\neg q$	$\neg q \vee r$	$p \rightarrow (\neg q \vee r)$
T	T	T	F	T	T
T	T	F	F	F	F
T	F	T	T	T	T
T	F	F	T	T	T
F	T	T	F	T	T
F	T	F	F	F	T
F	F	T	T	T	T
F	F	F	T	T	T

Example. *Construct the truth table for the following :- $[(p \vee q) \wedge (\sim r)] \leftrightarrow (q \rightarrow r)$*

[JNTU, February, 2007 M.C.A.]

P	q	r	$\neg r$	$p \vee q$	$(p \vee q) \wedge (\neg r)$	$q \rightarrow r$	$((p \vee q) \wedge (\neg r)) \leftrightarrow (q \rightarrow r)$
T	T	T	F	T	F	T	F
T	T	F	T	T	T	F	F
T	F	T	F	T	F	T	F
T	F	F	T	T	T	T	T
F	T	T	F	T	F	T	F
F	T	F	T	T	T	F	F
F	F	T	F	F	F	T	F
F	F	F	T	F	F	T	F

1.5 TAUTOLOGIES AND CONTRADICTIONS

We know how to construct the truth table of a given formula. In general, the final column of a truth table of a given formula contains both T and F. There are some formulas whose truth values are always T or always F regardless of the truth value assignments to the variables. Consider, for examples, the statement formulas $P \vee \neg P$ and $P \wedge \neg P$ in the following table.

Table : Tautology and contradiction.

Table 1.8.

P	$\neg$P	P $\vee$ $\neg$P	P $\wedge$ $\neg$P
T	F	T	F
F	T	T	F

The truth values of $P \vee \neg P$ and $P \wedge \neg P$ which are T and F respectively, are independent of the statement by which the variable P may be replaced.

Definition. *A statement formula which is true regardless of the truth values of the statements which replace the variables in it, is called a* ***universally valid formula*** *or a* ***tautology*** *or a* ***logical truth****.*

Definition. *A statement formula which is false regardless of the truth values of the statements which replace the variables in it is called a* ***contradiction****.*

Note that the negation of a contradiction is a tautology and vice-versa.

We may call a statement formula which is a tautology is identically true and a formula which is a contradiction is identically false.

We can easily observe the following :

1. Conjunction of two tautologies (contradictions) is also tautology (contradiction)
2. Conjunction of tautology, contradiction is contradiction.
3. Disjunction of two tautologies (contradiction) is also tautology (contradiction).
4. Disjunction of tautology, contradiction is tautology.

To determine whether a given formula is a Tautology, we have two methods.

I. Truth Table Method :

In this method, first we construct truth table.

If all the entries in the last column of table are T, then given formula is a Tautology.

Example. *Check whether the following formulas are tautologies (or) not.*

(i) $(P \vee Q) \rightarrow P$ *(ii)* $Q \rightarrow (P \vee Q)$

Solution.

(i)

P	Q	$P \vee Q$	$(P \vee Q) \rightarrow P$
T	T	T	T
T	F	T	T
F	T	T	F
F	F	F	T

Since all the entries in the last column of the truth table of $(P \vee Q) \rightarrow P$ are not T, the formula is not a tautology.

(ii)

P	Q	$P \vee Q$	$Q \rightarrow (P \vee Q)$
T	T	T	T
T	F	T	T
F	T	T	T
F	F	F	T

Since all the entries in the last column of the truth table of $Q \rightarrow (P \vee Q)$ are T, the formula is a tautology.

Example. *Show that the following propositions are tautologies.*

(i) $(P \wedge Q) \rightarrow P$ *(ii)* $P \vee \neg (P \wedge Q)$

(iii) $\neg (P \wedge Q) \rightleftarrows (\neg P \vee \neg Q)$.

Solution. We denote A : $(P \wedge Q) \rightarrow P$, B : $P \vee \neg (P \wedge Q)$, C : $\neg (P \wedge Q) \rightleftarrows (\neg P \vee \neg Q)$

P	Q	$\neg P$	$\neg Q$	$\neg P \vee \neg Q$	$P \wedge Q$	$\neg (P \wedge Q)$	A	B	C
T	T	F	F	F	T	F	T	T	T
T	F	F	T	T	F	T	T	T	T
F	T	T	F	T	F	T	T	T	T
F	F	T	T	T	F	T	T	T	T

Since all the entries in the last columns of the truth tables of A, B, C formulas are T, the formulas A, B, C are tautologies.

Example. *Show that the following are contradictions :*

(*i*) $(P \wedge \neg Q) \wedge Q$ (*ii*) $(P \vee Q) \wedge (\neg P \wedge \neg Q)$

Solution. Write A : $(P \wedge \neg Q) \wedge Q$, B : $(P \vee Q) \wedge (\neg P \wedge \neg Q)$

P	Q	¬P	¬Q	P∨Q	¬P∧¬Q	P∧¬Q	A	B
T	T	F	F	T	F	F	F	F
T	F	F	T	T	F	T	F	F
F	T	T	F	T	F	F	F	F
F	F	T	T	F	T	F	F	F

Since all the entries in the last column of each formula A, B truth table are F, A, B are contradictions.

Example. *Using truth tables, prove the following are tautologies.*

(*i*) $(P \rightarrow Q) \rightleftarrows (\neg P \vee Q)$ (*ii*) $((P \rightarrow R) \wedge (Q \rightarrow R)) \rightarrow ((P \vee Q) \rightarrow R)$

[JNTU - MCA, February, 2007]

Solution. (*i*)

P	Q	P→Q	¬P	¬P∨Q	(P→Q)⇄(¬P∨Q)
T	T	T	F	T	T
T	F	F	F	F	T
F	T	T	T	T	T
F	F	T	T	T	T

∴ $(P \rightarrow Q) \rightleftarrows \neg P \vee Q)$ is a tautology.

(*ii*) Consider A : $(P \rightarrow R) \wedge (Q \rightarrow R)$, B : $(P \vee Q) \rightarrow R$

P	Q	R	P→R	Q→R	P∨Q	B:(P∨Q)→R	A:(P→R)∧(Q→R)	A→B
T	T	T	T	T	T	T	T	T
T	T	F	F	F	T	F	F	T
T	F	T	T	T	T	T	T	T
T	F	F	F	T	T	F	F	T
F	T	T	T	T	T	T	T	T
F	T	F	T	F	T	F	F	T
F	F	T	T	T	F	T	T	T
F	F	F	T	T	F	T	T	T

A → B is a tautology

Hence $(P \rightarrow R) \wedge (Q \rightarrow R) \rightarrow (P \vee Q) \rightarrow R$ is a tautology.

Example. *Examine whether* $((p \vee q \rightarrow r) \leftrightarrow (\sim r \rightarrow \sim (p \vee q))$ *is a tautology.*

Solution.

P	q	r	$p \vee q$	$(p \vee q) \rightarrow r$	$\neg r$	$\neg(p \vee q)$	$\neg r \rightarrow \neg(p \vee q)$
T	T	T	T	T	F	F	T
T	T	F	T	F	T	F	F
T	F	T	T	T	F	F	T
T	F	F	T	F	T	F	F
F	T	T	T	T	F	F	T
F	T	F	T	F	T	F	F
F	F	T	F	T	F	T	T
F	F	F	F	T	T	T	T

Since $(p \vee q) \rightarrow r$ and $\neg r \rightarrow \neg (p \vee q)$ have the same truth values, $((p \vee q) \rightarrow r) \leftrightarrow (\neg r \rightarrow \neg (p \vee q))$ is a tautology.

Example. *Prove that, for any propositions, p, q, r, the compound proposition* $(p \rightarrow (q \rightarrow r)) \rightarrow (p \rightarrow r) \rightarrow (p \rightarrow r))$ *is a tautology.*

Solution. Take A : $p \rightarrow (q \rightarrow r)$

B : $(p \rightarrow q) \rightarrow (p \rightarrow r)$

p	q	r	$q \rightarrow r$	$p \rightarrow q$	$p \rightarrow r$	A : $p \rightarrow (q \rightarrow r)$	B : $(p \rightarrow q) \rightarrow (p \rightarrow r)$	A $\rightarrow$ B
T	T	T	T	T	T	T	T	T
T	T	F	F	T	F	F	F	T
T	F	T	T	F	T	T	T	T
T	F	F	T	F	F	T	T	T
F	T	T	T	T	T	T	T	T
F	T	F	F	T	T	T	T	T
F	F	T	T	T	T	T	T	T
F	F	F	T	T	T	T	T	T

$\therefore$ A $\rightarrow$ B is a tautology

Then $(p \rightarrow (q \rightarrow r)) \rightarrow ((p \rightarrow q) \rightarrow (p \rightarrow r))$ is a tautology.

Example. *Show that* $(P \rightarrow Q) \rightarrow (P \rightarrow (P \wedge Q))$ *by using truth table.*

Solution.

P	Q	P $\rightarrow$ Q	P $\wedge$ Q	$p \rightarrow$ (P $\wedge$ Q)	(P $\rightarrow$ Q) $\rightarrow$ (P $\rightarrow$ (P $\rightarrow$ Q))
T	T	T	T	T	T
T	F	F	F	F	T
F	T	T	F	T	T
F	F	T	F	T	T

$\therefore$ (P $\rightarrow$ Q) $\rightarrow$ (P $\rightarrow$ (P $\wedge$ Q)) is a tautology.

Note: $p \Rightarrow$ Q means "$p \rightarrow$ Q is a tautology".

Example. *Show that the following implications*

(i) $P \wedge Q \Rightarrow P$ (ii) $P \wedge Q \Rightarrow Q$

Solution. (*i*), and (*ii*)

P	Q	$P \wedge Q$	$(p \wedge Q) \to p$	$(p \wedge Q) \to Q$
T	T	T	T	T
T	F	F	T	T
F	T	F	T	T
F	F	F	T	T

Since $(p \wedge Q) \to p$ is a tautology, $P \wedge Q \Rightarrow P$.

Since $(p \wedge Q) \to Q$ is a tautology, $P \wedge Q \Rightarrow Q$.

Example. *Show that the following logical implications.*

(a) $P \wedge Q \Rightarrow (P \to Q)$ (b) $P \Rightarrow (Q \to P)$

(c) $(P \to (Q \to R)) \Rightarrow ((P \to Q) \to (P \to R))$.

Solution. (*a*)

P	Q	$P \wedge Q$	$P \to Q$	$(P \wedge Q) \to (P \to Q)$
T	T	T	T	T
T	F	F	F	T
F	T	F	T	T
F	F	F	T	T

$\therefore$ $(P \wedge Q) \to (P \to Q)$ is a tautology

Hence $(P \wedge Q) \Rightarrow (P \to Q)$

(*b*)

P	Q	$Q \to P$	$P \to (Q \to P)$
T	T	T	T
T	F	T	T
F	T	F	T
F	F	T	T

$\therefore$ $P \to (Q \to P)$ is a tautology

Hence $P \Rightarrow (Q \to P)$

(*c*) We have $(P \to (Q \to R)) \to ((P \to Q) \to (P \to R))$ is a tautology (See example on page No. 17)

$\therefore$ $(P \to (Q \to R)) \Rightarrow ((P \to Q) \to (P \to R))$.

Example. *Show the following implications without constructing the truth tables.*

(a) $(P \to Q) \Rightarrow P \to (P \wedge Q)$ (b) $((P \to Q) \to Q) \Rightarrow P \vee Q$

Solution.

(*a*) Let us assume truth value of $P \to Q$ is T.

$\therefore$ Truth value of P, Q are either T or F.

Case (*i*) If the truth values of P, Q are T then $P \wedge Q$ is T

Then the truth value of $P \to (P \wedge Q)$ is T.

Case (*ii*) If the truth values of P, Q are F then $P \wedge Q$ is F.

Then the truth value of $P \to (P \wedge Q)$ is T.

In both cases, the truth value of $P \rightarrow (P \wedge Q)$ is T.

(*b*) Let us assume truth value of $(P \rightarrow Q) \rightarrow Q$ is T.

$\Rightarrow$ Truth values of $P \rightarrow Q$, Q are both T or F.

Case (*i*) If Truth value of Q is T, truth value of $P \rightarrow Q$ is T

$\Rightarrow$ Truth value of P is T

$\Rightarrow$ Truth value of $P \vee Q$ is T.

Case (*ii*) If truth value of Q is F and truth value of $P \rightarrow Q$ is F.

$\Rightarrow$ Truth value of Q is F and truth value of P is T.

$\Rightarrow$ Truth value of $P \vee Q$ is T.

From above two cases, the truth value of $P \vee Q$ is T.

$\therefore$ $((P \rightarrow Q) \rightarrow Q) \Rightarrow P \vee Q$.

Example. *Show that the truth values of the following formulas are independent of their components :*

(*i*) $(P \wedge (P \rightarrow Q)) \rightarrow Q$ (*ii*) $(P \rightarrow Q) \rightleftarrows (\neg P \vee Q)$

(*iii*) $((P \rightarrow Q) \wedge (Q \rightarrow R)) \rightarrow (P \rightarrow R)$

Solution.

(*i*)

P	Q	$P \rightarrow Q$	$P \wedge (P \rightarrow Q)$	$P \wedge (P \rightarrow Q) \rightarrow Q$
T	T	T	T	T
T	F	F	F	T
F	T	T	F	T
F	F	T	F	T

$\therefore$ Truth value of $(P \wedge (P \rightarrow Q)) \rightarrow Q$ is independent of the truth values of P, Q.

(*ii*)

P	Q	$P \rightarrow Q$	$\neg P$	$\neg P \vee Q$	$(P \rightarrow Q) \rightleftarrows (\neg P \vee Q)$
T	T	T	F	T	T
T	F	F	F	F	T
F	T	T	T	T	T
F	F	T	T	T	T

$\therefore$ Truth value of $(P \rightarrow Q) \rightleftarrows (\neg P \vee Q)$ is independent of the truth values of P, Q.

(*iii*)

P	Q	R	$P \rightarrow Q$	$Q \rightarrow R$	$(P \rightarrow Q) \wedge (Q \rightarrow R) : A$	$P \rightarrow R$	$A \rightarrow (P \rightarrow R)$
T	T	T	T	T	T	T	T
T	T	F	T	F	F	F	T
T	F	T	F	T	F	T	T
T	F	F	F	T	F	F	T
F	T	T	T	T	T	T	T
F	T	F	T	F	F	T	T
F	F	T	T	T	T	T	T
F	F	F	T	T	T	T	T

∴ Truth value of $((P \to Q) \wedge (Q \to R)) \to (P \to R)$ is independent of the truth values of P, Q, R.

Example. *Construct the truth table for (i) $(P \to Q) \to (P \wedge Q)$ (ii) $(P \wedge Q) \to (P \vee Q)$ (iii) $(P \to Q) \wedge (Q \to P)$.* [Nov. 05 set I - JNTU]

Solution.

(*i*)

P	Q	$P \to Q$	$P \wedge Q$	$(P \to Q) \to (P \wedge Q)$
T	T	T	T	T
T	F	F	F	T
F	T	T	F	F
F	F	T	F	F

(*ii*)

P	Q	$P \wedge Q$	$P \vee Q$	$(P \wedge Q) \to (P \vee Q)$
T	T	T	T	T
T	F	F	T	T
F	T	F	T	T
F	F	F	F	T

(*iii*)

P	Q	$P \to Q$	$Q \to P$	$(P \to Q) \wedge (Q \to P)$
T	T	T	T	T
T	F	F	T	F
F	T	T	F	F
F	F	T	T	T

Example. *Given the truth values of P and Q as T and those of R and S as F, find the truth values of the following :*

(*i*) $P \vee (Q \wedge R)$

(*ii*) $(P \wedge (Q \wedge R)) \vee \neg ((P \vee Q) \wedge (R \vee S))$

(*iii*) $(\neg (P \wedge Q) \vee \neg R) \vee (((\neg P \wedge Q) \vee \neg R) \wedge S)$

Solution.

(*i*) Truth value of $P \vee (Q \wedge R)$ is $T \vee (T \wedge F) = T \vee F = T$

(*ii*) Truth value of given formula is $(T \wedge (T \wedge F)) \vee \neg ((T \vee T) \wedge (F \vee F)) = (T \wedge F) \vee \neg (T \wedge F)$ $= F \vee \neg F = F \vee T = T$

(*iii*) Truth value of given formula is $(\neg (T \wedge T) \vee \neg F) \vee (((\neg T \wedge T) \vee \neg F) \wedge F) = (\neg T \vee T) \vee (((F \wedge T) \vee T) \wedge F) = (F \vee T) \vee ((F \vee T) \wedge F) = T \vee (T \wedge F) = T \vee F = T$.

Example. *Given the truth values of P and Q as T and those of R and S as F, find the truth values of the following :*

(*i*) $(\neg (P \wedge Q) \vee \neg R) \vee ((Q \rightleftarrows \neg P) \to (R \vee \neg S))$

(*ii*) $(P \rightleftarrows R) \wedge (\neg Q \to S)$

(*iii*) $(P \vee (Q \to (R \wedge \neg P))) \rightleftarrows (Q \vee \neg S)$

Solution.

(*i*) Statement truth value is $(\neg (T \wedge T) \vee \neg F) \vee ((T \rightleftarrows \neg T) \rightarrow (F \vee \neg F))$

$$= (\neg T \vee T) \vee ((T \rightleftarrows F) \rightarrow (F \vee T))$$

$$= (F \vee T) \vee (F \rightarrow T) = T \vee T = T$$

(*ii*) Truth value of statement is $(T \rightleftarrows F) \wedge (\neg T \rightarrow F)$

$$= F \wedge (F \rightarrow F) = F \wedge T = F$$

(*iii*) Truth value of statement is $(T \vee (T \rightarrow (F \wedge \neg T))) \rightleftarrows (T \vee \neg F)$

$$= (T \vee (T \rightarrow (F \wedge F))) \rightleftarrows (T \vee T)$$

$$= (T \vee (T \rightarrow F)) \rightleftarrows T = (T \vee F) \rightleftarrows T$$

$$= T \rightleftarrows T = T.$$

II. Alternative Method (without Using Truth Table) :

Recall that the numbers of rows in a truth table is 2^n, where n is the number of distinct variables in the formula. Therefore, this process of determining whether a given formula is a tautology is tedious, particularly when the number of distinct variables is large or when the formula is complicated. Hence alternative method will be developed that will be able to determine whether a statement formula is tautology without having to construct its truth table.

A simple fact about tautologies is that the conjunction of two tautologies is also a tautology.

Definition. *A formula A is called a* ***substitution instance*** *of another formula B if A can be obtained from B by substituting formulae for some variables of B, with the condition that the same formula is substituted for the same variable each time it occurs.*

We now demonstrate this concept.

Let $B : P \rightarrow (E \wedge P)$

Substitute $R \rightleftarrows S$ for P in B, we get

$$A : (R \rightleftarrows S) \rightarrow (E \wedge (R \rightleftarrows S))$$

Then A is a substitution instance of B.

Note that $C : (R \rightleftarrows S) \rightarrow (E \wedge P)$ is not a substitution instance of B because the variable P in $E \wedge P$ was not replaced by $R \rightleftarrows S$.

It is possible to substitute more than one variable by other formulae, provided that all substitutions are considered to occur simultaneously.

Example. *Substitution instances of* $P \rightarrow \neg Q$ *are*

(1) $(R \wedge \neg S) \rightarrow \neg (E \vee M)$

(2) $(R \vee \neg S) \rightarrow \neg (R \vee \neg S)$

(3) $Q \rightarrow \neg P$

(4) $Q \rightarrow \neg (P \wedge \neg Q)$.

Note that in (1), P is replaced by $R \wedge \neg S$, Q is replaced by $E \vee M$.

In (2), P and Q are replaced by $R \vee \neg S$.

In (3), P is replaced Q and Q is replaced by P.

In (4), P is replaced by Q and Q is replaced by $P \wedge \neg Q$.

Example. Consider the following formulae from $P \to \neg Q$

1. Substitute $P \vee Q$ for P and R for Q to get the substitution instance $(P \vee Q) \to \neg R$
2. First substitute $P \vee Q$ for P to obtain the substitution instance $(P \vee Q) \to \neg Q$. Next, substitute R for Q in $(P \vee Q) \to \neg Q$, and we get $(P \vee R) \to \neg R$.

This formula is a substitution instance of $(P \vee Q) \to \neg Q$, but it is not a substitution instance of $P \to \neg Q$ under the substitution $(P \vee Q)$ for P and R for Q, because we did not substitute simultaneously as we did in (1).

Note that in constructing substitution instances of a formula, substitutions are made for the atomic formula and never for the compound formula. Thus $P \to Q$ is not a substitution instance of $P \to \neg R$, because it is R which must be replaced and not $\neg R$.

The importance of substitution instance lies in the fact that any substitution of a tautology is a tautology. For example, consider the tautology $P \vee \neg P$. Regardless of what is substituted for P, the truth value of $P \vee \neg P$ is always T. Therefore, if we substitute any statement formula for P, the resulting formula will be a tautology. Hence the following substitution instances of $P \vee \neg P$ are tautologies.

1. $(A \to B) \vee \neg (A \to B)$
2. $((A \vee C) \wedge B) \vee \neg ((A \vee C) \wedge B)$
3. $(((P \vee \neg S) \to R) \rightleftarrows S) \vee \neg (((P \vee \neg S) \to R) \rightleftarrows S)$

Thus, if it is possible to detect whether a given formula is a substitution instance of a tautology, then it is immediately known that the given formula is also a tautology.

Example. *Produce the substitution instances of the following formulae for the given substitutions.*

(i) $(((P \to Q) \to P) \to P)$ *; substitute* $(P \to Q)$ *for P and* $((P \wedge Q) \to R)$ *for Q.*

(ii) $((P \to Q) \to (Q \to P))$ *; substitute Q for P and* $(P \wedge \neg P)$ *for Q.*

Solution.

(i) $((((P \to Q) \to ((P \wedge Q) \to R)) \to (P \to Q)) \to (P \to Q))$

(ii) $((Q \to (P \wedge \neg P)) \to ((P \wedge \neg P) \to Q))$

Example. *Determine the formulae which are substitution instances of other formulae in the list and give the substitutions.*

(i) $(P \to (Q \to P))$

(ii) $(((P \to Q) \wedge (R \to S)) \wedge (P \vee R)) \to (Q \vee S))$

(iii) $(Q \to ((P \to P) \to Q))$

(iv) $(P \to ((P \to (Q \to P)) \to P))$

(v) $((((R \to S) \wedge (Q \to P)) \wedge (R \vee Q)) \to (S \vee P))$.

Solution.

(v) is substitution instance of *(ii)*.

Here, substitutions are R for P, S for Q, Q for R, P for S.

(iii) is substitution instance of *(i)*

Here substitutions are Q for P, P → P for Q.

(*iv*) is substitution instance of (*i*)

Here substitution is P → (Q → P) for Q.

1.6 EQUIVALENCE OF FORMULAS

Definition. *Two formulas A and B are said to be equivalent to each other if and only if A $\rightleftarrows$ B is a tautology.*

We shall represent the equivalence of two formulas A and B, by writing A ⇔ B, which is read as A is equivalent to B.

Note 1. ⇔ is only symbol, but not connective.

2. A ⇔ B if and only if the truth values of A, B are same *i.e.,* the final columns in the truth tables of A, B are same *i.e.,* the truth tables of A, B are same.

3. Equivalence relation is symmetric and transitive.

Example 1. ¬¬P ⇔ P

Example 2. P ∨ ¬P is equivalent to Q ∨ ¬Q.

Method I (Truth Table Method) :

One method to determine whether any two statement formulas are equivalent is to construct their truth tables.

Example. Prove the following :

(*i*) (P ∧ ¬P) ∨ Q ⇔ Q (*ii*) (P → Q) ⇔ (¬P ∨ Q)

(*i*)

P	Q	¬P	P ∧ ¬P	(P ∧ ¬P) ∨ Q
T	T	F	F	T
T	F	F	F	F
F	T	T	F	T
F	F	T	F	F

Since truth values (columns) of A, B are same, (P ∧ ¬P) ∨ Q ⇔ Q

(*ii*)

P	Q	¬P	¬P ∨ Q	P → Q	(P → Q) $\rightleftarrows$ (¬P ∨ Q)
T	T	F	T	T	T
T	F	F	F	F	T
F	T	T	T	T	T
F	F	T	T	T	T

∴ (P → Q) $\rightleftarrows$ (¬P ∨ Q) is a tautology, hence (P → Q) ⇔ (¬P ∨ Q)

Example. Prove the following :

P ∨ Q ⇔ ¬ (¬P ∧ ¬Q)

(*i*)

P	Q	$\neg$P	$\neg$Q	$\neg P \wedge \neg Q$	$\neg(\neg P \wedge \neg Q)$	$P \vee Q$	$(P \vee Q) \leftrightarrow \neg(\neg P \wedge \neg Q)$
T	T	F	F	F	T	T	T
T	F	F	T	F	T	T	T
F	T	T	F	F	T	T	T
F	F	T	T	T	F	F	T

$\therefore$ $(P \vee Q) \leftrightarrow \neg(\neg P \wedge \neg Q)$ is a tautology

Hence $(P \vee Q) \Leftrightarrow \neg(\neg P \wedge \neg Q)$.

Example. Prove that $(P \wedge Q) \wedge R \Leftrightarrow P \wedge (Q \wedge R)$ (Associative law)

(*ii*)

P	Q	R	$P \wedge Q$	$(P \wedge Q) \wedge R$	$Q \wedge R$	$P \wedge (Q \wedge R)$
T	T	T	T	T	T	T
T	T	F	T	F	F	F
T	F	T	F	F	F	F
T	F	F	F	F	F	F
F	T	T	F	F	T	F
F	T	F	F	F	F	F
F	F	T	F	F	F	F
F	F	F	F	F	F	F

Since the truth values (columns) of $(P \wedge Q) \wedge R$, $P \wedge (Q \wedge R)$ are same, $(P \wedge Q) \wedge R \Leftrightarrow P \wedge (Q \wedge R)$.

Example. *Show the following equivalences :*

(*i*) $\neg(P \wedge Q) \Leftrightarrow \neg P \vee \neg Q$

(*ii*) $\neg(P \vee Q) \Leftrightarrow \neg P \wedge \neg Q$

(*iii*) $\neg(P \rightarrow Q) \Leftrightarrow P \wedge \neg Q$

(*iv*) $\neg(P \rightleftarrows Q) \Leftrightarrow (P \wedge \neg Q) \vee (\neg P \wedge Q)$

Solution.

(*i*)

P	Q	$P \wedge Q$	$\neg(P \wedge Q)$	$\neg$P	$\neg$Q	$\neg P \vee \neg Q$	$\neg(P \wedge Q) \rightleftarrows (\neg P \vee \neg Q)$
T	T	T	F	F	F	F	T
T	F	F	T	F	T	T	T
F	T	F	T	T	F	T	T
F	F	F	T	T	T	T	T

$\therefore$ $\neg(P \wedge Q) \rightleftarrows (\neg P \vee \neg Q)$ is a tautology.

Hence $\neg(P \wedge Q) \Leftrightarrow (\neg P \vee \neg Q)$

(*ii*)

P	Q	$P \vee Q$	$\neg$P	$\neg$Q	$\neg(P \vee Q)$	$\neg P \wedge \neg Q$	$\neg(P \vee Q) \rightleftarrows \neg P \wedge \neg Q$
T	T	T	F	F	F	F	T
T	F	T	F	T	F	F	T
F	T	T	T	F	F	F	T
F	F	F	T	T	T	T	T

$\therefore$ $\neg(P \vee Q) \rightleftarrows (\neg P \wedge \neg Q)$ is a tautology

Hence $\neg(P \vee Q) \Leftrightarrow (\neg P \wedge \neg Q)$

(iii)

P	Q	$P \to Q$	$\neg(P \to Q)$	$\neg Q$	$P \wedge \neg Q$	$\neg(P \to Q) \rightleftarrows (P \wedge \neg Q)$
T	T	T	F	F	F	T
T	F	F	T	T	T	T
F	T	T	F	F	F	T
F	F	T	F	T	F	T

$\therefore \quad \neg (P \to Q) \rightleftarrows (P \wedge \neg Q)$ is a tautology.

Hence $\neg (P \to Q) \Leftrightarrow (P \wedge \neg Q)$

(iv)

P	Q	$\neg P$	$\neg Q$	$P \wedge \neg Q$	$\neg P \wedge Q$	A: $(P \wedge \neg Q) \vee (\neg P \wedge Q)$	$(P \rightleftarrows Q)$	B: $\neg(P \rightleftarrows Q)$	$A \rightleftarrows B$
T	T	F	F	F	F	F	T	F	T
T	F	F	T	T	F	T	F	T	T
F	T	T	F	F	T	T	F	T	T
F	F	T	T	F	F	F	T	F	T

$A \rightleftarrows B$ is a tautology then $A \Leftrightarrow B$

Therefore $\neg (P \rightleftarrows Q) \Leftrightarrow (P \wedge \neg Q) \vee (\neg P \wedge Q)$.

Example. *Show that P is equivalent to the following formulas*

(i) $\neg\neg P$ (ii) $P \wedge P$

(iii) $P \vee P$ (iv) $P \vee (P \wedge Q)$

(v) $P \wedge (P \vee Q)$ (vi) $(P \wedge Q) \vee (P \wedge \neg Q)$

(vii) $(P \vee Q) \wedge (P \vee \neg Q)$.

Solution.

P	$\neg P$	$\neg\neg P$	$P \wedge P$	$P \vee P$
T	F	T	T	T
F	T	F	F	F

$\therefore \quad P \Leftrightarrow \neg\neg P \Leftrightarrow P \wedge P \Leftrightarrow P \vee P$ and

Consider $A : P \vee (P \wedge Q)$; $B : P \wedge (P \vee Q)$; $C : (P \wedge Q) \vee (P \wedge \neg Q)$; and $D : (P \vee Q) \wedge (P \vee \neg Q)$.

P	Q	$P \wedge Q$	$P \vee Q$	$\neg P$	$\neg Q$	$P \wedge \neg Q$	$P \vee \neg Q$	A	B	C	D
T	T	T	T	F	F	F	T	T	T	T	T
T	F	F	T	F	T	T	T	T	T	T	T
F	T	F	T	T	F	F	F	F	F	F	F
F	F	F	F	T	T	F	T	F	F	F	F

$\therefore \quad P \Leftrightarrow A \Leftrightarrow B \Leftrightarrow C \Leftrightarrow D$

$\therefore \quad P \Leftrightarrow P \vee (P \wedge Q) \Leftrightarrow P \wedge (P \vee Q) \Leftrightarrow (P \wedge Q) \vee (P \wedge \neg Q) \Leftrightarrow (P \vee Q) \wedge (P \vee \neg Q)$

A list of some basic equivalent formulas which will be found useful is given in Table 1.9. In order to make the list complete, we use T and F as special variables in the sense that T can be replaced by only a tautology and F by only a contradiction.

Table 1.9. Equivalent Formulae

$P \vee P \Leftrightarrow P,$	$P \wedge P \Leftrightarrow P$	*(Idempotent laws)*
$(P \vee Q) \vee R \Leftrightarrow P \vee (Q \vee R),$	$(P \wedge Q) \wedge R \Leftrightarrow P \wedge (Q \wedge R)$	(Associative laws)
$P \vee Q \Leftrightarrow Q \vee P,$	$P \wedge Q \Leftrightarrow Q \wedge P$	(Commutative laws)
$P \vee (Q \wedge R) \Leftrightarrow (P \vee Q) \wedge (P \vee R),$	$P \wedge (Q \vee R) \Leftrightarrow (P \wedge Q) \vee (P \wedge R)$	(Distributive laws)
$P \vee (P \wedge Q) \Leftrightarrow P,$	$P \wedge (P \vee Q) \Leftrightarrow P$	(Absorption laws)
$P \vee F \Leftrightarrow P,$	$P \wedge T \Leftrightarrow P$	
$P \vee T \Leftrightarrow T,$	$P \wedge F \Leftrightarrow F$	
$\neg (P \vee Q) \Leftrightarrow \neg P \wedge \neg Q,$	$\neg (P \wedge Q) \Leftrightarrow \neg P \vee \neg Q$	(Demorgan's laws)

Method II. (Replacement Process) :

In constructing substitution instances of a statement formula, we are allowed to substitute only for the variables appearing in the formula. Furthermore, the same formula is to be substituted for every occurrence of a particular variable. This rule ensures that substitution instances of a tautology are also tautologies.

Consider now another process, called a **replacement process**, in which we replace any part of a statement formula which is itself a formula and it may be atomic or molecular, by any other formula.

We now demonstrate this process.

Consider the formula A : $P \to (Q \to R)$. The formula $Q \to R$ is a part of the formula A. If we replace $Q \to R$ by an equivalent formula $\neg Q \vee R$ in A, we get another formula B : $P \to (\neg Q \vee R)$. Here, we can easily verify that the formulas A and B are equivalent to each other. This process of obtaining B from A is known as the replacement process.

Example. *Show that* $P \to (Q \to R) \Leftrightarrow P \to (\neg Q \vee R) \Leftrightarrow (P \wedge Q) \to R$.

[JNTU - November, 2004 - Set 1]

Solution. We know that $Q \to R \Leftrightarrow \neg Q \vee R$.

Replacing $Q \to R$ by $\neg Q \vee R$, we get $P \to (\neg Q \vee R)$, which is equivalent to $\neg P \vee (\neg Q \vee R)$ by same rule.

Now, $\neg P \vee (\neg Q \vee R) \Leftrightarrow (\neg P \vee \neg Q) \vee R \Leftrightarrow \neg (P \wedge Q) \vee R \Leftrightarrow (P \wedge Q) \to R.$

by using associativity of $\vee$, Demorgan's law and the previous used rule.

Therefore $P \to (Q \to R) \Leftrightarrow P \to (\neg Q \vee R) \Leftrightarrow (P \wedge Q) \to R.$

Example. *Show the following equivalences :*

(i) $P \to (Q \to P) \Leftrightarrow \neg P \to (P \to Q)$.

(ii) $P \to (Q \vee R) \Leftrightarrow (P \to Q) \vee (P \to R)$.

Solution.

(*i*) $P \to (Q \to P) \Leftrightarrow P \to (\neg Q \vee P) \Leftrightarrow \neg P \vee (\neg Q \vee P)$

$$\Leftrightarrow (\neg P \vee \neg Q) \vee P$$

$$\Leftrightarrow P \vee (\neg P \vee \neg Q) \Leftrightarrow (P \vee \neg P) \vee \neg Q$$

$$\Leftrightarrow T \vee \neg Q$$

$$\Leftrightarrow T$$

and $\neg P \to (P \to Q) \Leftrightarrow \neg P \to (\neg P \vee Q)$

$$\Leftrightarrow \neg\neg P \vee (\neg P \vee Q)$$

$$\Leftrightarrow (P \vee \neg P) \vee Q \Leftrightarrow T \vee Q \Leftrightarrow T$$

Hence $P \to (Q \to P) \Leftrightarrow \neg P \to (P \to Q)$

(*ii*) $P \to (Q \vee R) \Leftrightarrow \neg P \vee (Q \vee R)$

$$\Leftrightarrow (\neg P \vee \neg P) \vee (Q \vee R)$$

$$\Leftrightarrow (\neg P \vee Q) \vee (\neg P \vee R)$$

$$\Leftrightarrow (P \to Q) \vee (P \to R)$$

Example. *Show that $(\neg P \wedge (\neg Q \wedge R)) \vee (Q \wedge R) \vee (P \wedge R) \Leftrightarrow R$ by using replacement process.*

[JNTU - November, 2004 -Set-1]

Solution. $(\neg P \wedge (\neg Q \wedge R)) \vee (Q \wedge R) \vee (P \wedge R)$

$\Leftrightarrow$ $((\neg P \wedge \neg Q) \wedge R) \vee (Q \wedge R) \vee (P \wedge R)$ (Associative law)

$\Leftrightarrow$ $((\neg P \wedge \neg Q) \wedge R) \vee ((Q \vee P) \wedge R)$ (Distributive law)

$\Leftrightarrow$ $(\neg (P \vee Q) \wedge R) \vee ((P \vee Q) \wedge R)$ (Demorgan's law, commutative)

$\Leftrightarrow$ $((\neg (P \vee Q) \vee (P \vee Q)) \wedge R)$ (Distributive law)

$\Leftrightarrow$ $(T \wedge R)$ since $\neg A \vee A = T$

$\Leftrightarrow$ R since $T \wedge R \Leftrightarrow R$.

Example. Show that $\neg (P \rightleftarrows Q) \Leftrightarrow (P \vee Q) \wedge \neg (P \wedge Q)$

Solution. $\neg(P \rightleftarrows Q) \Leftrightarrow \neg ((P \to Q) \wedge (Q \to P))$

$\Leftrightarrow$ $\neg ((\neg P \vee Q) \wedge (\neg Q \vee P))$

$\Leftrightarrow$ $\neg (((\neg P \vee Q) \wedge \neg Q) \vee ((\neg P \vee Q) \wedge P))$ (Distributive law)

$\Leftrightarrow$ $\neg ((\neg P \wedge \neg Q) \vee (Q \wedge \neg Q)) \vee ((\neg P \wedge P) \vee (Q \wedge P))$ (Distributive law)

$\Leftrightarrow$ $\neg ((\neg P \wedge \neg Q) \vee (Q \wedge \neg Q) \vee (\neg P \wedge P) \vee (P \wedge Q))$ (Commutative law)

$\Leftrightarrow$ $\neg (\neg (P \vee Q) \vee F \vee F \vee (P \wedge Q))$ (Demorgan law)

$\Leftrightarrow$ $\neg (\neg (P \vee Q) \vee F \vee (P \wedge Q))$

$\Leftrightarrow$ $\neg (\neg (P \vee Q) \vee (P \wedge Q))$

$\Leftrightarrow$ $(P \vee Q) \wedge \neg (P \wedge Q)$ (Demorgan law)

Example. *Show that $((P \vee Q) \wedge \neg (\neg P \wedge (\neg Q \vee \neg R))) \vee (\neg P \wedge \neg Q) \vee (\neg P \wedge \neg R)$ is a tautology.*

[JNTU - November, 2004 - Set 3]

Solution. By Demorgan's laws, we have

$$\neg P \wedge \neg Q \Leftrightarrow \neg (P \vee Q)$$

$$\neg P \wedge \neg R \Leftrightarrow \neg (P \vee R)$$

$$(\neg P \wedge \neg Q) \vee (\neg P \wedge \neg R) \Leftrightarrow \neg (P \vee Q) \vee \neg (P \vee R)$$

$\Leftrightarrow$ $\neg ((P \vee Q) \wedge (P \vee R))$

Also $\neg (\neg P \wedge (\neg Q \vee \neg R)) \Leftrightarrow \neg (\neg P \wedge \neg (Q \wedge R))$

$\Leftrightarrow$ $P \vee (Q \wedge R)$

$\Leftrightarrow$ $(P \vee Q) \wedge (P \vee R)$

Consequently, the given formula is equivalent to $((P \vee Q) \wedge (P \vee Q) \wedge (P \vee R)) \vee \neg ((P \vee Q) \wedge (P \vee R))$

$$\Leftrightarrow \quad ((P \vee Q) \wedge (P \vee R)) \vee \neg ((P \vee Q) \wedge (P \vee R))$$

which is a substitution instance of $P \vee \neg P$.

$\therefore$ Given formula is a tautology.

Example. *Are $(p \to q) \to r$ and $p \to (q \to r)$ logically equivalent ? Justify your answer by using the rules of logic to simply both expressions and also by using truth tables.*

[JNTU - November, 2006 - Set 4, February, 2007 - Set 1]

Solution. $(p \to q) \to r$ and $p \to (q \to r)$ are not logically equivalent because

Method I :

Consider
$$\begin{aligned}(p \to q) \to r &\Leftrightarrow (\neg p \vee q) \to r\\ &\Leftrightarrow \neg(\neg p \vee q) \vee r\\ &\Leftrightarrow (p \wedge \neg q) \vee r\\ &\Leftrightarrow (p \wedge r) \vee (\neg q \wedge r)\end{aligned}$$

and
$$\begin{aligned}p \to (q \to r) &\Leftrightarrow p \to (\neg q \vee r)\\ &\Leftrightarrow \neg p \vee (\neg q \vee r)\\ &\Leftrightarrow \neg p \vee \neg q \vee r.\end{aligned}$$

Method II : (Truth Table Method)

p	q	r	$p \to q$	$(p \to q) \to r$	$q \to r$	$p \to (q \to r)$
T	T	T	T	T	T	T
T	T	F	T	F	F	F
T	F	T	F	T	T	T
T	F	F	F	T	T	T
F	T	T	T	T	T	T
F	T	F	T	F	F	T
F	F	T	T	T	T	T
F	F	F	T	F	T	T

Here the truth values (columns) of $(p \to q) \to r$ and $p \to (q \to r)$ are not identical.

1.7 DUALITY LAW

In this section we shall consider formulas which contain the connectives $\wedge$, $\vee$ and $\neg$.

Definition. *Two formulas A and A* are said to be duals of each other if either one can be obtained from the other by replacing $\wedge$ by $\vee$ and $\vee$ by $\wedge$. The connectives $\wedge$ and $\vee$ are also called duals of each other. If the formulae, A contains the special variable T or F, then A*, its dual, is obtained by replacing T by F and F by T in addition to the above-mentioned interchanges.*

Example 1. *Write the duals of (i) $(P \vee Q) \wedge R$; (ii) $(P \wedge Q) \wedge T$*

Solution. Duals are (*i*) $(P \wedge Q) \vee R$; (*ii*) $(P \vee Q) \vee F$.

Example 2. *If the formula A is given by*

$$A : \neg (P \vee Q) \wedge (P \vee \neg (Q \wedge \neg R)), \text{ then find } A^*.$$

Solution. Dual of A is

$$A^* : \neg (P \wedge Q) \vee (P \wedge \neg (Q \vee \neg R)).$$

Theorem. Let A and A* be dual formulas and let $P_1, P_2, ..., P_n$ be the atomic variables that are in A and A*. That is A : A $(P_1, P_2, ..., P_n)$, A* : A* $(P_1, P_2, ..., P_n)$. Then through the use of Demorgan's laws $P \wedge Q \Leftrightarrow \neg (\neg P \vee \neg Q)$

$$P \vee Q \Leftrightarrow \neg (\neg P \wedge \neg Q)$$

1.8 TAUTOLOGICAL IMPLICATIONS

Definition. *Recall the definition of the conditional statement and its truth table. For any statement formula $P \to Q$, the statement formula $Q \to P$ is called its* ***Converse****, $\neg P \to \neg Q$ is called its* ***inverse*** *and $\neg Q \to \neg P$ is called its* ***contra-positive****.*

Example. *Show that the following :*

(i) $P \to Q \Leftrightarrow \neg Q \to \neg P$ *(ii)* $Q \to P \Leftrightarrow \neg P \to \neg Q$.

Solution.

P	Q	¬P	¬Q	P→Q	¬Q→¬P	Q→P	¬P→¬Q
T	T	F	F	T	T	T	T
T	F	F	T	F	F	T	T
F	T	T	F	T	T	F	F
F	F	T	T	T	T	T	T

$\therefore \quad P \to Q \Leftrightarrow \neg Q \to \neg P$ and $Q \to P \Leftrightarrow \neg P \to \neg Q$.

Definition. *A statement A is said to* ***tautologically imply*** *a statement B, if any and only if $A \to B$ is a tautology. We shall denote it by $A \Rightarrow B$, which is read as "A implies B".*

Note 1. $\Rightarrow$ is not a connective, $A \Rightarrow B$ is not a statement formula.

Note 2. By constructing the truth tables of A and B, we can determine whether $A \Rightarrow B$.

Example. *Define converse, contra positive and inverse of an implication.*

[JNTU – February, 2007 – Set 2 and 3, November, 2006 – Set 1, Set 2]

Solution. See the definition.

Example. *Prove that $(P \to Q) \Rightarrow (\neg Q \to \neg P)$.*

Solution.

P	Q	¬P	¬Q	P→Q	¬Q→¬P	(P→Q)→(¬Q→¬P)
T	T	F	F	T	T	T
T	F	F	T	F	F	T
F	T	T	F	T	T	T
F	F	T	T	T	T	T

Since all the entries in the last column are true, $(P \to Q) \to (\neg Q \to \neg P)$ is a tautology.

Hence $\quad (P \to Q) \Rightarrow (\neg Q \to \neg P)$.

Note. 3. In order to show any of the given implications, it is sufficient to show that an assignment of the truth value T to the antecedent of the corresponding conditional leads to the truth value T for the consequent. This procedure ensures that the conditional becomes a tautology, thereby proving the implication.

Example. *Prove that* $\neg Q \wedge (P \to Q) \Rightarrow \neg P$

Solution. Assume that the antecedent $\neg Q \wedge (P \to Q)$ has the truth value T, then both $\neg Q$ and $P \to Q$ have the truth value T, which means that Q has the value F. $P \to Q$ has the truth value T, and hence P must have the value F.

Therefore the consequent $\neg P$ must have the value T.

Note 4. Another method to show $A \Rightarrow B$ is to assume that the consequent B has the value F and then show that this assumption leads to *a's* having the value F.

Then $A \to B$ must have the value T.

Example. *Show that* $\neg (P \to Q) \Rightarrow P$.

Solution. Assume that P is false (F), when P is false $P \to Q$ has T, then $\neg (P \to Q)$ has F.

$\neg (P \to Q) \to P$ has T

Therefore $\neg (P \to Q) \Rightarrow P$.

Note 5. $A \Leftrightarrow B$ if and only if $A \Rightarrow B$ and $B \Rightarrow A$.

Note 6. If a formula is equivalent to a tautology, then it must be a tautology.

Note 7. If a formula is implied by a tautology then it must be a tautology.

Example. *Show that the implication is transitive.*

Solution. Assume that $A \Rightarrow B$ and $B \Rightarrow C$.

Then $A \to B$ and $B \to C$ are tautologies.

Hence $(A \to B) \wedge (B \to C)$ is also a tautology.

Therefore $(A \to B) \wedge (B \to C) \Rightarrow (A \to C)$

Since $(P \to Q) \wedge (Q \to R) \Rightarrow P \to R$

Hence $A \to C$ is a tautology.

Therefore $A \Rightarrow C$.

Therefore Implication is transitive.

Note 8. In order to show that $A \Rightarrow C$, it is convenient to introduce a series of formulas $B_1, B_2, ..., B_m$ such that $A \Rightarrow B_1, B_1 \Rightarrow B_2, ..., B_{m-1} \Rightarrow B_m$ and $B_m \Rightarrow C$.

Note 9. If $A \Rightarrow B$, $A \Rightarrow C$, then $A \Rightarrow (B \wedge C)$

(Assume if A is true, then B and C are both true. Thus $B \wedge C$ is true and hence $A \to (B \wedge C)$ is true.)

Theorem. If $H_1, H_2, ..., H_m$ and P imply Q, then $H_1, H_2, ..., H_m$ imply $P \to Q$.

Proof : Given that $(H_1 \wedge H_2 \wedge ... \wedge H_m \wedge P) \Rightarrow Q$.

This means $(H_1 \wedge H_2 \wedge ... \wedge H_m \wedge P) \to Q$ is a tautology.

Since $(P \wedge Q) \to R \Leftrightarrow P \to (Q \to R)$, $(H_1 \wedge H_2 \wedge ... \wedge H_m) \to (P \to Q)$ is a tautology.

Hence the theorem.

1.9 FUNCTIONALLY COMPLETE SETS OF CONNECTIVES

So far, we have defined the connectives $\wedge$, $\vee$, $\neg$, $\rightarrow$ and $\rightleftarrows$. Now we introduce some other connectives namely NAND and NOR which have useful applications in the design of computers. The word 'NAND' is a combination of 'NOT' and 'AND', while the word 'NOR' is a combination of 'NOT' and 'OR', where 'NOT' stands for negation, 'AND' stands for conjunction and 'OR' stands for disjunction.

Definition. *The connective 'NAND' is denoted by the symbol $\uparrow$, and it is defined as $P \uparrow Q \Leftrightarrow \neg (P \vee Q)$, for any two formulas P and Q.*

Definition. *The connective 'NOR' is denoted by the symbol $\downarrow$, and it is defined as $P \downarrow Q \Leftrightarrow \neg (P \vee Q)$, for any two formulas P and Q.*

Note: 1. Connectives $\uparrow$ and $\downarrow$ are duals of each other.

2. Connectives $\uparrow$ and $\downarrow$ truth tables are shown below.

Table 1.10

P	Q	$P \wedge Q$	$P \uparrow Q$	$P \vee Q$	$P \downarrow Q$
T	T	T	F	T	F
T	F	F	T	T	F
F	T	F	T	T	F
F	F	F	T	F	T

Definition. *Let P, Q be any two formulas and the connective exclusive OR (or) exclusive disjunction is denoted by the symbol $\bar{\vee}$, then truth value of $P \bar{\vee} Q$ is true whenever either P or Q is true, but not both true. Its truth table is given below :*

Table 1.11.

P	Q	$P \bar{\vee} Q$
T	T	F
T	F	T
F	T	T
F	F	F

The following equivalences follow from the definition of connective $\bar{\vee}$.

1. $P \bar{\vee} Q \Leftrightarrow Q \bar{\vee} P$.
2. $(P \bar{\vee} Q) \bar{\vee} R \Leftrightarrow P \bar{\vee} (Q \bar{\vee} R)$
3. $P \wedge (Q \bar{\vee} R) \Leftrightarrow (P \wedge Q) \bar{\vee} (P \wedge R)$
4. $(P \bar{\vee} Q) \Leftrightarrow (P \wedge \neg Q) \vee (\neg P \wedge Q)$
5. $(P \bar{\vee} Q) \Leftrightarrow \neg (P \rightleftarrows Q)$.

Definition. *A set of connectives is said to be* ***functionally complete*** *set of connectives if every formula can be expressed in terms of an equivalent formula containing the connectives only from this set.*

Note 1. Functionally complete set should not contain any redundant connectives *i.e.,* a connective which can be expressed in terms of the other connectives.

Already, we have

(*i*) $P \vee Q \Leftrightarrow \neg(\neg P \wedge \neg Q)$

(*ii*) $P \wedge Q \Leftrightarrow \neg(\neg P \vee \neg Q)$

(*iii*) $P \rightarrow Q \Leftrightarrow \neg P \vee Q$

(*iv*) $P \rightleftarrows Q \Leftrightarrow (\neg P \vee Q) \wedge (P \vee \neg Q)$.

In order to arrive at a functionally complete set, we first examine the following equivalence :

$$P \rightleftarrows Q \Leftrightarrow (P \rightarrow Q) \wedge (Q \rightarrow P).$$

This equivalence suggests that in any formula we can replace the part containing the bi-conditional by an equivalent formula not containing the bi-conditional. Thus all the bi-conditionals can be replaced in a formula.

Example 1. *Write an equivalent formula for $P \wedge (Q \rightleftarrows R) \vee (R \rightleftarrows P)$ which does not contain the bi-conditional.*

Solution. $P \wedge (Q \rightleftarrows R) \vee (R \rightleftarrows P) \Leftrightarrow P \wedge ((Q \rightarrow R) \wedge (R \rightarrow Q)) \vee ((R \rightarrow P) \wedge (P \rightarrow R))$.

Next we now consider the equivalence $P \rightarrow Q \Leftrightarrow \neg P \vee Q$. This equivalence suggests that the conditionals can also be eliminated by replacing those parts which contain conditionals by their equivalents.

Example 2. *Write an equivalent formula for $P \wedge (Q \rightleftarrows R)$ which contains neither the bi-conditional nor the conditional.*

Solution. $P \wedge (Q \rightleftarrows R) \Leftrightarrow P \wedge ((Q \rightarrow R) \wedge (R \rightarrow Q))$

$\Leftrightarrow \quad P \wedge ((\neg Q \vee R) \wedge (\neg R \vee Q))$

Hence we can first replace all bi-conditionals, then the conditionals and finally all the conjunctions or all the disjunctions in any formula, we can obtain an equivalent formula which contains either the negation and disjunction only or the negation and conjunction only. In other words, for every formula, we can find an equivalent formula containing the connectives $\vee$ and $\neg$ only or $\wedge$ and $\neg$ only. From the definition of functionally complete set of connectives, the sets of connectives $\{\wedge, \neg\}$ and $\{\vee, \neg\}$ are functionally complete sets.

Note 2. The set $\{\wedge, \vee\}$ is not functionally complete, as for the formula $\neg P$, it is not possible to find an equivalent formula containing connectives only from the set $\{\wedge, \vee\}$.

Example. Show that $\{\uparrow\}$, $\{\downarrow\}$ are functionally complete.

Proof : In order to prove, it is sufficient to show that the sets of connectives $\{\wedge, \neg\}$ and $\{\vee, \neg\}$ can be expressed either in terms of $\uparrow$ alone or in terms of $\downarrow$ alone.

(*i*) To show that $\{\vee, \neg\}$ is functionally complete, it is enough to show that $\neg$ and $\vee$ can be expressed in terms of $\downarrow$ alone. The following valid equalities help us in this direction. We have,

$$\neg P \Leftrightarrow \neg P \wedge \neg P \Leftrightarrow \neg(P \vee P) \Leftrightarrow P \downarrow P \text{ and}$$

$$P \vee Q \Leftrightarrow \neg(\neg P \wedge \neg Q) \Leftrightarrow \neg P \downarrow \neg Q \Leftrightarrow (P \downarrow P) \downarrow (Q \downarrow Q)$$

Then $\{\downarrow\}$ is a functionally complete set.

(*ii*) To show that $\{\wedge, \neg\}$ is functionally complete, it is enough to show that $\neg$ and $\wedge$ can be expressed in terms of $\uparrow$ alone.

$$\neg P \Leftrightarrow \neg P \vee \neg P \Leftrightarrow \neg (P \wedge P) \Leftrightarrow P \uparrow P \text{ and}$$

$$P \wedge Q \Leftrightarrow \neg (P \uparrow Q) \Leftrightarrow (P \uparrow Q) \uparrow (P \uparrow Q).$$

Then $\{\uparrow\}$ is a functionally complete set.

Note 3. We call each of the sets $\{\uparrow\}$ and $\{\downarrow\}$ a minimal functionally complete set or in short a minimal set, because a single operator NAND or NOR is functionally complete.

Note 4. The idea of minimal set of connectives is useful in the economics of the design of electronic circuits.

We now list some of the basic properties of the connectives NAND and NOR.

(*i*) $P \uparrow Q \Leftrightarrow Q \uparrow P$; $P \downarrow Q \Leftrightarrow Q \downarrow P$ (Commutative)

(*ii*) The connectives $\uparrow$ and $\downarrow$ are not associative.

Because $P \uparrow (Q \uparrow R) \Leftrightarrow P \uparrow \neg (Q \wedge R) \Leftrightarrow \neg (P \wedge \neg (Q \wedge R))$

$$\Leftrightarrow \neg P \vee (Q \wedge R)$$

while $(P \uparrow Q) \uparrow R \Leftrightarrow (P \wedge Q) \vee \neg R$

Thus the connective $\uparrow$ is not associative.

Similarly, we can easily observe that

$$P \downarrow (Q \downarrow R) \Leftrightarrow \neg P \wedge (Q \vee R), (P \downarrow Q) \downarrow R \Leftrightarrow (P \vee Q) \wedge \neg R.$$

Thus the connective $\downarrow$ is not associative.

(*iii*) $P \uparrow Q \uparrow R \Leftrightarrow \neg (P \wedge Q \wedge R)$

However, $P \uparrow Q \uparrow R$ is not equivalent to any of $P \uparrow (Q \uparrow R)$, $(P \uparrow Q) \uparrow R$, $Q \uparrow (P \uparrow R)$.

This non associatively of the connectives NAND and NOR creates some difficulty in using them.

(*iv*) $P \uparrow Q \Leftrightarrow \neg (P \wedge Q) \Leftrightarrow \neg P \vee \neg Q \Leftrightarrow (\neg P \wedge Q) \vee (P \wedge \neg Q) \vee (\neg P \wedge \neg Q)$.

$P \downarrow Q \Leftrightarrow \neg (P \vee Q) \Leftrightarrow \neg P \wedge \neg Q \Leftrightarrow (\neg P \vee Q) \wedge (P \vee \neg Q) \wedge (\neg P \vee \neg Q)$.

EXERCISE 1.1

1. State which of the following sentences are propositions. If so find its truth value.
   (*a*) A cow has four legs
   (*b*) $3 + 5 = 7$
   (*c*) $3 + 5 = 8$
   (*d*) $\sqrt{3}$ is irrational number
   (*e*) Moscow is the capital of Britain.
   (*f*) Open the door.
2. Write the negation of each of the following statements :
   (*a*) Arasavalli Surya Devalayam is in Andhra Pradesh.
   (*b*) 2 divides 5
   (*c*) Vijayawada is capital of Krishna District
   (*d*) Taj Mahal is in Agra.
3. Determine the truth value of each of the following :
   (*a*) Taj Mahal is in Agra and $5 + 4 = 8$

(*b*) If $3 + 5 = 8$, then $4 + 1 = 5$.

(*c*) $1 + 6 = 7$ or $2 + 4 = 7$.

4. State whether the following are well-formed formulas :

(*a*) $\neg (P \vee Q)$ (*b*) $\neg P \wedge Q$ (*c*) $(P \rightarrow Q) \rightarrow (\wedge Q)$

(*d*) $\sim p \vee (\sim q)$ (*e*) $p \rightarrow (q \vee r)$ (*f*) $(p \rightarrow (q \vee r))$

5. Given the truth values of p and q as 1 and those of r and s as 0, find the truth values of the following :

(*a*) $p \wedge (q \vee r)$ (*b*) $(p \leftrightarrow r) \wedge ((\neg q) \rightarrow s)$ O

(*c*) $(p \wedge (q \rightarrow (r \vee \sim q))) \leftrightarrow (r \vee \sim p)$

6. Let P be 'He is tall' and Q be 'He is handsome'. Write each of the following in symbolic form using P and Q.

(*a*) He is tall and handsome.

(*b*) He is tall but not handsome.

(*c*) He is neither tall nor handsome.

7. Translate the following sentences into propositional forms.

(*a*) Amar is rich and happy

(*b*) Amar is rich but not happy

(*c*) Amar is rich or happy

(*d*) Amar is neither rich nor happy.

8. Write the inverse, converse and contra positive of the following propositions :

(*a*) $P \rightarrow \neg Q$ (*b*) $P \rightarrow (Q \rightarrow R)$ (*c*) $(P \wedge (P \rightarrow Q)) \rightarrow Q$.

9. Construct the truth tables for the following :

(*a*) $(p \vee q) \wedge (\neg p)$ (*b*) $\neg (p \vee \neg q)$ (*c*) $p \rightarrow (q \rightarrow r)$

(*d*) $q \leftrightarrow (\neg p \vee \neg q)$ (*e*) $[(p \wedge q) \vee (\neg r)] \leftrightarrow p$.

10. Determine whether the following are tautologies :

(*a*) $p \rightarrow (q \rightarrow p)$ (*b*) $q \vee (\neg q \wedge p)$

(*c*) $p \rightarrow (p \vee q)$ (*d*) $p \vee (\sim (p \wedge q))$

11. Show that the truth values of the following formula is independent of their components.

$(P \rightleftarrows Q) \rightleftarrows ((P \wedge Q) \vee (\neg P \wedge \neg Q))$

12. Verify if the proposition $((p \vee \neg q) \rightarrow r) \leftrightarrow s \vee \neg (((p \vee \neg q) \rightarrow r) \leftrightarrow s)$ is a tautology.

13. Show that the following propositions are contradiction.

(*a*) $(p \wedge q) \wedge \neg (p \vee q)$ (*b*) $(\sim q \wedge p) \wedge q$

14. Show that $\neg (p \vee (\neg p \wedge q))$ and $\neg p \wedge \neg q$ are logically equivalent.

15. Show that the following :

(*a*) $((p \vee q) \rightarrow r) \Leftrightarrow ((p \rightarrow r) \wedge (q \rightarrow r))$

(*b*) $p \wedge (q \vee r) \Leftrightarrow (p \wedge q) \vee (p \wedge r)$

(*c*) $((p \rightarrow q) \wedge (p \rightarrow r)) \Leftrightarrow p \rightarrow (q \wedge r)$

(*d*) $((p \rightleftarrows q) \wedge (q \rightleftarrows r) \wedge (r \rightleftarrows p)) \rightleftarrows ((p \wedge q) \wedge (q \rightarrow r) \wedge (r \rightarrow p))$.

16. Prove the following logical equivalences without using truth tables.

(*a*) $p \vee (p \wedge (p \vee q)) \Leftrightarrow p$

(*b*) $((\neg p \vee \neg q) \rightarrow (p \wedge q \wedge r)) \Leftrightarrow p \wedge q$

(c) $(p \vee q \vee (\sim p \wedge \sim q \wedge r)) \Leftrightarrow (p \vee q \vee r)$.

17. Prove that each of the following is a tautology.

(a) $(\neg p \wedge (\neg q \wedge r)) \vee (q \wedge r) \vee (p \wedge r) \rightleftarrows r$ (b) $(p \wedge (p \rightarrow q)) \rightarrow q$.

18. Show that the following equivalences

(a) $(P \rightarrow Q) \wedge (R \rightarrow Q) \Leftrightarrow (P \vee R) \rightarrow Q$ (b) $\neg (P \rightleftarrows Q) \Leftrightarrow (P \vee Q) \wedge \neg (P \wedge Q)$.

[JNTU June, 2004 - Set-01]

19. Show the following implications without constructing the truth tables

(a) $((P \vee \neg P) \rightarrow Q) \rightarrow ((P \vee \neg P) \rightarrow R) \Rightarrow (Q \rightarrow R)$

(b) $\neg Q \wedge (P \rightarrow Q) \Rightarrow \neg P$

(c) $(P \rightarrow Q) \wedge (Q \rightarrow R) \Rightarrow (P \rightarrow R)$

(d) $(P \vee Q) \wedge (P \rightarrow R) \wedge (Q \rightarrow R) \Rightarrow R$.

20. Write down the duels of the following propositions.

(a) $\neg (p \vee q) \wedge (p \vee \neg (q \wedge \neg s))$ (b) $((p \wedge q) \vee T) \wedge ((p \vee q) \wedge F)$

21. Verify the principle of duality for the following logical equivalence : $(\neg (p \wedge q) \rightarrow (\neg p \vee (\neg p \vee q))) \Leftrightarrow (\neg p \vee q)$

22. Show that $(P \wedge Q) \vee (\neg P \wedge Q) \vee (P \wedge \neg Q) \vee (\neg P \wedge \neg Q)$ is a tautology.

[JNTU - November, 2005 - Set 1, November, 2003 - Set - 3]

23. Construct the truth table of the following formula $\neg (P \vee (Q \wedge R)) \wedge ((P \vee Q) \wedge (P \vee R))$

[JNTU - May, 05 - Set 4, November, 2003 - Set 1]

24. What is the minimal set of connections required for a well formed formula.

[JNTU - November, 2006 Set - 3, February, 2007 - Set - 4]

ANSWERS

1. (a), (b), (c), (d), (e) are propositions but (f) is not proposition. and (a), (c), (d) truth values are Ts and (b), (e) are Fs

2. (a) Arasavalli Surya Devalayam is not in A.P.

(b) 2 does not divide 5

(c) Vijayawada is not capital of Krishna district.

(d) It is not true that Taj Mahal is in Agra.

3. (a) F (b) T (c) T

4. (a) Yes (b) No (c) No (d) No

(e) No (f) Yes.

5. (a) 1 (b) 0 (c) 1

6. (a) $P \wedge Q$ (b) $P \wedge \neg Q$ (c) $\neg P \wedge \neg Q$

7. Take R : Amar is rich, H : Amar is happy.

(a) $R \wedge H$ (b) $R \wedge \neg H$ (c) $R \vee H$ (d) $\neg R \wedge \neg H$.

8. (a) $\sim P \rightarrow Q$, $\sim Q \rightarrow P$, $Q \rightarrow \sim P$

(b) $\sim P \rightarrow (Q \wedge \sim R)$, $(Q \rightarrow R) \rightarrow P$, $(Q \wedge \neg R) \rightarrow \neg P$

(c) $\neg P \vee (P \wedge Q) \rightarrow \neg Q$, $Q \rightarrow (P \wedge (P \rightarrow Q))$, $\neg Q \rightarrow (\neg P \vee (P \wedge \neg Q))$

9. (*a*),(*b*), (*d*)

p	q	$(p \vee q) \wedge \neg p$	$\neg(p \vee \neg q)$	$q \leftrightarrow (\neg p \vee \neg q)$
T	T	F	F	F
T	F	F	F	F
F	T	T	T	T
F	F	F	F	F

(*c*), (*e*)

p	q	r	$p \to (q \to r)$	$((p \wedge q) \vee \neg r) \leftrightarrow p$
T	T	T	T	T
T	T	F	F	T
T	F	T	T	F
T	F	F	T	T
F	T	T	T	F
F	T	F	T	F
F	F	T	T	T
F	F	F	T	F

10. (*a*) Yes (*b*) No (*c*) Yes (*d*) Yes.

20. (*a*) $\neg(p \wedge q) \vee (p \wedge \neg(q \vee \neg s))$

(*b*) $((p \vee q) \wedge \text{F}) \vee ((p \wedge q) \vee \text{T})$

23. Consider A : $\neg(\text{P} \vee (\text{Q} \wedge \text{R})) \wedge ((\text{P} \vee \text{Q}) \wedge (\text{P} \vee \text{R}))$

P	Q	R	A
T	T	T	F
T	T	F	F
T	F	T	F
T	F	F	F
F	T	T	F
F	T	F	F
F	F	T	F
F	F	F	F

1.10 NORMAL FORMS

Let A $(\text{P}_1, \text{P}_2, ..., \text{P}_n)$ be a statement formula where $\text{P}_1, \text{P}_2, ..., \text{P}_n$ are the atomic variables. If we consider all possible assignments of the truth values of $\text{P}_1, \text{P}_2, ..., \text{P}_n$ and obtain the resulting truth values of the formula A, then we get the truth table for A. Such a truth table contains 2^n rows. But this is very tedious, and the construction of truth tables may not be practical, even with the aid of a computer. We therefore, consider other procedures known as reduction to normal forms.

In this section, we will denote the word **"product"** in place of **"conjunction"** and **"sum"** in place of **"disjunction"**.

Definition. *A product of the variables and their negations in a formula is called an **elementary product**.*

Example. Let P, Q be any two atomic variables. Then P, ¬P, ¬P ∧ Q, ¬Q ∧ P ∧ ¬P, P ∧ ¬P, Q ∧ ¬P, ¬P ∧ ¬Q are some examples of elementary products.

Definition. *A sum of the variables and their negations in a formula is called an **elementary sum**.*

Example. Let P, Q be any two atomic variables. Then P, ¬P, ¬P ∨ Q, ¬Q ∨ P ∨ ¬P, P ∨ ¬P, Q ∨ ¬P, ¬P ∨ ¬Q are some examples of elementary sums.

Definition. *Any part of an elementary sum (product) which is itself an elementary sum (product) is called a **factor** of the original elementary sum (product).*

Example. Q, Q ∨ ¬P, ¬P ∨ ¬Q are some of the factors of Q ∨ ¬P ∨ ¬Q and ¬Q, P ∧ ¬P, ¬Q ∧ P are some of the factors of ¬Q ∧ P ∧ ¬P.

The following statements hold for elementary sums and products.

Result 1. A necessary and sufficient condition for an elementary product to be identically false is that it contains atleast one pair of factors in which one is the negation of the other.

Explanation. We know that for any variable P, P ∧ ¬P is identically false. Hence if P ∧ ¬P appears in the elementary product, then the product is identically false.

Result 2. A necessary and sufficient condition for an elementary sum to be identically true is that it contain atleast one pair of factors in which one is the negation of the other.

Explanation. We know that for any variable, P, P ∨ ¬P is identically true. Hence if P ∨ ¬P appears in the elementary sum, then the sum is identically true.

Example 1. P ∧ ¬Q ∧ ¬P is identically false.

Example 2. P ∨ ¬Q ∨ ¬P is identically true.

1.10.1 Disjunctive Normal Form (d.n.f.)

Definition. *A formula which is equivalent to a given formula and which consists of a sum of elementary products is called a **disjunctive normal form** of the given formula.*

Procedure to obtain a disjunctive normal form of a given formula.

Step 1. If the connectives "→" and "⇄" appear in the given formula, obtain an equivalent formula in which "→" and "⇄" do not appear, *i.e.,* an equivalent formula can be obtained in which "→" and "⇄" are replaced by ∧, ∨ and ¬.

Example. P → Q is replaced by ¬P ∨ Q and P ⇄ Q is replaced by either (P ∧ Q) ∨ (¬P ∧ ¬Q) or (¬P ∨ Q) ∧ (¬Q ∨ P).

Therefore there is no loss of generality in assuming that the given formula contains the connectives ∧, ∨ and ¬ only.

Step 2. If the negation is applied to the formula or to a part of the formula and not to the variables appearing in it, then by using Demorgan's laws an equivalent formula can be obtained in which the negation is applied to the variables only.

Step 3. Apply the distributive law until we get a sum of elementary products is obtained and apply idempotent law if necessary. Thus, we obtain disjunctive normal form of a given formula.

Note 1. The *d.n.f.* of a given formula is not unique, because different *d.n.f. s* can be obtained for a given formula if the distributive laws are applied in different ways.

Note 2. A given formula is identically false if every elementary product appearing in its disjunctive normal form is identically false.

Example. *Obtain disjunctive normal forms of the following formulas :*

(*i*) $P \wedge (P \to Q)$ (*ii*) $\neg (P \vee Q) \rightleftarrows (P \wedge Q)$.

Solution. (*i*) $P \wedge (P \to Q) \Leftrightarrow P \wedge (\neg P \vee Q) \Leftrightarrow (P \wedge \neg P) \vee (P \wedge Q)$

which is *d.n.f.* of given formula.

(*ii*) $\neg (P \vee Q) \rightleftarrows (P \wedge Q) \Leftrightarrow (\neg (P \vee Q) \wedge (P \wedge Q)) \vee ((P \vee Q) \wedge \neg (P \wedge Q))$ (Elimination of bi-conditional)

$\Leftrightarrow (\neg P \wedge \neg Q \wedge P \wedge Q) \vee ((P \vee Q) \wedge (\neg P \vee \neg Q))$

$\Leftrightarrow (\neg P \wedge \neg Q \wedge P \wedge Q) \vee (P \wedge (\neg P \vee \neg Q)) \vee (Q \wedge (\neg P \vee \neg Q))$

$\Leftrightarrow (\neg P \wedge \neg Q \wedge P \wedge Q) \vee (P \wedge \neg P) \vee (P \wedge \neg Q) \vee (Q \wedge \neg P) \vee (Q \wedge \neg Q)$ (Distributive law)

which is the required d.n.f. of given formula.

Example. *Find disjunctive normal form of*

(*i*) $Q \wedge (P \vee \neg Q)$ (*ii*) $(Q \to P) \wedge (\neg P \wedge Q)$

Solution. (*i*) $Q \wedge (P \vee \neg Q) \Leftrightarrow (Q \wedge P) \vee (Q \wedge \neg Q)$

which is *d.n.f.* of given formula.

(*ii*) $(Q \to P) \wedge (\neg P \wedge Q) \Leftrightarrow (\neg Q \vee P) \wedge (\neg P \wedge Q)$

$\Leftrightarrow (\neg Q \wedge \neg P \wedge Q) \vee (P \wedge \neg P \wedge Q)$ (Distributive law)

which is *d.n.f.* of given formula.

Example. *Obtain disjunctive normal form of* $P \to ((P \to Q) \wedge \neg (\neg Q \vee \neg P))$

Solution. $P \to ((P \to Q) \wedge \neg (\neg Q \vee \neg P))$

$\Leftrightarrow \neg P \vee ((P \to Q) \wedge \neg (\neg Q \vee \neg P))$

$\Leftrightarrow \neg P \vee ((\neg P \vee Q) \wedge (Q \wedge P))$

$\Leftrightarrow \neg P \vee ((\neg P \wedge Q \wedge P) \vee (Q \wedge Q \wedge P))$

$\Leftrightarrow \neg P \vee (\neg P \wedge P \wedge Q) \vee (Q \wedge P)$

$\Leftrightarrow \neg P \vee (\neg P \wedge P \wedge Q) \vee (P \wedge Q)$

which is required *d.n.f.*

1.10.2 Conjunctive Normal Form (*c.n.f.*)

Definition. *A formula which is equivalent to a given formula and which consists of a product of elementary sums is called a* ***conjunctive normal form*** *of the given formula.*

The method for finding a conjunctive normal form of a given formula is similar to the one given for disjunctive normal form.

Note 1. The conjunctive normal form is not unique.

For example, consider $P \wedge (Q \vee R)$. This is already in the *c.n.f.*

We may write

$$P \wedge (Q \vee R) \Leftrightarrow (P \wedge Q) \vee (P \wedge R)$$
$$\Leftrightarrow ((P \wedge Q) \vee P) \wedge ((P \wedge Q) \vee R)$$
$$\Leftrightarrow (P \vee P) \wedge (Q \vee P) \wedge (P \vee R) \wedge (Q \vee R)$$

which is another equivalent *c.n.f.*

Note 2. A given formula is identically true (or) a tautology if every elementary sum in its conjunctive normal form is identically true.

Example. *Obtain conjunctive normal forms of the following formulas :*

(i) $P \wedge (P \rightarrow Q)$ (ii) $\neg (P \vee Q) \rightleftarrows (P \wedge Q)$.

Solution. (*i*) $P \wedge (P \rightarrow Q) \Leftrightarrow P \wedge (\neg P \vee Q)$

which is required *c.n.f.*

(*ii*) $\neg (P \vee Q) \rightleftarrows (P \wedge Q) \Leftrightarrow (\neg (P \vee Q) \rightarrow (P \wedge Q)) \wedge ((P \wedge Q) \rightarrow \neg (P \vee Q))$

$\because \quad (R \rightleftarrows S) \Leftrightarrow (R \rightarrow S) \wedge (S \rightarrow R))$

$$\Leftrightarrow ((P \vee Q) \vee (P \wedge Q)) \wedge (\neg (P \wedge Q) \vee \neg (P \vee Q))$$
$$\Leftrightarrow ((P \vee Q \vee P) \wedge (P \vee Q \vee Q)) \wedge ((\neg P \vee \neg Q) \vee (\neg P \wedge \neg Q))$$
$$\Leftrightarrow (P \vee Q \vee P) \wedge (P \vee Q \vee Q) \wedge (\neg P \vee \neg Q \vee \neg P) \wedge (\neg P \vee \neg Q \vee \neg Q)$$

which is required *c.n.f.*

$$\Leftrightarrow (P \vee Q) \wedge (P \vee Q) \wedge (\neg P \vee \neg Q) \wedge (\neg P \vee \neg Q)$$
$$\Leftrightarrow (P \vee Q) \wedge (\neg P \vee \neg Q)$$

which is also required *c.n.f.*

Example. *Obtain c.n.f. of the following formulas :*

(i) $((p \rightarrow q) \wedge \neg Q) \rightarrow \neg P$ (ii) $((p \rightarrow q) \wedge \neg P) \rightarrow \neg Q$

and find which of the above formulas are tautologies :

Solution. (*i*) $((p \rightarrow q) \wedge \neg Q) \rightarrow \neg P \Leftrightarrow ((\neg P \vee q) \wedge \neg q) \rightarrow \neg P$

$$\Leftrightarrow \quad \neg ((\neg P \vee q) \wedge \neg q) \vee \neg P$$
$$\Leftrightarrow \quad ((\neg\neg p \wedge \neg q) \vee \neg\neg q) \vee \neg P$$
$$\Leftrightarrow \quad ((p \wedge \neg q) \vee q) \vee \neg P$$
$$\Leftrightarrow \quad ((p \vee q) \wedge (\neg q \vee q))0 \vee \neg P$$
$$\Leftrightarrow \quad (p \vee q \vee \neg P) \wedge (\neg q \vee q \vee \neg P)$$

which is required *c.n.f.*

Since each of the elementary sums is a tautology, the given formula is a tautology.

(*ii*) $((p \to q) \wedge \neg P) \to \neg Q \Leftrightarrow ((\neg P \vee q) \wedge \neg P) \to \neg q$

$\Leftrightarrow \quad \neg((\neg P \vee q) \wedge \neg P) \vee \neg q$

$\Leftrightarrow \quad \neg(\neg P \vee q) \vee \neg\neg p \vee \neg q$

$\Leftrightarrow \quad ((p \wedge \neg Q) \vee p) \vee \neg q$

$\Leftrightarrow \quad ((p \vee p) \wedge (\neg q \vee p)) \vee \neg q$

$\Leftrightarrow \quad (p \vee p \vee \neg q) \wedge (\neg q \vee p \vee \neg q).$

which is required *c.n.f.*

Since no one elementary sum is tautology, the given formula is not a tautology.

1.10.3 Principal Disjunctive Normal Form (p.d.n.f.)

Definition. *A **min term** consists of conjunctions in which each variable or its negation, but not both, appears only once.*

Example 1. For two variables P, Q, there are 2^2 min terms given by $P \wedge Q$, $P \wedge \neg Q$, $\neg P \wedge Q$ and $\neg P \wedge \neg Q$ which are also called Boolean conjunctions of P, Q.

Example 2. For three variables P, Q and R, there are 2^3 min terms given by $P \wedge Q \wedge R$, $P \wedge Q \wedge \neg R$, $P \wedge \neg Q \wedge R$, $P \wedge \neg Q \wedge \neg R$, $\neg P \wedge Q \wedge R$, $\neg P \wedge Q \wedge \neg R$, $\neg P \wedge \neg Q \wedge R$, $\neg P \wedge \neg Q \wedge \neg R$.

Note 1. From the following truth tables of min terms for P, Q, it is clear that no two min terms are equivalent.

Table 1.12

P	Q	$P \wedge Q$	$P \wedge \neg Q$	$\neg P \wedge Q$	$\neg P \wedge \neg Q$
T	T	T	F	F	F
T	F	F	T	F	F
F	T	F	F	T	F
F	F	F	F	F	T

Note 2. Each min term has the truth value T for exactly one combination of the truth values of the variables P and Q. This fact is shown in above table.

We assert that if the truth table of any formula containing only the variables P and Q is known, then one can easily obtain an equivalent formula which consists of a disjunction of some of the min terms.

Definition. *An equivalent formula consisting of disjunctions of min terms only is known as its **principal disjunctive normal form** (Sum-of-products canonical form).*

Simply *p.d.n.f.*

Methods to obtain *p.d.n.f.* of a given formula.

I. Truth Table Method :

For every truth value T of the given formula, select the min term which also has the value T for the same combination of the truth values of the variables.

Example. *Obtain p.d.n.f. of each of the following :*

(i) $P \to Q$ *(ii)* $P \vee Q$ *(iii)* $\neg (P \wedge Q)$.

Solution.

Table 1.13.

P	Q	$P \to Q$	$P \vee Q$	$P \wedge Q$	$\neg (P \wedge Q)$
T	T	T	T	T	F
T	F	F	T	F	T
F	T	T	T	F	T
F	F	T	F	F	T

Hence

(i) $P \to Q \Leftrightarrow (P \wedge Q) \vee (\neg P \wedge Q) \vee (\neg P \wedge \neg Q)$

(ii) $P \vee Q \Leftrightarrow (P \wedge Q) \vee (P \wedge \neg Q) \vee (\neg P \wedge Q)$

(iii) $\neg (P \wedge Q) \Leftrightarrow (P \wedge \neg Q) \vee (\neg P \wedge Q) \vee (\neg P \wedge \neg Q)$.

which are *p.d.n.* forms of given formulas.

Note. The number of min terms appearing in the normal form is the same as the number of entries with the truth value T in the truth table of the given formula. Thus, every formula (which is not a contradiction) has an equivalent *p.d.n.f.* Such a normal form is unique.

Example. *Obtain the p.d.n.f. of the following formula.*

$P \vee (\neg P \to (Q \vee (\neg Q \to R)))$. [JNTU - May, 2005, Set 3]

Solution. Consider $A : \neg P \to (Q \vee (\neg Q \to R))$, $B : P \vee (\neg P \to (Q \vee (\neg Q \to R)))$.

P	Q	R	$\neg P$	$\neg Q$	$\neg Q \to R$	$Q \vee (\neg Q \to R)$	A	B	Necessary minterms
T	T	T	F	F	T	T	T	T	$P \wedge Q \wedge R$
T	T	F	F	F	T	T	T	T	$P \wedge Q \wedge \neg R$
T	F	T	F	T	T	T	T	T	$P \wedge \neg Q \wedge R$
T	F	F	F	T	F	F	T	T	$P \wedge \neg Q \wedge \neg R$
F	T	T	T	F	T	T	T	T	$\neg P \wedge Q \wedge R$
F	T	F	T	F	T	T	T	T	$\neg P \wedge Q \wedge \neg R$
F	F	T	T	T	T	T	T	T	$\neg P \wedge \neg Q \wedge R$
F	F	F	T	T	F	F	F	F	

$\therefore$ The *p.d.n.f.* of $P \vee (\neg P \to (Q \vee (\neg Q \to R)))$ is $(P \wedge Q \wedge R) \vee (P \wedge Q \wedge \neg R) \vee (P \wedge \neg Q \wedge R) \vee (P \wedge \neg Q \wedge \neg R) \vee (\neg P \wedge Q \wedge R) \vee (\neg P \wedge Q \wedge \neg R) \vee (\neg P \wedge \neg Q \wedge R)$.

II. Replacement Method (Without using truth table)

To obtain the *p.d.n.f.* of a given formula without using its truth table, we use the following steps :

1. First replace the conditionals and bi-conditionals by their equivalent formulas containing only $\wedge$, $\vee$ and $\neg$.
2. The negations are applied to the variables by using Demorgan's laws followed by the application of distributive laws (as was done earlier in obtaining *d.n.f.* or *c.n.f.*)

3. Any elementary product which is a contradiction is dropped.
4. Min terms are obtained in the disjunctions by introducing the missing factors. Identical min terms appearing in the disjunctions are deleted.

Note. This method can be used to determine whether two given formulas are equivalent.

Example. *Obtain the p.d.n.f. for the following formulas, without using truth tables.*

(i) $P \rightleftarrows Q$ (ii) $\neg P \vee Q$

(iii) $(P \wedge Q) \vee (\neg P \wedge R) \vee (Q \wedge R)$

Solution. (*i*) $P \rightleftarrows Q \Leftrightarrow (\neg P \vee Q) \wedge (P \vee \neg Q)$

$$\Leftrightarrow (\neg P \wedge P) \vee (\neg P \wedge \neg Q) \vee (Q \wedge P) \vee (Q \wedge \neg Q)$$
$$\Leftrightarrow F \vee (\neg P \wedge \neg Q) \vee (P \wedge Q) \vee F$$
$$\Leftrightarrow (P \wedge Q) \vee (\neg P \wedge \neg Q)$$

which is required *p.d.n.f.*

(*ii*) $\neg P \vee Q \Leftrightarrow (\neg P \wedge T) \vee (T \wedge Q)$

$$\Leftrightarrow (\neg P \wedge (Q \vee \neg Q)) \vee ((P \vee \neg P) \wedge Q)$$
$$\Leftrightarrow (\neg P \wedge Q) \vee (\neg P \wedge \neg Q) \vee (P \wedge Q) \vee (\neg P \wedge Q)$$
$$\Leftrightarrow (\neg P \wedge Q) \vee (\neg P \wedge \neg Q) \vee (P \wedge Q)$$
$$\Leftrightarrow (P \wedge Q) \vee (\neg P \wedge Q) \vee (\neg P \wedge \neg Q)$$

which is required *p.d.n.f.*

(*iii*) $(P \wedge Q) \vee (\neg P \wedge R) \vee (Q \wedge R)$

$$\Leftrightarrow ((P \wedge Q) \wedge T) \vee ((\neg P \wedge R) \wedge T) \vee ((Q \wedge R) \wedge T)$$
$$\Leftrightarrow ((P \wedge Q) \wedge (R \vee \neg R)) \vee ((\neg P \wedge R) \wedge (Q \vee \neg Q)) \wedge ((Q \wedge R) \wedge (P \vee \neg P))$$
$$\Leftrightarrow (P \wedge Q \wedge R) \vee (P \wedge Q \wedge \neg R) \vee (\neg P \wedge R \wedge Q) \vee (\neg P \wedge R \wedge \neg Q) \vee (Q \wedge R \wedge P) \vee (Q \wedge R \wedge \neg P)$$
$$\Leftrightarrow (P \wedge Q \wedge R) \vee (P \wedge Q \wedge \neg R) \vee (\neg P \wedge Q \wedge R) \vee (\neg P \wedge \neg Q \wedge R)$$

which is required *p.d.n.f.*

1.10.4 Principal Conjunctive Normal Form (*p.c.n.f.*)

Definition. *A **max term** consists of disjunctions in which each variable or its negation, but not both, appears only once.*

Example 1. For two variables P, Q, there are 2^2 max terms given by $P \vee Q$, $P \vee \neg Q$, $\neg P \vee Q$ and $\neg P \vee \neg Q$.

Example 2. For three variables P, Q and R, there are 2^3 max terms given by $P \vee Q \vee R$, $P \vee Q \vee \neg R$, $P \vee \neg Q \vee R$, $P \vee \neg Q \vee \neg R$, $\neg P \vee Q \vee R$, $\neg P \vee Q \vee \neg R$, $\neg P \vee \neg Q \vee R$, $\neg P \vee \neg Q \vee \neg R$.

Note 1. The max terms are duals of min terms.

2. Either from the duality principle or directly from the truth tables it can be concluded that each of the max terms has the truth value F for exactly one combination of the truth values of the variables. Different max terms have the truth value F for different combinations of the truth values of the variables.

Definition. ***Principal Conjunctive Normal Form (p.c.n.f.) :*** *For a given formula, an equivalent formula consisting of conjunctions of the max terms only is known as its principal conjunctive normal form. It is also called the* ***product-of-sums canonical form****.*

Note 1. Every formula, which is not a tautology has an equivalent principal conjunctive normal form which is unique except for the rearrangement of the factors in the max terms and their conjunctions.

Note 2. All the assertions made for the principal disjunctive normal forms can also be made for the principal conjunctive normal forms by the duality principle.

Methods to obtain *p.c.n.f.* of a given formula.

1. The method for obtaining the principal conjunctive normal form for a given formula is similar to the method for the principal disjunctive normal form.
2. If the principal disjunctive (conjunctive) normal form of a given formula A containing n variables is known, then the principal disjunctive (conjunctive) normal form of $\neg A$ will consist of the disjunction (conjunction) of the remaining min terms (max terms) which do not appear in the principal disjunctive (conjunctive) normal form of A. From $A \Leftrightarrow \neg\neg A$ one can obtain the principal conjunctive (disjunctive) normal form of A by repeated applications of Demorgan's laws to the principal disjunctive (conjunctive) normal form of $\neg A$.

Example. *Obtain the principal conjunctive normal form of the formula A given by $(\neg P \to R) \wedge (Q \rightleftarrows P)$ and hence find disjunctive normal form of A.*

Solution. A: $(\neg P \to R) \wedge (Q \rightleftarrows P) \Leftrightarrow (\neg\neg P \vee R) \wedge ((Q \to P) \wedge (P \to Q))$

$\Leftrightarrow \quad (P \vee R) \wedge ((\neg Q \vee P) \wedge (\neg P \vee Q))$

$\Leftrightarrow \quad (P \vee R \vee (Q \wedge \neg Q)) \wedge (\neg Q \vee P \vee (R \wedge \neg R)) \wedge (\neg P \vee Q \vee (R \wedge \neg R))$

Since $\quad P \vee F \Leftrightarrow P$

$\Leftrightarrow \quad (P \vee R \vee Q) \wedge (P \vee R \vee \neg Q) \wedge (\neg Q \vee P \vee R) \wedge (\neg Q \vee P \vee \neg R) \wedge (\neg P \vee Q \vee R) \wedge (\neg P \vee Q \vee \neg R).$

$\Leftrightarrow \quad (P \vee Q \vee R) \wedge (P \vee \neg Q \vee R) \wedge (P \vee \neg Q \vee R) \wedge (P \vee \neg Q \vee \neg R) \wedge (\neg P \vee Q \vee R) \wedge (\neg P \vee Q \vee \neg R)$ which is *p.c.n.f.* of A.

Now $\quad \neg A : (P \vee Q \vee \neg R) \wedge (\neg P \vee \neg Q \vee R) \wedge (\neg P \vee \neg Q \vee \neg R)$ (Remaining max terms) is *p.c.n.f.* of $\neg A$.

$\therefore \quad \neg\neg A : \neg ((P \vee Q \vee \neg R) \wedge (\neg P \vee \neg Q \vee R) \wedge (\neg P \vee \neg Q \vee \neg R))$

$A \Leftrightarrow \neg (P \vee Q \vee \neg R) \vee \neg (\neg P \vee \neg Q \vee R) \vee \neg (\neg P \vee \neg Q \vee \neg R)$

$\Leftrightarrow \quad (\neg P \wedge \neg Q \wedge R) \vee (P \wedge Q \wedge \neg R) \vee (P \wedge Q \wedge R)$

(Since $\neg\neg A = A$)

which is *p.d.n.f.* of A by Demorgan's law.

Example. *Obtain p.c.n.f. of $(P \wedge Q) \vee (\neg P \wedge Q) \vee (P \wedge \neg Q)$*

Solution. $(P \wedge Q) \vee (\neg P \wedge Q) \vee (P \wedge \neg Q) \Leftrightarrow (P \wedge Q) \vee (P \wedge \neg Q) \vee (\neg P \wedge Q)$

$\Leftrightarrow \quad [(P \wedge (Q \vee \neg Q)] \vee (\neg P \wedge Q)$

$\Leftrightarrow \quad (P \wedge T) \vee (\neg P \wedge Q)$

$\Leftrightarrow \quad P \vee (\neg P \wedge Q)$

$\Leftrightarrow \quad (P \vee \neg P) \wedge (P \vee Q)$

$\Leftrightarrow \quad T \wedge (P \vee Q)$

$\Leftrightarrow \quad P \vee Q.$

Alternative method : Let A : $(P \wedge Q) \vee (\neg P \wedge Q) \vee (P \wedge \neg Q)$ be *p.d.n.f.*.

$\neg A : (\neg P \wedge \neg Q)$ (remaining terms) is *p.d.n.f.* of $\neg A$.

$\neg\neg A : \neg (\neg P \wedge \neg Q) \Leftrightarrow P \vee Q$ is *p.c.n.f.* of A

$A : P \vee Q$ is *p.c.n.f.* of A.

Example. *Find p.c.n.f. of* $(P \wedge Q) \vee (\neg P \wedge Q \wedge R)$.

Solution. Let S : $(P \wedge Q) \vee (\neg P \wedge Q \wedge R) \Leftrightarrow ((P \wedge Q) \wedge T) \vee (\neg P \wedge Q \wedge R)$

$\Leftrightarrow \quad ((P \wedge Q) \wedge (R \vee \neg R)) \vee (\neg P \wedge Q \wedge R)$

$\Leftrightarrow \quad (P \wedge Q \wedge R) \vee (P \wedge Q \wedge \neg R) \vee (\neg P \wedge Q \wedge R)$ which is *p.d.n.f.* of S.

p.d.n.f. of $\neg S$ is $(P \wedge \neg Q \wedge R) \vee (P \wedge \neg Q \wedge \neg R) \vee (\neg P \wedge \neg Q \wedge R) \vee (\neg P \wedge Q \wedge \neg R) \vee (\neg P \wedge \neg Q \wedge \neg R)$.

p.c.n.f. of $\neg\neg S \Leftrightarrow S$ is $\neg (P \wedge \neg Q \wedge R) \wedge \neg (P \wedge \neg Q \wedge \neg R) \wedge \neg (\neg P \wedge \neg Q \wedge R) \wedge \neg (\neg P \wedge Q \wedge \neg R) \wedge \neg (\neg P \wedge \neg Q \wedge \neg R)$ (Demorgan's law)

$\Leftrightarrow (\neg P \vee Q \neg R) \wedge (\neg P \vee Q \vee R) \wedge (P \vee Q \vee \neg R) \wedge (P \vee \neg Q \vee R) \wedge (P \vee Q \vee R)$

which is *p.c.n.f.* of S.

Example. *Obtain the product of sums canonical form (or) p.c.n.f. of the following formula.*

$A : (\neg S \wedge \neg P \wedge R \wedge Q) \vee (S \wedge P \wedge \neg R \wedge \neg Q) \vee (\neg S \wedge P \wedge R \wedge \neg Q) \vee (Q \wedge \neg P \wedge \neg R \wedge S) \vee (P \wedge \neg S \wedge \neg R \wedge Q).$

Solution. Given A : $(\neg S \wedge \neg P \wedge R \wedge Q) \vee (S \wedge P \wedge \neg R \wedge \neg Q) \vee (\neg S \wedge P \wedge R \wedge \neg Q) \vee (Q \wedge \neg P \wedge \neg R \wedge S) \vee (P \wedge \neg S \wedge \neg R \wedge Q)$ is *p.d.n.f.* of A.

p.d.n.f. of $\neg A$ is $(\neg S \wedge \neg P \wedge R \wedge \neg Q) \vee (\neg S \wedge \neg P \wedge \neg R \wedge Q) \vee (\neg S \wedge \neg P \wedge \neg R \neg Q) \vee (\neg S \wedge P \wedge R \wedge Q) \vee (\neg S \wedge P \wedge \neg R \wedge \neg Q) \vee (S \wedge P \wedge R \wedge Q) \vee (S \wedge P \wedge \neg R \wedge Q) \vee (S \wedge P \wedge R \wedge \neg Q) \vee (S \wedge \neg P \wedge R \wedge Q) \vee (S \wedge \neg P \wedge R \wedge \neg Q) \vee (S \wedge \neg P \wedge \neg R \wedge \neg Q)$.

Therefore *p.c.n.f.* of $\neg\neg A \Leftrightarrow A$ is $(S \vee P \vee \neg R \vee Q) \wedge (S \vee P \vee R \vee \neg Q) \wedge (S \vee P \vee R \vee Q) \wedge (S \vee \neg P \vee \neg R \vee \neg Q) \wedge (S \vee \neg P \vee R \vee Q) \wedge (\neg S \vee \neg P \vee \neg R \vee \neg Q) \wedge (\neg S \vee \neg P \vee R \vee \neg Q) \wedge (\neg S \vee \neg P \vee \neg R \vee Q) \wedge (\neg S \vee P \vee \neg R \vee Q) \wedge (\neg S \vee P \vee \neg R \vee Q) \wedge (\neg S \vee P \vee R \vee Q)$.

Example. *Write equivalent forms for the following formulas and hence obtain the p.c.n.f.*

(*i*) $\neg (P \vee Q)$ (*ii*) $\neg (P \rightarrow Q)$

Solution.

(*i*) A : $\neg (P \vee Q) \Leftrightarrow \neg P \wedge \neg Q$ which is *p.d.n.f.*

p.d.n.f. of $\neg A : (P \wedge Q) \vee (\neg P \wedge Q) \vee (P \wedge \neg Q)$

p.c.n.f. of $\neg\neg A : (\neg P \vee \neg Q) \wedge (P \vee \neg Q) \wedge (\neg P \vee Q)$

which is *p.c.n.f.* of A (since $\neg\neg A \Leftrightarrow A$)

(*ii*) A : $\neg (P \rightarrow Q) \Leftrightarrow \neg (\neg P \vee Q) \Leftrightarrow P \wedge \neg Q$ which is *p.d.n.f.*

p.d.n.f. of $\neg A : (\neg P \wedge Q) \vee (P \wedge Q) \vee (\neg P \wedge \neg Q)$

$p.c.n.f.$ of $\neg\neg A : (P \vee \neg Q) \wedge (\neg P \vee \neg Q) \wedge (P \vee Q)$

which is $p.c.n.f.$ of A (since $\neg\neg A \Leftrightarrow A$)

Example. *Obtain P.D.N.F., P.C.N.F. for the following and which of the formulas are tautologies ?*

(a) $Q \wedge (P \vee \neg Q)$ (b) $(Q \to P) \wedge (\neg P \wedge Q)$

Solution.

(a) Let A: $Q \wedge (P \vee \neg Q)$

$\Leftrightarrow \quad (Q \vee (P \wedge \neg P)) \wedge (P \vee \neg Q)$ ($\because R \vee F \Leftrightarrow R$)

$\Leftrightarrow \quad (Q \vee P) \wedge (Q \vee \neg P) \wedge (P \vee \neg Q)$

which is P.C.N.F.

$$\neg A \Leftrightarrow \neg P \vee \neg Q$$

$$\neg\neg A \Leftrightarrow \neg (\neg P \vee \neg Q)$$

$A \Leftrightarrow P \wedge Q$, which is P.D.N.F. and not a tautology.

(b) Consider A : $(Q \to P) \wedge (\neg P \wedge Q)$

$\therefore \quad A \Leftrightarrow (\neg Q \vee P) \wedge (\neg P \wedge Q)$

$\Leftrightarrow \quad (\neg Q \vee P) \wedge (\neg P \vee (Q \wedge \neg Q)) \wedge (Q \vee (P \wedge \neg P))$

$\Leftrightarrow \quad (\neg Q \vee P) \wedge (\neg P \vee Q) \wedge (\neg P \vee \neg Q) \wedge (Q \vee P) \wedge (Q \vee \neg P)$

$\Leftrightarrow \quad (P \vee Q) \wedge (P \vee \neg Q) \wedge (\neg P \vee Q) \wedge (\neg P \vee \neg Q)$

which is $p.c.n.f.$ and is not a tautology.

And P.D.N.F. does not exist.

Example. *Find the min term normal form (p.d.n.f.) of* $\neg ((P \vee Q) \wedge R) \wedge (P \vee R)$

Solution. $\neg ((P \vee Q) \wedge R) \wedge (P \vee R) \Leftrightarrow (\neg (P \vee Q) \vee \neg R) \wedge (P \vee R)$

$\Leftrightarrow \quad ((\neg P \wedge \neg Q) \vee \neg R) \wedge (P \vee R)$

$\Leftrightarrow \quad ((\neg P \wedge \neg Q) \wedge (P \vee R)) \vee (\neg R \wedge (P \vee R))$

$\Leftrightarrow \quad (\neg P \wedge \neg Q \wedge P) \vee (\neg P \wedge \neg Q \wedge R) \vee (\neg R \wedge P) \vee (\neg R \wedge R)$

$\Leftrightarrow \quad (F \wedge \neg Q) \vee (\neg P \wedge \neg Q \wedge R) \vee (\neg R \wedge P) \vee F$

$\Leftrightarrow \quad F \vee (\neg P \wedge \neg Q \wedge R) \vee (\neg R \wedge P) \vee F.$

$\Leftrightarrow \quad (\neg P \wedge \neg Q \wedge R) \vee (\neg R \wedge P)$

$\Leftrightarrow \quad (\neg P \wedge \neg Q \wedge R) \vee (P \wedge Q \wedge \neg R) \vee (P \wedge \neg Q \wedge \neg R)$

$(\because \neg R \wedge P \Leftrightarrow (\neg R \wedge P \wedge Q) \vee (\neg R \wedge P \wedge \neg Q))$

which $p.d.n.f.$

Example. *Obtain the product of sums canonical form for* $(P \wedge Q \wedge R) \vee (\neg P \wedge R \wedge Q) \vee (\neg P \wedge \neg Q \wedge \neg R)$.

Solution. Let S: $(P \wedge Q \wedge R) \vee (\neg P \wedge R \wedge Q) \vee (\neg P \wedge \neg Q \wedge \neg R)$

Then $\neg S \Leftrightarrow (P \wedge \neg Q \wedge R) \vee (P \wedge Q \wedge \neg R) \vee (P \wedge \neg Q \wedge \neg R) \vee (\neg P \wedge \neg Q \wedge R) \vee (\neg P \wedge Q \wedge \neg R)$

$\therefore \neg\neg S \Leftrightarrow \neg (P \wedge \neg Q \wedge R) \wedge \neg (P \wedge Q \wedge \neg R) \wedge \neg (P \wedge \neg Q \wedge \neg R) \wedge$
$\neg (\neg P \wedge \neg Q \wedge R) \wedge \neg (\neg P \wedge Q \wedge \neg R)$

$S \Leftrightarrow (\neg P \vee Q \vee \neg R) \wedge (\neg P \vee \neg Q \vee R) \wedge (\neg P \vee Q \vee R) \wedge (P \vee Q \vee \neg R) \wedge (P \vee \neg Q \vee R)$

Which is the product of sums canonical form.

Example. *Obtain the p.d.n.f. and p.c.n.f. of the formula.*

$P \vee (\neg P \to (Q \vee (\neg Q \to R)))$ [JNTU, May, 2005 - Set 3]

Solution. Consider S : $P \vee (\neg P \to (Q \vee (\neg Q \to R)))$

$\therefore \quad S \Leftrightarrow P \vee (P \vee (Q \vee (Q \vee R))) \quad (\because \; P \to Q \Leftrightarrow \neg P \vee Q, \neg\neg P \Leftrightarrow P)$

$\Leftrightarrow \quad P \vee (P \vee Q \vee R)$

$\Leftrightarrow \quad P \vee Q \vee R$, which is *p.c.n.f.* of S.

$$\neg S \Leftrightarrow (P \vee Q \vee \neg R) \wedge (P \vee \neg Q \vee R) \wedge (P \vee \neg Q \vee \neg R) \wedge (\neg P \vee Q \vee R) \wedge (\neg P \vee Q \vee \neg R) \wedge (\neg P \vee \neg Q \vee R) \wedge (\neg P \vee \neg Q \vee \neg R)$$

$$\neg\neg S \Leftrightarrow \neg (P \vee Q \vee \neg R) \vee \neg (P \vee \neg Q \vee R) \vee \neg (P \vee \neg Q \vee \neg R) \vee \neg (\neg P \vee Q \vee R) \vee \neg (\neg P \vee Q \vee \neg R) \vee \neg (\neg P \vee \neg Q \vee R) \vee \neg (\neg P \vee \neg Q \vee \neg R)$$

$$\therefore \quad S \Leftrightarrow (\neg P \wedge \neg Q \wedge R) \vee (\neg P \wedge Q \wedge \neg R) \vee (\neg P \wedge Q \wedge R) \vee (P \wedge \neg Q \wedge \neg R) \vee (P \wedge \neg Q \wedge R) \vee (P \wedge Q \wedge \neg R) \vee (P \wedge Q \wedge R)$$

which is *p.d.n.f.* of S.

Example. *Explain the difference between the principal disjunctive and conjunctive normal forms.*

(JNTU - March - 06 - Set 2, November - 05 - Set 2, November - 04 - Set 2.

Solution. Principal disjunctive normal form (*p.d.n.f.*)

D_1 Definition. *An equivalent formula consisting of disjunctions of min terms only is known as its principal disjunctive normal form (sum of products canonical form).*

C_1 Definition. *An equivalent formula consisting of conjunctions of max terms only is known as its principal conjunctive normal form (product of sums canonical form).*

D_2 : For two variables P, Q, there are 2^2 Min terms given by $P \wedge Q$, $P \wedge \neg Q$, $\neg P \wedge Q$ and $\neg P \wedge \neg Q$ which are also called Boolean conjunctions of P, Q.

C_2 : For two variables, P, Q, there are 2^2 Max terms given by $P \vee Q$, $P \vee \neg Q$, $\neg P \vee Q$, and $\neg P \vee \neg Q$ which are also called Boolean disjunctions of P, Q.

D_3 : For the three variables P, Q and R, there are $2^3 = 8$ min terms given by $P \wedge Q \wedge R$, $P \wedge Q \wedge \neg R$, $P \wedge \neg Q \wedge R$, $P \wedge \neg Q \wedge \neg R$, $\neg P \wedge Q \wedge R$, $\neg P \wedge Q \wedge \neg R$, $\neg P \wedge \neg Q \wedge R$, $\neg P \wedge \neg Q \wedge \neg R$.

C_3 : For the three variables P, Q and, R there are $2^3 = 8$ max terms given by $P \vee Q \vee R$, $P \vee Q \vee \neg R$, $P \vee \neg Q \vee R$, $P \vee \neg Q \vee \neg R$, $\neg P \vee Q \vee R$, $\neg P \vee Q \vee \neg R$, $\neg P \vee \neg Q \vee R$, $\neg P \vee \neg Q \vee \neg R$.

Methods to obtain *p.d.n.f.* and *p.c.n.f.* of a given formula.

D_4 : **Truth table Method** (i). For every truth value T of the given formula, select the min terms which also has the value T for the same combination of the truth values of the variables.

Example. *p.d.n.f.* of $P \to Q$ is $(P \wedge Q) \vee (\neg P \wedge Q) \vee (\neg P \wedge \neg Q)$

C_4 : *p.c.n.f.* of any formula is obtained by the following method (2).

D_5 : **Without constructing truth table.**

(*i*) First replace the conditionals and bi-conditional by their equivalent formulas containing only $\wedge$, $\vee$ and $\neg$.

(ii) Apply other known equivalent formulas until we get all min terms.

Hence obtained form is called *p.d.n.f.*

C_5 : If A $(P_1, P_2, ..., P_n)$ is any statement formula and if an equivalent *p.d.n.f.* is given then its *p.c.n.f.* is obtained by $\neg\neg A\ (P_1, P_2, ..., P_n)$.

EXERCISE 1.2.

1. Obtain the principal disjunctive normal forms of

   (a) $\neg p \vee q$ (b) $p \to ((p \to q) \wedge \neg(\neg q \vee \neg p))$ (c) $(p \wedge q) \vee (\neg p \wedge r) \vee (q \wedge r)$

2. Obtain the principal conjunctive normal forms of

   (a) $(p \wedge q) \vee (\neg p \wedge q) \vee (p \wedge \neg q)$

   (b) $(p \wedge q) \vee (\neg p \wedge q \wedge r)$

   (c) $\neg(p \vee q)$

3. Obtain *p.d.n.f.*, *p.c.n.f.* for the following formulae :

   (a) $(\neg P \to R) \wedge (Q \leftrightarrow P)$

   (b) $P \to (P \wedge (Q \to P))$

   (c) $P \to ((P \to Q) \wedge \neg(\neg Q \vee \neg P))$

4. Obtain the product of sums canonical form of the following formula.

   $(\neg P \to R) \wedge (Q \to P)$ [JNTU - March 06 - Set 2, November 05 - Set 2.]

5. Obtain the sum-of-products canonical form of the following formula.

   $(P \wedge Q) \vee (\neg P \wedge R) \vee (Q \wedge R)$

6. Show that the following are equivalent formulas

   (a) $P \vee (P \wedge Q) \Leftrightarrow P$ (b) $P \vee (\neg P \wedge Q) \Leftrightarrow P \vee Q$

7. Obtain the *p.d.n.f.* of $(\neg P \vee \neg Q) \to (P \to \neg Q)$ (JNTU, May 2005, Set-3)

ANSWERS

1. (a) $(\neg p \wedge q) \vee (\neg p \wedge \neg q) \vee (p \wedge q)$

   (b) $(\neg p \wedge q) \vee (\neg p \wedge \neg q) \vee (p \wedge q)$

   (c) $(p \wedge q \wedge r) \vee (p \wedge q \wedge \neg r) \vee (\neg p \wedge q \wedge r) \vee (\neg p \wedge \neg q \wedge r)$

2. (a) $p \vee q$

   (b) $(\neg p \vee q \vee r) \wedge (\neg p \vee q \vee r) \wedge (p \vee q \vee \neg r) \wedge (p \vee \neg q \vee r) \wedge (p \vee q \vee r)$.

   (c) $(p \vee \neg q) \wedge (\neg p \vee q) \wedge (\neg p \vee \neg q)$

3. (a) *p.d.n.f.* is $(\neg P \wedge \neg Q \wedge R) \vee (P \wedge Q \wedge R) \vee (P \vee Q \vee \neg R)$

   p.c.n.f. is $(P \vee Q \vee R) \wedge (P \vee \neg Q \vee R) \wedge (P \vee \neg Q \vee \neg R) \wedge (\neg P \vee Q \vee R) \wedge (\neg P \vee Q \vee \neg R)$

   (b) *p.d.n.f.* is $(\neg P \wedge Q) \vee (\neg P \wedge \neg Q) \vee (P \wedge \neg Q) \vee (P \wedge Q)$ *p.c.n.f.* does not exist..

   (c) *p.d.n.f.* is $(\neg P \wedge Q) \vee (\neg P \wedge \neg Q) \vee (P \wedge Q)$ *p.c.n.f.* $\neg P \vee Q$.

4. $(P \vee Q \vee R) \wedge (P \vee \neg Q \vee R) \wedge (P \vee \neg Q \vee R) \wedge (P \vee \neg Q \vee \neg R)$

5. $(P \wedge Q \wedge R) \vee (P \wedge Q \wedge \neg R) \vee (\neg P \wedge Q \wedge R) \vee (\neg P \wedge \neg Q \wedge R)$

7. $(P \wedge Q) \vee (\neg P \wedge Q) \vee (\neg P \wedge \neg Q) \vee (\neg Q \wedge P)$

CHAPTER

2

Statement Calculus & Predicate Logic

2.1 THE THEORY OF INFERENCE IN STATEMENT CALCULUS

Definition. *The main aim of logic is to provide rules of inference to infer a conclusion from certain premises. The theory associated with rules of inference is known as* ***inference theory.***

Definition. *If a conclusion is derived from a set of premises by using the accepted rules of reasoning, then such a process of derivation is called a* ***deduction*** *or a* ***formal proof*** *and the argument is called a valid argument or conclusion is called a valid conclusion.*

Note : Premises means set of assumptions, axioms, hypotheses.

Definition. *Let A, B be two statement formulas. We say that "B logically follows from A" (or) "B is a valid conclusion (consequence) of the premise A" iff $A \to B$ is a tautology i.e., $A \Rightarrow B$.*

By extending the above definition, we say that from a set of premises $\{H_1, H_2, ..., H_m\}$ a conclusion C follows logically iff $H_1 \wedge H_2 \wedge ... \wedge H_m \Rightarrow C$.

Note. To determine whether the conclusion logically follows from the given premises, we use following methods :

1. Truth table (technique) method.
2. Without using truth table method.

I. Method : Truth Table Method :

1. Let $P_1, P_2, ..., P_n$ be all the atomic variables appearing in the premises $H_1, H_2, ..., H_m$ and the conclusion C. If all possible combinations of truth values are assigned to $P_1, P_2, ..., P_n$ and if the truth values of $H_1, H_2, ..., H_m$ and C are entered in a table, then it is easy to see from such a table whether $H_1 \wedge H_2 \wedge ..., \wedge H_m \Rightarrow C$ is true.
2. Look for the rows in which all $H_1, H_2, ..., H_m$ have the value T. If for every such row, C also has the value T, then $H_1 \wedge H_2 \wedge ... \wedge H_m \Rightarrow C$ holds.

3. Alternatively, look for the rows in which C has the value F. If in every such row, atleast one of the values of $H_1, H_2, ..., H_m$ is F, then $H_1 \wedge H_2 \wedge ... \wedge H_m \Rightarrow C$ also holds.

Example. *Determine whether the conclusion C follows logically from the hypotheses H_1 and H_2.*

(a) $H_1 : P \rightarrow Q$ $\quad H_2 : P$ $\quad C : Q$
(b) $H_1 : P \rightarrow Q$ $\quad H_2 : \neg P$ $\quad C : Q$
(c) $H_1\ P \rightarrow Q$ $\quad H_2 : \neg (P \wedge Q)$ $\quad C : \neg P$
(d) $H_1 : \neg P$ $\quad H_2\ P \rightleftarrows Q$ $\quad C : \neg (P \wedge Q)$
(e) $H_1 : P \rightarrow Q$ $\quad H_2 : Q$ $\quad C : P$

Solution. We first construct the appropriate truth table, as shown below :

Table 2.1.

P	Q	$P \rightarrow Q$	$\neg P$	$\neg Q$	$\neg (P \wedge Q)$	$P \rightleftarrows Q$
T	T	T	F	F	F	T
T	F	F	F	T	T	F
F	T	T	T	F	T	F
F	F	T	T	T	T	T

For (*a*), We observe that the first row is the only row in which both the premises have the value T. The conclusion also has the value T in that row. Hence it is valid.

For (*b*), We observe that the third and fourth rows. The conclusion Q is true only in the third row, but not in the fourth, and hence the conclusion is not valid.

In (*c*), Observe the third and fourth rows. Thc conclusion $\neg P$ is true in these two rows. Hence it is valid.

In (*d*), Observe that the fourth row is the only row in which both the premises have the value T. The conclusion $\neg(P \wedge Q)$ has the value T in that row. Hence it is valid.

In (*e*), Observe that the first and third rows. The conclusion P is true only in first row, but not in third row and hence the conclusion is not valid.

Example. *Show that the conclusion C follows from the premises $H_1, H_2, ...,$ in the following Cases:*

(a) $H_1 : P \rightarrow Q$ $\quad C : P \rightarrow (P \wedge Q)$
(b) $H_1 : \neg P$ $\quad H_2 : P \vee Q$ $\quad C : Q$
(c) $H_1 : \neg P \vee Q$ $\quad H_2 : \neg (Q \wedge \neg R)$ $\quad H_3 : \neg R$ $\quad C : \neg P.$

Table 2.2.

Solution. (*a*)

P	Q	$H_1 : P \rightarrow Q$	$P \wedge Q$	$C : P \rightarrow (P \wedge Q)$
T	T	T	T	T
T	F	F	F	F
F	T	T	F	T
F	F	T	F	T

Therefore, C follows from the premise H_1 because in first, third, fourth rows H_1 value is T and C value T.

Table 2.3.

(*b*)

P	Q	P ∨ Q	¬P
T	T	T	F
T	F	T	F
F	T	T	T
F	F	F	T

Therefore, C follows from the premises H_1 and H_2 because in third row H_1, H_2 values are T and C value is T.

Table 2.4.

(*c*)

P	Q	R	¬P	¬R	¬P ∨ Q	Q ∧ ¬R	¬(Q ∧ ¬R)
T	T	T	F	F	T	F	T
T	T	F	F	T	T	T	F
T	F	T	F	F	F	F	T
T	F	F	F	T	F	F	T
F	T	T	T	F	T	F	T
F	T	F	T	T	T	T	F
F	F	T	T	F	T	F	T
F	F	F	T	T	T	F	T

Therefore, C follows from H_1, H_2, H_3 because in last row H_1, H_2, H_3 have value T and C value T.

Example. *Determine whether the conclusion C is valid in the following when H_1, H_2, ..., are the premises.*

(*a*) $H_1 : p \to q$, $H_2 : \neg q$, $C : p$

(*b*) $H_1 : p \vee q$, $H_2 : p \to r$, $H_3 : q \to r$, $C : r$.

Table 2.5.

Solution. (*a*)

p	*q*	*p* → *q*	¬*q*
T	T	T	F
T	F	F	T
F	T	T	F
F	F	T	T

Therefore, C does not follow H_1, H_2 because in fourth row H_1 value is T, H_2 value is T and C value is F.

Table 2.6.

(*b*)

p	q	r	$p \vee q$	$p \to r$	$q \to r$
T	T	T	T	T	T
T	T	F	T	F	F
T	F	T	T	T	T
T	F	F	T	F	T
F	T	T	T	T	T
F	T	F	T	T	F
F	F	T	F	T	T
F	F	F	F	T	T

∴ C logically follows premises H_1, H_2, H_3 because in first, third, fifth rows, H_1, H_2, H_3 values T and C value T.

2.2 RULES OF INFERENCE

II Method (Without Using Truth Table) :

The truth table technique becomes tedious when the number of atomic variables present in all the formulae representing the premises and the conclusion is large. To overcome this disadvantage, we need to investigate other possible methods, without using the truth table.

We now describe the process of derivation by which one demonstrates that a particular formula is a valid consequence of a given set of premises. Before we do this, we give two rules of inference which are called rules P and T.

Rule P : A premise may be introduced at any point in the derivation.

Rule T : A formula S may be introduced in a derivation if S is a tautologically implied by any one or more of the preceding formulas in the derivation.

Before we proceed with the actual process of derivation, we list some important implications and equivalences that will be referred to frequently.

Implications Table 2.7.

I_1	$P \wedge Q \Rightarrow P$	(simplification)
I_2	$P \wedge Q \Rightarrow Q$	
I_3	$P \Rightarrow P \vee Q$	(addition)
I_4	$Q \Rightarrow P \vee Q$	
I_5	$\neg P \Rightarrow P \to Q$	
I_6	$Q \Rightarrow P \to Q$	
I_7	$\neg (P \to Q) \Rightarrow P$	
I_8	$\neg (P \to Q) \Rightarrow \neg Q$	

I_9	$P, Q \Rightarrow P \wedge Q$	
I_{10}	$\neg P, P \vee Q \Rightarrow Q$	(disjunctive syllogism)
I_{11}	$P, P \rightarrow Q \Rightarrow Q$	(modus ponens)
I_{12}	$\neg Q, P \rightarrow Q \Rightarrow \neg P$	(modus tollens)
I_{13}	$P \rightarrow Q, Q \rightarrow R \Rightarrow P \rightarrow R$	(hypothetical syllogism)
I_{14}	$P \vee Q, P \rightarrow R, Q \rightarrow R \Rightarrow R$	(dilemma)

Equivalences Table 2.8.

E_1	$\neg\neg P \Leftrightarrow P$	(double negation)
E_2	$P \wedge Q \Leftrightarrow Q \wedge P$	(commutative law)
E_3	$P \vee Q \Leftrightarrow Q \vee P$	
E_4	$(P \wedge Q) \wedge R \Leftrightarrow P \wedge (Q \wedge R)$	(associative law)
E_5	$(P \vee Q) \vee R \Leftrightarrow P \vee (Q \vee R)$	
E_6	$P \wedge (Q \vee R) \Leftrightarrow (P \wedge Q) \vee (P \wedge R)$	(distributive law)
E_7	$P \vee (Q \wedge R) \Leftrightarrow (P \vee Q) \wedge (P \vee R)$	
E_8	$\neg (P \wedge Q) \Leftrightarrow \neg P \vee \neg Q$	(De Morgans law)
E_9	$\neg (P \vee Q) \Leftrightarrow \neg P \wedge \neg Q$	
E_{10}	$P \vee P \Leftrightarrow P$	
E_{11}	$P \wedge P \Leftrightarrow P$	
E_{12}	$R \vee (P \wedge \neg P) \Leftrightarrow R$	
E_{13}	$R \wedge (P \wedge \neg P) \Leftrightarrow R$	
E_{14}	$R \vee (P \vee \neg P) \Leftrightarrow T$	
E_{15}	$R \wedge (P \wedge \neg P) \Leftrightarrow F$	
E_{16}	$P \rightarrow Q \Leftrightarrow \neg P \vee Q$	
E_{17}	$\neg (P \rightarrow Q) \Leftrightarrow P \wedge \neg Q$	
E_{18}	$P \rightarrow Q \Leftrightarrow \neg Q \rightarrow \neg P$	
E_{19}	$P \rightarrow (Q \rightarrow R) \Leftrightarrow (P \wedge Q) \rightarrow R$	
E_{20}	$\neg \left(P \rightleftarrows Q\right) \Leftrightarrow P \rightleftarrows \neg Q$	
E_{21}	$P \rightleftarrows Q \Leftrightarrow (P \rightarrow Q) \wedge (Q \rightarrow P)$	
E_{22}	$\left(P \rightleftarrows Q\right) \Leftrightarrow (P \wedge Q) \vee (\neg P \wedge \neg Q)$	

Example. *Demonstrate that R is a valid inference from the premises $P \rightarrow Q$, $Q \rightarrow R$ and P.*

Solution.

{1}	(1)	$P \rightarrow Q$	Rule P.
{2}	(2)	P	Rule P
{1, 2}	(3)	Q	Rule T, (1), (2), and I_{11}
{4}	(4)	$Q \rightarrow R$	Rule P
{1, 2, 4}	(5)	R	Rule T, (3), (4), and I_{11}

Therefore, R is a valid inference.

There are five lines in this derivation. The second column of numbers designates the formula as well as the line of derivation in which it occurs. The set of numbers in braces (the first column) for each line shows the premises on which the formula in the line depends. On the right, P or T represents the rule of inference, followed by a comment showing from which formulas and tautology that particular formula has been obtained. For example, if we follow this notation, the third line shows that the formula in this line is numbered (3) and has been obtained from premises in (1) and (2). The comment on the right says that the formula Q has been introduced using rule T and also indicates the details of the application of rule T.

Example. *Show that* $\neg P$ *follows logically from the premises*

$$\neg (P \wedge \neg Q),\ \neg Q \vee R,\ \neg R$$

Solution.

{1}	1.	$\neg (P \wedge \neg Q)$	Rule P.
{1}	2.	$\neg P \vee Q$	Rule T, (1), E_8, E_1.
{1, 2}	3.	$P \rightarrow Q$	Rule T, (2), E_1, E_{16}.
{4}	4.	$\neg Q \vee R$	Rule P.
{4}	5.	$Q \rightarrow R$	Rule T, (4), E_1, E_{16}.
{1, 2, 4}	6.	$P \rightarrow R$	Rule T, (3), (5), I_{13}.
{1, 2, 4}	7.	$\neg R \rightarrow \neg P$	Rule T, (6), E_{18}.
{8}	8.	$\neg R$	Rule P.
{1, 2, 4, 8}	9.	$\neg P$	Rule T, (7), (8), I_{11}.

Example. *Show that* $R \vee S$ *follows logically from the premises* $C \vee D$, $(C \vee D) \rightarrow \neg H$, $\neg H \rightarrow (A \wedge \neg B)$, *and* $(A \wedge \neg B) \rightarrow (R \vee S)$. *[JNTU - March, 2006 - Set 3, November - 05 Set 4, June - 04 Set 3].*

Solution.

{1}	(1)	$(C \vee D) \rightarrow \neg H$	Rule P.
{2}	(2)	$\neg H \rightarrow (A \wedge \neg B)$	Rule P.
{1, 2}	(3)	$(C \vee D) \rightarrow (A \wedge \neg B)$	Rule T, (1), (2) and I_{13}.
{4}	(4)	$(A \wedge \neg B) \rightarrow R \vee S$	Rule P.
{1, 2, 4}	(5)	$(C \vee D) \rightarrow (R \vee S)$	Rule T, (3), (4), and I_{13}.
{6}	(6)	$C \vee D$	Rule P.
{1, 2, 4, 6}	(7)	$R \vee S$	Rule T, (5), (6) and I_{11}.

The two tautologies frequently used in the above derivations are I_{13}, known as hypothetical syllogism, and I_{11}, known as modus ponens.

Example. *Show that* $S \vee R$ *is tautologically implied by* $(P \vee Q) \wedge (P \rightarrow R) \wedge (Q \rightarrow S)$
[JNTU - March 06 - Set 4, November 05 - Set 3, June 04 - Set 2]

Solution.

{1}	1.	$P \vee Q$	Rule P.
{1}	2.	$\neg P \rightarrow Q$	Rule T, (1), E_1 and E_{16}.
{3}	3.	$Q \rightarrow S$	Rule P.
{1, 3}	4.	$\neg P \rightarrow S$	Rule T, (2), (3) and I_{13}.
{1, 3}	5.	$\neg S \rightarrow P$	Rule T, (4), E_{18} and E_1.

{6}	6. $P \to R$	Rule P.
{1, 3, 6}	7. $\neg S \to R$	Rule T, (5), (6), and I_{13}.
{1, 3, 6}	8. $S \vee R$	Rule T, (7), E_{16} and E_1.

Example. *Show that $M \vee N$ is a valid argument from the premises $\neg J \to (M \vee N)$, $(H \vee G) \to \neg J$, $H \vee G$.*

Solution.

{1}	1. $(H \vee G) \to \neg J$	Rule P.
{2}	2. $H \vee G$	Rule P.
{1, 2}	3. $\neg J$	Rule T, (1), (2), I_{11}.
{4}	4. $\neg J \to (M \vee N)$	Rule P.
{1, 2, 4}	5. $M \vee N$	Rule T, (3), (4), I_{11}.

Example. *Show that $\neg S$ is a valid argument from the premises $P \to Q$, $(\neg Q \vee R) \wedge (\neg R)$, $\neg (\neg P \wedge S)$.*

Solution.

{1}	1. $P \to Q$	Rule P.
{2}	2. $(\neg Q \vee R) \wedge (\neg R)$	Rule P.
{2}	3. $\neg Q \vee R$	Simplification, (2).
{2}	4. $Q \to R$	Rule T, (3), E_1, E_{16}.
{1, 2}	5. $P \to R$	Rule T, (1), (4), I_{13}.
{6}	6. $\neg (\neg P \wedge S)$	Rule P.
{6}	7. $\neg (\neg P \wedge \neg\neg S)$	Rule T, E_1.
{6}	8. $\neg (\neg (\neg P \to \neg S))$	Rule T, (7), E_{17}.
{6}	9. $\neg (\neg (\neg S))$	Rule T, (8), I_8
{6}	10. $\neg S$	Rule T, E_1, (8).

Example. *Show that $R \wedge (P \vee Q)$ is a valid conclusion from the premises $P \vee Q$, $Q \to R$, $P \to M$ and $\neg M$.* [JNTU June - 04 - Set 3]

Solution.

{1}	1. $P \to M$	Rule P.
{2}	2. $\neg M$	Rule P.
{1, 2}	3. $\neg P$	Rule T, (1), (2) and I_{12}
{4}	4. $P \vee Q$	Rule P.
{1, 2, 4}	5. Q	Rule T, (3), (4) and I_{10}
{6}	6. $Q \to R$	Rule P.
{1, 2, 4, 6}	7. R	Rule T, (5), (6) and I_{11}.
{1, 2, 4,6}	8. $R \wedge (P \vee Q)$	Rule T, (4), (7) and I_9.

Example. *Show that $I_{12} : \neg Q, P \to Q \Rightarrow \neg P$.*

Solution.

{1}	1. $P \to Q$	Rule P.
{1}	2. $\neg Q \to \neg P$	Rule T, (1) and E_{18}.
{3}	3. $\neg Q$	Rule P.
{1, 3}	4. $\neg P$	Rule T, (2), (3) and I_{11}.

We shall now introduce a third inference rule, known as rule CP or rule of conditional proof.

Rule CP : If we can derive S from R and a set of premises, then we can derive $R \to S$ from the set of premises alone.

The general idea of this rule is that we may introduce a new premise R conditionally and use it in the conjunction with the original premises to derive a conclusion S, and then assert that the implication $R \to S$ follows from the original premises alone. If S is a valid inference from premises $P_1, P_2, ..., P_n$ and R, Then $R \to S$ is a valid inference from premises $P_1, P_2, ..., P_n$.

Rule CP is not new for our purpose here because it follows from the equivalence E_{19} which states that $(P \wedge R) \to S \Leftrightarrow P \to (R \to S)$.

Where P denote the conjunction of the set of premises say $P_1, P_2, ..., P_n$ and let R be any formula. The above equivalence states that if R is included as an additional premise and S is derived from $P \wedge R$, then $R \to S$ can be derived from the premises P alone. Rule CP is also called the **deduction theorem.**

Rule CP is generally used if the conclusion is of the form $R \to S$. In such cases, R is taken as an additional premise and S is derived from the given premises and R.

Example. *Show that $R \to S$ can be derived from the premises $P \to (Q \to S)$, $\neg R \vee P$ and Q.*

[JNTU - March 06 - Set 3, November - 05 - Set 4, June -04 - Set 3 and 4]

Solution. Instead of deriving $R \to S$, we shall include R as an additional premise and show S first.

{1}	1. $\neg R \vee P$	Rule P.
{2}	2. R	Rule P (additional premise)
{1, 2}	3. P	Rule T, (1), (2), and I_{10}
{4}	4. $P \to (Q \to S)$	Rule P.
{1, 2, 4}	5. $Q \to S$	Rule T, (3), (4), and I_{11}.
{6}	6. Q	Rule P.
{1, 2, 4, 6}	7. S	Rule T, (5), (6), and I_{11}.
{1, 4, 6}	8. $R \to S$	Rule CP.

Example. *Show that $P \to S$ can be derived from the premises $\neg P \vee Q$, $\neg Q \vee R$, $R \to S$.*

Solution. We include P as an additional premise and derive S.

{1}	1. $\neg P \vee Q$	Rule P.
{2}	2. P	Rule P (additional premise).
{1, 2}	3. Q	Rule T, (1), (2) and I_{10}.
{4}	4. $\neg Q \vee R$	Rule P.
{1, 2, 4}	5. R	Rule T, (3), (4), and I_{10}.
{6}	6. $R \to S$	Rule P.
{1, 2, 4, 6}	7. S	Rule T, (5), (6), and I_{11}.
{1, 4, 6}	8. $P \to S$	Rule CP.

2.3 CONSISTENCY OF PREMISES AND INDIRECT METHOD OF PROOF

Definition. *A set of formulas $H_1, H_2, ..., H_m$ is said to be* ***consistent*** *if their conjunction has the truth value T for some assignment of the truth values to the atomic variables appearing in $H_1, H_2, ..., H_m$.*

Definition. *A set of formulas $H_1, H_2, ..., H_m$ is said to be* ***inconsistent*** *if at least one of the formulas $H_1, H_2, ..., H_m$ is false for every assignment of the truth values to the atomic variables appearing in $H_1, H_2, ..., H_m$.*

(or) A set of formulas $H_1, H_2, ..., H_m$ is inconsistent if $H_1 \wedge H_2 \wedge ... \wedge H_m \Rightarrow R \wedge \neg R$ where R is any formula.

Note : That $R \wedge \neg R$ is a contradiction, and it is necessary and sufficient for the implication that $H_1 \wedge H_2 \wedge ... \wedge H_m$ be a contradiction.

The notion of inconsistency is used in the following method.

In direct method of proof :

The notion of inconsistency is used in a procedure called **proof by contradiction** (or) **indirect method of proof** (or) **reduction *ad* absurdum.**

The technique of indirect method of proof is as follows :

(*i*) In order to show that a conclusion C follows logically from the premises $H_1, H_2, ..., H_m$, we assume that C is false and consider ¬C as an additional premise

(*ii*) If the new set of premises is inconsistent then they imply a contradiction. Therefore C is true when-ever $H_1 \wedge H_2 \wedge ... \wedge H_m$ is true.

(*iii*) Thus, C follows logically from the premises $H_1, H_2, ..., H_m$.

Example. *Show that $\neg (P \wedge Q)$ follows from $\neg P \wedge \neg Q$.*

Solution. *We introduce $\neg\neg (P \wedge Q)$ as an additional premise and show that this additional premise leads to a contradiction.*

{1}	1. $\neg\neg (P \wedge Q)$	Rule P (assumed)
{1}	2. $P \wedge Q$	Rule T, (1) and E_1.
{1}	3. P	Rule T, (2), and I_1.
{4}	4. $\neg P \wedge \neg Q$	Rule P.
{4}	5. $\neg P$	Rule T, (4), I_1.
{1, 4}	6. $P \wedge \neg P$	Rule T, (3), (5), I_9.

Example. *Show that the following premises are inconsistent.*

1. *If the contract is valid, then John is liable for penalty.*
2. *If John is liable for penalty, he will go bankrupt.*
3. *If the bank will loan him money, he will not go bankrupt.*
4. *As a matter of fact, the contract is valid and the bank will loan him money.*

Solution. We indicate the given statements as follows :

V : The contract is valid.

L : John is liable for penalty.

M : Bank will loan him money.

B : He will go bankrupt.

Then the given premises are

$V \to L$, $L \to B$, $M \to \neg B$, $V \wedge M$.

{1}	1. $V \to L$	Rule P.
{2}	2. $L \to B$	Rule P.
{1, 2}	3. $V \to B$	Rule T, (1), (2), and I_{13}.
{4}	4. $V \wedge M$	Rule P.
{4}	5. V	Rule T, (4).
{4}	6. M	Rule T, (4).
{1, 2, 4}	7. B	Rule T, (3), (5), modus ponens
{8}	8. $M \to \neg B$	Rule P.
{4, 8}	9. $\neg B$	Rule T, (6), (8).
{1, 2, 4, 8}	10. $B \wedge \neg B$	Rule T, (7), (9).

Thus, the given set of premises leads to a contradiction and hence it is inconsistent.

Example. *Show that the following premises are inconsistent.*

1. *If Jack misses many classes through illness, then he fails high school.*
2. *If Jack fails high school, then he is uneducated.*
3. *If Jack reads a lot of books, then he is not uneducated.*
4. *Jack misses many classes through illness and reads a lot of books.*

[JNTU - November - 04 - Set - 4]

Solution. We indicate the given statements as follows :

E : Jack misses many classes.

S : Jack fails high-school.

A : Jack reads a lot of books.

H : Jack is uneducated.

The given premises are $E \to S$, $S \to H$, $A \to \neg H$ and $E \wedge A$

{1}	1. $E \to S$	Rule P.
{2}	2. $S \to H$	Rule P.
{1, 2}	3. $E \to H$	Rule T, (1), (2) and I_{13}.
{4}	4. $A \to \neg H$	Rule P.
{4}	5. $H \to \neg A$	Rule T, (4), E_{18}.
{1, 2, 4}	6. $E \to \neg A$	Rule T, (3), (5), I_{13}.
{1, 2, 4}	7. $\neg E \vee \neg A$	Rule T, (6), E_{16}.
{1, 2, 4}	8. $\neg (E \wedge A)$	Rule T, (7), E_8.
{9}	9. $E \wedge A$	Rule P.
{1, 2, 4, 9}	10. $(E \wedge A) \wedge \neg (E \wedge A)$	Rule T, (8), (9), I_9.

Thus, the given set of premises leads to a contradiction and hence it is inconsistent.

Proof by contradiction is sometimes convenient. However, it can always be eliminated and replaced by a conditional proof (CP). Observe that,

$$P \to (Q \wedge \neg Q) \Rightarrow \neg P \quad \text{... (1)}$$

In the proof by contradiction we show that $H_1, H_2, ..., H_m \Rightarrow C$ by showing

$$H_1, H_2, ..., H_m, \neg C \Rightarrow R \wedge \neg R \quad \text{... (2)}$$

Now (2) can be converted to the following by using rule CP

$$H_1, H_2, ..., H_m \Rightarrow \neg C \to (R \wedge \neg R) \quad \text{... (3)}$$

From (3) and (1) and E_1, we can show

$$H_1, H_2, ..., H_m \Rightarrow C$$

Which is the required derivation.

Example. *Show that the following set of premises are inconsistent.*

$P \to Q, P \to R, Q \to \neg R, P$ Hence show that $P \to Q, P \to R, Q \to \neg R, P \Rightarrow M$.

Solution.

{1}	1. $P \to Q$	Rule P.
{2}	2. P	Rule P.
{1, 2}	3. Q	Rule T, (1), (2), I_{11}.
{4}	4. $Q \to \neg R$	Rule P.
{1, 2, 4}	5. $\neg R$	Rule T, (3), (4), I_{11}.
{6}	6. $P \to R$	Rule P.
{1, 2, 4, 6}	7. $\neg P$	Rule T, (5), (6), I_{12}.
{8}	8. P	Rule P.
{1, 2, 4, 6, 8}	9. $P \wedge \neg P$	Rule T, (7), (8), I_9.

Hence the given premises are in consistent.

And take $\neg M$ as additional premise. Since $\neg M \wedge (P \wedge \neg P) \Leftrightarrow F$,

$P \to Q, P \to R, Q \to \neg R, P \Rightarrow M$.

Example. *Prove by indirect method that*

$\neg Q, P \to Q, P \vee R \Rightarrow R$.

Solution. Assume new prmise $\neg R$.

{1}	1. $P \vee R$	Rule P.
{2}	2. $\neg R$	Rule P, (additional premise)
{1, 2}	3. P	Rule T, (1), (2).
{4}	4. $P \to Q$	Rule P.
{1, 2, 4}	5. Q	Rule T, (3), (4), modus ponens
{6}	6. $\neg Q$	Rule P.
{1, 2, 4, 6}	7. $Q \wedge \neg Q$	Rule T, (5), (6).

The set of given premises, including $\neg R$, leads to a contradiction. Hence $\neg Q, P \rightarrow Q, P \vee R \Rightarrow R$.

Example. *By indirect proof method, show that*

$P \rightarrow Q, Q \rightarrow R, P \vee R \Rightarrow R$

Solution.

{1}	1. $Q \rightarrow R$	Rule P.
{2}	2. $\neg R$	Rule P (additional premise).
{1, 2}	3. $\neg Q$	Rule T, (1), (2).
{4}	4. $P \rightarrow Q$	Rule P.
{1, 2, 4}	5. $\neg P$	Rule T, (3), (4), modus ponens
{6}	6. $P \vee R$	Rule P.
{1, 2, 4, 6}	7. R	Rule T, (5), (6).
{1, 2, 4, 6}	8. $R \wedge \neg R$	Rule T, (2), (7).

Thus, we get a contradiction

Therefore, $P \rightarrow Q, Q \rightarrow R, P \vee R \Rightarrow R$.

Example. *Show that the following (use indirect method if needed).*

$(R \rightarrow \neg Q), R \vee S, S \rightarrow \neg Q, P \rightarrow Q \Rightarrow \neg P$.

Solution.

{1}	1. P	Rule P (additional premise).
{2}	2. $P \rightarrow Q$	Rule P.
{1, 2}	3. Q	Rule T, (1), (2), I_{11}.
{4}	4. $S \rightarrow \neg Q$	Rule P.
{4}	5. $Q \rightarrow \neg S$	Rule T, (4), E_{18}.
{1, 2, 4}	6. $\neg S$	Rule T, (3), (5), I_{11}.
{7}	7. $R \vee S$	Rule P.
{7}	8. $\neg R \rightarrow S$	Rule T, (7), E_{16}.
{7}	9. $\neg S \rightarrow R$	Rule T_1 (8), E_{18}.
{1, 2, 4, 7}	10. R	Rule T, (6), (9), I_{11}.
{11}	11. $R \rightarrow \neg Q$	Rule P.
{1, 2, 4, 7, 11}	12. $\neg Q$	Rule T, (10), (11), I_{11}.
{1, 2, 4, 7, 11}	13. $Q \wedge \neg Q$	Rule T, (3), (12), I_9.

Thus, we get a contradiction.

By indirect method, $(R \rightarrow \neg Q), (R \vee S), (S \rightarrow \neg Q), (P \rightarrow Q) \Rightarrow \neg P$.

Example. *Establish the validity of $A \vee C$ from $A \rightleftarrows (B \rightarrow C)$, $B \rightleftarrows (\neg A \vee \neg C)$, $C \rightleftarrows (A \vee \neg B)$ and B.* [JNTU - March 2006 - Set 1, June 03-Set 1]

Solution.

{1}	1. $A \rightleftarrows (B \to C)$	Rule P.
{2}	2. $B \rightleftarrows (\neg A \vee \neg C)$	Rule P.
{3}	3. $C \rightleftarrows (A \vee \neg B)$	Rule P.
{1}	4. $A \to (B \to C)$	Rule T, (1), E_{21}.
{1}	5. $(A \wedge B) \to C$	Rule T, (4), E_{19}.
{3}	6. $C \to (A \vee \neg B)$	Rule T, (3), E_{21}.
{1, 3}	7. $(A \wedge B) \to (A \vee \neg B)$	Rule T, (5), (6), I_{13}.
{1, 3}	8. $\neg (A \wedge B) \vee (A \wedge \neg B)$	Rule T, (7), E_{16}.
{1, 3}	9. $(\neg A \vee \neg B) \vee (A \vee \neg B)$	Rule T, (8), E_8.
{1, 3}	10. $\neg B$	Rule (9), E_{13}.
{11}	11. B	Rule P.
{1, 3, 11}	12. $B \wedge \neg B$	Rule T, (10), (11).

Which is a contradiction.

Hence the given premises are inconsistent.

Example. *Without constructing the truth table, find whether $A \wedge E$ is valid or not for the following :*

$A \rightleftarrows B, B \rightleftarrows (C \wedge D), C \rightleftarrows (A \vee E), A \vee E$ [JNTU - March - 06 - Set 1, June 03 - Set]

Solution.

{1}	1. $A \rightleftarrows B$	Rule P.
{2}	2. $B \rightleftarrows (C \wedge D)$	Rule P.
{1}	3. $A \to B$	Rule T, (1), E_{21}.
{2}	4. $B \to (C \wedge D)$	Rule T, (2), E_{21}.
{1, 2}	5. $A \to (C \wedge D)$	Rule T, (3), (4), I_{13}.
{6}	6. $C \rightleftarrows (A \vee E)$	Rule P.
{6}	7. $C \to (A \vee E)$	Rule T, (6), E_{21}.
{8}	8. $A \vee E$	Rule P.

There is a falt in the anticedent, hence $A \wedge E$ is not a valid consequence.

Example. *Test whether the following is a valid argument.*

If Sachin hits a century, then he gets a free car. Sachin does not get a free car.

Therefore, Sachin has not hit a century.

Solution. Let p : Sachin hits a century.

q : Sachin gets a free car.

Then the given premises are

$p \to q, \neg q$

and conclusion is $\neg p$.

{1} 1. $p \rightarrow q$ Rule P.

{2} 2. $\neg q$ Rule P.

{1, 2} 3. $\neg p$ Rule T, (1), (2).

Therefore, conclusion $\neg P$ is valid from the premises $p \rightarrow q$, $\neg q$.

2.4 AUTOMATIC THEOREM

The method of derivation just discussed provides only a partial solution to the decision problem, because if an argument is valid, then it is possible to show by this method that the argument is valid. On the other hand, if an argument is not valid, then it is very difficult to decide, after a finite number of steps, that this is the case. Also the formulation or process of derivation given earlier could not be used for this purpose because the construction of derivation depends upon the skill, experience and ingenuity of the person to make the right decision at every step. Let us first examine the limitations of the earlier procedure of derivation.

Rule P permits the introduction of a premise at any point in the derivation, but does not suggest either the premise or the step at which it should be introduced.

Rule T allows us to introduce any formula which follows from the previous steps. However, there is no definite choice of such a formula.

Similarly, Rule CP does not tell anything about the stages at which an antecedent is to be introduced as an assumed premise.

At every step, such decisions are taken from a large number of alternatives, with the ultimate aim of reaching the conclusion. Such a procedure is far from mechanical.

Now we describe a set of rules and a procedure which allow one to construct each step of derivation in a specified manner without recourse to any ingenuity and finally to arrive at a last step which clearly indicates whether a given conclusion follows from the premises. Our procedure is mechanical, also it is a full decision process for validity.

Our system of derivation of consisting of 10 rules an axiom schema and rules of well formed sequents and formulas :

1. **Variables :** The capital letters A, B, C, ..., P, Q, R, ... are used as statement variables and statement formulas.
2. **Connectives :** The connectives, $\neg$, $\wedge$, $\vee$ $\rightarrow$ and $\rightleftarrows$ appear in the formulas with the order of precedence as given.
3. **String of Formulas :** A string of formula is defined as follows :
   (*a*) Any formula is a string of formulas.
   (*b*) If α and β are strings of formulas, then α, β and β, α are strings of formulas.
   (*c*) Only those strings which are obtained by steps (*a*) and (*b*) are strings of formulas, with the exception of the empty string which is also a string of formulas.

Note : The order in which the formulas appear in any string is not important and so the strings A, B, C ; B, C, A ; A, C, B ; etc., ..., are the same.

4. **Sequents :** If α and β are strings of formulas, then $\alpha \xrightarrow{s} \beta$ is called a sequent in which α is denoted the antecedent and β the consequent of the sequent.

Thus, A, B, C $\xrightarrow{s}$ D, E, F is true iff $A \wedge B \wedge C \rightarrow D \vee E \vee F$ is true *i.e.*, A sequent $\alpha \xrightarrow{s} \beta$ is true iff either at least one of the formulas of the antecedent is false or at least one of the formulas of the consequent is true.

In this sense, the symbol "$\xrightarrow{s}$" is a generalization of the connection $\rightarrow$ to strings of formulas. Similarly, we use the symbol "$\overset{s}{\Rightarrow}$" applied to strings of formulas as a generalization of the symbol "$\Rightarrow$". Thus, $A \Rightarrow B$ means "A implies B" or $A \rightarrow B$ is a tautology while $\alpha \overset{s}{\Rightarrow} \beta$ means that $\alpha \overset{s}{\rightarrow} \beta$ is true.

Example. P, Q, R $\overset{s}{\Rightarrow}$ P, N.

The empty antecedent is interpreted as the logical constant "true" (T) and the empty consequent is interpreted as the logical constant "false" (F).

5. **Axiom Schema :** If α and β are strings of formulas such that every formula in both α and β is a variable only, then the sequent $\alpha \overset{s}{\rightarrow} \beta$ is an axiom iff α and β have at least one variable in common.

Example A, B, C $\overset{s}{\rightarrow}$ P, B, R is an axiom, where A, B, C, P and R are variables.

Note that if $\alpha \overset{s}{\rightarrow} \beta$ is an axiom, then $\alpha \overset{s}{\rightarrow} \beta$.

6. **Theorem :** The following sequent are theorems of our system.
   (*a*) Every axiom is a theorem.
   (*b*) If a sequent α is a theorem and a sequent β results from α through the use of one of the 10 rules of the system which are given below, then β is a theorem.
   (*c*) Sequent obtained by (*a*) and (*b*) are the only theorems.
7. **Rules :** The following rules are used to combine formulas within strings by introducing connectives. Corresponding to each of the connectives there are two rules, one for the introduction of the connective in the antecedent and the other for its introduction in the consequent. In the description of these rules, $\alpha, \beta, \gamma, \ldots$, are strings of formulas while X and Y are formulae to which the connectives are applied.

Antecedent Rules :

Rule $\neg \Rightarrow$: If $\alpha, \beta \overset{s}{\Rightarrow} X, \gamma$, then $\alpha, \neg X, \beta \overset{s}{\Rightarrow} \gamma$

Rule $\wedge \Rightarrow$: If $X, Y, \alpha, \beta \overset{s}{\Rightarrow} \gamma$, then $\alpha, X \wedge Y, \beta \overset{s}{\Rightarrow} \gamma$

Rule $\vee \Rightarrow$: If $X, \alpha, \beta \overset{s}{\Rightarrow} \gamma$ and $Y, \alpha, \beta \overset{s}{\Rightarrow} \gamma$, then $\alpha, X \vee Y, \beta \overset{s}{\Rightarrow} \gamma$.

Rule $\rightarrow \Rightarrow$: If $Y, \alpha, \beta \overset{s}{\Rightarrow} \gamma$ and $\alpha, \beta \overset{s}{\Rightarrow} X, \gamma$ then $\alpha, X \rightarrow Y, \beta \overset{s}{\Rightarrow} \gamma$.

Rule $\rightleftarrows \Rightarrow$: If $X, Y, \alpha, \beta \overset{s}{\Rightarrow} \gamma$ and $\alpha, \beta \overset{s}{\Rightarrow} X, Y, \gamma$, then $\alpha, X \rightleftarrows Y, \beta \overset{s}{\Rightarrow} \gamma$.

Consequent Rules :

Rule $\Rightarrow \neg$: If X, $\alpha \overset{s}{\Rightarrow} \beta, \gamma$ then $\alpha \overset{s}{\Rightarrow} \beta, \neg X, \gamma$.

Rule $\Rightarrow \wedge$: If $\alpha \overset{s}{\Rightarrow}$ X, β, γ, and $\alpha \overset{s}{\Rightarrow}$ Y, β, γ, then $\alpha \overset{s}{\Rightarrow} \beta, X \wedge Y, \gamma$.

Rule $\Rightarrow \vee$: If $\alpha \overset{s}{\Rightarrow}$ X, Y, β, γ, then $\alpha \overset{s}{\Rightarrow} \beta, X \vee Y, \gamma$.

Rule $\Rightarrow \rightarrow$: If X, $\alpha \overset{s}{\Rightarrow}$ Y, β, γ, then $\alpha \overset{s}{\Rightarrow} \beta, X \rightarrow Y, \gamma$.

Rule $\Rightarrow \rightleftarrows$: If X, $\alpha \overset{s}{\Rightarrow}$ Y, β, γ, and Y, $\alpha \overset{s}{\Rightarrow}$ X, β, γ then $\alpha \overset{s}{\Rightarrow} \beta, X \rightleftarrows Y, \gamma$.

The system described here is equivalent to the one described earlier except that the procedures and techniques of derivation are different. This difference does not effect the validity of an argument.

The description of our system is complete. Now, we describe the procedure used in practice. In order to show that the conclusion C follows from $H_1, H_2, ..., H_m$, we establish that

$$\overset{s}{\rightarrow} H_1 \rightarrow (H_2 \rightarrow (H_3 \ldots (H_m \rightarrow C) \ldots)) \qquad \ldots(I)$$

is a theorem. We must show that

$$\overset{s}{\Rightarrow} H_1 \rightarrow (H_2 \rightarrow (H_3 \ldots (H_m \rightarrow C) \ldots)) \qquad \ldots(II)$$

Our procedure involves showing (I) to be a theorem.

For this, we first assume (II) and then show that this assumption is or is not justified.

This is accomplished by working backward from (II), using the rules and showing that (II) holds if some simpler sequent is a theorem (By a simpler sequent we mean a sequent in which some connective is eliminated in one of the formulas appearing in the antecedent or the consequent). We continue working backward until we arrive at the simplest possible sequents (*i.e.,*) those which do not have any connectives. If these sequents are axioms, then we have justified our assumption of (II). If atleast one of the simplest sequents is not an axiom, then the assumption of (II) is not justified and C does not follow from $H_1, H_2, ..., H_m$. In case when C follows from $H_1, H_2, ..., H_m$ the derivation of (II) is easily constructed by simply working through the same steps, starting from the axioms obtained, we now demonstrate this procedure by means of examples.

Example. *Show that P ∨ Q follows from P.*

Solution. We need to show that

(1) $\overset{s}{\Rightarrow} P \rightarrow (P \vee Q)$

(1) If (2) $P \overset{s}{\Rightarrow} P \vee Q$ $(\Rightarrow \rightarrow)$

(2) If (3) $P \overset{s}{\Rightarrow} P, Q$ $(\Rightarrow \vee)$

We first eliminate the connective $\rightarrow$ in (1).

Using the rule $\Rightarrow\,\rightarrow$ we have "If $P \overset{s}{\Rightarrow} P \vee Q$ then $\overset{s}{\Rightarrow} P \rightarrow (P \vee Q)$". Here we have named $P \overset{s}{\Rightarrow} P \vee Q$ by (2). Each line of derivation thus introduces the name as well as gives a rule. Note also that "(1) if (2)" means "if (2) then (1)". The chain of arguments is then given by (1) holds if (2) and (2) holds if (3).

Finally (3) is a theorem because it is an axiom. The actual derivation is simply a reversal of those steps in which (3) is an axiom that leads to $\overset{s}{\Rightarrow} P \rightarrow (P \vee Q)$ as shown.

(*a*) $P \overset{s}{\Rightarrow} P, Q$ Axiom.

(*b*) $P \overset{s}{\Rightarrow} P \vee Q$ Rule $(\Rightarrow \vee)$, (*a*).

(*c*) $\overset{s}{\Rightarrow} P \rightarrow (P \vee Q)$ Rule $(\Rightarrow \rightarrow)$, (*b*).

Example. *Show that* $\overset{s}{\Rightarrow} (\neg Q \wedge (P \rightarrow Q)) \rightarrow \neg P$.

Solution. (1) $\overset{s}{\Rightarrow} (\neg Q \wedge (P \rightarrow Q)) \rightarrow \neg P$.

(1) If (2) $\quad \neg Q \wedge (P \rightarrow Q) \overset{s}{\Rightarrow} \neg P \; (\Rightarrow \rightarrow)$

(2) If (3) $\quad \neg Q, P \rightarrow Q \overset{s}{\Rightarrow} \neg P \; (\wedge \Rightarrow)$

(3) If (4) $\quad P \rightarrow Q \overset{s}{\Rightarrow} \neg P, Q \; (\neg \Rightarrow)$

(4) If (5) $\quad Q \overset{s}{\Rightarrow} \neg P, Q$ and (6) $\overset{s}{\Rightarrow} P, \neg P, Q \; (\rightarrow \Rightarrow)$

(5) If (7) $\quad P, Q \overset{s}{\Rightarrow} Q \; (\Rightarrow \neg)$.

(6) If (8) $\quad P \overset{s}{\Rightarrow} P, Q \; (\Rightarrow \neg)$

Now (7) and (8) are axioms, hence the theorem (1) follows derivation, easily obtained by starting with the axioms (7) and (8) and retracing the steps.

Example. *Does P follow from* $P \vee Q$*?*

Solution. We investigate whether $\overset{s}{\rightarrow} (P \vee Q) \rightarrow P$ is a theorem,

Assume (1) $\overset{s}{\Rightarrow} (P \vee Q) \rightarrow P$

(1) If (2) $P \vee Q \overset{s}{\Rightarrow} P$. $\quad (\Rightarrow \rightarrow)$

(2) If (3) $P \overset{s}{\Rightarrow} P$ and (4) $Q \overset{s}{\Rightarrow} P \; (\vee \Rightarrow)$

Here (3) is an axiom, but (4) is not.

Hence P does not follow from P ∨ Q.

Note : That in some cases the derivation is longer if this procedure is used. The reason is that for every connective appearing there is at least one step introduced. In some steps a branching appears and then we have to pursue two steps.

Example. *Show that $S \vee R$ is tautologically implied by $(P \vee Q) \wedge (P \to R) \wedge (Q \to S)$.*

Solution. To show

(1) $\overset{s}{\Rightarrow} ((P \vee Q) \wedge (P \to R) \wedge (Q \to S)) \to (S \vee R)$

(1) If (2) $(P \vee Q) \wedge (P \to R) \wedge (Q \to S) \overset{s}{\Rightarrow} (S \vee R)$ $(\Rightarrow \to)$

(2) If (3) $(P \vee Q) \wedge (P \to R) \wedge (Q \to S) \overset{s}{\Rightarrow} S, R$ $(\Rightarrow \vee)$

(3) If (4) $(P \vee Q), (P \to R), (Q \to S) \overset{s}{\Rightarrow} S, R$ $(\wedge \Rightarrow$ twice$)$

(4) If (5) $P, P \to R, Q \to S \overset{s}{\Rightarrow} S, R$ and (6) $Q, P \to R, Q \to S \overset{s}{\Rightarrow} S, R$ $(\vee \Rightarrow)$

(5) If (7) $P, R, Q \to S \overset{s}{\Rightarrow} S, R$ and (8) $P, Q \to S \overset{s}{\Rightarrow} P, S, R$ $(\to \Rightarrow)$

(7) If (9) $P, R, S \overset{s}{\Rightarrow} S, R$ and (10) $P, R \overset{s}{\Rightarrow} S, R, Q$ $(\to \Rightarrow)$

(8) If (11) $P, S \overset{s}{\Rightarrow} P, S, R$ and (12) $P \overset{s}{\Rightarrow} P, S, R, Q$ $(\to \Rightarrow)$

(6) If (13) $Q, R, Q \to S \overset{s}{\Rightarrow} S, R$ and (14) $Q, Q \to S \overset{s}{\Rightarrow} S, R, P$ $(\to \Rightarrow)$

(13) If (15) $Q, R, S \overset{s}{\Rightarrow} S, R$ and (16) $Q, R \overset{s}{\Rightarrow} S, R, Q$ $(\to \Rightarrow)$

(14) If (17) $Q, S \overset{s}{\Rightarrow} S, R, P$ and (18) $Q \overset{s}{\Rightarrow} S, R, P, Q$ $(\to \Rightarrow)$

Now (9) to (12) and (15) to (18) are all axioms ; therefore the result follows. The nesting of steps shows how the branching occurs.

EXERCISE 2.1.

1. Show that the conclusion C follows from the premises $H_1, H_2, \ldots,$ in the following cases :
   (*a*) $H_1 : \neg P \vee Q, H_2 : \neg(Q \wedge \neg R), H_3 : \neg R, C : \neg P$
   (*b*) $H_1 : P \to Q, H_2 : Q \to R, C : P \to R$
   (*c*) $H_1 : R, H_2 : P \vee \neg P, C : R.$
2. Determine whether the conclusion C follows logically from the premises H_1 and H_2.
   (*a*) $H_1 : P \to Q, H_2 : \neg (P \wedge Q), C : \neg P$
   (*b*) $H_1 : \neg P, H_2 : P \rightleftarrows Q, C : \neg (P \wedge Q)$
   (*c*) $H_1 : P \to Q, H_2 : \neg P, C : Q$
   (*d*) $H_1 : \neg Q, H_2 : P \to Q, C : \neg P.$

3. Show that the $L \vee M$ follows from $P \wedge Q \wedge R$, $(Q \rightleftarrows R) \rightarrow (L \vee M)$.
4. Show that D follows logically from the premises $(A \rightarrow B) \wedge (A \rightarrow C)$, $\neg (B \wedge C)$, $D \vee A$.
5. Show that $R \rightarrow S$ can be derived from the premises $P \rightarrow (Q \rightarrow S)$, $\neg R \vee P$ and Q, using rule CP.
6. Derive the following, using rule CP if necessary
   (*a*) $P, P \rightarrow (Q \rightarrow (R \wedge S)) \Rightarrow (Q \rightarrow S)$
   (*b*) $P \rightarrow Q \Rightarrow P \rightarrow (P \wedge Q)$
   (*c*) $(P \vee Q) \rightarrow R \Rightarrow (P \wedge Q) \rightarrow R$.
7. "If there was a ball game, then travelling was difficult. If they arrived on time, then travelling was not difficult. They arrived on time. Therefore, there was no ball game". Show that these statements constitute a valid argument.
8. Show that the following by using indirect proof (method). $S \rightarrow \neg Q, S \vee R, \neg R, \neg R \rightleftarrows Q \Rightarrow \neg P$.
9. Show that the following :
   (*a*) $P \Rightarrow (\neg P \rightarrow Q)$
   (*b*) $P \wedge \neg P \wedge Q \Rightarrow R$
   (*c*) $P \wedge \neg P \vee (P \wedge Q) \Rightarrow Q$.
10. Determine the validity of the following argument. If two sides of a triangle are equal, then two opposite angles are equal.
    Two sides of a triangle are not equal.
    Therefore, the opposite angles are not equal.

ANSWERS

2. (*a*) Yes (*b*) Yes (*c*) No (*d*) Yes.
10. Not valid.

2.5 THE PREDICATE LOGIC

So far our discussion of symbolic logic has been limited to the consideration of statements and statement formulas. The inference theory was also restricted in the sense that the premises and conclusions were all statements. The symbols P, Q, R, ..., P_1, Q_1, R_1, ... were used for statements or statement variables. We now introduce the concept of a predicate in an atomic statement. The logic based upon the analysis of predicates in any statement is called predicate logic.

Predicates :

Let us consider the two statements :

Ravi is a bachelor.

Kishore is a bachelor.

If we introduce some symbol to denote "is a bachelor" and a method to join it with symbols denoting the names of individuals, then we will have a symbolism to denote statements about any individual's being a bachelor. The part "is a bachelor" is called a predicate.

We shall symbolize a predicate by a capital letter and the names of individuals or objects in general by small letters. Every predicate describes something about one or more objects (the word "object" is being used in a very general sense to include individuals as well). Therefore, a statement could be written symbolically in terms of the predicate letter followed by the name or names of the objects to which the predicate is applied.

We again consider the statements :

(1) Ravi is a bachelor.

(2) Kishore is a bachelor.

Denote the predicate "is a bachelor" symbolically by the predicate letter B, "Ravi" by *r*, and "Kishore" by *k*. The statements (1) and (2) can be written as B (r), and B (k) respectively.

In general, any statement of the type "*p* is Q" where Q is a predicate and *p* is the subject can be denoted by Q (*p*).

Further, the connectives described earlier can now be used to form compound statements such as "Ravi is a bachelor, and this painting is red", which can be written as $B(r) \wedge R(p)$ where R (*p*) : This painting is red. Other connectives can also be used to form statements such as $B(r) \to R(p)$, $\neg R(p)$, $B(r) \vee R(p)$, etc.

A predicate requiring *m* ($m > 0$) names or objects is called an *m*-place predicate.

Consider the examples

(*i*) Indu is a student.

The predicate S : is a student is a 1-place predicate because it is related to one object (or) name, say Indu (*i*). This statement can be translated as S (*i*).

(*ii*) Phaneendhar is taller than Mohan.

The predicate "is taller than" is a 2-place predicate. This statement can be translated as T (*p*, *m*) where T symbolizes "is taller than", *p* denotes "Phaneendhar" and m denotes "Mohan".

Note : That the order in which the names or objects appear in the statement as well as in the predicate is important.

In order to extend our definition to $m = 0$, we shall call a statement a 0-place predicate because no names are associated with a statement.

Examples of 3-place predicates and 4-place predicates are :

1. Suman sits between Rahul and Nithin.
2. Venkat and Karthik played bridge against Rohith and Raj.

In general, an n-place predicate requires *n* names of objects to be inserted in fixed positions in order to obtain a statement. The position of these names or objects is important. If S is an *n*-place letter and $a_1, a_2, ..., a_n$ are the names of objects, then $S(a_1, a_2, ..., a_n)$ is a statement.

The statement function, variables and quantifiers.

Let H be the predicate "is a mortal", *b* the name "Brown", *c* the name "Canada" and *s* the name "A shirt". Then H (*b*), H (*c*), and H (*s*) all denote statements. In fact, these statements have a common form. If we write H (*x*) for "x is mortal", then H (*b*), H (*c*), H (*s*), and others having the same form can be obtained from H (x) by replacing x by an appropriate name. Note that H (*x*) is not a statement, but it results in a statement when x is replaced by the name of an object. The letter x used here is a place holder. From now on we shall use small letters as individual or object variables as well as names of objects.

A simple statement function of one variable is defined to be an expression consisting of a predicate symbol and an individual variable. Such a statement function becomes a statement when the variable is replaced by the name of any object. We can form compound statement functions by combining one or more simple statement functions and the logical connectives. For example, if we let M (x) be "x is a man" and H (x) be "x is a mortal", then we can form compound statement functions such as

$$M(x) \wedge H(x),\ M(x) \rightarrow H(x),\ \neg H(x),\ M(x) \vee \neg H(x), \text{ etc.}$$

An extension of this idea to the statement functions of two or more variables is straight forward. Consider, for example, the statement function of two variables.

$$G(x, y) : x \text{ is taller than y.}$$

If both x and y are replaced by the names of objects, we get a statement. If m represents Mr. Madhu and g represents Mr. Gopi, then we have

G (m, g) : Mr. Madhu is taller than Mr. Gopi ; and

G (g, m) : Mr. Gopi is taller than Mr. Madhu.

Quantifiers :

Consider the following example :

All Roses are red. This can be understood as for every x. If x is a Rose, then x is red. If we denote R (x) : x is a Rose and P (x) : x is a red, then we can write above statement as $(x)\ (R(x) \rightarrow P(x))$. Some times $(x)\ (R(x) \rightarrow P(x))$ is also written as $(\forall\ x)\ (R(x) \rightarrow P(x))$. The symbols (x) or $(\forall\ x)$ are called **Universal Quantifiers.**

Examples :

1. All men are mortal.
2. Every apple is red.
3. Any integer is positive or negative.

Let us paraphrase these in the following manner.

1*a*. For all x, if x is a man, then x is a mortal.

2*a*. For all x, if x is an apple, then x is red.

3*a*. For all x, if x is an integer, then x is either positive or negative

and consider statement functions are as follows :

M (x) : x is a man, H (x) : x is a mortal.

A (x) : x is an apple, R (x) : x is a red.

N (x) : x is an integer, P (x) : x is either positive or negative.

We write (1*a*), (2*a*), (3*a*) as

$$1b: (x)\ (M(x) \rightarrow H(x))$$

$$2b: (x)\ (A(x) \rightarrow R(x))$$

$$3b: (x)\ (N(x) \rightarrow P(x))$$

Note that statement (x) $(M(x))$ can be translated as

4. For all x, x is a man

4*a*. For every x, x is a man

4*b*. Every thing is a man

4*c*. For each x, x is a man.

Note : That the particular variable appearing in the statements involving a quantifier is not important because the statements remain unchanged if x is replaced by y throughout. Thus, the statements (x) $(M(x) \rightarrow H(x))$ and (y) $(M(y) \rightarrow H(y))$ are equivalent.

Sometimes it is necessary to use more than one universal quantifier in a statement. For example consider

$$G(x, y) : x \text{ is taller than } y.$$

We can state that "for any x and any y, if x is taller than y, then y is not taller than x" or "for any x and y, if x is taller than y, then it is not true that y is taller than x". This statement can now be symbolized as (x) (y) $(G(x, y) \rightarrow \neg G(y, x))$.

Another quantifier will now be introduced to symbolize expressions such as "for some", "there is at least one" (or) "there exists some".

Consider the following statements :

5. There exists a man.
6. Some men are clever.
7. Some real numbers are rational.

The first statement can be expressed in various ways, say

5*a*. There exists an x such that x is a man.

5*b*. There is at least one x such that x is a man.

Similarly, (6) can be written as

6*a*. There exists an x such that x is a man and x is clever.

6*b*. There exists at least one x such that x is a man and x is clever.

Here we introduce the symbol "$(\exists x)$", called the **existential quantifier**, which symbolizes expressions such as "there is at least one x such that" or "there exists an x such that" or "for some x".

Denoting $M(x) : x$ is a man

$C(x) : x$ is clever

$R_1(x) : x$ is a real number

$R_2(x) : x$ is rational

and using the existential quantifier, we can write the statements (5), (6), (7) as

5*c*. $(\exists x)$ $(M(x))$

6*c*. $(\exists x)$ $(M(x) \wedge C(x))$

7*c*. $(\exists x)$ $(R_1(x) \wedge R_2(x))$

It may be noted that a conjunction has been used in writing the statements of the form "Some A are B", while a conditional was used in writing statements of the form "All A are B".

Example. *Write the following statements in symbolic form.*

1. *Something is good.*
2. *Everything is good.*
3. *Nothing is good.*
4. *Something is not good.*

Solution. Statement (1) means "there is at least one x such that x is good".

Statement (2) means "for all x, x is good"

Statement (3) means "for all x, x is not good"

Statement (4) means "there is at least one x such that x is not good".

We denote G (x) : x is good, then

Statement (1) can be denoted by $(\exists\ x)\ (G\ (x))$

Statement (2) can be denoted by $(\forall\ x)\ (G\ (x))$

Statement (3) can be denoted by $(\forall\ x)\ (\neg G\ (x))$

Statement (4) can be denoted by $(\exists\ x)\ (\neg G\ (x))$

Example. *Write each of the following in symbolic form.*

(*a*) *All men are good.*

(*b*) *No men are good.*

(*c*) *Some men are good.*

(*d*) *Some men are not good.*

Solution. We assume that the universe consists some of which are not men.

Let M (x) : x is a man and

G (x) : x is good.

(*a*): Statement (*a*) means 'for all x, if x is a man, then x is good'. So (*a*) is $(\forall\ x)\ (M\ (x) \rightarrow G\ (x))$.

(*b*): Statement (*b*) means "for all x, if x is a man, then x is not good" and (*b*) is $(\forall\ x)\ (M\ (x) \rightarrow \neg G\ (x))$.

(*c*): Statement (*c*) means "there is an x, such that x is a man and x is good" and it is represented by $(\exists\ x)\ (M\ (x) \wedge G\ (x))$

(*d*): Statement (*d*) means "there is an x, such that x is a man and x is not good". So it is $(\exists\ x)\ (M\ (x) \wedge \neg G\ (x))$.

2.6 STATEMENT FORMULAS IN PREDICATE LOGIC

P $(x_1, x_2, \ldots\ldots, x_n)$ denotes an n-place predicate formula in which the letter P is an n-place predicate an $x_1, x_2, \ldots\ldots, x_n$ are object or individual variables. In general P $(x_1, x_2, \ldots\ldots, x_n)$ will be called an **atomic formula** of predicate calculus.

The following are some examples of atomic formulas R, Q (x), P (x, y), A (x, y, z), P (a, y), and A (x, a, z). A well-formed formula of predicate calculus is obtained by using the following rules.

1. An atomic formula is a well-formed formula.
2. If A is a well-formed formula, then $\neg$A is a well-formed formula.
3. If A and B are well-formed **formula,** then $(A \wedge B)$, $(A \vee B)$, $(A \rightarrow B)$, and $(A \leftrightarrows B)$ are also well-formed formulas.
4. If A is a well-formed formula and x is any variable, then (x) A and $(\exists\ x)$ A are well-formed formulas.
5. Only those formulas obtained by using rules (1) to (4) are well-formed formulas.

Free and Bound Variables

Given a formula containing a part of the form (x) P (x) or $(\exists\ x)$ P (x), such a part is called an x-bound part of the formula. Any occurrence of x is an x-bound part of a formula is called a **bound occurrence** of x, while any occurrence of x or of any varaible that is not a bound occurrence is called a **free occurrence** and the formula P (x) either in (x) P (x) or in $(\exists\ x)$ P (x) is described as the **scope of the quantifier**. In other words, the **scope of a quantifier** is the formula immediately following the quantifier.

Consider the following formulas :

(1) (x) P (x, y)
(2) (x) (P $(x) \rightarrow$ Q (x))
(3) (x) (P $(x) \rightarrow (\exists\ y)$ R (x, y))
(4) $(\exists\ x)$ (P $(x) \wedge$ Q (x))
(5) $(\exists\ x)$ P $(x) \vee$ Q (x).

In (1), P (x, y) is the scope of the quantifier, and occurrence of x is bound occurrence, while the occurrence of y is a free occurrence.

In (2), the scope of the universal quantifier is P $(x) \rightarrow$ Q (x), and all occurrences of x are bound.

In (3), the scope of (x) is P $(x) \rightarrow (\exists\ y)$ R (x, y), while the scope of $(\exists\ y)$ is R (x, y). All occurrences of both x and y are bound occurrences.

In (4), the scope of $(\exists\ x)$ is P $(x) \wedge$ Q (x).

and In (5), the scope of $(\exists\ x)$ is P (x) and the last occurrence of x in Q (x) is free.

Example. *Indicate the variables that are free and bound and write the scope of the quantifiers :*

(a) (x) (P $(x) \wedge$ R (x)) $\rightarrow (x)$ P $(x) \wedge$ Q (x)
(b) (x) (P $(x) \wedge (\exists x)$ Q (x)) $\vee$ ((x) P $(x) \rightarrow$ Q (x))
(c) (x) (P $(x) \leftrightarrows$ Q $(x) \wedge (\exists x)$ R (x)) $\wedge$ S (x).

Solution. In *(a)*, the scope of the first quantifier (x) is P $(x) \wedge$ R (x), the scope of the second quantifier (x) is P (x) and the last occurrence of x in Q (x) is free while the first two occurrences of x are bound occurrences.

In *(b)*, the scope of the first quantifier (x) is P $(x) \wedge (\exists\ x)$ Q (x) while the scope of $(\exists\ x)$ is Q (x), the scope of the second quantifier (x) is P (x) and the last occurrence of x in Q (x) is free while the other occurrences of x are bound occurrences.

In (*c*), the scope of the quantifier (x) is $P(x) \leftrightarrows Q(x) \wedge (\exists x) R(x)$ while the scope of the quantifier $(\exists x)$ is $R(x)$ and the last occurrence of x in $S(x)$ is free while the other occurrences of x are bound occurrences.

Example. *Symbolize the expression "x is the father of the mother of y".*

Solution. Write $P(x)$: x is a person.

$F(x, y)$: x is the father of y.

$M(x, y)$: x is the mother of y.

First we name a person called z as the mother of y.

Hence x is the father of z and z is the mother of y.

We symbolize the given predicate as

$(\exists z)(P(z) \wedge F(x, z) \wedge M(z, y))$.

Example. *Symbolize the expression "All the world loves a lover".* [JNTU - Nov 04 - Set 3]

Solution. First we note that the quotation really means that everybody loves a lover.

Let $P(x)$: x is a person.

$L(x)$: x is lover.

$R(x, y)$: x loves y.

The given expression can be written in symbolical form as $(x)(P(x) \rightarrow (y)(P(y) \wedge L(y) \rightarrow R(x, y)))$.

The Universe of Discourse

Some restriction can be introduced by limiting the class of objects under consideration. This limitation means that the variables which are mentioned stand for only those objects which are members of a particular set or class. Such a restricted class is called the **universe of discourse** (or) **the domain of individuals** (or) simply the **universe**. If the discussion refers to human beings only, then the universe of discourse is the class of human beings. In elementary algebra or number theory the universe of discourse could be numbers (real, rational, complex, etc.)

Example. *Symbolize the statement : "All mean are giants".* [JNTU - Nov 4 - Set 3]

Solution. We denote $G(x)$: x is a giant.

$M(x)$: x is a man.

Then the given statement can be symbolized as $(x)(M(x) \rightarrow G(x))$.

(or) If we restrict the variable x to the universe which is the class of man, then the statement is $(x) G(x)$.

Example. *Consider the statement "Given any positive integer, there is a greater positive integer". Symbolize this statement with and without using the set of positive integers as the universe of discourse.*

Solution.

(*i*) Let the variables x and y be restricted to the set of positive integers. Then given statement can be written as follows : For all x, there exists a y such that y is greater than x. Let $G(x, y)$: x is a greater than y.

Then the given statement is $(x)(\exists y) G(y, x)$.

(*ii*) If we do not impose the restriction on the universe of discourse and if we write P (x) : x is a positive integer then the given statement is symbolic form as

(x) (P $(x) \rightarrow (\exists\, y)$ (P $(y) \wedge$ G (y, x))).

The universe of discourse, if any, must be explicitly stated, because the truth value of a statement depends upon it.

Consider the predicate Q (x) : x is less than 5 and statements (x) Q (x) and $(\exists\, x)$ Q (x). If the universe of discourse is given by the sets

1. $\{-2, 0, 1, 3, 4\}$
2. $\{4, -2, 6, -3\}$
3. $\{10, 15, 26\}$

Then (x) Q (x) is true for the universe of discourse (1), false for the universe of discourse (2), (3) and the statement $(\exists\, x)$ Q (x) is true for both (1), (2) but false for (3).

2.7 VALID FORMULAS AND EQUIVALENCES

Let A and B be any two predicate formulas defined over a common universe E. If, for ever assignment of object names from the universe E to each of the variables appearing in A and B, the resulting statements have the same truth values, then the predicate formulas A and B are said to be equivalent to each other over E. This idea is symbolized by writing A $\Leftrightarrow$ B over E. If E is arbitrary, we say A and B are equivalent and write A $\Leftrightarrow$ B. Similarly, a formula A is said to be valid in E if, for every assignment of object names from E to the corresponding variables in A and for every assignment of statements to statement variables, the resulting statements have the truth value T.

We write $\models$ A in E. If A is valid for an arbitrary E, then we write $\models$ A.

In this discussion we derive several valid formulas, which will be useful in the inference theory of predicate logic.

Formulas of predicate calculus that involve quantifiers and no free variables are also formulas of the statement calculus. Therefore, substitution instances of all the tautologies by these formulas yield any number of special tautologies.

For example, consider the tautologies of the statement calculus given by

$$P \vee \neg P \quad (P \rightarrow Q) \rightleftarrows (\neg P \vee Q)$$

and substitute the formulas (x) R (x) and $(\exists\, x)$ S (x) for P and Q respectively. It is assumed that (x) R (x) and $(\exists\, x)$ S (x) do not contain any free variables. The following tautologies are obtained.

$$((x)\, R\, (x) \vee \neg((x)\, R\, (x))$$

$$((x)\, R\, (x)) \rightarrow ((\exists\, x)\, S\, (x)) \rightleftarrows \neg\, ((x)\, R\, (x)) \vee ((\exists\, x)\, S\, (x)).$$

These tautologies form only a very special class of valid formulas of the predicate calculus. Another class of valid formulas that follow from the tautologies of the statement calculus will now be discussed.

Recall that any substitution instance of a tautology is also a tautology in the statement calculus. A substitution instance is one in which any variable in a formula is consistently replaced by any other formula throughout. A similar situation exists for predicate formulas, we shall call a predicate formula a prime (atomic, simple) formula if no sentential connectives appear in it. In general, a tautology of the statement calculus remains a valid formula of the predicate calculus when prime formulas are substituted for statement variables throughout the formula. Thus all the implications and equivalences of the statement calculus given in section 2.2 can be considered as implications and equivalences of the predicate calculus in which the statement variables are replaced by prime predicate formulas.

As an illustration, Let A (x), B (x), and C (x, y) denote any prime formula of the predicate calculus. Then the following are valid formulas of the predicate calculus.

$$\neg\neg A(x) \rightleftarrows A(x) \quad E_1$$

$$C(x, y) \wedge B(x) \rightleftarrows B(x) \wedge C(x, y) \quad E_2$$

$$A(x) \to B(x) \rightleftarrows \neg A(x) \vee B(x) \quad E_{16}$$

2.8 SOME VALID FORMULAS OVER FINITE UNIVERSES

In this section we denote predicate formulas by capital letters such as A, B, C, ..., followed by object variables x, y, ... Thus A (x), A (x, y), B (y), and C (x, y, z) are examples of predicate formulas. Some classification is necessary at this stage. In the formula A (x), we wish to say that A is predicate formula in which x is one of the free variables. This variable x is of interest to us, and we want to emphasize the dependence of A on it. For example, we may write B (x) for (y) P $(y) \vee$ Q (x).

If in a formula A (x) we replace each free occurrence of the variable x by another variable y, then we say that y is substituted for x in the formula, and the resulting formula is denoted by A (y). For such a substitution, the formula A (x) must be free for y. A formula A (x) is said to be free for y if no free occurrence of x is in the scope of the quantifiers (y) or $(\exists\, y)$. If A (x) is not free for y, then it is necessary to change the variable y, appearing as a bound variable, to another variable before substituting y for x. If y is to be substituted, then it is usually a good idea to make all the bound variables different from y. The following examples show what A (y) is for a given A (x).

A (x)	A (y)
$P(x, y) \wedge (\exists x)\, Q(y)$	$P(y, y) \wedge (\exists y)\, Q(y)$ or $P(y, y) \wedge (\exists z)\, Q(z)$
$(S(x) \wedge S(y)) \vee (x)\, R(x)$	$(S(y) \wedge S(y)) \vee (x)\, R(x)$ or $(S(y) \wedge S(y)) \vee (z)\, R(z)$

The following formulas are not free for y.

$P(x, y) \wedge (y)\, Q(x, y)$ $\qquad\qquad$ $(y)\,(S(y) \to S(x))$.

In order to substitute y in place of the variable x in these formulas, it is necessary to first make them free for y as follows :

A (x)	A (y)
$P(x, y) \wedge (z)\, Q(x, z)$	$P(y, y) \wedge (z)\, Q(y, z)$
$(z)\,(S(z) \to S(x))$	$(z)\,(S(z) \to S(y))$

If the universe of discourse is a finite set, then all possible substitutions of the object variables can be enumerated. However, it is not possible to enumerate all possible substitutions if the universe of discourse is infinite we shall now give some equivalences which hold for a finite universe. Later we show that these equivalences also hold for an arbitrary universe.

Let the universe of discourse be denoted by a finite set S given by $S = \{a_1, a_2, \ldots\ldots, a_n\}$. From the meaning of the quantifiers and by simple enumeration of all the objects in S, it is easy to see that

$$(x)\, A(x) \Leftrightarrow A(a_1) \wedge A(a_2) \wedge \ldots\ldots \wedge A(a_n) \qquad \ldots(1)$$

$$(\exists x)\, A(x) \Leftrightarrow A(a_1) \vee A(a_2) \vee \ldots\ldots \vee A(a_n) \qquad \ldots(2)$$

Equivalences (1) and (2) over S show that the quantifiers can be dispensed with if the universe of discourse is finite.

Example. Prove the following De Morgan's Laws

1. $\neg ((x)\, A(x)) \Leftrightarrow (\exists x) \neg A(x)$
2. $\neg ((\exists x)\, A(x)) \Leftrightarrow (x) \neg A(x)$

Proof :

1. $\neg ((x)\, A(x)) \Leftrightarrow \neg (A(a_1) \wedge A(a_2) \wedge \ldots\ldots \wedge A(a_n))$
$\Leftrightarrow \neg A(a_1) \vee \neg A(a_2) \vee \ldots\ldots \vee \neg A(a_n)$
$\Leftrightarrow (\exists x) \neg A(x)$
2. $\neg((\exists x)\, A(x)) \Leftrightarrow \neg(A(a_1) \vee A(a_2) \vee \ldots\ldots \vee A(a_n))$
$\Leftrightarrow \neg A(a_1) \wedge \neg A(a_2) \wedge \ldots\ldots \wedge \neg A(a_n)$
$\Leftrightarrow (x) \neg A(x)$.

If the universal and existential quantifiers are called duals of each other, then the above equivalences can be summarized by saying that the negation of a quantified formula is equivalent to a formula in which the quantifier is replaced by its dual and the scope of the quantifier by its negation.

Example. *Negate and simplify each of the following :*

(i) $(\exists x)\, (P(x) \vee Q(x))$ *(ii)* $(x)\, (P(x) \wedge \neg Q(x))$

(iii) $(x)\, (p(x) \rightarrow q(x))$ *(iv)* $(\exists x)\, ((P(x) \vee Q(x)) \rightarrow R(x))$

Solution.

(i) $\neg(\exists x)\, (P(x) \vee Q(x)) \Leftrightarrow (x)\, (\neg (P(x) \vee Q(x))$
$\Leftrightarrow (x)\, (\neg P(x) \wedge \neg Q(x))$

(ii) $\neg (x)\, (P(x) \wedge \neg Q(x)) \Leftrightarrow (\exists x)\, (\neg (P(x) \wedge \neg Q(x)))$
$\Leftrightarrow (\exists x)\, (\neg P(x) \vee Q(x))$

(iii) $\neg (x)\, (p(x) \rightarrow q(x)) \Leftrightarrow (\exists x)\, (\neg (p(x) \rightarrow q(x)))$
$\Leftrightarrow (\exists x)\, (\neg (\neg p(x) \vee q(x)))$
$\Leftrightarrow (\exists x)\, (p(x) \wedge \neg q(x))$

(iv) $\neg (\exists x)\, ((P(x) \vee Q(x)) \rightarrow R(x))$
$\Leftrightarrow (x)\, (\neg ((P(x) \vee Q(x)) \rightarrow R(x)))$
$\Leftrightarrow (x)\, (\neg (\neg (P(x) \vee Q(x)) \vee R(x)))$
$\Leftrightarrow (x)\, (P(x) \vee Q(x) \wedge \neg R(x))$

Example. *Write down the symbolic forms of each of the following statements and write negation of each.*

(*i*) For all integers *n*, if *n* is not divisible by 2, then *n* is odd.

(*ii*) If *k*, *m*, *n* are any integers where *k-m* and *m-n* are odd, then *k-n* is even.

Solution. Let Z denote the set of all integers and R denote the set of all real numbers.

(*i*) Take $p(n)$: *n* is divisible by 2, $q(n)$: *n* is odd.

Then given statement is $(n)(\neg p(n) \rightarrow q(n))$

Negation of the statement is $(\exists n)(\neg(\neg p(n) \rightarrow q(x))$

$\Leftrightarrow (\exists n)(\neg(p(n) \vee q(n))$

$\Leftrightarrow (\exists n)(\neg p(n) \wedge \neg q(n))$.

In words, for some integer *n*, *n* is not divisible by 2 and *n* is not odd.

(*ii*) Take

$p(x)$: *k-m* is odd, $q(x)$: *m-n* is odd, $r(x)$: *k-n* is even.

Then given statement is $(x)((p(x) \wedge q(x)) \rightarrow r(x))$

Negation of this is $(\exists x)(\neg((p(x) \wedge q(x)) \rightarrow r(x)))$

$\Leftrightarrow (\exists x)(\neg(\neg(p(x) \wedge (q(x)) \vee r(x)))$

$\Leftrightarrow (\exists x)((p(x) \wedge q(x)) \wedge \neg r(x))$

In words, there exists integers *k*, *m*, *n* such that $k - m$, $m - n$ are odd and $k - n$ is not even.

Example. *Let P (x) denote the statement "x is a professional athlete" and let Q (x) denote the statement "x plays soccer". The domain is the let of all people. Write each of the following proposition in English. (i) $(\forall x)(P(x) \rightarrow Q(x))$; (ii) $(\exists x)(P(x) \wedge Q(x))$; and (iii) $(\forall x)(P(x) \vee Q(x))$ and also write the negation of each of the above proportions both in symbols and in words.*

[JNTU - Nov 06 - Set 2 and 4, Feb 0¬ - Set 1 and 4].

Solution.

(*i*) If *x* is a professional athlete. Then *x* plays soccer.

(*ii*) For some professional athlete plays soccer.

(*iii*) For any *x*, *x* is a professional athlete or plays soccer.

And negation of (*i*) is

$$\neg(\forall x)(P(x) \rightarrow Q(x)) \Leftrightarrow (\exists x)(\neg(\neg P(x) \vee Q(x)))$$
$$\Leftrightarrow (\exists x)(P(x) \wedge \neg Q(x))$$

In words, for some *x*, *x* is a professional athlete and *x* not play soccer.

Negation of (*ii*) is $\neg(\exists x)(P(x) \wedge Q(x)) \Leftrightarrow (\forall x)(\neg P(x) \vee \neg Q(x))$

In words, for all *x*, *x* is not professional athlete or *x* is not plays soccer.

Negation of (*iii*) is $\neg(\forall x)(P(x) \vee Q(x)) \Leftrightarrow (\exists x)(\neg P(x) \wedge \neg Q(x))$

In words, for some *x*, *x* is not professional athlete and *x* is not plays soccer.

2.9 SPECIAL VALID FORMULAS INVOLVING QUANTIFIERS

In this section, we give four implications. These will be used in the theory of inference of the predicate calculus and will permit us to either remove or add quantifiers during the course of a derivation.

Let A (*x*) be a predicate formula where *x* is a particular object variable of interest.

Then $(x)\,A(x) \Rightarrow A(y)$ ...(1)

Where y is substituted for x in A (x) to obtain A (y)

In order to show (1), we assume that (x) A (x) is true. Obviously, A (y) must also be true. Hence the implication (1) holds. In case (x) A (x) is false, nothing need be proved. This implication can be written in a more convenient form as

$$(x)\ A\ (x) \Rightarrow A\ (x) \qquad \ldots(2)$$

Implication (2) will be called the rule of universal specification (or instantiation) and will be denoted by US in the theory of inference. It states that from a universally valid premise (x) A (x), we can conclude a specific case A (x), so that one can drop the quantifier in the derivation.

For the next implication, let x be any object variable, B be any formula not containing any free occurrence of x, and A (x) be any formula.

Then $$(B \rightarrow A\ (x)) \Rightarrow (B \rightarrow (x)\ A\ (x)) \qquad \ldots(3)$$

Implication (3) states that if A (x) follows logically from B, then we can conclude (x) A (x) from B.

For the truth of (3), note that if B $\rightarrow$ A (x) is assumed to be true for any variable x, then changing x does not change the truth value of B. Thus B $\rightarrow$ (x) A (x) must also be true. It is important to note that the variable x in A (x) is arbitrary in the sense that nothing that appears in B affects the arbitrariness of x. Certain restrictions are discussed in previous section which guarantee this statement.

As a special case of (3), let us replace B by P $\vee \neg$ P, where P is any statement variable. Since P $\vee \neg$ P is a tautology, from

$$(P \wedge \neg P) \rightarrow A\ (x) \Rightarrow (P \vee \neg P) \rightarrow (x)\ A\ (x)$$

we get the implication

$$A\ (x) \Rightarrow (x)\ A\ (x) \qquad \ldots(4)$$

Implication (4) permits us to conclude (x) A (x) from A (x) ; *i.e.,* it permits us to add the universal quantifier in the conclusion during the course of derivation. This rule is therefore called the rule of universal generalization and will be denoted by UG.

Corresponding to the rules in (2) and (4) which permit the universal quantifier to be removed or added during the course of a derivation, we also have two more rules which will permit us to remove or add the existential quantifier during the course of a derivation.

The following implications, derivable under assumptions similar to those used for (2) and (4), establish such rules :

$$(\exists\ x)\ A\ (x) \Rightarrow A\ (y) \qquad \ldots(5)$$

$$A\ (y) \Rightarrow (\exists\ x)\ A\ (x) \qquad \ldots(6)$$

Implication (5) is known as existential specification, abbreviated here as ES, while (6) is known as existential generalization, or EG. That some care is necessary in the interpretation of these rules will be discussed in the next section.

Some other important implications and equivalences involving the quantifiers are listed in the following.

$$(\exists x)(A(x) \vee B(x)) \Leftrightarrow (\exists x) A(x) \vee (\exists x) B(x) \qquad E_{23}$$

$$(x)(A(x) \wedge B(x)) \Leftrightarrow (x) A(x) \wedge (x) B(x) \qquad E_{24}$$

$$\neg (\exists x) A(x) \Leftrightarrow (x) \neg A(x) \qquad E_{25}$$

$$\neg (x) A(x) \Leftrightarrow (\exists x) \neg A(x) \qquad E_{26}$$

These can be proved by using the definitions of the quantifiers and arguments similar to the one given above. In certain other cases it is more convenient to prove them by using the method of derivation given in the next section. If a formula A does not depend upon the variable *x*, then $(x)\, A \Leftrightarrow A$ and $(\exists x)\, A(x) \Leftrightarrow A$, and the implications I_{15} and I_{16} become E_{27} and E_{28}. Using E_{23}, we can easily prove $(\exists x)(A(x) \rightarrow B(x)) \Leftrightarrow (x) A(x) \rightarrow (\exists x) B(x)$ E_{33} similarly, from I_{15} and I_{16} we can prove

2.10 THEORY OF INFERENCE FOR THE PREDICATE LOGIC

The method derivation involving predicate formulas uses the rules of inference given for the statement calculus and also certain additional rules which are required to deal with the formulas involving quantifiers. The rules P and T regarding the introduction of a premise at any stage of derivation and the introduction of any formula which follows logically from the formulas already introduced, remains the same. If the conclusion is given in the form of a conditional, we shall also use the rule CP. Certain additional rules and required to deal with the formulas involving quantifiers. The elimination of quantifiers can be done by rules of specification called US and ES. To prefix the correct quantifier, we need the rules of generalization called UG and EG.

The rules of generalization and specification follow

Rule US (Universal Specification)

From $(x)\, P(x)$ one can conclude $P(y)$

Rule ES (Existential Specification)

From $(\exists x)\, P(x)$ one can conclude the value of $P(y)$ provided that y is not free in any given premise and also not free in any prior step of the derivation. These requirements can easily be met by choose a new variable each time ES is used.

Rule EG (Existential Generalization)

From $P(y)$ one can conclude $(\exists x)\, P(x)$

Rule UG (Universal Generalization)

From $P(y)$ one can conclude $(x)\, P(x)$ provided that *y* is not free in any of the given premises and provided that if *y* is free in the prior step which resulted from the use of ES, then no variables introduced by that use of ES appear free in $P(y)$.

We shall now show, by means of an example, how an invalid conclusion could be arrived at if the second restriction on rule UG were not imposed. The other restrictions on ES and UG are easy to understand.

Let $D(x, y)$: *x* is divisible by *y*. Assume that the universe of discourse is {5, 7, 10, 11}, so that the statement $(\exists x)\, D(x, 5)$ is true because both $D(5, 5)$ and $D(10, 5)$ are true. On the other hand, $(u)\, D(u, 5)$ is false because $D(7, 5)$ and $D(11, 5)$ are false.

Consider now the following derivation :

{1}	1.	$(\exists x)\ D(x, 5)$	Rule P.
{1}	2.	$D(a, 5)$	ES, (1)
{1}	3.	$(u)\ D(4, 5)$	UG, (2), (neglecting second restriction)

In step 3 we have obtained from $D(a, 5)$, the conclusion $(u)\ D(4, 5)$. Obviously a is not free in the premise, and so the first restriction is satisfied. But a is free in step 2 which resulted by use of ES, and that a has been introduced by the use of ES and appears free in $D(a, 5)$. Hence it cannot be generalized. This is the reason why we obtained a false conclusion from a true premise.

Example. *Show that* $(x)\ (H(x) \to M(x)) \wedge H(s) \Rightarrow M(s)$

Solution.

{1}	1.	$(x)\ (H(x) \to M(x))$	Rule P
{1}	2.	$H(s) \to M(s)$	US, (1)
{3}	3.	$H(s)$	Rule P.
{1, 3}	4.	$M(s)$	Rule T, (2), (3), I_{11}.

Example. *Verify the validity of the following argument.*

All men are mortal.

Socrates is a man.

Therefore Socrates is a mortal.

Solution. We denote $H(x)$: x is a man

$M(x)$: x is a mortal.

s: Socrates.

We need to show $(\forall x)\ (H(x) \to M(x)) \wedge H(s) \Rightarrow M(s)$

It is already verified in previous example

Example. *Show that* $(x)\ (P(x) \to Q(x)) \wedge (x)\ (Q(x) \to R(x))$

$\Rightarrow \qquad (x)\ (P(x) \to R(x))$

Solution.

{1}	1.	$(x)\ (P(x) \to Q(x))$	Rule P.
{1}	2.	$P(y) \to Q(y)$	US, (1)
{3}	3.	$(x)\ (Q(x) \to R(x))$	Rule P.
{3}	4.	$Q(y) \to R(y)$	US, (3)
{1, 3}	5.	$P(y) \to R(y)$	Rule T, (2), (4), I_{13}.
{1, 3}	6.	$(x)\ (P(x) \to R(x))$	UG, (5).

Example. *Given an argument which will establish the validity of the following inference.*

All integers are rational numbers.

Some integers are powers of 5.

Therefore, some rational numbers are powers of 5.

Solution. We denote $P(x)$: x is an integer

$R(x)$: x is a rational number.

$S(x)$: x is a power of 5.

Next we shall show

$(x)(P(x) \rightarrow R(x)) \wedge (\exists x)(P(x) \wedge S(x)) \Rightarrow (\exists x)(R(x) \wedge S(x))$

{1} 1. $(\exists x)(P(x) \wedge S(x))$ Rule P.
{1} 2. $P(b) \wedge S(b)$ ES, (1).
{1} 3. $P(b)$ Rule T, (2).
{1} 4. $S(b)$ Rule T, (2).
{2} 5. $(x)(P(x) \rightarrow R(x))$ Rule P.
{2} 6. $P(b) \rightarrow R(b)$ US, (5)
{1, 2} 7. $R(b)$ Rule T, (3), (6).
{1, 2} 8. $R(b) \wedge S(b)$ Rule T, (4), (7).
{1, 2} 9. $(\exists x)(R(x) \wedge S(x))$ EG, (8).

Example. *Show that $(\exists x) M(x)$ follows logically from the premises $(x)(H(x) \rightarrow M(x))$ and $(\exists x) H(x)$*

Solution. {1} 1. $(\exists x) H(x)$ Rule P.
{1} 2. $H(y)$ ES, (1)
{3} 3. $(x)(H(x) \rightarrow M(x))$ P
{3} 4. $H(y) \rightarrow M(y)$ US, (3)
{1, 3} 5. $M(y)$ T, (2), (4), I_{11}
{1, 3} 6. $(\exists x) M(x)$ EG, (5).

Example. *Prove that $(\exists x)(P(x) \wedge Q(x)) \Rightarrow (\exists x) P(x) \wedge (\exists x) Q(x)$.*

Solution. {1} 1. $(\exists x)(P(x) \wedge Q(x))$ P
{1} 2. $P(y) \wedge Q(y)$ ES, (1), y fixed
{1} 3. $P(y)$ T, (2), I_1
{1} 4. $Q(y)$ T, (2), I_2
{1} 5. $(\exists x) P(x)$ EG, (3)
{1} 6. $(\exists x) Q(x)$ EG, (4)
{1} 7. $(\exists x) P(x) \wedge (\exists x) Q(x)$ T, (4), (5), I_9.

Example. *Show that $(x)(P(x) \vee Q(x)) \Rightarrow (x) P(x) \vee (\exists x) Q(x)$.*

Solution. We shall use the indirect method of proof by assuming $\neg((x) P(x) \vee (\exists x) Q(x))$ as an additional premise.

{1} 1. $\neg((x) P(x) \vee (\exists x) Q(x))$ P (assumed)
{1} 2. $\neg(x) P(x) \wedge (\exists x) Q(x)$ T, (1), E_9
{1} 3. $\neg(x) P(x)$ T, (2), I_1
{1} 4. $(\exists x) \neg P(x)$ T, (3), E_{26}
{1} 5. $\neg(\exists x) Q(x)$ T, (2), I_2
{1} 6. $(x) \neg Q(x)$ T, (5), E_{25}

$\{1\}$ 7. $\neg P(y)$ ES, (4).

$\{1\}$ 8. $\neg Q(y)$ US, (6)

$\{1\}$ 9. $\neg P(y) \wedge \neg Q(y)$ T, (7), (8), I_9

$\{1\}$ 10. $\neg(P(y) \vee Q(y))$ T, (9), E_9.

$\{11\}$ 11. $(x)(P(x) \vee Q(x))$ P

$\{11\}$ 12. $P(y) \vee Q(y)$ US, (11)

$\{1, 11\}$ 13. $\neg(P(y) \vee Q(y)) \wedge (P(y) \vee Q(y))$ T, (10), (12), I_9 contradiction.

Example. *Show that from*

(a) $(\exists x)(F(x) \wedge S(x)) \rightarrow (y)(M(y) \rightarrow W(y))$

(b) $(\exists y)(M(y) \wedge \neg W(y))$

the conclusion $(x)(F(x) \rightarrow \neg S(x))$ *follows :*

Solution. $\{1\}$ 1. $(\exists y)(M(y) \wedge \neg W(y))$ P

$\{1\}$ 2. $M(z) \wedge \neg W(z)$ ES, (1)

$\{1\}$ 3. $\neg(M(z) \rightarrow W(z))$ T, (2), E_{17}

$\{1\}$ 4. $(\exists y) \neg(M(y) \rightarrow W(y))$ EG, (3)

$\{1\}$ 5. $\neg(y)(M(y) \rightarrow W(y))$ E_{26}, (4)

$\{6\}$ 6. $(\exists x)(F(x) \wedge S(x)) \rightarrow (y)(M(y) \rightarrow W(y))$ P

$\{1, 6\}$ 7. $\neg(\exists x)(F(x) \wedge S(x))$ T, (5), (6), I_{12}

$\{1, 6\}$ 8. $(x) \neg(F(x) \wedge S(x))$ T, (7), E_{25}

$\{1, 6\}$ 9. $\neg(F(x) \wedge S(x))$ US, (8)

$\{1, 6\}$ 10. $F(x) \rightarrow \neg S(x)$ T, (9), E_9, E_{16}, E_{17}.

$\{1, 6\}$ 11. $(x)(F(x) \rightarrow \neg S(x))$ UG, (10).

Example. *Using CP or otherwise, obtain the following implication*

$(x)(P(x) \rightarrow Q(x)), (x)(R(x) \rightarrow \neg Q(x)) \Rightarrow (x)(R(x) \rightarrow \neg P(x))$

Solution. $\{1\}$ 1. $(x)(P(x) \rightarrow Q(x))$ Rule P

$\{2\}$ 2. $(x)(R(x) \rightarrow \neg Q(x))$ Rule P

$\{2\}$ 3. $R(x) \rightarrow \neg Q(x)$ US, (2)

$\{4\}$ 4. $R(x)$ P (assumed)

$\{2, 4\}$ 5. $\neg Q(x)$ Rule T, (3), (4).

$\{1\}$ 6. $P(x) \rightarrow Q(x)$ US, (1)

$\{1, 2, 4\}$ 7. $\neg P(x)$ Rule T, (5), (6)

$\{1, 2, 4\}$ 8. $R(x) \rightarrow \neg P(x)$ CP, (4), (7)

$\{1, 2, 4\}$ 9. $(x)(R(x) \rightarrow \neg P(x))$ UG, (8).

Example. *Are the following conclusions validly derivable from the premises given?*

(a) $(x)(P(x) \rightarrow Q(x)), (\exists y) P(y)$ $C: (\exists z) Q(z)$

(b) $(\exists x)(P(x) \wedge Q(x))$ $\qquad$ $C: (x)P(x)$.

Solution. *(a)*

{1}	1. $(x)(P(x) \rightarrow Q(x))$	Rule P.
{1}	2. $P(z) \rightarrow Q(z)$	US, (1)
{3}	3. $(\exists y) P(y)$	Rule P
{3}	4. $P(z)$	ES, (3)
{1, 3}	5. $Q(z)$	Rule T, I_1, (2), (4).
{1, 3}	6. $(\exists z) Q(z)$	EG, (5).

Therefore conclusion C is valid from premises.

(b) {1}	1. $(\exists x)(P(x) \wedge Q(x))$	Rule P.
{1}	2. $P(z) \wedge Q(z)$	EG, (1)
{1}	3. $P(z)$	Rule T, I_1, (2).
{1}	4. $(x) P(x)$	UG, (3).

Therefore conclusion C is valid from premises.

Example. *There is a mistake in the following derivation. Find it. Is the conclusion valid ? If so, obtain a correct derivation.*

{1}	1. $(x)(P(x) \rightarrow Q(x))$	*Rule P.*
{1}	2. $P(y) \rightarrow Q(y)$	*US, (1)*
{3}	3. $(\exists x)(P(x))$	*Rule P.*
{3}	4. $P(y)$	*ES, (3)*
{1, 3}	5. $Q(y)$	*Rule T, (2), (4),* I_{11}.
{1, 3}	6. $(\exists x) Q(x)$	*EG, (5).*

Solution. The y introduced in step (2) is free, it should not be introduced again by ES in fourth step. We try to eliminate this mistake by obtaining a correct derivation.

{1}	1. $(\exists x) P(x)$	Rule P.
{1}	2. $P(y)$	ES, (1)
{3}	3. $(x) P(x) \rightarrow Q(x))$	Rule P.
{3}	4. $P(y) \rightarrow Q(y)$	US, (3)
{1, 3}	5. $Q(y)$	Rule T, (2), (4), I_{11}.
{1, 3}	6. $(\exists x) Q(x)$	EG, (5).

2.11 FORMULAS INVOLVING MORE THAN ONE QUANTIFIER

So far we have considered only those formulas in which the universal and existential quantifiers appear singly. We shall now consider cases in which the quantifiers occur in combinations. These combinations are possible even in the case of 1-place predicates, and they become particularly important in the case of n-place predicates ($n \geq 2$). For example, if $P(x, y)$ is a 2-place predicate formula, then the following possibilities exist.

$(x)\ (y)\ P\ (x, y)$, $(x)\ (\exists\ y)\ P\ (x, y)$, $(\exists\ x)\ (y)\ P\ (x, y)$, $(\exists\ x)\ (\exists\ y)\ P\ (x, y)$, $(y)\ (x)\ P\ (x, y)$, $(\exists\ y)\ (x)\ P\ (x, y)$, $(y)\ (\exists\ x)\ P\ (x, y)$, $(\exists\ y)\ (\exists\ x)\ P\ (x, y)$.

Note 1. $(x)\ (y)\ P\ (x, y)$ means $(x)\ ((y)\ P\ (x, y))$ and $(\exists\ x)\ (y)\ P\ (x, y)$ means $(\exists\ x)\ ((y)\ P\ (x, y))$.

Note 2. The following formulas can be obtained from the meaning of the quantifiers.

1. $(x)\ (y)\ P\ (x, y) \Leftrightarrow (y)\ (x)\ P\ (x, y)$
2. $(x)\ (y)\ P\ (x, y) \Rightarrow (\exists\ y)\ (x)\ P\ (x, y)$
3. $(y)\ (x)\ P\ (x, y) \Rightarrow (\exists\ x)\ (y)\ P\ (x, y)$
4. $(\exists\ y)\ (x)\ P\ (x, y) \Rightarrow (x)\ (\exists\ y)\ P\ (x, y)$
5. $(\exists\ x)\ (y)\ P\ (x, y) \Rightarrow (y)\ (\exists\ x)\ P\ (x, y)$
6. $(x)\ (\exists\ y)\ P\ (x, y) \Rightarrow (\exists\ y)\ (\exists\ x)\ P\ (x, y)$
7. $(y)\ (\exists\ x)\ P\ (x, y) \Rightarrow (\exists\ x)\ (\exists\ y)\ P\ (x, y)$
8. $(\exists\ x)\ (\exists\ y)\ P\ (x, y) \Leftrightarrow (\exists\ y)\ (\exists\ x)\ P\ (x, y)$.

From the following figure shows implications (2) to (7) and equivalences (1) and (8). We can also prove these implications and equivalences using the method of derivation given in the previous section.

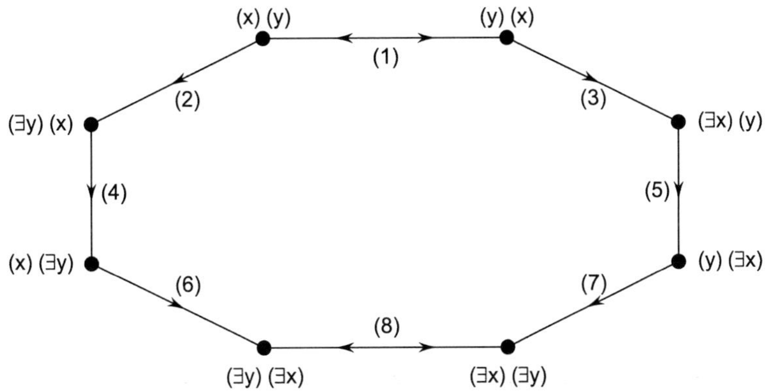

Figure 2.1 *Graphical representation of relationships among formulas involving two quantifiers :*

The negation of any of the above formulas can be obtained by repeated applications of the equivalences E_{25} and E_{26} of previous section.

For example,

$\neg\ (\exists\ y)\ (x)\ P\ (x, y) \Leftrightarrow (y)\ (\neg\ (x)\ P\ (x, y) \Leftrightarrow (y)\ (\exists\ x) \neg\ P\ (x, y)$ and $\neg\ (y)\ (\exists\ x)\ P\ (x, y) \Leftrightarrow (\exists\ y)$ $(\neg\ (\exists\ x)\ P\ (x, y)) \Leftrightarrow (\exists\ y)\ (x) \neg\ P\ (x, y)$.

The negations of other formulas of this type are obtained in a similar manner.

The inference rules and the method of derivation are given in previous section also apply to n-place predicate formulas. Obviously, some special care would now be needed in the use of the rules UG, EG, US and ES. In the case of US and ES, the specific variable should be chosen in such away that it is different from the bound variable used elsewhere.

To illustrate this discussion, we consider the formula $(x)\ (\exists\ y)\ P\ (x, y)$. Using US, we can write any of the formulas $(\exists\ y)\ P\ (x, y)$, $(\exists\ y)\ P\ (z, y)$ but we should not write $(\exists\ y)\ P\ (y, y)$ because the variable y is used as a bound variable. *i.e.,* $(\exists\ y)\ P\ (x, y)$ is not free for y. Similarly, in using EG, one should be careful. For example, from $(x)\ P(x, y)$ we can generalize $(\exists y)\ (x)\ P(x, y)$ (or $(\exists z)$ $(x)\ P(x, z)$ but not $(\exists x)\ (x)\ P(x, x)$.

Similar care is required in the use of UG and ES.

Example. *Show that $\neg P(a, b)$ follows logically from $(x)\ (y)\ (P(x, y) \rightarrow W(x, y))$ and $\neg W(a, b)$.*

Solution.

{1}	1.	$(x)\ (y)\ (P(x, y) \rightarrow W(x, y))$	Rule P.
{1}	2.	$(y)\ (P(a, y) \rightarrow W(a, y))$	US, (1)
{1}	3.	$P(a, b) \rightarrow W(a, b)$	US, (2)
{1}	4.	$\neg W(a, b)$	Rule P.
{1, 4}	5.	$\neg P(a, b)$	Rule T, (3), (4), I_{12}.

Example. *Negate each of the following expressions :*

(a) $\neg (x)\ (y)\ R(x, y)$

(b) $\neg (x)\ (\exists y)\ R(x, y)$

(c) $\neg ((\exists x)\ (\exists y) \neg (R(x, y)) \wedge (x)\ (y)\ P(x, y))$

Solution. *(a)* $(\exists x)\ (\exists y) \neg R(x, y)$

(b) $(\exists x)\ (y) \neg R(x, y)$

(c) $((x)\ (y)\ R(x, y) \vee (\exists x)\ (\exists y) \neg P(x, y))$.

Example. *Let $P(x)$: x is a weak student.*

$Q(x)$: x is logical.

$R(x)$: x is able to pass an examination.

Suppose that the universe of discourse is the set of all students. Express each of the following statements using quantifiers :

(a) Weak students are illogical.

(b) Weak students are not able to pass an examination.

Solution. *(a)* $(x)\ (P(x) \rightarrow \neg Q(x))$.

(b) $(x)\ (P(x) \rightarrow \neg R(x))$.

Example. *Translate the statement $(\forall x)\ (P(x) \vee (\exists y)\ (P(y) \wedge Q(x, y))$ into English, where $P(x)$: x has a book,*

$Q(x, y)$: x and y are friends, and the universe of discourse for both x and y is the set of all students in a school.

Solution. The statement says that, for every student *x* in a school, *x* has a book or there is a student *y* such that *y* has a book and *x* and *y* are friends.

Example. *Express the statements "Some student in the class study computers" and "Every student in the class study either mathematics or computers" using quantifiers.*

Solution. Let the universe of discourse for the variables '*x*' be the set of students in a class. Let $P(x)$ be the statement "*x* study computers" and $Q(x)$ the statement "*x* study mathematics".

Now "Some student in the class study computers" can be written as $(\exists x)\ P(x)$. The statement "Every student in the class study either Mathematics or Computers" can be written as $(x)\ (P(x) \vee Q(x))$.

CHAPTER

3

Set Theory

3.1 SETS

The concept of a set underlies most of modern mathematics and much of Computer Science. To use sets as a foundation for all the other structures in this text, we first need to understand both the language used to describe sets and the operations normally associated with sets. The Language of Sets is very precise. When we use this language carefully, we gain precision in expressing problems and describing solutions to problems. Understanding basic operations on sets and the properties of these operations is a model for the approach that is used to introduce most other discrete structures in this text. In extending our understanding of operations on sets, we will learn proof techniques to explore other discrete mathematical topics, such as relations, functions and graphs. We will use these proof techniques, for example, to prove that algorithms are correct and to determine how well we have chosen an algorithm for a given task.

Definition. *A **set** is a well-defined collection or class of distinct objects.*

By a well defined collection we mean that there exists a rule with the help of which we should be able to say whether any given object or entity belongs to or not to the collection under consideration.

Example 1. The odd positive integers less than 6.

2. The vowels in English alphabet.
3. Students in a college who studying Computer Science.
4. Rivers in India.

Definition. *The objects in a set are called its **elements** (or) **members.** The elements in the set must be distinct and distinguishable.*

Generally, capital letters A, B, C, X, Y, Z etc., are used to denote sets, set symbol is { } and its elements by small letters a, b, c, x, y, z, etc.

Example. A = {*a*, *b*, *c*, *x*, *y*}, B = {1, 2, 3, 4, 5}, C = {*a*, *e*, *i*, *o*, *u*}, D = {Red, Black, White, Green}

Standard Sets and Symbols

The choice of the name for a set is more like the choice of an identifier name in programming. There are few sets of numbers that we will use frequently in the text and so we use standard symbols for them, such as

$\mathbf{N}\ (\mathbf{Z}^+) = \{1, 2, 3, 4, ...\}$, is the set of natural numbers.

$\mathbf{W} = \{0, 1, 2, 3, 4, ...\}$, is the set of non-negative integers.

$\mathbf{Z} = \{..., -3, -2, -1, 0, 1, 2, 3, ...\}$, is the set of integers.

$\mathbf{Z}^- = [-1, -2, -3, ...\}$, is the set of negative integers.

$\mathbf{Q} = \{x \mid x = \frac{p}{q}$ where $p, q \in Z, q \neq 0\}$, is the set of rational numbers.

Examples. $1, \frac{2}{3}, -\frac{7}{5}, -2, 0, ...$

$\mathbf{Q}^+\ (\mathbf{Q}_+) = \{x \mid x > 0, x \in \mathbf{Q}\}$, is the set of positive rational numbers.

$\mathbf{Q}^-\ (\mathbf{Q}_-) = \{x \mid x < 0, x \in \mathbf{Q}\}$, is the set of negative rational numbers.

$\mathbf{Q}_0\ (\mathbf{Q}^0) = \{x \mid x \neq 0, x \in \mathbf{Q}\}$, is the set of non-zero rational numbers.

$\mathbf{Q}' = \{x \mid x \in \mathbf{R}, x \notin \mathbf{Q}\}$, is the set of irrational numbers.

Example. $\sqrt{2}, \sqrt{3}, \sqrt{5}, \sqrt{6}, ...$

$$\mathbf{R} = \mathbf{Q} \cup \mathbf{Q}'$$

$= \{x \mid x \in \mathbf{Q}$ or $x \in \mathbf{Q}'\}$, is the set of real numbers.

$\mathbf{R}^+\ (\mathbf{R}_+) = \{x \mid x > 0, x \in \mathbf{R}\}$, is the set of positive real numbers.

$\mathbf{R}^-\ (\mathbf{R}_-) = \{x \mid x < 0, x \in \mathbf{R}\}$, is the set of negative real numbers.

$\mathbf{R}_0\ (\mathbf{R}^0) = \mathbf{R} - \{0\}$, is the set of non-zero real numbers.

$C = \{a + ib\}\ a, b, \in \mathbf{R}, i^2 = -1\}$, is the set of complex numbers.

3.2 TYPES OF SET

1. **Finite Set.** A set is said to be finite if it contains finite number of different elements.

   Example. A = {1, 2, 5, 7}

   B = {*x* | *x* is country in the world}

2. **Infinite Set.** A set is said to be infinite if it contains infinite distinct elements.

   Example. N = {1, 2, 3, 4, 5, ...}

   A = {*x* | *x* is the point in the plane}.

3. **Singleton Set.** A set which contains only one element is called a singleton set.

   Example. A = {*x* | 2 < *x* < 4, *x* is an integer} = {3}

   B = {0}.

4. **Null Set.** A set which contains no element is called an empty (null or void) set. This set is denoted by ϕ (*phi*)

   Example. $\phi = \{x \mid 2 < x < 3, x \text{ is an integer}\}$.

   Example. Set of all integers whose square is $3 = \phi$.

   A set which is not a null set is called non-empty set.

5. **Equality of Sets.** Two sets A and B are said to be equal if every element of A is an element of B and also every element of B is an element of A. The equality of two sets A and B is denoted by A = B.

$$A = B \quad \text{iff } x \in A \Leftrightarrow x \in B.$$

   Example. A = {1, 2, 3, 4}, B = {1, 3, 2, 4}.

6. **Equivalent Sets.** If the elements of one set can be put into one-to-one correspondence with the elements of another set, then the two sets are called equivalent sets.

   The symbol ~ (or) = is used to denote equivalent sets.

   Example. Let A = {*a*, *b*, *c*, *d*}, B = {1, 2, 3, 4} be two sets.

   The elements of A can be put into one-to-one correspondence with those of B. Then A ~ B.

7. **Subset.** Let A, B be two non-empty sets. The set A is subset of B (or A is contained in B) iff every element of A is an element of B.

   (*i.e.,*) $A \subseteq B$ iff $x \in A \Rightarrow x \in B$.

   Here B is called the superset of A if $A \subseteq B$.

   Example 1. Let A = {1, 5, 7}, B = {1, 3, 5, 7} then $A \subseteq B$.

   2. $N \subseteq W \subseteq Z \subseteq Q \subseteq R \subseteq C$

   3. Null set ϕ is subset of every set.

   Note. 1. $A \subseteq B, B \subseteq A \Rightarrow A = B$.

   2. $A \subseteq B, B \subseteq C \Rightarrow A \subseteq C$.

8. **Proper Subset.** The set A is a proper subset of the set B (or) A is properly contained in B iff (1) every element of A is also an element of B (*i.e.,*) $A \subseteq B$.

   (2) There is at least one element in the set B, which is not in A (*i.e.,*) $A \neq B$.

   If A is a proper subset of B, then we write $A \subset B$.

   Example 1. Let A = {1, 5, 7}, B = {1, 3, 5, 7} then $A \subset B$.

   Note : 2. $\mathbf{N} \subset \mathbf{W} \subset \mathbf{Z} \subset \mathbf{Q} \subset \mathbf{R} \subset C$.

9. **Power Set.** If S is any set, then the family of all the subsets of S is called the power set of S and is denoted by P (*s*) (*i.e.,*) $P(S) = \{A \mid A \subseteq S\}$.

   Clearly ϕ and S are both members of P (S).

   Note. If A is a finite set of *n* elements then the power set of A P (A) contains 2^n elements.

   Example 1. Let A = {*a*}, then P (A) = {ϕ, {a}}.

   2. Let A = {*a*, *b*}, then P (A) = {ϕ, {*a*}, {*b*}, {*a*, *b*}}.

   3. Let A = {1, 2, 3}, then P (A) = {ϕ, {1}, {2}, {3}, {1, 2}, {1, 3}, {2, 3}, {1, 2, 3}}

10. **Universal Set.** In applications of Set Theory all the sets under consideration are assumed to be the subset of the fixed set. This set is called the universal set (or) universal of discourse. This set is usually denoted by U.

Example. All the people in the world constitute the universal set in any study of human population.

3.3 VENN-EULER DIAGRAMS

A device known as **Venn-Euler** diagram (or) simply **Venn** diagram is a pictorial representation of sets. In Venn diagrams, a universal set U is represented by the interior of rectangle and each subset of U is represented by the circle inside the rectangle.

If few elements are common both in A and B, then the sets A and B are represented by the Fig. 3.1(*i*) if the sets A, B are disjoint (*i.e.,*) they have no common elements, then circles A, B are represented by the Fig. 3.1(*ii*) and if A is subset of B (*i.e.,*) all elements in A are also in B, then circles A and B are represented by the Figure 3.1(*iii*).

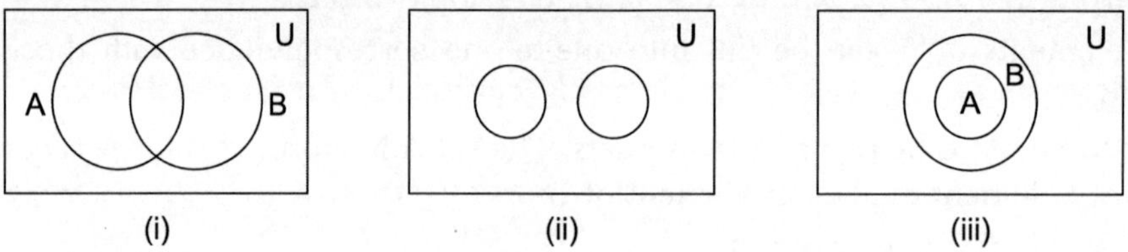

Fig. 3.1.

Example. *Write down the following sets :*

(*i*) $A = \{x \mid x^2 = 9\}$ (*ii*) $B = \{x \mid x^2 + 4 = 0\}$

Solution. (*i*) $A = \{-3, 3\}$

(*ii*) $x^2 + 4 = 0 \Rightarrow x^2 = -4 = 4i^2 \Rightarrow x = \pm 2i$

$\therefore \quad B = \{-2i, 2i\}.$

Example. *Which of the following sets are equal ?*

(*i*) $A = \{1, 2, 2, 3\}$ (*ii*) $B = \{x \mid x^2 - 3x + 2 = 0\}$

(*iii*) $C = \{1, 2, 3\}$ (*iv*) $D = \{x \mid x^3 - 6x^2 + 11x - 6 = 0\}$

(*v*) $E = \{-1, 2, 3\}$

Solution. $A = \{1, 2, 3\}$

$B = \{x \mid (x-1)(x-2) = 0\} = \{1, 2\}$

$C = \{1, 2, 3\}$

$D = \{x \mid (x-1)(x-2)(x-3) = 0\} = \{1, 2, 3\}$

$E = \{-1, 2, 3\}$

$\therefore \quad A = C = D.$

Example. *Find null and singleton sets from the following :*

(*i*) $A = \{x \mid x^2 = 7, x \text{ is an integer}\}$ (*ii*) $B = \{x \mid x + 2 = 2\}$

(iii) $C = \{x \mid x^2 = 7, 3x = 5\}$ *(iv)* $D = \{x \mid 0 < x < 2, x \text{ is an integer}\}$

Solution. (*i*) $x^2 = 7 \Rightarrow x = \sqrt{7}$ which is not an integer $\Rightarrow A = \{\ \} = \phi$

(*ii*) $x + 2 = 2 \Rightarrow x = 0 \Rightarrow B = \{0\}$

(*iii*) $x^2 = 7 \Rightarrow x = \pm\sqrt{7} \Rightarrow 3x = 3x(\pm\sqrt{7}) = \pm 3\sqrt{7} \neq 5$

$\therefore$ $C = \{\ \} = \phi.$

(*iv*) $D = \{1\}$

Hence A, C are null sets and B, D are singleton sets.

3.4 OPERATIONS ON SETS AND PROPERTIES OF SET ALGEBRA

In this section we shall define certain basic set operations on sets so as to yield new sets with the given sets.

1. **Union of Sets :** Let A, B be any two non-empty sets. The union of A and B is the set of all elements which are either in A (or) in B (or) in the both A and B. The union of A, B is denoted by $A \cup B$, we read it as A union B.

   Symbolically, $A \cup B = \{x \mid x \in A \text{ (or) } x \in B\}$.

   Venn diagram of $A \cup B$ in Fig. 3.2.

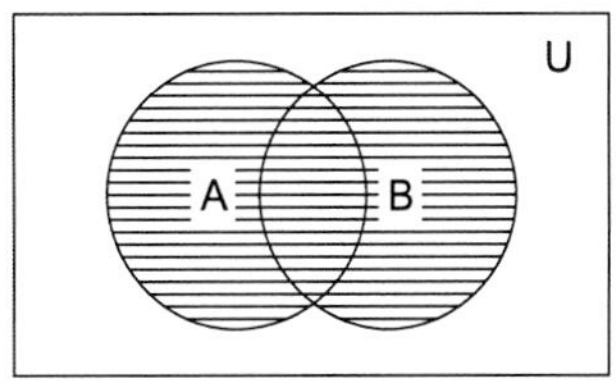

$A \cup B$ (Shaded area)

Figure. 3.2

   Example 1. Let $A = \{1, 2, 3\}$, $B = \{2, 3, 6\}$

   Then $A \cup B = \{1, 2, 3, 6\}$

   2. Let $A = \{x \mid x \in \mathbf{Z}, 0 < x < 5\}$, $B = \{y \mid y \in \mathbf{Z}, 4 < x < 8\}$

   Then $A = \{1, 2, 3, 4\}$, $B = \{5, 6, 7\} \Rightarrow A \cup B = \{1, 2, 3, 4, 5, 6, 7\}$.

2. **Intersection of Sets :** Let A and B be two non-empty sets. The intersection of A and B is the set of all elements which are in both A and B. Intersection of A, B is denoted by $A \cap B$. We read it as A intersection B.

   Symbolically $A \cap B = \{x \mid x \in A \text{ and } x \in B\}$.

   Venn diagram of $A \cap B$ is Fig. 3.3.

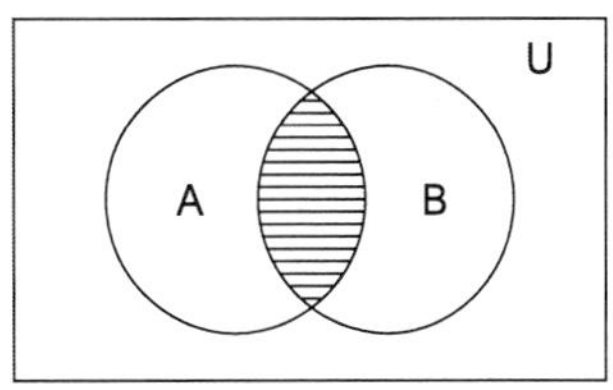

$A \cap B$ (Shaded part)

Figure. 3.3.

   Example 1. If $A = \{1, 2, 3, 4, 5, 6\}$, $B = \{5, 6, 7, 8\}$

   Then $A \cap B = \{5, 6\}$

   2. If $A = \{x \mid x \in Z, 0 \leq x < 6\}$,

   $B = \{y \mid y \in Z, 3 < y < 8\}$

   Then $A = \{0, 1, 2, 3, 4, 5\}$, $B = \{4, 5, 6, 7\}$

   Hence $A \cap B = \{4, 5\}$.

3.4.1 Properties of Union of Sets and Intersection of Sets

1. $A \cup B = B \cup A$, $A \cap B = B \cap A$ (Commutative Law)
2. $A \cup (B \cup C) = (A \cup B) \cup C$, $A \cap (B \cap C) = (A \cap B) \cap C$ (Associative Law)
3. $A \cup A = A$, $A \cap A = A$ (Idempotent Law)
4. $A \cup (B \cap C) = (A \cup B) \cap (A \cup C)$,

   $A \cap (B \cup C) = (A \cap B) \cup (A \cap C)$ (Distributive Law)

5. $A \cup \phi = A, A \cap \phi = \phi$

6. $A \cup U = U, A \cap U = A$.

Disjoint Sets. Let A and B be two non-empty sets. These two sets A, B are said to be disjoint if they have no common elements *i.e.,* $A \cap B = \phi$.

Venn diagram of disjoint sets A, B is Fig. 3.4.

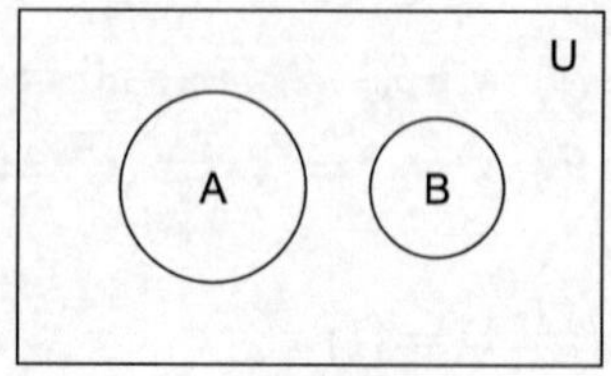

$A \cap B = \phi$

Figure. 3.4.

Example. If A = {1, 2, 3, 4}, B = {5, 6, 7}

Then $A \cap B = \phi$ (*i.e.,*) A, B are disjoint.

Difference of Two Sets. If A, B are any sets, then difference of A and B is the set containing elements which belong to A and does not belong to B. The difference of A and B is denoted by A – B (or) A ~ B. We read it as A difference B.

Symbolically, $A - B = \{x \mid x \in A \text{ and } x \notin B\}$

$B - A = \{x \mid x \in B \text{ and } x \notin A\}$

Venn diagram of A – B is Fig. 3.5.

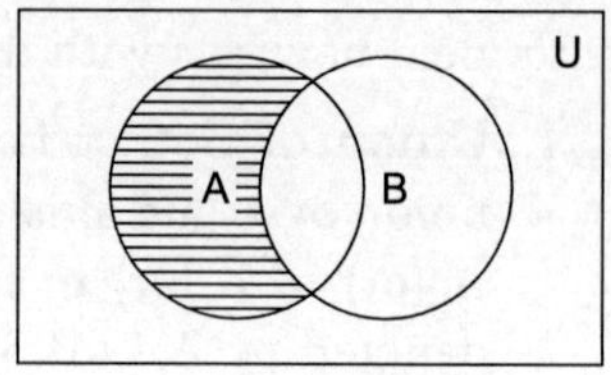

A – B (Shaded part)

Figure. 3.5.

Example. A = {2, 4, 6, 8}, B = {6, 8, 10, 12}

Then A – B = {2, 4}, B – A = {10, 12}

Note that $A - B \neq B - A$.

Complement of a Set : Let A be any set. The complement of A defined as the set of elements that are in the universal set but not in A. It is denoted by U – A (or) A′ (or) A^C (or) ^{-}A.

Symbolically, $A' = U - A = \{x \mid x \in U \text{ and } x \notin A\}$

Venn diagram of A^C is Fig. 3.6.

Example. If {1, 2, 3, 4, ...} is the universal set, and Let A = {2, 4, 6, 8, ...}

Then U – A = {1, 3, 5, 7, ...}

Note that $U^C = \phi, \phi^C = U$.

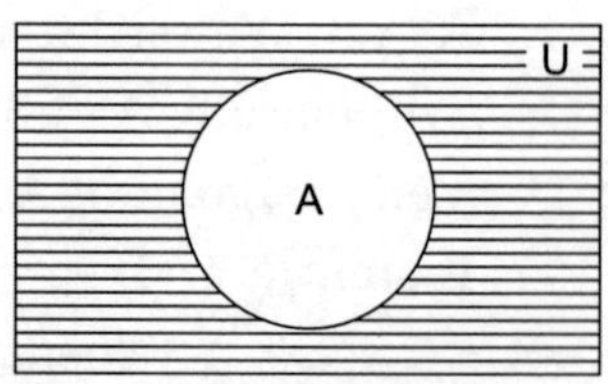

Figure. 3.6.

Complementary properties :

1. $A \cup A^C = U$
2. $A \cap A^C = \phi$
3. $(A^C)^C = A$.
4. $(A \cup B)^C = A^C \cap B^C$, $(A \cap B)^C = A^C \cup B^C$ (Demorgan's laws)

Symmetric Difference of Sets

Definition. *Let A, B be two non-empty sets. Then the symmetric difference of A, B denoted by* $A \oplus B$ *or* $A \,\Delta\, B$, *is defined as the set containing elements which either belongs to A or B but not to both.*

$$\therefore \quad A \oplus B = (A \cup B) - (A \cap B)$$

Example. Let A = {1, 2, 3}, B = {3, 4, 5}, then

$$A \oplus B = (A \cup B) - (A \cap B) = \{1, 2, 3, 4, 5\} - \{3\} = \{1, 2, 4, 5\}$$

Venn diagram of $A \oplus B$ is given in Figure 3.7.

Properties :

1. $A \oplus B = B \oplus A$ (Commutative)
2. $(A \oplus B) \oplus C = A \oplus (B \oplus C)$ (Associative)
3. $A \oplus B = (A - B) \cup (B - A)$
4. $A \oplus A = \phi$
5. $A \oplus \phi = A$.

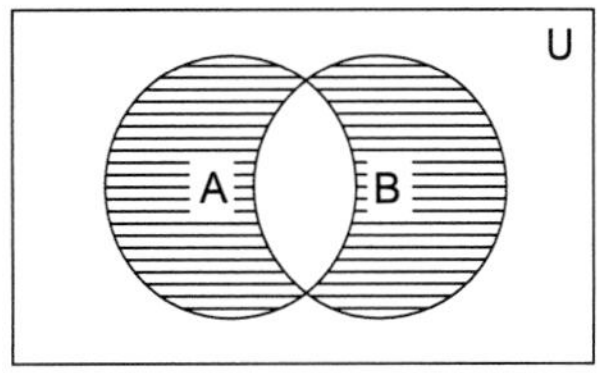

A⊕B (Shaded part)

Figure. 3.7.

Demorgan's Laws :

For any two sets A and B,

(*i*) $(A \cup B)' = A' \cap B'$

(*ii*) $(A \cap B)' = A' \cap B'$.

Demorgan's laws can be extended as follows :

If $A_1, A_2, ..., A_n$ are sets, then

(*i*) $(A_1 \cup A_2 \cup ..., \cup A_n)' = A_1' \cap A_2' \cap \ldots \cap A_n'$

i.e., $$\left(\bigcup_{i=1}^{n} A_i\right)' = \bigcap_{i=1}^{n} A_i'$$

(*ii*) $(A_1 \cap A_2 \cap ... \cap A_n)' = A_1' \cup A_2' \cup \ldots \cup A_n'$

i.e., $$\left(\bigcap_{i=1}^{n} A_i\right)' = \bigcup_{i=1}^{n} A_i'$$

3.4.2 Properties of the Difference Operation

Let A, B, C are sets and U be the universal set, then

(*i*) $A' = U - A$

(*ii*) $A - B = A \cap B'$

(*iii*) $A - A = \phi$

(*iv*) $A - \phi = A$

(*v*) $A - B = B - A \Leftrightarrow A = B$.

(*vi*) $A - B = A \Leftrightarrow A \cap B = \phi$

(*vii*) $A - B = \phi \Leftrightarrow A \subseteq B$

(*viii*) $A - (B \cup C) = (A - B) \cap (A - C)$

(*ix*) $A - (B \cap C) = (A - B) \cup (A - C)$

(*x*) $A \cap (B - A) = \phi$

(*xi*) $(A - B) \cap B = \phi$

(*xii*) $(A - B) \cup B = A \cup B$

(*xiii*) $A - B = B' - A'$

(*xiv*) $A - (A - B) = A \cap B$

(*xv*) $(A - C) \cap (B - C) = (A \cap B) - C$

(*xvi*) $A \cap (B - C) = (A \cap B) - (A \cap C)$

(*xvii*) $A \cup (B - C) \neq (A \cup B) - (A \cup C)$.

Example. *Illustrate the distributive law $A \cap (B \cup C) = (A \cap B) \cup (A \cap C)$ with Venn diagrams.*

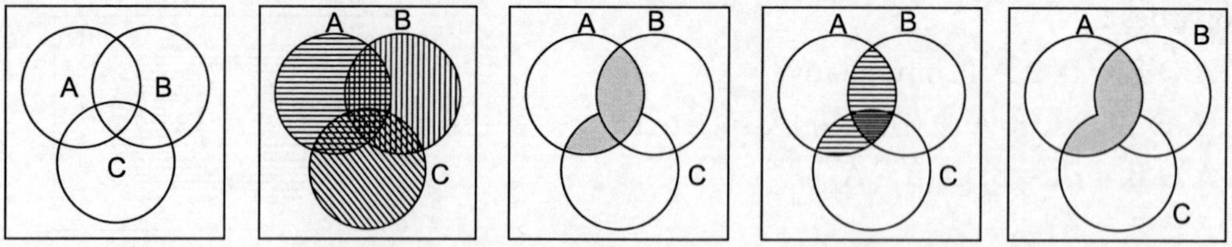

(i) A, $B \cup C$ are shaded (ii) $A \cap (B \cup C)$ is shaded (iii) $A \cap B$, $A \cup C$ are shaded (iv) $(A \cap B) \cup (A \cup C)$ is shaded (v)

$\therefore \quad A \cap (B \cup C) = (A \cap B) \cup (A \cap C).$

Figure. 3.8.

Example. Show that $A \cap (B \cap C) = (A \cap B) \cap C$

Solution. First we shall show that $A \cap (B \cap C) \subset (A \cap B) \cap C$

Let $x \in A \cap (B \cap C) \Rightarrow x \in A$ and $x \in B \cap C$

$\Rightarrow$ $x \in A$ and ($x \in B$ and $x \in C$)

$\Rightarrow$ ($x \in A$ and $x \in B$) and $x \in C$

$\Rightarrow$ $x \in A \cap B$ and $x \in C$

$\Rightarrow$ $x \in (A \cap B) \cap C$

$\therefore$ $A \cap (B \cap C) \subseteq (A \cap B) \cap C.$

Similarly we can prove $(A \cap B) \cap C \subseteq A \cap (B \cap C)$

$\therefore$ $A \cap (B \cap C) = (A \cap B) \cap C.$

Example. Show thatA $\cup$ (B $\cap$ C) = A $\cup$ B) $\cap$ (A $\cup$ C)

Solution. First we shall show that $A \cup (B \cap C) \subseteq (A \cup B) \cap (A \cup C)$

Let $x \in A \cup (B \cap C) \Rightarrow x \in A$ or $x \in B \cap C$

$\Rightarrow$ $x \in A$ or ($x \in B$ and $x \in C$)

$\Rightarrow$ ($x \in A$ or $x \in B$) and ($x \in A$ or $x \in C$).

$\Rightarrow$ ($x \in A \cup B$) and ($x \in A \cup C$)

$\Rightarrow$ $x \in (A \cup B) \cap (A \cup C)$

$\therefore$ $A \cup (B \cap C) \subseteq (A \cup B) \cap (A \cup C)$

Similarly we can prove $(A \cup B) \cap (A \cup C) \subseteq A \cup (B \cap C)$

$\therefore$ $A \cup (B \cap C) = (A \cup B) \cap (A \cup C).$

Theorem. Let A, B be any two sets then prove that

(*i*) $(A \cup B)^C = A^C \cap B^C$.

(*ii*) $(A \cap B)^C = A^C \cup B^C$.

Proof : (*i*) Let x be any arbitrary element in $(A \cup B)^C$

$\Leftrightarrow x \in U$ and $x \notin A \cup B$

$\Leftrightarrow x \in U$ and ($x \notin A$ and $x \notin B$)

$\Leftrightarrow$ ($x \in U$ but $x \notin A$) and ($x \in U$ but $x \notin B$)

$\Leftrightarrow x \in A^C$ and $x \in B^C$.

$$\Leftrightarrow x \in A^C \cap B^C.$$

$\therefore \qquad (A \cup B)^C = A^C \cap B^C.$

(*ii*) Let x be any arbitrary element in $(A \cap B)^C$.

$$\Leftrightarrow x \in U \text{ and } x \notin A \cap B$$
$$\Leftrightarrow x \in U \text{ and } (x \notin A \text{ or } x \notin B)$$
$$\Leftrightarrow (x \in U \text{ and } x \notin A) \text{ or } (x \in U \text{ and } x \notin B)$$
$$\Leftrightarrow x \in A^C \text{ or } x \in B^C$$
$$\Leftrightarrow x \in A^C \cup B^C.$$

$\therefore \qquad (A \cap B)^C = A^C \cup B^C.$

Theorem. If A, B are any two sets, then prove that

(*i*) $A - B = A \cap B^C$

(*ii*) $A \cap (B - A) = \phi$

(*iii*) $(A - B) \cap B = \phi$.

Proof (*i*) Let $x \in A - B \Leftrightarrow x \in A \text{ and } x \notin B$

$$\Leftrightarrow x \in A \text{ and } x \in B^C$$
$$\Leftrightarrow x \in A \cap B^C$$

Hence $\qquad A - B = A \cap B^C.$

(*ii*) Let $\quad x \in A \cap (B - A) \Leftrightarrow x \in A \text{ and } x \in B - A$

$$\Leftrightarrow x \in A \text{ and } (x \in B \text{ and } x \notin A)$$
$$\Leftrightarrow (x \in A \text{ and } x \notin A) \text{ and } x \in B.$$
$$\Leftrightarrow (x \in A \cap A^C = \phi) \text{ and } x \in B$$
$$\Leftrightarrow x \in \phi.$$

$\therefore \qquad A \cap (B - A) = \phi.$

(*iii*) Let $x \in (A - B) \cap B \Leftrightarrow x \in A - B \text{ and } x \in B$

$$\Leftrightarrow (x \in A \text{ and } x \notin B) \text{ and } x \in B$$
$$\Leftrightarrow x \in A \text{ and } (x \notin B \text{ and } x \in B)$$
$$\Leftrightarrow x \in A \text{ and } (x \in B^C \cap B = \phi)$$
$$\Leftrightarrow x \in \phi$$

$\therefore \qquad (A - B) \cap B = \phi.$

Example. *Let* $S_1 = \{1, 2, 3\}$, $S_2 = \{3, 4, 5, 6\}$, *then find*

(*i*) $S_1 \cup S_2$ (*ii*) $S_1 \cap S_2$

(*iii*) $S_1 - S_2$ (*iv*) $S_2 - S_1$.

Solution. (*i*) $S_1 \cup S_2 = \{1, 2, 3, 4, 5, 6\}$

(*ii*) $S_1 \cap S_2 = \{3\}$

(*iii*) $S_1 - S_2 = \{1, 2\}$

(*iv*) $S_2 - S_1 = \{4, 5, 6\}$

Example. *If $A = \{2, 3, 4\}$, $B = \{3, 4, 5, 6\}$, $C = \{2, 4, 6, 8\}$, then verify $A \cap (B - C) = (A \cap B) - (A \cap C)$.*

Solution.

$$B - C = \{3, 5\}$$
$$A \cap (B - C) = \{3\}$$
$$A \cap B = \{3, 4\}$$
$$A \cap C = \{2, 4\}$$
$$(A \cap B) - (A \cap C) = \{3\}$$
$$\therefore \quad A \cap (B - C) = (A \cap B) - (A \cap C).$$

Example. *Let $U = \{1, 2, 3, 4, 5, 6, 7, 8, 9, 10\}$, $A = \{1, 3, 5\}$, $B = \{2, 4, 6, 8\}$ and $C = \{2, 5, 10\}$ then verify (i) $(A \cap B)^C = A^C \cup B^C$; (ii) $(A \cup B)^C = A^C \cap B^C$; (iii) $A \cap (B \cup C) = (A \cap B) \cup (A \cap C)$; and (iv) $A \cup (B \cap C) = (A \cup B) \cap (A \cup C)$.*

Solution. Given $U = \{1, 2, 3, 4, 5, 6, 7, 8, 9, 10\}$

$$A \cup B = \{1, 2, 3, 4, 5, 6, 8\} \Rightarrow (A \cup B)^C = \{7, 9, 10\}$$
$$A^C = \{2, 4, 6, 7, 8, 9, 10\}, B^C = \{1, 3, 5, 7, 9, 10\}$$
$$A^C \cap B^C = \{7, 9, 10\} \mid A^C \cup B^C = \{1, 2, 3, 4, 5, 6, 7, 8, 9, 10\}$$
$$A \cap B = \phi \Rightarrow (A \cap B)^C = \{1, 2, 3, 4, 5, 6, 7, 8, 9, 10\}$$
$$\therefore \quad (A \cap B)^C = A^C \cup B^C \text{ and}$$
$$(A \cup B)^C = A^C \cap B^C.$$

and
$$B \cup C = \{2, 4, 5, 6, 8, 10\}, A \cap (B \cup C) = \{5\}.$$
$$A \cap B = \phi, A \cap C = \{5\} \Rightarrow (A \cap B) \cup (A \cap C) = \{5\}$$
$$\therefore \quad A \cap (B \cup C) = (A \cap B) \cup (A \cap C).$$
$$B \cap C = \{2\}$$
$$A \cup (B \cap C) = \{1, 2, 3, 5\}$$
$$A \cup B = \{1, 2, 3, 4, 5, 6, 8\}$$
$$A \cup C = \{1, 2, 3, 5, 10\}$$
$$(A \cup B) \cap (A \cup C) = \{1, 2, 3, 5\}$$
$$\therefore \quad A \cup (B \cap C) = (A \cup B) \cap (A \cup C).$$

Example. *If $U = \{x \mid x < 12, x \in N\}$, $A = \{x \mid x = 1, 2, 3, 4\}$, $B = \{x \mid x < 11$, and x is even natural number$\}$ and $C = \{x \mid x \in N, 2 < x < 7\}$*

Find (i) A^C ; (ii) B^C ; and (iii) $(B - C)^C$.

Solution. Given

$$U = \{1, 2, 3, 4, 5, 6, 7, 8, 9, 10, 11\}$$
$$A = \{1, 2, 3, 4\}$$
$$B = \{2, 4, 6, 8, 10\} \text{ and } C = \{3, 4, 5, 6\}$$

(*i*) $A^C = U - A = \{5, 6, 7, 8, 9, 10, 11\}$

(*ii*) $B^C = U - B = \{1, 3, 5, 7, 9, 11\}$

(*iii*) $B - C = \{2, 8, 10\} \Rightarrow (B - C)^C = \{1, 3, 4, 5, 6, 7, 9, 11\}$

Example. *Let A, B, C* $\subseteq R^2$ *where* $A = \{(x, y) \mid y = 2x + 1\}$, $B = \{(x, y)/ y = 3x\}$, *and* $C = \{(x, y) / x - y = 7\}$

Determine each of the following :

(i) $A \cap B$ (ii) $B \cap C$

(iii) $\overline{\overline{A} \cup \overline{C}}$ (iv) $\overline{B} \cup \overline{C}$. (JNTU, Feb 2007, Set 2 and 3)

Solution. (*i*) $A \cap B = \{(x, y) / y = 2x + 1 \text{ and } y = 3x\}$

$$= \{(x, y) / y = 2x + 1 = 3x\}$$

$$\therefore \quad 2x + 1 = 3x \Rightarrow x = 1$$

$$\Rightarrow \quad y = 3$$

$$\therefore \quad A \cap B = \{(1, 3)\}$$

(*ii*) $B \cap C = \{(x, y) / y = 3x, y = x - 7\}$

$$\therefore \quad 3x = x - 7 \Rightarrow x = -\frac{7}{2}$$

$$\Rightarrow \quad y = -\frac{21}{2}$$

$$\therefore \quad B \cap C = \left\{\left(-\frac{7}{2}, -\frac{21}{2}\right)\right\}$$

(*iii*) $\overline{\overline{A} \cup \overline{C}} = \overline{\overline{A}} \cap \overline{\overline{C}} = A \cap C$

$$= \{(x, y) / y = 2x + 1 \text{ and } y = x - 7\}$$

$$\therefore \quad 2x + 1 = x - 7 \Rightarrow x = -8$$

$$\Rightarrow \quad y = -15.$$

$$\therefore \quad A \cap C = \{(-8, -15)\}$$

$$\Rightarrow \quad \overline{\overline{A} \cup \overline{C}} = \{(-8, -15)\}$$

(*iv*) We have $\overline{B \cap C} = \overline{B} \cup \overline{C}$ and $B \cap C = \left\{\left(-\frac{7}{2}, -\frac{21}{2}\right)\right\}$

$$\therefore \quad \overline{B} \cup \overline{C} = \overline{B \cap C} = \left\{(x, y) \,\middle|\, (x, y) \neq \left(-\frac{7}{2}, -\frac{21}{2}\right)\right\}$$

Example. *Determine the sets A and B, given that*

$A - B = \{1, 2, 4\}$, $B - A = \{7, 8\}$ *and* $A \cup B = \{1, 2, 4, 5, 7, 8, 9\}$

Solution. We have $A = (A \cup B) - (B - A)$

$$= \{1, 2, 4, 5, 9\} \text{ and}$$

$$B = (A \cup B) - (A - B) = \{5, 7, 8, 9\}.$$

Duality. Consider a set P formed by the intersections and/or unions of sets which may include the null set ϕ and a universal set U. With this set in hand, suppose we form another set by replacing.

(*i*) each intersection operation by the union operation.

(*ii*) each union operation by the intersection operation and by replacing U by ϕ and ϕ by U.

The new set formed in this way is called the dual of the original set P and is denoted by P^d. For example if,

$$p = ((A \cup B) \cap \phi) \cup ((C \cap D) \cup U)$$

Then
$$P^d = ((A \cap B) \cup C) \cap ((C \cup D) \cap \phi).$$

Note. We can prove that if P and Q are sets such that P = Q. Then $P^d = Q^d$, this is known as the **Principle of Duality** for the equality of sets.

Example. *Write down the dual statement for each of the following statements :*

(*i*) $A = (A \cup B) \cap (A \cup \phi)$; (*ii*) $U = (A \cap B) \cup (A \cap \bar{B}) \cup (\bar{A} \cap \bar{B}) \cup (\bar{A} \cap B)$.

Solution. (*i*) $A = (A \cap B) \cup (A \cup U)$

(*ii*) $\phi = (A \cup B) \cap (\bar{A} \cup \bar{B}) \cap (\bar{A} \cup \bar{B}) \cap (\bar{A} \cup B)$.

3.5 A COMPUTER REPRESENTATION FOR SETS

Let U = {1, 2, 3, 4, 5, 6} be a set, and let $X \subseteq U$, A bit representation for X is a Six-digit binary number $x_1\, x_2\, x_3\, x_4\, x_5\, x_6$ with bit x_i for $1 \le i \le 6$ defined as

$$x_i = \begin{cases} 1 & \text{if } i \in X \\ 0 & \text{if } i \notin X \end{cases}.$$

For example, if B = {2, 3, 6}, then B = 0 1 1 0 0 1. The operations of union, intersection, and complement can be carried out using operators UNION, INTER, DIFF and COMP that operate on binary numbers bit-by-bit. Let B, $C \subseteq U$ with $B = b_1\, b_2\, b_3\, b_4\, b_5\, b_6$; and $C = c_1\, c_2\, c_3\, c_4\, c_5\, c_6$. Define the union as UNION (B, C) $= x_1\, x_2\, x_3\, x_4\, x_5\, x_6$ where for $1 \le i \le 6$,

$$x_i = \begin{cases} 1 & \text{if } b_i = 1 \text{ or } c_i = 1 \\ 0 & \text{otherwise} \end{cases}.$$

Define the intersection as INTER (B, C) $= x_1\, x_2\, x_3\, x_4\, x_5\, x_6$, where for $1 \le i \le 6$,

$$x_i = \begin{cases} 1 & \text{if } b_i = 1 \text{ and } c_i = 1 \\ 0 & \text{otherwise} \end{cases}.$$

Define the complement as COMP (B) $= x_1\, x_2\, x_3\, x_4\, x_5\, x_6$, where for $1 \le i \le 6$,

$$x_i = \begin{cases} 1 & \text{if } b_i = 0 \\ 0 & \text{otherwise} \end{cases}.$$

Define the relative difference as DIFF (B, C) $= x_1\, x_2\, x_3\, x_4\, x_5\, x_6$, where for $1 \le i \le 6$,

$$x_i = \begin{cases} 1 & \text{if } b_i = 1 \text{ and } c_i = 0 \\ 0 & \text{otherwise} \end{cases}.$$

Example. Let B = {1, 2, 3, 4, 5} and C = {3, 4, 5, 6, 7, 8} be subsets of the universal set U = {1, 2, ..., 9}. Find UNION (B, C), INTER (B, C), COMP (C), and DIFF (B, C).

Solution. $B \cup C = \{1, 2, 3, 4, 5, 6, 7, 8\}$,

$$B \cap C = \{3, 4, 5\}$$

$$\overline{C} = \{1, 2, 9\}$$

$$B - C = \{1, 2\}.$$

Therefore, UNION (B, C) = 1 1 1 1 1 1 1 1 0

INTER (B, C) = 0 0 1 1 1 0 0 0 0

COMP (C) = 1 1 0 0 0 0 0 0 1

DIFF (B, C) = 1 1 0 0 0 0 0 0 0.

EXERCISE 3.1.

1. Define the following, give examples
   (*i*) Union of sets (*ii*) Intersectionof sets
   (*iii*) Complement of a set (*iv*) Power set
2. If $U = \{1, 2, 3, 4, 5, 6, 7, 8\}$, $A = \{2, 4, 6, 8\}$, $B = \{1, 3, 5, 7\}$ and $C = \{3, 6\}$ then find (*i*) $A \cup B$; (*ii*) $B \cap C$; (*iii*) C^C ; (*iv*) $A - B$; (*v*) $B - A$; (*vi*) $B - C$; (*vii*) $C - B$; and (*viii*) $B \oplus C$.
3. Let $U = \{x \mid x \in \mathbf{N}, 1 \le x \le 12\}$ be the universal set and $A = \{1, 9, 10\}$, $B = \{3, 4, 6, 11, 12\}$ and $C = \{2, 5, 6\}$ are subsets of U then find the following :
   (*i*) $A \cup B$ (*ii*) $A \cup C$ (*iii*) $(A \cup B) \cap (A \cup C)$
   (*iv*) $B \cap C$ (*v*) $A \cup (B \cap C)$ (*vi*) $A \cup B \cup C$
   (*vii*) $(A \cup B \cup C)^C$.
   and verify $A \cup (B \cap C) = (A \cup B) \cap (A \cup C)$.
4. Let $U = \{x \mid |x| < 5, x \in \mathbf{Z}\}$ be universal set.
   $A = \{x \mid x^3 + x^2 - 2x = 0\}$ and $B = \{x \mid x \text{ is even integer}\}$
   Find (*i*) $A - B$; and
   (*ii*) $B - A$.
5. If $U = \{1, 2, 3, 4, 5, 6, 7, 8, 9\}$, $A = \{1, 2, 4, 6, 8\}$ and $B = \{2, 4, 5, 9\}$, then find the following :
   (*i*) $\overline{A}$ (*ii*) $\overline{B}$ (*iii*) $\overline{A} \cup \overline{B}$
   (*iv*) $\overline{A \cup B}$ (*v*) $\overline{A} \cap \overline{B}$ (*vi*) $\overline{A \cap B}$
   (*vii*) $B - A$ (*viii*) $A - B$ (*ix*) $A \oplus B$.
6. Determine the sets A and B, given that $A - B = \{1, 3, 7, 11\}$, $B - A = \{2, 6, 8\}$ and $A \cap B = \{4, 9\}$.
7. If $A = \{a, b, c\}$, find the power set of A.
8. If $A = \{a, b, c, d\}$, find the power set of A.
9. Prove that $A - B = A \cap \overline{B}$.
10. Use Venn diagram to show that :
    (*i*) $A \cap (B \cup C) = (A \cap B) \cup (A \cap C)$
    (*ii*) $A \cup (B \cap C) = (A \cup B) \cap (A \cup C)$.

11. For any sets A, B, C, prove the associative laws

(*i*) $A \cup (B \cup C) = (A \cup B) \cap C$ (*ii*) $A \cap (B \cap C) = (A \cap B) \cap C$.

12. Prove the following by Venn diagrams :

(*i*) $(A^C)^C = A$ (*ii*) $(A \cap B)^C = A^C \cup B^C$

(*iii*) $(A \cup B)^C = A^C \cap B^C$.

13. Write down the dual statement for each of the following statements :

(*i*) $A = A \cap (A \cup B)$ (*ii*) $A \cup B = (A \cap B) \cup (A \cap \bar{B}) \cup (\bar{A} \cap B)$.

ANSWERS

2. (*i*) {1, 2, 3, 4, 5, 6, 7, 8} (*ii*) ϕ (*iii*) {1, 2, 4, 5, 7, 8}

(*iv*) {2, 4, 6, 8} (*v*) {1, 3, 5, 7} (*vi*) {1, 5, 7}

(*vii*) {6} (*viii*) {1, 5, 6, 7}.

3. (*i*) {1, 3, 4, 6, 9, 10, 11, 12} (*ii*) {1, 2, 5, 6, 9, 10} (*iii*) {1, 6, 9, 10}

(*iv*) {6} (*v*) {1, 6, 9, 10}

(*vi*) {1, 2, 3, 4, 5, 6, 9, 10, 11, 12} (*vii*) {7, 8}.

4. (*i*) $A - B = \{1\}$ (*ii*) $B - A = \{-4, 2, 4\}$

5. (*i*) {3, 5, 7, 9} (*ii*) {1, 3, 6, 7, 8} (*iii*) {1, 3, 5, 6, 7, 8, 9}

(*iv*) {3, 7} (*v*) {3, 7} (*vi*) {1, 3, 5, 6, 7, 8, 9}

(*vii*) {5, 9} (*viii*) {1, 6, 8} (*ix*) {1, 5, 6, 8, 9}.

6. A = {1, 3, 4, 7, 9, 11}, B = {2, 4, 6, 8, 9}

7. P (A) = {ϕ, {*a*}, {*b*}, {*c*}, {*a*, *b*}, {*a*, *c*}, {*b*, *c*}, {*a*, *b*, *c*}}

8. P (A) = {ϕ, {*a*}, {*b*}, {*c*}, {*d*}, {*a*, *b*}, {*a*, *c*}, {*a*, *d*}, {*b*, *c*}, {*b*, *d*}, {*c*, *d*}, {*a*, *b*, *c*}, {*a*, *b*, *d*}, {*a*, *c*, *d*}, {*b*, *c*, *d*}, A}

13. (*i*) $A = A \cup (A \cap B)$ (*ii*) $A \cap B = (A \cup B) \cap (A \cup \bar{B}) \cap (\bar{A} \cup B)$

3.6 CARTESIAN PRODUCT

Definition. *Let A, B be two sets. Then the* **Cartesian product** *of* **A, B** *denoted by A × B and it is defined as set of all ordered pairs (a, b) where a ∈ A, b ∈ B.*

In symbols $A \times B = \{(a, b) \mid a \in A, b \in B\}$

More generally the cartesian product of *n* sets $A_1, A_2, ..., A_n$ is defined as $A_1 \times A_2 \times ... \times A_n = \{(a_1, a_2, ..., a_n) \mid a_i \in A_i, i = 1, 2, ..., n\}$. The expression $(a_1, a_2, ..., a_n)$ is called an ordered *n*-tuple.

Note : 1. A × B is read as A cross B

2. $R \times R = \{(a, b) \mid a, b \in R\} = R^2$

3. $R \times R \times ... \times R$ (*n* times) $= \{(a_1, a_2, ..., a_n) \mid a_i \in R, i = 1, 2, ..., n\} = R^n$.

4. If $n(A) = m$, $n(B) = n$, then $n(A \times B) = mn$ and $P(A \times B) = 2^{mn}$.

5. $A \times B \neq B \times A$.

Example 1. If A = {*a*, *b*, *c*}, B = {1, 2}, then

$$A \times B = \{(a, 1), (a, 2), (b, 1), (b, 2), (c, 1), (c, 2)\}$$
$$A \times A = \{(a, a), (a, b), (a, c), (b, a), (b, b), (b, c), (c, a), (c, b), (c, c)\}$$
$$B \times B = \{(1, 1), (1, 2), (2, 1), (2, 2)\}$$
$$B \times A = \{(1, a), (1, b), (1, c), (2, a), (2, b), (2, c)\}$$

We observed that $A \times B \neq B \times A$.

Example 2. Let R be the set of real numbers. Then the Cartesian product $R \times R = \{(x, y) \mid x, y \in R\}$.

Example. *Find x and y in each of the following cases :*

(i) $(2x - 3, 3y + 1) = (5, 7)$ *(ii)* $(x + 2, 4) = (5, 2x + y)$ *(iii)* $(x, y) = (x^2, y^2)$

Solution.

(*i*) **Given :** $(2x - 3, 3y + 1) = (5, 7)$

$\Rightarrow \quad 2x - 3 = 5, \quad 3y + 1 = 7$

$\Rightarrow \quad x = 4, \quad y = 2.$

(*ii*) **Given :** $(x + 2, 4) = (5, 2x + y)$

$\Rightarrow \quad x + 2 = 5, \quad 2x + y = 4$

$\Rightarrow \quad x = 3, \quad y = 4 - 2x = 4 - 6 = -2$

(*iii*) **Given :** $(x, y) = (x^2, y^2)$

$\Rightarrow \quad x = x^2 ; \quad y = y^2$

$\Rightarrow \quad x = 0, 1 ; \quad y = 0, 1.$

3.7 THE IMPORTANT RESULTS ON CARTESIAN PRODUCT

Theorem 3.7.1. *If A, B and C are sets and $A \subseteq B$ then prove that $(A \times C) \subseteq (B \times C)$.*

Proof : Let (a, c) be an arbitrary element of $A \times C$.

$\Rightarrow \quad a \in A$ and $c \in C$

$\Rightarrow \quad a \in B$ and $c \in C \Rightarrow (a, c) \in B \times C$

Thus $A \times C \subseteq B \times C$.

Theorem 3.7.2. *If $A \subseteq B$ and $C \subseteq D$ then prove that $(A \times C) \subseteq (B \times D)$*

Proof : Let (a, c) be an arbitrary element of $A \times C$

$\Rightarrow \quad a \in A$ and $c \in C$

$\Rightarrow \quad a \in B$ and $c \in D \Rightarrow (a, c) \in B \times D$

$\Rightarrow$ Thus $A \times C \subseteq B \times D$.

Theorem 3.7.3. *If $A \subseteq B$, show that $A \times A = (A \times B) \cap (B \times A)$*

Proof : Let (x, y) be an arbitrary element of $A \times A$, then

$(x, y) \in A \times A \Leftrightarrow x \in A$ and $y \in A$

$\Leftrightarrow (x \in A$ and $y \in A)$ and $(x \in A$ and $y \in A)$

$\Leftrightarrow (x \in A$ and $y \in B)$ and $(x \in B$ and $y \in A)$

$$\Leftrightarrow (x, y) \in A \times B \text{ and } (x, y) \in B \times A.$$
$$\Leftrightarrow (x, y) \in (A \times B) \cap (B \times A)$$

Therefore, $A \times A = (A \times B) \cap (B \times A)$.

Theorem 3.7.4. *If A, B, C are sets, prove that $A \times (B \cap C) = (A \times B) \cap (A \times C)$.*

Proof : Let (x, y) be an arbitrary element of $A \times (B \cap C)$, then

$$\begin{aligned}(x, y) \in A \times (B \cap C) &\Leftrightarrow x \in A \text{ and } y \in (B \cap C)\\ &\Leftrightarrow x \in A \text{ and } (y \in B \text{ and } y \in C)\\ &\Leftrightarrow (x \in A \text{ and } y \in B) \text{ and } (x \in A \text{ and } y \in C)\\ &\Leftrightarrow (x, y) \in A \times B \text{ and } (x, y) \in A \times C\\ &\Leftrightarrow (x, y) \in (A \times B) \cap (A \times C)\end{aligned}$$

Therefore, $A \times (B \cap C) = (A \times B) \cap (A \times C)$.

Theorem 3.7.5. *Prove that $A \times (B \cup C) = (A \times B) \cup (A \times C)$.*

Proof : Let (x, y) be an arbitrary element of $A \times (B \cup C)$

$$\begin{aligned}(x, y) \in A \times (B \cup C) &\Leftrightarrow x \in A \text{ and } y \in (B \cup C)\\ &\Leftrightarrow x \in A \text{ and } (y \in B \text{ or } y \in C)\\ &\Leftrightarrow (x \in A \text{ and } y \in B) \text{ or } (x \in A \text{ and } y \in C)\\ &\Leftrightarrow (x, y) \in A \times B \text{ or } (x, y) \in A \times C\\ &\Leftrightarrow (x, y) \in (A \times B) \cup (A \times C)\end{aligned}$$

Therefore, $A \times (B \cup C) = (A \times B) \cup (A \times C)$

Theorem 3.7.6. *Prove that $(A \times B) \cap (S \times T) = (A \cap S) \times (B \cap T)$*

Proof. Let (x, y) be an arbitrary element of $(A \times B) \cap (S \times T)$

$$\begin{aligned}(x, y) \in (A \times B) \cap (S \times T) &\Leftrightarrow (x, y) \in A \times B \text{ and } (x, y) \in S \times T\\ &\Leftrightarrow (x \in A \text{ and } y \in B) \text{ and } (x \in S \text{ and } y \in T)\\ &\Leftrightarrow (x \in A \text{ and } x \in S) \text{ and } (y \in B \text{ and } y \in T)\\ &\Leftrightarrow x \in A \cap S \text{ and } y \in B \cap T\\ &\Leftrightarrow (x, y) \in (A \cap S) \times (B \cap T)\end{aligned}$$

$\therefore$ $(A \times B) \cap (S \times T) = (A \cap S) \times (B \cap T)$.

Theorem 3.7.7. *Let A, B, C are any three sets, prove that $A \times (B - C) = (A \times B) - (A \times C)$*

Proof. Let (x, y) be an arbitrary element of $A \times (B - C)$

$$\begin{aligned}(x, y) \in A \times (B - C) &\Leftrightarrow x \in A \text{ and } y \in B - C\\ &\Leftrightarrow x \in A \text{ and } (y \in B \text{ and } y \notin C)\\ &\Leftrightarrow (x \in A \text{ and } y \in B) \text{ and } (x \in A \text{ and } y \notin C)\\ &\Leftrightarrow (x, y) \in A \times B \text{ and } (x, y) \notin A \times C\\ &\Leftrightarrow (x, y) \in (A \times B) - (A \times C)\end{aligned}$$

$\therefore$ $A \times (B - C) = (A \times B) - (A \times C)$.

Theorem 3.7.8. *Let A, B be two non-empty sets, prove that $A \times B = B \times A \Leftrightarrow A = B$.*

Proof : Let A = B, then $A \times B = A \times A$ and $B \times A = A \times A$

$\Rightarrow$ $A \times B = B \times A$

Conversely suppose $A \times B = B \times A$

let $x \in A \Rightarrow (x, y) \in A \times B$ for $y \in B$

$\Rightarrow$ $(x, y) \in B \times A$

$\Rightarrow$ $x \in B$

$\therefore$ $A \subseteq B$.

Similarly, we can prove that $B \subseteq A$.

Then $A = B$.

Hence $A \times B = B \times A \Leftrightarrow A = B$.

3.8 RELATIONS

Human language has many words and phrases to describe relationships between or among objects. It may be that for two people A and B, that A is a parent of B, that A is an ancestor of B, that A is taller than B, or that A is infront of B. In algebra, it may be that the value of variable x is less than the value of variable y. In set theory, it may be that a set X is a subset of *a* set Y (or) that X is disjoint from Y. At a particular moment while a computer program is running, it may be that the value of X is less than the value of Y. All these notions are special instances of a relation.

Definition. *Let A, B are two sets. A subset $A \times B$ is called a* ***binary relation*** *(or)* ***relation*** *from A to B.*

Note 1. If $R \subseteq A \times B$, then R is a relation on from A to B.

2. If B = A, then R is a relation on A.

3. If $(a, b) \in R$ is also can be written as aRb, we read it as a relates b.

Definition. *Let R be a relation from A to B then*

(*i*) ***domain of R*** *is denoted by dom R and is defined as dom $R = \{a \mid a \in A, (a, b) \in R$ for some $b \in B\}$.*

(*ii*) ***Range of R*** *is denoted by ran R and is defined as*

$$\text{ran } R = \{b \mid b \in B, (a, b) \in R \text{ for some } a \in A\}.$$

Note. Domain of $R \subseteq A$, Range of $R \subseteq B$.

Example. *Let $A = \{2, 3, 4\}$ and $B = \{3, 4, 5, 6, 7\}$.*

Define a relation R from A to B by $(a, b) \in R$ if a divides b (*i.e.,*) a/b

Then $R = \{(2, 4), (2, 6), (3, 3), (3, 6), (4, 4)\}$

Hence domain of R is {2, 3, 4}, and range of R is {3, 4, 6}.

Definition. *Let R be a relation from A to B. Then inverse of relation R from B to A is denoted by R^{-1} and it is defined as $R^{-1} = \{(b, a) \mid (a, b) \in R\}$.*

Example. If R = {(2, 4), (2, 6), (3, 3), (3, 6), (4, 4)} is a relation from A to B. Then R^{-1} = {(3, 3), (4, 2), (4, 4), (6, 2), (6, 3)} is inverse relation from B to A.

3.9 PROPERTIES OF RELATIONS

Definition. *A relation R on a set A is said to be*

(i) Reflexive on A if (a, a) ∈ R (i.e.,) aRa ∀ a ∈ A

(ii) Symmetric on A if (a, b) ∈ R ⇒ (b, a) ∈ R ∀ a, b ∈ A

(iii) Transitive on A if (a, b) ∈ R, (b, c) ∈ R ⇒ (a, c) ∈ R ∀ a, b, c ∈ A.

Example 1. Let **Z** be the set of integers, For $a, b \in \mathbf{Z}$, define a relation R by a*R*b iff 2 divides $a - b$, then R is reflexive, symmetric and transitive on **Z**.

Example 2. Let N be the set of natural numbers set. A relation R defined by aRb if $a + b = 7$ is symmetric, but not reflexive, not transitive on **N**.

Example 3. Let A = {1, 2, 3, 4} and Let R = {(1, 3), (4, 2), (2, 4), (2, 3), (3, 1)} then the relation R is not reflexive because (1, 1) ∉ R,

R is not symmetric because (2, 3) ∈ R but (3, 2) ∉ R and R is not transitive since (2, 3) ∈ R, (3, 1) ∈ R, but (2, 1) ∉ R.

Example 4. Let A = {1, 2, 3, 4} and S = {(1, 1), (2, 2), (3, 3), (4, 4), (2, 3), (3, 1), (2, 1)} is reflexive and transitive but not symmetric relation on A.

Example 5. The relation '<' is not reflexive on Real numbers set R.

Example 6. The relation '≤' is reflexive, transitive but not symmetric on real numbers set R.

3.10 EQUIVALENCE RELATIONS

Definition. *A relation R on a set A is said to be an* ***equivalence*** *relation on A if*

(i) R is reflexive on A

(ii) R is symmetric on A

(iii) R is transitive on A.

Example 1. *Let R be a relation on N defined as*

$$R = \{(a, b) \mid a, b \in \text{N and } a + b \text{ is even}\}$$
$$= \{(a, b) \mid a, b \in \text{N}, 2 \text{ divides } a + b\}$$

then R is an equivalence relation on **N**.

Because (*i*) $a + a$ is even for any $a \in \mathbf{N}$

(*ii*) If $a + b$ is even ⇒ $b + a$ is also even ∀ $a, b \in \mathbf{N}$.

(*iii*) If $a + b$ is even, $b + c$ *is even* ⇒ $a + b + b + c = a + c$ is also even ∀ $a, b, c, \in \mathbf{N}$.

Example 2. *A relation R defined in* **Z** *as aRb* = *a* – *b is a multiple of n is an equivalence relation on* **Z**.

Example. *Each of the following defines a relation on* **N**.

(i) "x is greater than y" *(ii) x + y = 10.*

Determine which relations are (a) Reflexive (b) Symmetric (c) Transitive.

Solution. Let R be a relation on N

(*i*) **Given :** $x \text{ R} y$ iff $x > y \ \forall\ x, y \in \mathbf{N}$.

(*a*) For any $x \in \mathbf{N}$, $x > x \Rightarrow$ R is not reflexive on **N**.

(*b*) For any $x, y \in \mathbf{N}$, $x > y \Rightarrow y \not> x$

$\Rightarrow$ R is not symmetric on **N**.

(c) For any $x, y, z \in \mathbf{N}$, $x > y, y > z \Rightarrow x > z$

$\Rightarrow$ R is transitive on **N**.

(*ii*) Given relation R defined as $x \text{R} y \Leftrightarrow x + y = 10 \ \forall\ x, y \in \mathbf{N}$.

(*a*) For any $x \in \mathbf{N}$, $x + x \neq 10$ $\qquad (\because \ 2 + 2 = 4 \neq 10)$

$\Rightarrow$ R is not reflexive on **N**.

(*b*) For any $x, y \in \mathbf{N}$, $x \text{R} y \Rightarrow x + y = 10$

$\Rightarrow \qquad y + x = 10$

$\Rightarrow \qquad y \text{R} x \ \forall\ x, y \in \mathbf{N}$.

$\therefore$ R is symmetric on **N**.

(*c*) R is not transitive, because for $2, 8 \in \mathbf{N}$,

$$2 + 8 = 10, 8 + 2 = 10 \quad \Rightarrow \quad 2 + 2 \neq 10.$$

Example. Let A = {1, 2, 3, ..., 9} and let ~ be the relation on A × A defined by $(a, b) \sim (c, d)$ iff $a + d = b + c$.

Then prove that ~ is an equivalence relation on A × A

Proof. (*i*) Let $(a, b) \in A \times A$, then $a + b = b + a$

$\Rightarrow \quad (a, b) \sim (a, b)$

$\therefore \quad \sim$ is reflexive on A × A.

(*ii*) Let $(a, b), (c, d) \in A \times A$, $(a, b) \sim (c, d)$

$\Rightarrow \qquad a + d = b + c \ \Rightarrow \ d + a = c + b$

$\Rightarrow \qquad c + b = d + a \ \Rightarrow \ (c, d) \sim (a, b)$.

$\therefore \quad \sim$ is symmetric on A × A.

(*iii*) Let $(a, b), (c, d), (e, f) \in A \times A$,

$(a, b) \sim (c, d), (c, d) \sim (e, f) \Rightarrow a + d = b + c, c + f = d + e$

$\Rightarrow \qquad a + d = b + c, d + e = c + f.$

$\Rightarrow \qquad a + b + c + f = b + c + d + e$

$\Rightarrow \qquad a + f + d + c = b + e + d + c$

$\Rightarrow \qquad a + f = b + e$

$\Rightarrow \qquad (a, b) = (e, f)$

$\therefore \quad \sim$ is transitive on A × A.

So ~ is an equivalence relation on A × A.

Example. *Show that the relation R is defined N × N by (a, b) R (c, d) if and only if a + d = b + c is an equivalence relation.*

Solution. $N \times N = \{(a, b) \mid a, b \in N\}$

(*i*) Clearly, $(a, b)\ R\ (a, b)\ \forall\ (a, b) \in N \times N$ since $a + d = b + c$.

$\therefore$ R is reflexive on N × N.

(*ii*) Let $(a, b)\ R\ (c, d)\ \forall\ (a, b), (c, d) \in N \times N$

$\Rightarrow \quad a + d = b + c$

i.e., $\quad c + b = d + a$

$\therefore \quad (c, d)\ R\ (a, b)$

Hence R is symmetric on N × N.

(*iii*) Let $(a, b)\ R\ (c, d), (c, d)\ R\ (e, f)\ \forall\ (a, b), (c, d), (e, f) \in N \times N$.

$\Rightarrow \quad a + d = b + c$ and $c + f = d + e$

$\Rightarrow \quad a + d + c + f = b + c + d + e$

$\Rightarrow \quad a + f = b + e$

$\Rightarrow \quad (a, b)\ R\ (e, f)$

$\Rightarrow$ R is transitive on N × N.

Hence R is an equivalence relation on N × N.

Example. *Let R be the relation on N × N which is defined by (a, b) is related to (c, d) which can be written as (a, b) ≃ (c, d) if and only if ad = bc. Prove that R is an equivalence relation.*

Solution. Since $ab = ba$, $(a, b) \simeq (a, b)\ \forall\ (a, b) \in N \times N$

$\therefore$ R is reflexive on N × N.

Let $(a, b) \simeq (c, d)\ \forall\ (a, b), (c, d) \in N \times N$

$\Rightarrow \quad ad = bc$

$\Rightarrow \quad cb = da$

$\Rightarrow \quad (c, d) \simeq (a, b)$

$\Rightarrow$ R is symmetric on N × N.

Now suppose $(a, b) \simeq (c, d)$ and $(c, d) \simeq (e, f)\ \forall\ (a, b), (c, d), (e, f) \in N \times N$

$\Rightarrow \quad ad = bc$ and $cf = de$.

$\Rightarrow \quad (ad)(cf) = (bc)(de)$

$\Rightarrow \quad a(dc)f = b(cd)e$

$\Rightarrow \quad a(cd)f = b(cd)e$ since $dc = cd$.

$\Rightarrow \quad af = be$ by cancelling cd on both sides.

$\Rightarrow \quad (a, b) \simeq (e, f) \Rightarrow$ R is transitive on N × N

$\therefore$ R is equivalence relation on N × N.

Example 1. *Let R denote a relation on the set of all ordered pairs of positive integers by (x, y) R (u, v) iff xv = yu.*

Show that R is an equivalence relation.

Solution. If x, y, u, v are positive integers, it is given that (x, y) R (u, v) iff $xv = yu$.

(*i*) As $xy = yx$ $\forall$ positive integers x, y, we have (x, y) R (x, y) for all ordered pair (x, y) of positive integers. R is reflexive.

(*ii*) (x, y) R (u, v) $\forall$ ordered pairs (x, y), (u, v) of positive integers.

$$\Rightarrow \quad xv = yu \quad \Rightarrow \quad yu = xv$$

$$\Rightarrow \quad uy = vx$$

$\Rightarrow$ (u, v) R (x, y).

$\therefore$ R is symmetric.

(*iii*) (x, y) R (u, v) and (u, v) R (m, n) $\forall$ ordered pairs (x, y), (u, v), (m, n) of positive integers.

$$\Rightarrow \quad xv = yu \quad \text{and} \quad un = vm$$

$$\Rightarrow \quad xv\,un = yu\,vm$$

$$\Rightarrow \quad x\,(vu)\,n = y\,(uv)\,m$$

$$\Rightarrow \quad x\,(uv)\,n = y\,(uv)\,m \text{ since } vu = uv.$$

$$\Rightarrow \quad xn = ym \text{ by cancelling } uv \text{ on both sides}$$

$$\Rightarrow \quad (x, y)\ R\ (m, n)$$

$\Rightarrow$ R is transitive.

$\therefore$ R is an equivalence relation.

Example. *Give an example of relation which is symmetric but neither reflexive nor anti-symmetric nor transitive.*

Solution. Take A = {1, 2, 3}. Consider the relation

$$R = \{(1, 1), (1, 2), (2, 1), (3, 2), (2, 3)\} \text{ on A}$$

Here $(2, 2) \notin R \Rightarrow R$ is not reflexive.

$$(1, 2) \notin R \Rightarrow (2, 1) \in R, (3, 2) \in R \Rightarrow (2, 3) \in R, (1, 1) \in R \Rightarrow (1, 1) \in R$$

Then R is symmetric on A.

$$(1, 2) \in R, (2, 1) \in R \Rightarrow 2 \neq 1$$

$\therefore$ R is not anti-symmetric relation on A.

and $(1, 2) \in R, (2, 3) \in R \Rightarrow (1, 3) \notin R$

$\therefore$ R is not transitive on A.

Example. *Give one example of a relation which is transitive but neither reflexive nor symmetric nor anti-symmetric.*

Solution. Let A = {1, 2, 3}, R = {(1, 1), (2, 2), (1, 2), (1, 3), (2, 1), (2, 3)} be a relation on A.

As $(3, 3) \notin$ R, R is not reflexive on A.

If $(1, 3) \in R \Rightarrow (3, 1) \notin R \Rightarrow R$ is not symmetric on A.

If $(1, 2), (2, 1) \in R \Rightarrow 1 \neq 2 \Rightarrow$ R is not anti-symmetric and R is transitive on A because $(1, 2), (2, 3) \in R \Rightarrow (1, 3) \in R$.

Example. *Give an example of relation which is symmetric but neither reflexive nor anti-symmetric nor transitive.*

Solution. Take A = {1, 2, 3}, consider the relation R = {(1, 1), (1, 2), (2, 1), (3, 2), (2, 3)} on A.

Here (*i*) $(2, 2) \notin R \Rightarrow$ R is not reflexive.

(*ii*) $(1, 2) \in R \Rightarrow (2, 1) \in R$

$(3, 2) \in R \Rightarrow (2, 3) \in R$

$(1, 1) \in R \Rightarrow (1, 1) \in R$, then R is symmetric on A.

(*iii*) $(1,2) \in R, (2, 1) \in R \Rightarrow 1 \neq 2$

$\therefore$ R is not anti-symmetric relation on A.

(*iv*) $(1, 2) \in R, (2, 3) \in R \Rightarrow (1, 3) \notin R$

$\therefore$ R is not transitive on A.

Example. *Give one example of a relation which is transitive but neither reflexive nor symmetric nor anti-symmetric.*

Solution. Let A = {1, 2, 3}, R = {(1, 1), (2, 2), (1, 2), (1, 3), (2, 1), (2, 3)} be a relation on A.

As $(3, 3) \notin R$, R is not reflexive.

As $(1, 3) \in R$ but $(3, 1) \notin R$, R is not symmetric.

Also $(1, 2), (2, 1) \in R$ but $1 \neq 2 \Rightarrow$ R is not anti-symmetric and clearly R is transitive on A because $(1, 2), (2, 3) \in R \Rightarrow (1, 3) \in R$.

Example. *Give an example of a relation which is anti-symmetric but neither reflexive nor symmetric nor transitive.*

Solution. Let $R = \{(x, y) \in \mathbf{R} \times \mathbf{R} : x, y$ are integers and $x + 3y = 12\}$.

Clearly, R is a relation on R.

Since $(1, 1) \notin \mathbf{R}$, R is not reflexive.

$(6, 2) \in R$ but $(2, 6) \notin R \Rightarrow$ R is not symmetric.

Now suppose $(x, y), (y, x) \in R$.

Then $x + 3y = 12 = y + 3x \Rightarrow 2x = 2y \Rightarrow x = y$

Hence R is anti-symmetric.

and R is not transitive, because $(-6, 6), (6, 2) \in R$ but $(-6, 2) \notin R$.

Example. *Let N be the set of all natural numbers and let R be the relation on N defined by $x\,Ry$ iff x divides y (i.e.,) $x \mid y$ (i.e.,) $\exists z \in N \in y = xz$.*

Then R is reflexive, transitive, anti-symmetric, not symmetric relation.

Solution.

(*i*) For any $x \in N$, $x \mid x \Rightarrow$ R is reflexive on N

(*ii*) For any $x, y, z \in N$, $x \mid y, y \mid z \exists p, q \in N \ni y = px, z = qy$

$\therefore \qquad z = qy = q\,(px) = (pq)\,x \Rightarrow x \mid z$

Then R is transitive on N.

(*iii*) For any $x, y \in \mathbf{N}$, $x \mid y$, $y \mid x \;\exists\; p, q, \in \text{N} \ni y = px$, $x = qy$

$\Rightarrow \qquad y = px = p\,(qy) = (pq)\,y \Rightarrow pq = 1 \Rightarrow p = 1,\ q = 1$

$\therefore \qquad x = y \Rightarrow$ R is anti-symmetric on N.

(*iv*) R is not symmetric on N because $\frac{2}{4}$ but 4 + 2 for 2, 4 ∈ N.

Example. *Let A be the set of integers and n is fixed positive integer. Define the relation R_n on A by $a\,R_n\,b$ iff $n \mid a - b$ (i.e.,) n divides $a - b$. In other words $a\,R_n\,b$ iff $a - b = kn$ where k is some integer. The most common notation is to write $a \equiv b \bmod n$ instead of $a\,R_n\,b$ and we say that a is congruent to b module n. Then prove that R_n is an equivalence relation on A.*

Solution. (*i*) Let $a \in \text{A}$, $a - a = 0 = 0\,n \Rightarrow a \equiv a \pmod{n} \Rightarrow a\,\text{R}_n\,a$

$\therefore$ R_n is reflexive on A.

(*ii*) Let $a, b \in \text{A}$, $a\,\text{R}_n\,b \Rightarrow a \equiv b \pmod{n} \Rightarrow a - b = kn \Rightarrow b - a = (-k)\,n$

$\Rightarrow \qquad b \equiv a \pmod{n} \Rightarrow b\,\text{R}_n\,a \Rightarrow \text{R}_n$ is symmetric on A.

(*iii*) Let $a, b, c \in \text{A}$, $a\,\text{R}_n\,b$, $b\,\text{R}_n\,c \Rightarrow a \equiv b \pmod{n}$, $b \equiv c \pmod{n}$

$\Rightarrow \qquad a - b = kn,\ b - c = mn$

$\Rightarrow \qquad a - b + b - c = (k + m)\,n \Rightarrow a - c = (k + m)\,n$

$\Rightarrow \qquad a \equiv c \pmod{n}$

$\therefore$ R_n is transitive on A.

Hence R_n is an equivalence relation on A.

Definition. *A relation R on a set A is said to be (i) irreflexive if $(a, a) \notin \mathbf{R}$ (i.e.,) a Ra $\forall\, a \in A$.*

(*ii*) *Anti-symmetric on A if a Rb, b Ra $\Rightarrow a = b \;\forall\, a, b \in A$.*

(*iii*) *Assymmetric on A if a Rb $\Rightarrow$ b* **R** *a.*

(*i.e.,*) $(a, b) \in \text{R} \Rightarrow (b, a) \notin \text{R} \;\forall\; a, b \in \text{A}$.

Example

1. The relation '<' is irreflexive on **R**.
2. The relation "divides" is assymmetric on **R** and antisymmetric on **R**.
3. Let A = {1, 2, 3, 4} and relation R = {(1, 3), (2, 3), (2, 4), (3, 1), (4, 2)}, then the relation **R** is irreflexive, assymmetric, not transitive and not anti-symmetric on A.
4. Let A = {1, 2, 3, 4} and relation S = {(1, 1), (2, 2), (3, 3), (4, 4), (2, 3), (3, 1), (2, 1)} is reflexive and transitive but not symmetric on A.
5. Let A = {1, 2, 3, 4} and relation T = {(1, 1), (2, 2), (4, 4), (2, 4), (4, 2)} is symmetric, transitive but not reflexive on A.
6. Let A = {1, 2, 3, 4}, relation {(2, 2), (2, 3)} is transitive but neither reflexive nor symmetric on A.
7. Let A = {1, 2, 3, 4}, relation {(1, 2), (2, 4)} is not reflexive, not symmetric, and not transitive on A.

3.11 OPERATIONS ON RELATIONS

Definition. *Let R, S are relations from A to B.*

Then,

(*i*) $R \cup S = \{(a, b) \in A \times B \mid (a, b) \in R \text{ (or) } (a, b) \in S\}$ is the union of the relations R, S.

(*i.e.,*) $(a, b) \in R \cup S \Leftrightarrow (a, b) \in R \text{ (or) } (a, b) \in S$.

(*ii*) $R \cap S = \{(a, b) \in A \times B \mid (a, b) \in R \text{ and } (a, b) \in S\}$ is the intersection of the relations R, S.

(*i.e.,*) $(a, b) \in R \cap S \Leftrightarrow (a, b) \in R, (a, b) \in S$.

(*iii*) $R - S = \{(a, b) \in A \times B \mid (a, b) \in R, (a, b) \notin S\}$ is the difference of the relations R, S.

(*i.e.,*) $(a, b) \in R - S \Leftrightarrow (a, b) \in R, (a, b) \notin S$

(*iv*) $R^c = R' = \{(a, b) \in A \times B \mid (a, b) \notin R\}$ is the complement of the relation R.

(*i.e.,*) $(a, b) \in R^c \Leftrightarrow (a, b) \notin R$.

Definition. *Let R be a relation from A to B and S be a relation from B to C. Then* ***composite relation*** *of R and S is denoted by ROS (or) RS and it is defined as*

$$ROS = \{(a, c) \mid a \in A, c \in C \ \exists\ b \in B \ni (a, b) \in R, (b, c) \in S\}.$$

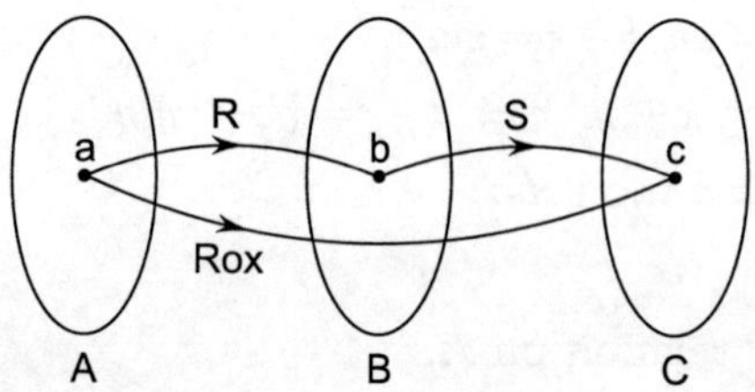

Figure. 3.9. *Composition of Relations*

Note 1. $R \subseteq A \times B$

2. $S \subseteq B \times C$

3. $ROS \subseteq A \times C$

4. $SOR \neq ROS$

5. So R may not have meaning when A, B, C are different sets.

6. If R is a relation on A, then ROR is again a relation on A. This composition relation is denoted by R^2.

7. $(ROR)\ OR = R^3$, and soon $R^{n-1}\ OR = R^n$.

Example. *Let R and S be two relations on*

$$A = \{1, 2\} \text{ and } R = \{(1, 1), (2, 2\}, S = \{(1, 2), (2, 1)\}$$

Then $\quad ROS = \{(1, 2), (2, 1)\}$

Theorem. *Let R, S are equivalence relations on A. Then prove that $R \cap S$ is an equivalence relation on A.*

(or) *Let R, S are relations on A. Then ⊕*

(*i*) *$R \cap S$ is reflexive if R, S are reflexive*

(*ii*) *$R \cap S$ is symmetric if R, S are symmetric.*

(*iii*) $R \cap S$ is transitive if R, S are transitive.

Proof : $(x, y) \in R \cap S$ for any $x, y \in A$

$\Leftrightarrow \quad (x, y) \in R$ and $(x, y) \in S$ for any $x, y \in A$.

(*i*) Since R, S are reflexive on A, $(x, x) \in R, (x, x) \in S \ \forall \ x \in A$

$\Rightarrow \quad (x, x) \in R \cap S$

$\therefore \quad R \cap S$ is reflexive on A.

(*ii*) Let $(x, y) \in R \cap S \Rightarrow (x, y) \in R$ and $(x, y) \in S$.

$\Rightarrow \quad (y, x) \in R$ and $(y, x) \in S \qquad (\because$ R, S are symmetric on A)

$\Rightarrow \quad (y, x) \in R \cap S \ \forall \ x, y \in A$

$\therefore \quad R \cap S$ is symmetric on A.

(*iii*) Let $(x, y), (y, z) \in R \cap S \Rightarrow (x, y), (y, z) \in R$ and $(x, y), (y, z) \in S$

$\Rightarrow \quad (x, z)$ R, $(x, z) \in S \ \forall \ x, y, z \in A \qquad (\because$ R, S are transitive on A)

$\Rightarrow \quad (x, z) \in R \cap S$

$\Rightarrow \quad R \cap S$ is transitive on A.

$\therefore \quad R \cap S$ is an equivalence relation on A.

Theorem. *Let R, S are relations on A. Then,*

(*i*) *$R \cup S$ is reflexive if R, S are reflexive.*

(*ii*) *$R \cup S$ is symmetric if R, S are symmetric.*

Proof : (*i*) Since R, S are reflexive,

$(x, x) \in R, (x, x) \in S \ \forall \ x \in A$

$\therefore \quad (x, x) \in R \cup S \ \forall \ x \ni A.$

$\therefore \quad R \cup S$ is reflexive on A.

(*ii*) Let $(x, y) \in R \cup S \ \forall \ x, y \in A$.

$\Rightarrow \quad (x, y) \notin R$ (or) $(x, y) \in S$.

$\Rightarrow \quad (y, x) \in R$ (or) $(y, x) \in S \qquad (\because$ R, S are symmetric on A)

$\Rightarrow (y, x) \in R \cup S \ \forall \ x, y \in A$.

$\therefore \quad R \cup S$ is symmetric on A.

Theorem. *A relation R on A is reflexive iff $I_A \subset R$.*

Proof : Suppose R is reflexive on A and $(a, a) \in I_A$ for any $a \in A$. Since R is reflexive on A, $(a, a) \in R$.

$\Rightarrow \quad I_A \subset R.$

Conversely suppose $\quad I_A \subset R.$

For any $\quad a \in A, (a, a) \in I_A$

$\Rightarrow \quad (a, a) \in R \ \Rightarrow$ R is reflexive on A.

Theorem. *A relation R on a set A is symmetric, iff $R^{-1} = R$.*

Proof. Let relation R is symmetric on A.

Then $(x, y) \in R \Leftrightarrow (y, x) \in R$

$\Leftrightarrow$ $(x, y) \in R^{-1}$

$\therefore$ $R^{-1} = R$

Conversely suppose $R^{-1} = R$.

For any $x, y \in A$, $(x, y) \in R \Rightarrow (x, y) \in R^{-1}$

$\Rightarrow$ $(y, x) \in (R^{-1})^{-1} = R$

$\Rightarrow$ $(y, x) \in R$.

$\therefore$ R is symmetric on A.

Theorem. *A relation R on a set A is transitive iff ROR $\subset$ R (i.e.,) $R^2 \subset R$.*

Proof. Let relation R is transitive on A and let $(a, c) \in ROR$

$\exists\, b \in A \ni (a, b) \in R, (b, c) \in R$.

Since R is transitive on

A, $(a, c) \in R$

$\therefore$ $ROR \subset R$

Conversely suppose $ROR \subset R$.

Let $(a, c) \in ROR$, let us suppose $(a, b) \in R, (b, c) \in R$

$\forall\, a, b, c \in A$.

$\Rightarrow$ $(a, c) \in ROR$ for some $b \in A$.

$\Rightarrow$ $(a, c) \in R$ since $ROR \subset R$.

$\therefore$ R is transitive on A.

Theorem. *The relation R on a set A is transitive iff $R^n \subseteq R$ for n = 1, 2, 3, ...*

Proof. Suppose $R^n \subseteq R$ for n = 1, 2, 3, ...

In particular $R^2 \subseteq R$.

by above theorem, R is transitive relation.

Conversely suppose R is transitive.

Next we shall show that $R^n \subseteq R$ by induction. When $n = 1$, the result is true.

Suppose $R^n \subseteq R$,

Now we prove $R^{n+1} \subseteq R$.

Let $(a, b) \in R^{n+1}$, since $R^{n+1} = R^n OR$, there is an element x with $x \in A$ such that $(a, x) \in R$ and $(x, b) \in R^n$.

Since $R^n \subseteq R$, $(x, b) \in R$.

Since R is transitive, $(a, x) \in R, (x, b) \in R \Rightarrow (a, b) \in R$

$\therefore$ $R^{n+1} \subseteq R$.

Hence the result, by induction on n.

Theorem. *If a relation R on a set A is both reflexive and transitive then ROR = R.*

Proof : Since R is reflaxive, $I_A \subset R$

Since R is transitive, $ROR \subset R$

Let $\quad a \in A, (a, a) \in I_A \Rightarrow (a, a) \in R$

$\Rightarrow \quad (a, a) \in ROR$

$\therefore \quad R \subset ROR$

Hence $\quad ROR = R.$

Theorem. *Let R be a relation on A, then prove that*

(*i*) $(R^{-1})^{-1} = R$ (*ii*) $ROI_A = I_A OR = R$

Proof : (*i*) Let R be a relation on A.

$\Rightarrow \quad R = \{x, y) \mid x, y \in A\}$

$\Rightarrow \quad R^{-1} = \{(y, x) \mid x, y \in A\}$

$\Rightarrow \quad (R^{-1})^{-1} = \{(x, y) \mid x, y \in A\} = R$

$\therefore \quad (R^{-1})^{-1} = R.$

(*ii*) Let $R = \{(x, y) \mid x, y \in A\}$ and

take $\quad I_A = \{(y, y) \mid y \in A\}$

$\Rightarrow \quad I_A OR = \{(x, y) \mid x, y \in A\} = R \quad$ ...(1)

and take $\quad I_A = \{(x, x) \mid x \in A\}$

$\rightarrow \quad ROI_A - \{(x, y) \mid x, y \subset A\} = R \quad$ ...(2)

From (1), (2), $\quad I_A OR = R = ROI_A$

Theorem. Let R be a relation from A to B and S be a relation from B to C, then prove that $(ROS)^{-1} = S^{-1}OR^{-1}$.

Proof : Since R is relation from A to B, $R \subseteq A \times B$.

Since S is relation from B to C, $S \subseteq B \times C$

$\Rightarrow \quad ROS \subseteq A \times C, SOR \subseteq C \times A.$

and $\quad R^{-1} \subseteq B \times A, S^{-1} \subseteq C \times B.$

$\Rightarrow \quad S^{-1}OR^{-1} \subseteq C \times A, (ROS)^{-1} \subseteq C \times A.$

$\therefore \quad R = \{(a, b) \mid a \in A, b \in B\},$

$\quad S = \{(b, c) \mid b \in B, c \in C\}.$

$\Rightarrow \quad ROS = \{(a, c) \mid a \in A, c \in C\}$

$\Rightarrow \quad (ROS)^{-1} = \{(c, a) \mid c \in C, a \in A\} \quad$ ...(1)

and $\quad S^{-1} = \{(c, b) \mid b \in B, c \in C\}$

$\quad R^{-1} = \{(b, a) \mid a \in A, b \in B\}$

$\Rightarrow \quad S^{-1}OR^{-1} = \{(c, a) \mid a \in A, c \in C\} \quad$ ...(2)

From (1), (2), $(ROS)^{-1} = S^{-1}OR^{-1}$.

Theorem. *If R is an equivalence relation on A, then R^{-1} is also an equivalence relation on A.*

Proof : $R = \{(x, y) \mid x, y \in A\} \Leftrightarrow R^{-1} = \{(y, x) \mid x, y \in A\} = \{(y, x) \mid (x, y) \in R\}$

(*i*) Let R is reflexive on A.

Let $x \in A \Rightarrow xRx \Rightarrow (x, x) \in R \Rightarrow (x, x) \in R^{-1} \Rightarrow R^{-1}$ is reflexive on A.

(*ii*) Let R is symmetric on A.

For any $x, y \in A, (x, y) \in R \Rightarrow (y, x) \in R$

If $(x, y) \in R^{-1} \Rightarrow (y, x) \in R \Rightarrow (x, y) \in R$ since R is symmetric

$\Rightarrow (y, x) \in R^{-1}$

$\Rightarrow R^{-1}$ is symmetric on A.

(*iii*) Let R is transitive on A.

For any $x, y, z \in A, (x, y) \in R^{-1}, (y, z) \in R^{-1}$

$\Rightarrow (y, x) \in R, (z, y) \in R$

$\Rightarrow (z, x) \in R$ since R is transitive on A

$\Rightarrow (x, z) \in R^{-1}$.

$\therefore$ R^{-1} is transitive on A.

Hence R^{-1} is an equivalence relation on A.

Example. Let R, S be two relations on **Z** so that

$$R = \{(x, 2x) \mid x \in \mathbf{Z}\}, S = \{(x, 7x) \mid x \in \mathbf{Z}\}$$

Then (*i*) $ROS = \{(x, 14x) \mid x \in \mathbf{Z}\} = SOR$

(*ii*) $ROR = \{(x, 4x) \mid x \in \mathbf{Z}\}$.

(*iii*) $ROROR = \{(x, 8x) \mid x \in \mathbf{Z}\}$.

(*iv*) $ROSOR = \{(x, 28x) \mid x \in \mathbf{Z}\}$.

Example. *Let R, S be two relations on A = {1, 2, 3} and R = {(1, 1), (1, 2), (2, 3), (3, 1), (3, 3)}, S = {(1, 2), (1, 3), (2, 1), (3, 3)}, then find the following :*

(*i*) $R \cup S$ (*ii*) $R \cap S$. (*iii*) $R - S$

(*iv*) R^c (*v*) ROS (*vi*) S^2 *or* SOS

(*vii*) $R^2 = ROR$ (*viii*) SOR (*ix*) $S - R$.

Solution.

(*i*) $R \cup S = \{(1, 1), (1, 2), (1, 3), (2, 1), (2, 3), (3, 1), (3, 3)\}$

(*ii*) $R \cap S = \{(1, 2), (3, 3)\}$

(*iii*) $R - S = \{(1, 1), (2, 3), (3, 1)\}$

(*iv*) $R^c = \{(1, 3), (2, 1), (2, 2), (3, 2)\}$.

(*v*) $ROS = \{(1, 2), (1, 3), (1, 1), (2, 3), (3, 2), (3, 3)\}$

(*vi*) $S^2 = SOS = \{(1, 1), (1, 3), (2, 2), (2, 3), (3, 3)\}$

(*vii*) $R^2 = ROR = \{(1, 1), (1, 2), (1, 3), (2, 1), (2, 3), (3, 1), (3, 2), (3, 3)\}$

(*viii*) $SOR = \{(1, 1), (1, 3), (2, 1), (2, 2) (3, 1), (3, 3)\}$

(*ix*) $S - R = \{(1, 3), (2, 1)\}$.

Example. *Let S, T are relations on* $\mathbf{Z}$ *which are defined by* $S = \{(1, n) \mid n \in \mathbf{Z}\}$, $T = \{(n, 1) \mid n \in Z\}$, *what are ST, TS relations ?*

Solution.
$$ST = SOT = \{(1, 1)\}$$
$$TS = TOS = \{(n, n) \mid n \in \mathbf{Z}\} = \mathbf{Z} \times \mathbf{Z}.$$

Example. *Let R be a relation on the set N of positive integers defined by the equation* $x^2 + 2y = 100$. *Find the domain of R, range of R and* R^{-1}. (JNTU April 04 - Set 3)

Solution. We are given $x^2 + 2y = 100$, and also R is a relation on the set **N** of positive integers.

Therefore, from the above conditions we can find the relation R = {(2, 48), (4, 42), (6, 32), (8, 18)}

Hence domain of R is {2, 4, 6, 8} and Range of R is {18, 32, 42, 48}

We can find R^{-1} from R as

$$R^{-1} = \{(18, 8), (32, 6), (42, 4), (48, 2)\}$$

Range of $R^{-1} = \{8, 6, 4, 2\} = \{2, 4, 6, 8\}$.

Example. *Let* $A = \{1, 2, 3, 4, 5\}$, *define relation R on A by aR b iff* $a + 1 = b$. *We define* $R^2 = ROR$, $R^3 = R^2R$. *Then compute R,* R^2, R^3.

Solution.
$$R = \{(1, 2), (2, 3), (3, 4), (4, 5)\}$$
$$R^2 = \{(1, 3), (2, 4), (3, 5)\}$$
$$R^3 = \{(1, 4), (2, 5)\}.$$

Example. *Let* $A = \{1, 2, 3\}$, $B = \{1, 2, 3, 4\}$, $C = \{1, 2, 3, 4, 5\}$ *be sets and relation* $R = \{(1, 2), (3, 4), (2, 2)\}$ *from A to B.*

Relation $S = \{(1, 3), (2, 5), (3, 1), (4, 2)\}$ from B to C.

Then compute (*i*) ROS ; (*ii*) SOR ; (*iii*) RO (SOR) ; (*iv*) (ROS) OR ; (*v*) ROR ; (*vi*) SOS ; and (*vii*) ROROR.

Solution.

(*i*) $ROS = \{(1, 5), (2, 5), (3, 2)\}$

(*ii*) $SOR = \{(1, 4), (3, 2), (4, 2)\}$

(*iii*) $(ROS)\ OR = \{(3, 2)\}$

(*iv*) $RO\ (SOR) = \{(3, 2)\}$.

(*v*) $ROR = \{(1, 2), (2, 2)\}$

(*vi*) $SOS = \{(1, 1), (3, 3), (4, 5)\}$

(*vii*) $ROROR = \{(1, 2), (2, 2)\}$.

Definition. *Let R be an equivalence relation on a set A, $x \in A$. The equivalence class of a is given by $[a]_R = \{x \in A \mid (a, x) \in R\}$. Here a is called generator of equivalence class $[a]_R$.*

Note. $[a]_R \neq \phi$ since $a \in [a]$.

Definition. ***Quotient Set :*** *The collection of all equivalence classes of elements of A determined by the* ***equivalence relation*** *R on A is denoted by $\frac{A}{R}$ and is defined as $\frac{A}{R} = \{[a] \mid a \in A\}$.*

Definition. *Let S be a given set and $A = \{A_1, A_2, ..., A_m\}$ where each A_i $(i = 1, 2, ..., m)$ is a subset of S and $\bigcup_{i=1}^{m} A_i = S$, then the set A is called a covering of S and the sets $A_1, A_2, ..., A_m$ are said to cover S.*

If, in addition the elements of A which are subsets of S are mutually disjoint, then A is called a ***partition*** *of S and the sets $A_1, A_2, ..., A_m$ are called the* ***blocks*** *of the partition.*

(or)

Partition Set : Let A be a non-empty set and $A_1, A_2, ..., A_n$ are the subsets of A. A set denoted by π is called a partition set of A if

(*i*) $A_i \cap A_j = \phi,\ i \neq j$ (*ii*) $\bigcup_{i=1}^{n} A_i = A$.

Example. Let **Z** be the set of all integers. Consider a relation $R = \{(a, b) \mid a - b$ is divisible by $3\}$. The relation **R** is an equivalence relation on **Z**. This relation partitions the set **Z** into following three mutually disjoint equivalence classes. The equivalence classes determined by 0, 1, 2, are the sets consisting of all integers related to 0, 1, 2 respectively.

$$\therefore \quad [0] = \{a \mid a \in Z,\ a R 0\} = \{a \mid a - 0 = 3m,\ m \in \mathbf{Z}\}$$

$$= \{......, -6, -3, 0, 3, 6, ...\}$$

Similarly,

$$[1] = \{......, -5, -2, 1, 4, 7, ...\}$$

$$[2] = \{......, -4, -1, 2, 5, 8, ...\}$$

$$[3] = \{......, -6, -3, 0, 3, 6, 9, ...\}$$

$$[4] = \{......, -5, -2, 1, 4, 7, 10, ...\}$$

$$[5] = \{......, -4, -1, 2, 5, 8, 11, ...\} \text{ and so on.}$$

$$\therefore \quad [0] = [3] = ...$$

$$[1] = [4] = ...$$

$$[2] = [5] = ...$$

$$\therefore \quad \frac{\mathbf{Z}}{R} = \{[0], [1], [2]\}.$$

Example. *Show that the relation of congruence module m has m distinct equivalence classes.*

(JNTU March 2006 - Set 1, June 04 - Set 2).

Solution. From previous example, we have

$$\frac{\mathbf{Z}}{R} = \{[0], [1], [2]\}$$

In a similar manner, we can find the equivalence classes generated by a relation "Congruence modulo *m*" for any integer *m*. The equivalence classes of the relation congruence modulo *m* are called the Congruence classes modulo *m*. The congruence class of an integer '*a* modulo *m*' is denoted $[a]_m$, so that

$$[a]_m = \{......, a - 2m, a - m, a + m, a + 2m, ...\}$$

There are *m* different Congruence classes modulo *m* denoted by $[0]_m$, $[1]_m$, ..., $[m - 1]_m$. They form a partition of the set of integers.

Example 1. *Let $A = \{a, b, c\}$, consider $A_1 = \{a\}$, $A_2 = \{b\}$, $A_3 = \{c\}$, $A_4 = \{a, b\}$, $A_5 = \{b, c\}$, $A_6 = \{a, c\}$, partitions of A are*

$$\pi_1 = \{A_1, A_2, A_3\}$$
$$\pi_2 = (A_1, A_5\}$$
$$\pi_3 = \{A_2, A_6\}$$
$$\pi_4 = \{A_3, A_4\}$$

Examples. Let A = {1, 2, 3, 4, 5, 6, 7, 8, 9, 10}

Consider the subsets of A are $A_1 = \{1, 2, 3, 4, 5\}$, $A_2 = \{6, 7, 8\}$, $A_3 = \{9, 10\}$, $A_4 = \{1, 2, 3, 4\}$ and $A_5 = \{3\}$.

∴ $\pi_1 = \{A_1, A_2, A_3\}$ is a partition because $A_1 \cap A_2 = \phi$, $A_1 \cap A_3 = \phi$, $A_2 \cap A_3 = \phi$, $A_1 \cup A_2 \cup A_3 = A$.

But $\pi_2 = \{A_1, A_4, A_5\}$ is not a partition because $A_1 \cap A_4 \neq \phi$ and $\pi_3 = \{A_2, A_4, A_5\}$ is not a partition because $A_4 \cap A_5 \neq \phi$, $A_2 \cup A_4 \cap A_5 \neq A$.

Examples. *Let $\mathbf{Z}^+$ be the set of all positive integers, A_1 be the set of all positive odd integers, A_2 be the set of all positive even integers,*

Then $\pi = \{A_1, A_2\}$ is a partition of $\mathbf{Z}^+$.

Theorem. Let P be a partition of the set. Define a relation R on A as $x \, R y \Leftrightarrow x$ and y are members of the same block then R is an equivalence relation on A.

Proof : (*i*) If $x \in A$ then *x* and *x* are in the same block,

$$\Leftrightarrow x \, Rx \Rightarrow R \text{ is reflexive.}$$

(*ii*) If $x \, Ry \Rightarrow x$ and *y* are in the same block

⇒ *y* and *x* are in the same block ⇒ $y \, Rx$.

⇒ R is symmetric.

(*iii*) If $x \, Ry$, $y \, Rz \Rightarrow x, y$ and y, z are in the same block

⇒ *x* and *z* are in the same block

⇒ $x \, Rz \Rightarrow$ R is transitive.

Hence R is an equivalence relation on A.

Theorem. Let R be an equivalence relation defined on A then R induces a partition on A.

Proof : First we shall show that any two equivalence classes are identical (or) disjoint.

Let R be an equivalence relation on set A.

Let $a, b \in A$ so that $(a, b) \in R$

Let $x \in [a]_R \Leftrightarrow (x, a) \in R$

$\Leftrightarrow$ $(x, b) \in R$ since $(x, a) \in R$, $(a, b) \in R \Rightarrow (x, b) \in R$

$\Leftrightarrow$ $x \in [b]_R$

$\therefore$ $[a]_R = [b]_R$

any if $[a]_R \cap [b]_R = \phi$, then nothing to prove.

Suppose $[a]_R \cap [b]_R \neq \phi$, then $x \in [a]_R \cap [b]_R$

$\Rightarrow$ $x \in [a]_R$ and $x \in [b]_R$

$\Rightarrow$ $(x, a) \in R$ and $(x, b) \in R$

$\Rightarrow$ $[x]_R = [a]_R$ and $[x]_R = [b]_R$

$\Rightarrow$ $[a]_R = [b]_R$

Hence any two equivalence classes are identical (or) disjoint.

Thus R induces a partition P of A procedure for construction of $\frac{A}{R}$. See below of this page.

Example. *Let* $A = \{a, b, c, d\}$ *and let* $R = \{(a, a), (a, b), (b, a), (b, b), (c, d), (d, c), (c, c), (d, d)\}$ *be an equivalence relation on R. Find* $\frac{A}{R}$.

Solution. Here $[a] = \{a, b\} = [b]$, $[c] = \{c, d\} = [d]$

The partition of A is $\{(a, b), (c, d)\}$

Hence $\frac{A}{R} = \{[a], [c]\}$

Example. *Let* $A = \{1, 2, 3, 4\}$ *and* $P = \{\{1, 2, 3\}, \{4\}\}$ *be a partition of A. Find the equivalence relation determined by P.*

Solution. Since each element in the block is related to every other element in the same block and only to these elements,

$$R = \{(1, 1), (1, 2), (1, 3), (2, 1), (2, 2), (2, 3), (3, 1), (3, 2), (3, 3), (4, 4)\}.$$

Example. *Let R be the equivalence relation on the set* $A = \{1, 2, 3, 4, 5, 6\}$ *where* $R = \{(1, 1), (1, 5), (2, 2), (2, 3), (2, 6), (3, 2), (3, 3), (3, 6), (4, 4), (5, 1), (5, 5), (6, 2), (6, 3), (6, 6)\}$. *Find the partition of A induced by R (i.e.,) equivalence classes of R.*

Solution. The elements related to 1 are 1 and 5 $\Rightarrow [1] = \{1, 5\}$

Take 2 which is not in [1]

Here 2 is related to 2, 3, 6 $\Rightarrow [2] = \{2, 3, 6\}$

Take 4 which is not belongs to [1] or [2].

Here 4 is related to 4 itself $\Rightarrow [4] = \{4\}$.

Hence partition of A induced by

$$R\ ;\ \frac{A}{R} = \{[1], [2], [4]\} = \{\{1, 5\}, \{2, 3, 6\}, \{4\}\}.$$

Procedure :

1. Choose $a \in$ A and find $[a]$.

2. If $[a] \neq$ A, choose $b \in$ A and $b \notin [a]$ and find $[b]$.

3. If $[a] \cup [b] \neq$ A, choose C $\in$ A and C $\notin [a] \cup [b]$ and find $[c]$.

4. Repeat steps until all elements of are included in the computed equivalence classes.

Example. *If $P = \{\{1, 3, \}, \{2, 4, \}\}$ is a partition set of $A = \{1, 2, 3, 4, 5\}$. Determine the corresponding equivalence relation.*

Solution.

$$\begin{aligned} R &= \{\{1, 3\} \times \{1, 3\}, \{2, 4, 5\} \times \{2, 4, 5\}\} \\ &= \{(1, 1), (1, 3), (3, 1), (3, 3), (2, 2), (2, 4), (2, 5), (4, 2), (4, 4), (4, 5), (5, 2), (5, 4), (5, 5)\}. \end{aligned}$$

Which is equivalence relation corresponding to the given partition set P.

Example. *Let $A = \{x, y, z, w\}$ and $R = \{(x, x), (x, y), (x, z), (y, x), (y, y), (y, z), (z, x), (z, y), (z, z), (w, w)\}$*

Then show that R is an equivalence relation and find partition of A induced by R.

Solution. Since $(x, x), (y, y), (z, z), (w, w) \in$ R, R is reflexive on A.

Let $(x, y) \in R \Rightarrow (y, x) \in R.$

$(y, z) \in R \Rightarrow (z, y) \in R.$

$(x, z) \in R \Rightarrow (z, x) \in R$

$\Rightarrow$ R is symmetric on A.

and let $(x, y), (y, z) \in R \Rightarrow (x, z) \in R$

$(x, z), (z, x) \in R \Rightarrow (x, x) \in R$

$\Rightarrow$ R is transitive on A.

Hence R is an equivalence relation on A and partition of A induced by R is $\frac{R}{A} = \{\{x, y, z\}, w\}$.

Example. *Let $A = \{1, 2, 3, 4, 5\}$, $P = A \times A$ and R be a relation so that $(x, y)\ R\ (x', y') \Leftrightarrow xy' = x'y$. Show that R is an equivalence relation and find $\frac{A}{R}$.*

Proof : We know that R is an equivalence relation (Refer page 104)

and

$$\begin{aligned} \frac{A}{R} = \{&(1, 1), (2, 2), (3, 3), (4, 4), (5, 5)\}, \{(1, 2), (2, 4)\}, \{(2, 1), (4, 2)\}, \\ &\{(1, 3)\}, \{(1, 4)\}, \{(2, 3)\}, \{(2, 5)\}, \{(3, 1)\}, \{(3, 2)\}, \{(3, 4)\}, \{(1, 5)\}, \\ &\{(3, 5)\}, \{(4, 1)\}, \{(4, 3)\}, \{(4, 5)\}, \{(5, 1)\}, \{(5, 2)\}, \{(5, 3)\}, \{(5, 4)\}\}. \end{aligned}$$

Example. *If A is a set with m elements and B is a set with n elements. Find the number of relations from A to B.*

Solution. Since a relation from A to B is precisely a subset of A × B, the set of all relations from A to B is precisely the set of all subsets of A × B. Therefore, the number of relations from A to B is equal to the number of subsets of A × B.

Since n (A) = m and n (B) = n, we have n (A × B) = mn. Therefore, A × B has 2^{mn} number of subsets. This implies that there are 2^{mn} relations from A to B.

Example 1. *Let A, B are sets with n (A) = 2 and n (B) = 3, then the number of relations from A to B are* $2^6 = 64$.

Example 2. *Let A and B be two sets such that | B | = 3. If there are 32768 relations from A to B, then find n (A).*

Solution. Let n (A) = m, n (B) = n, there are 2^{mn} relations from A to B. We have $2^{3m} = 32768 = 8 \times 64 \times 64 = 2^{15}$

$$\Rightarrow \qquad m = 5$$

(or)
$$3m \log_e 2 = \log_e 32768 \quad \Rightarrow \quad m = \frac{\log_e 32768}{3 \log_e 2} = 5$$

3.12 REPRESENTATION OF RELATIONS

In this section we will discuss two alternative methods for representing relations. These are (*i*) matrix method ; (*ii*) directed graph method.

(*i*) **Matrix Method :**

Definition. *A binary relation R from a set A with n elements to a set B with m elements is represented by n × m matrix, called* ***Relation matrix***, *denoted by* $M_R = [a_{ij}]$ *where*

$$a_{ij} = \begin{cases} 1 & \text{if } i^{\text{th}} \text{ element of A is related to } j^{\text{th}} \text{ element of B } (i.e.,) \left(a_i, a_j\right) \in \text{R} \\ 0 & \text{otherwise } (i.e.,) \left(a_i, a_j\right) \notin \text{R} \end{cases}$$

Example. *Let A = {1, 2, 3}, B = {a, b, c, d}, the relation R from A to B is given by R = {(1, a), (1, d), (2, b), (2, c), (3, a), (3, c)}, then the relation matrix*

$$M_R = \begin{bmatrix} 1 & 0 & 0 & 1 \\ 0 & 1 & 1 & 0 \\ 1 & 0 & 1 & 0 \end{bmatrix}$$

Example. *Let A = {1, 2, 3, 4, 5, 6}, Define Relation R as < on A then find Relation matrix on A.*

Solution. Relation R = {(1, 2), (1, 3), (1, 4), (1, 5), (1, 6), (2, 3), (2, 4), (2, 5), (2, 6), (3, 4), (3, 5), (3, 6), (4, 5), (4, 6), (5, 6)}

$$\Rightarrow \qquad M_R = \begin{bmatrix} 0 & 1 & 1 & 1 & 1 & 1 \\ 0 & 0 & 1 & 1 & 1 & 1 \\ 0 & 0 & 0 & 1 & 1 & 1 \\ 0 & 0 & 0 & 0 & 1 & 1 \\ 0 & 0 & 0 & 0 & 0 & 1 \\ 0 & 0 & 0 & 0 & 0 & 0 \end{bmatrix}$$

Example. *Let* $A = \{a_1, a_2\}$, $B = \{b_1, b_2, b_3, b_4, b_5\}$ *and relation matrix* $M_R = \begin{bmatrix} 0 & 1 & 0 & 1 & 0 \\ 1 & 0 & 1 & 1 & 0 \end{bmatrix}$, *then write the relation R.*

Solution. $R = \{(a_1, b_2), (a_1, b_4), (a_2, b_1), (a_2, b_3), (a_2, b_4)\}$

A relation matrix reflects some of the properties of a relation. Further the matrix of a relation on a set, which is square matrix, can be used to determine whether the relation has certain properties :

(*i*) R is reflexive iff all the elements in the main diagonal of M_R are equal to 1.

(*ii*) R is symmetric if $a_{ij} = a_{ji}\ \forall\ i, j$ (*i.e.,*) $M_R = (M_R)^T$.

(*iii*) R is anti-symmetric of $a_{ij} = 1$ with $i \neq j$, then $a_{ji} = 0$.

In other words either $a_{ij} = 0$ (or) $a_{ji} = 0$, when $i \neq j$.

Example. *Suppose the relation R on a set is represented by the matrix,*

$$M_R = \begin{bmatrix} 1 & 0 & 0 \\ 0 & 1 & 1 \\ 0 & 1 & 1 \end{bmatrix}, \text{ is R reflexive, symmetric, anti-symmetric ?}$$

Solution.

(*i*) Since all the diagonal elements of this matrix are equal to 1, R is reflexive.

(*ii*) Since $M_R = (M_R)^T$, R is symmetric.

(*iii*) Clearly R is not anti-symmetric.

Second Method

2. **Digraph of a Relation (or) Graph of a Relation);** A relation can also be represented pictorially by drawing its **digraph**. Although we shall introduce some of the concepts of graph theory which are discussed in chapter 7. Here we shall use graphs only as a tool to represent relations. The procedure for drawing digraph of a given relation.

   Let R be a relation in a set

$$X = \{x_1, x_2, ..., x_m\}.$$

The elements of X are represented by points (or) circles called nodes (or) vertices.

An arrow is drawn from the vertex x_i to vertex x_j iff x_i R x_j. This is called an (directed) edge.

This pictorial representation of R is called a **directed graph** (or) **digraph** of R.

Note that an element of the form (x, x) in a relation corresponds to a directed edge from x to x, such an edge is called **loop**.

Example. *Draw the graph for the following relations :*

(i) $R = \{(1, 1), (2, 2), (1, 2)\}$ *on* $X = \{1, 2\}$

(ii) $S = \{(1, 1), (1, 3), (2, 1), (2, 3), (2, 4), (3, 1), (3, 2), (4, 1)\}$ *on the set* $X = \{1, 2, 3, 4\}$

(iii) $R = \{(1, 1), (1, 2), (1, 3), (2, 2), (2, 3), (3, 3)\}$ *on* $Y = \{1, 2, 3\}$.

Solution.

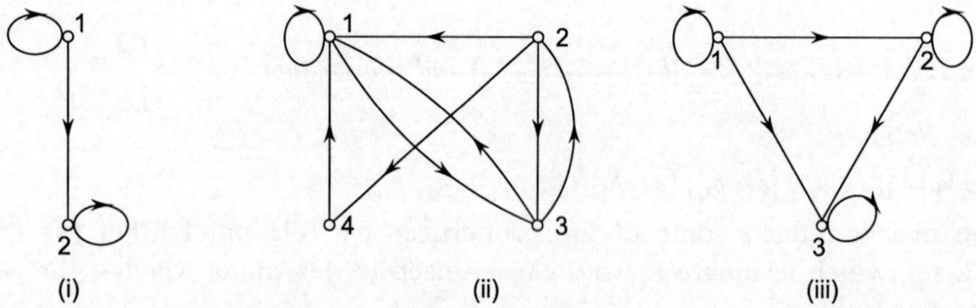

Figure 3.10.

Example. *Let* $X = \{1, 2, 3, 4\}$ *and relation* $R = \{(x, y) \mid x > y\}$. *Draw the graph of R and also give its matrix.*

Solution. Relation R on X = {1, 2, 3, 4} is R = {(2, 1), (3, 1), (3, 2), (4, 1), (4, 2), (4, 3)} and graph, matrix are below (or) Fig. 3.11.

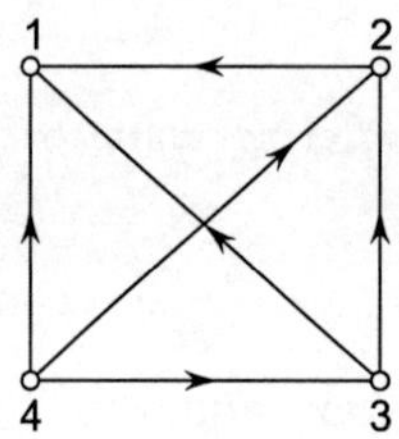

$$M_R = \begin{bmatrix} 0 & 0 & 0 & 0 \\ 1 & 0 & 0 & 0 \\ 1 & 1 & 0 & 0 \\ 1 & 1 & 1 & 0 \end{bmatrix}$$

Figure 3.11.

Example. *Find whether the relations for the directed graphs shown in the following figures are reflexive, symmetric, anti-symmetric and/or transitive and also write the corresponding relation matrices.*

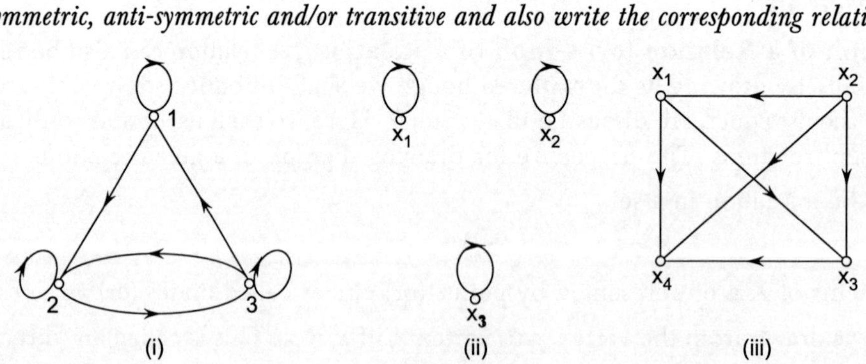

Figure 3.12.

Solution. (*i*) Since there are loops at each vertex of the relation graph, R is reflexive.

R is neither symmetric nor anti-symmetric since there is an edge between 1 and 2 but not from 2 to 1, but there are edges connecting 2 and 3 in both directions. And R is not transitive since there is an edge from 1 to 2 and 2 to 3, but no edge from 1 to 3, and

$$\text{Matrix } M_R = \begin{bmatrix} 1 & 1 & 0 \\ 0 & 1 & 1 \\ 1 & 1 & 1 \end{bmatrix}.$$

(*ii*) R is reflexive since there are loops at each vertex of the graph and matrix

$$M_R = \begin{bmatrix} 1 & 0 & 0 \\ 0 & 1 & 0 \\ 0 & 0 & 1 \end{bmatrix}.$$

(*iii*) R is not reflexive since graph has no loops.

R is not symmetric since there is an edge between x_1, x_4, but not from x_4 to x_1.

and R is transitive since there are edges from x_1 to x_3 and x_3 to x_4, then x_1 to x_4, and this property is true for all x_1, x_2, x_3, x_4 and matrix

$$M_R = \begin{bmatrix} 0 & 0 & 1 & 1 \\ 1 & 0 & 1 & 1 \\ 0 & 0 & 0 & 1 \\ 0 & 0 & 0 & 0 \end{bmatrix}.$$

Example. *Determine the properties of the relations given by the graphs shown below and also write the corresponding relation matrices.*

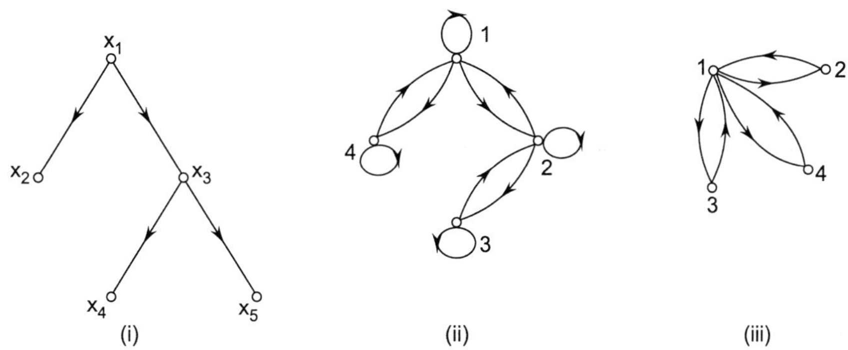

Figure 3.13.

Solution.

(*i*) Clearly relation is anti-symmetric but not reflexive, not symmetric not transitive.

(*ii*) Relation is reflexive, symmetric but not transitive.

(*i*) Matrix, $M_R = \begin{bmatrix} 0 & 1 & 1 & 0 & 0 \\ 0 & 0 & 0 & 0 & 0 \\ 0 & 0 & 0 & 1 & 1 \\ 0 & 0 & 0 & 0 & 0 \\ 0 & 0 & 0 & 0 & 0 \end{bmatrix}$ (*ii*) $M_R = \begin{bmatrix} 1 & 1 & 0 & 1 \\ 1 & 1 & 1 & 0 \\ 0 & 1 & 1 & 0 \\ 1 & 0 & 0 & 1 \end{bmatrix}.$

(*iii*) Relation is not reflexive because loops are not present in graph. Further it is symmetric, not anti-symmetric and not transitive.

$$\text{Matrix } M_R = \begin{bmatrix} 0 & 1 & 1 & 1 \\ 1 & 0 & 0 & 0 \\ 1 & 0 & 0 & 0 \\ 1 & 0 & 0 & 0 \end{bmatrix}.$$

Composition of Two Relations :

Let R be a binary relation from the set A to the set B and S be a binary relation from the set B to C, then the ordered pair (R, S) is said to be composable. If (R, S) is a composable pair of binary relations, the composite ROS is a binary relation from A to C such that for $a \in A$, $c \in C$, a (ROS) C is for some $b \in B$, both a R b and b S c.

Example. Let A = {1, 2, 3, 4}, B = {1, 3, 9, 10}, C = {5, 6, 7, 8}, R = {(1, 1), (1, 3), (2, 9), (2, 10), (3, 3), (4, 10)}, S = {(1, 5), (3, 7), (9, 7), (10, 8}. Find ROS its relation graph and relation matrix.

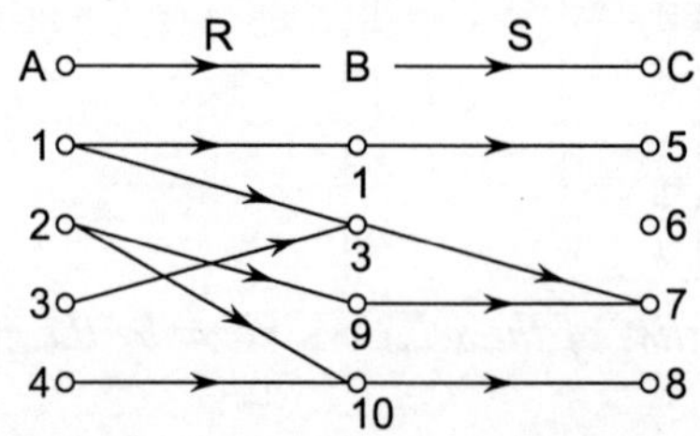

∴ ROS = {(1, 5), (1, 7), (2, 7), (2, 8), (3, 7), (4, 8)}

ROS Relation graph is

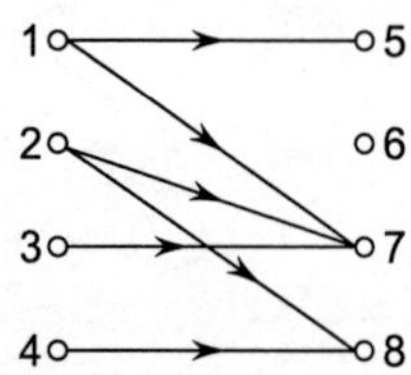

$$\text{and corresponding matrix is } M_{ROS} = \begin{bmatrix} 1 & 0 & 1 & 0 \\ 0 & 0 & 1 & 1 \\ 0 & 0 & 1 & 0 \\ 0 & 0 & 0 & 1 \end{bmatrix}.$$

We obtain the matrices of converse relation of R and composition of two relations as follows :

1. The relation matrix $M_{\tilde{R}}$ of $\tilde{R}$ can be obtained by simply interchanging the rows and columns of M_R. Such a matrix is called the transpose of M_R. Therefore, $M_{\tilde{R}} = (M_R)^T$. The graph of $\tilde{R}$ is also obtained from R by simply reversing the arrows on each arc.

Example. *Given the relation matrix* $M_R = \begin{bmatrix} 1 & 1 & 1 \\ 0 & 1 & 1 \\ 1 & 0 & 1 \end{bmatrix}$. *Then find* $M_{\tilde{R}}$.

Solution. $M_{\tilde{R}} = \text{Transpose of } M_R = \begin{bmatrix} 1 & 0 & 1 \\ 1 & 1 & 0 \\ 1 & 1 & 1 \end{bmatrix}$.

2. Boolean Arithmetic with Zero-One Matrices :

This arithmetic is based on the boolean operations $\vee$ and $\wedge$ which operate on pairs of bits defined by

$$a_1 \wedge a_2 = \begin{cases} 1 & \text{if } a_1 = a_2 = 1 \\ 0 & \text{otherwise} \end{cases}$$

and
$$a_1 \vee a_2 = \begin{cases} 1 & \text{if } a_1 = 1 \text{ or } a_2 = 1 \\ 0 & \text{otherwise} \end{cases}.$$

Definition. *Let $A = [a_{ij}]$ and $B = [b_{ij}]$ be $m \times n$ (zero-one) matrices. Then **the join of A and B** is the (zero-one) matrix with $(i, j)^{th}$ entry $a_{ij} \vee b_{ij}$. The join of A and B is denoted by $A \vee B$. The **meet of A and B** is the (zero-one) matrix with $(i, j)^{th}$ entry $a_{ij} \wedge b_{ij}$. The meet of A and B is denoted by $A \wedge B$.*

Example. *If* $A = \begin{bmatrix} 1 & 0 & 0 \\ 0 & 1 & 1 \\ 1 & 0 & 1 \end{bmatrix}$, $B = \begin{bmatrix} 0 & 1 & 0 \\ 1 & 0 & 0 \\ 0 & 1 & 1 \end{bmatrix}$ *are (zero-one) matrices.*

Then find $A \vee B$, $A \wedge B$.

Solution.
$$A \vee B = \begin{bmatrix} 1 \vee 0 & 0 \vee 1 & 0 \vee 0 \\ 0 \vee 1 & 1 \vee 0 & 1 \vee 0 \\ 1 \vee 0 & 0 \vee 1 & 1 \vee 1 \end{bmatrix} = \begin{bmatrix} 1 & 1 & 0 \\ 1 & 1 & 1 \\ 1 & 1 & 1 \end{bmatrix}$$

and
$$A \wedge B = \begin{bmatrix} 1 \wedge 0 & 0 \wedge 1 & 0 \wedge 0 \\ 0 \wedge 1 & 1 \wedge 0 & 1 \wedge 0 \\ 1 \wedge 0 & 0 \wedge 1 & 1 \wedge 1 \end{bmatrix} = \begin{bmatrix} 0 & 0 & 0 \\ 0 & 0 & 0 \\ 0 & 0 & 1 \end{bmatrix}.$$

Definition. *Let $A = [a_{ij}]$ be an $m \times r$ (zero-one) matrix and $B = [b_{ij}]$ be an $r \times n$ (zero-one) matrix. Then the Boolean product of A and B, denoted by $A \cdot B$, is the $m \times n$ matrix with $(i, j)^{th}$ entry $[c_{ij}]$ where $c_{ij} = (a_{i1} \wedge b_{1j}) \vee (a_{i2} \wedge b_{2j}) \vee \ldots \vee (a_{ir} \wedge b_{rj})$*

$$= \bigcup_{k=1}^{r} \left(a_{ik} \wedge b_{kj}\right), \ i = 1, 2, \ldots, m$$

$$j = 1, 2, \ldots, n$$

Example. *Find the Boolean product of A and B, where*

$$A = \begin{bmatrix} 1 & 0 \\ 0 & 1 \\ 1 & 0 \end{bmatrix} \quad and \quad B = \begin{bmatrix} 1 & 1 & 0 \\ 0 & 1 & 1 \end{bmatrix}.$$

Solution.

$$A \odot B = \begin{bmatrix} (1\wedge 1)\vee(0\wedge 0) & (1\wedge 1)\vee(0\wedge 1) & (1\wedge 0)\vee(0\wedge 1) \\ (0\wedge 1)\vee(1\wedge 0) & (0\wedge 1)\vee(1\wedge 1) & (0\wedge 0)\vee(1\wedge 1) \\ (1\wedge 1)\vee(0\wedge 0) & (1\wedge 1)\vee(0\wedge 1) & (1\wedge 0)\vee(0\wedge 1) \end{bmatrix}$$

$$= \begin{bmatrix} 1\vee 0 & 1\vee 0 & 0\vee 0 \\ 0\vee 0 & 0\vee 1 & 0\vee 1 \\ 1\vee 0 & 1\vee 0 & 0\vee 0 \end{bmatrix} = \begin{bmatrix} 1 & 1 & 0 \\ 0 & 1 & 1 \\ 1 & 1 & 0 \end{bmatrix}.$$

The Boolean operations join and meet can be used to find the matrices representing the union and the intersection of two relations. Suppose that R and S are relations on a set A represented by the matrices M_R and M_S respectively. The matrix representing the union of these relations has a 1 in the positions where either M_R or M_S has a 1. The matrix representing the intersection of these relations has a 1 in the positions where both M_R and M_S have a 1. Thus, the matrices representing the union and intersection of these relations are $M_{R\cup S} = M_R \vee M_S$ and $M_{R\cap S} = M_R \wedge M_S$.

Example. *Let R and S are relations on a set A represented by the matrices,* $M_R = \begin{bmatrix} 1 & 1 & 0 \\ 0 & 1 & 1 \\ 0 & 1 & 0 \end{bmatrix}$ *and* $M_S = \begin{bmatrix} 1 & 0 & 1 \\ 1 & 1 & 0 \\ 1 & 0 & 0 \end{bmatrix}$, *then find the matrices representing relations $R \cup S$ and $R \cap S$.*

Solution.

$$M_{R\cup S} = M_R \vee M_S = \begin{bmatrix} 1\vee 1 & 1\vee 0 & 0\vee 1 \\ 0\vee 1 & 1\vee 1 & 1\vee 0 \\ 0\vee 1 & 1\vee 0 & 0\vee 0 \end{bmatrix} = \begin{bmatrix} 1 & 1 & 1 \\ 1 & 1 & 1 \\ 1 & 1 & 0 \end{bmatrix}$$

and

$$M_{R\cap S} = M_R \wedge M_S = \begin{bmatrix} 1\wedge 1 & 1\wedge 0 & 0\wedge 1 \\ 0\wedge 1 & 1\wedge 1 & 1\wedge 0 \\ 0\wedge 1 & 1\wedge 0 & 0\wedge 0 \end{bmatrix} = \begin{bmatrix} 1 & 0 & 0 \\ 0 & 1 & 0 \\ 0 & 0 & 0 \end{bmatrix}.$$

We now find the matrix for the composite of relations. This matrix can be found using the Boolean product of the matrices for these relations.

Suppose R is a relation from A to B and S is a relation from B to C. Suppose that A, B and C have *m*, and *p*, and *n* elements respectively. Let the zero-one matrices for SOR, R and S be $M_{SOR} = [t_{ij}]$, $M_R = [r_{ij}]$ and $M_S = [s_{ij}]$ respectively (where $m \times p$, $m \times n$ and $n \times p$ are sizes of matrices M_{SOR}, M_R, M_S respectively). The ordered pair (a_i, c_j) belongs to SOR if and only if there is an element b_k such that (a_i, b_k) belongs to R and (b_k, c_j) belongs to S. It follows that $t_{ij} = 1$ if and only if $r_{ik} = s_{kj} = 1$ for some k.

From the definition of the Boolean product, $M_{SOR} = M_R \odot M_S$.

Example. *Find the matrix representing the relation SOR where the matrices representing R and S are*

$$M_R = \begin{bmatrix} 1 & 0 & 1 \\ 0 & 1 & 1 \\ 1 & 0 & 0 \end{bmatrix} \text{ and } M_S = \begin{bmatrix} 0 & 0 & 1 \\ 0 & 1 & 1 \\ 1 & 0 & 1 \end{bmatrix}$$

Solution. $M_{SOR} = M_R \odot M_S = \begin{bmatrix} 1 & 0 & 1 \\ 0 & 1 & 1 \\ 1 & 0 & 0 \end{bmatrix} \odot \begin{bmatrix} 0 & 0 & 1 \\ 0 & 1 & 1 \\ 1 & 0 & 1 \end{bmatrix} = \begin{bmatrix} 1 & 0 & 1 \\ 1 & 1 & 1 \\ 0 & 0 & 1 \end{bmatrix}.$

3. We now consider the converse of a composite relation. For this purpose, let R be a relation from A to B and S be a relation from B to C. Obviously $\tilde{R}$ is a relation from B to A, $\tilde{S}$ from C to B, SOR is a relation from A to C, and $S \,\tilde{o}\, R$ is a relation from C to A. Also the relation $\tilde{R} \circ \tilde{S}$ is from C to A.

Example (Theorem). Show that $S \,\tilde{o}\, R = \tilde{R} \circ \tilde{S}$, where R is a relation from A to B and S is a relation from B to C.

Proof : Given R is a relation from A to B, and S is a relation from B to C.

Let $a \in A$, $c \in C$ and let $(c, a) \in S \,\tilde{o}\, R \Leftrightarrow (a, c) \in S \text{ O } R$

$$\Leftrightarrow \exists\, b \in B \ni (a, b) \in R, (b, c) \in S.$$

$$\Leftrightarrow (c, b) \in \tilde{S}, (b, a) \in \tilde{R}.$$

$$\Leftrightarrow (c, a) \in \tilde{R} \circ \tilde{S}$$

$$\therefore \qquad S \,\tilde{o}\, R = \tilde{R} \circ \tilde{S}.$$

The same rule can be expressed in terms of the relation matrices by saying that the transpose of $M_{S \circ R}$ is the same as the matrix $M_{\tilde{R} \circ \tilde{S}}$. The matrix $M_{\tilde{R} \circ \tilde{S}}$ can be obtained from the matrices $M_{\tilde{R}}$ and $M_{\tilde{S}}$, which in turn can be obtained from the matrices M_R and M_S.

Example. *If* $M_R = \begin{bmatrix} 1 & 0 & 1 \\ 1 & 1 & 0 \\ 1 & 1 & 1 \end{bmatrix}$, $M_S = \begin{bmatrix} 1 & 0 & 0 & 1 \\ 1 & 0 & 1 & 0 \\ 0 & 1 & 0 & 1 \end{bmatrix}$, *then find* $M_{S \circ R}$, $M_{\tilde{R}}$, $M_{\tilde{S}}$, $M_{S \tilde{o} R}$ *and verify that* $M_{S \tilde{o} R} = M_{\tilde{R} \circ \tilde{S}}$.

Solution. $M_{\tilde{R}} = \begin{bmatrix} 1 & 1 & 1 \\ 0 & 1 & 1 \\ 1 & 0 & 1 \end{bmatrix}$, $M_{\tilde{S}} = \begin{bmatrix} 1 & 1 & 0 \\ 0 & 0 & 1 \\ 0 & 1 & 0 \\ 1 & 0 & 1 \end{bmatrix}.$

$$M_{S\circ R} = M_R \odot M_S = \begin{bmatrix} 1 & 0 & 1 \\ 1 & 1 & 0 \\ 1 & 1 & 1 \end{bmatrix} \odot \begin{bmatrix} 1 & 0 & 0 & 1 \\ 1 & 0 & 1 & 0 \\ 0 & 1 & 0 & 1 \end{bmatrix}$$

$$M_{S\circ R} = \begin{bmatrix} (1\wedge 1)\vee(0\wedge 1)\vee(1\wedge 0) & (1\wedge 0)\vee(0\wedge 0)\vee(1\wedge 1) & (1\wedge 0)\vee(0\wedge 1)\vee(1\wedge 0) & (1\wedge 1)\vee(0\wedge 0)\vee(1\wedge 1) \\ (1\wedge 1)\vee(1\wedge 1)\vee(0\wedge 0) & (1\wedge 0)\vee(1\wedge 0)\vee(0\wedge 1) & (1\wedge 0)\vee(1\wedge 1)\vee(0\wedge 0) & (1\wedge 1)\vee(1\wedge 0)\vee(0\wedge 1) \\ (1\wedge 1)\vee(1\wedge 1)\vee(1\wedge 0) & (1\wedge 0)\vee(1\wedge 0)\vee(1\wedge 1) & (1\wedge 0)\vee(1\wedge 1)\vee(1\wedge 0) & (1\wedge 1)\wedge(1\wedge 0)\vee(1\wedge 1) \end{bmatrix}$$

$$\therefore \quad M_{S\circ R} = \begin{bmatrix} 1\vee 0\vee 0 & 0\vee 0\vee 1 & 0\vee 0\vee 0 & 1\vee 0\vee 1 \\ 1\vee 1\vee 0 & 0\vee 0\vee 0 & 0\vee 1\vee 0 & 1\vee 0\vee 0 \\ 1\vee 1\vee 0 & 0\vee 0\vee 1 & 0\vee 1\vee 0 & 1\vee 0\vee 1 \end{bmatrix}$$

$$= \begin{bmatrix} 1 & 1 & 0 & 1 \\ 1 & 0 & 1 & 1 \\ 1 & 1 & 1 & 1 \end{bmatrix}.$$

$$M_{S\tilde{\circ}R} = \text{Transpose of } M_{S\circ R} = \begin{bmatrix} 1 & 1 & 1 \\ 1 & 0 & 1 \\ 0 & 1 & 1 \\ 1 & 1 & 1 \end{bmatrix}.$$

and

$$M_{\tilde{R}\circ\tilde{S}} = M_{\tilde{S}} \odot M_{\tilde{R}} = \begin{bmatrix} 1 & 1 & 0 \\ 0 & 0 & 1 \\ 0 & 1 & 0 \\ 1 & 0 & 1 \end{bmatrix} \odot \begin{bmatrix} 1 & 1 & 1 \\ 0 & 1 & 1 \\ 1 & 0 & 1 \end{bmatrix}$$

$$= \begin{bmatrix} 1 & 1 & 1 \\ 1 & 0 & 1 \\ 0 & 1 & 1 \\ 1 & 1 & 1 \end{bmatrix} = M_{S\tilde{\circ}R}.$$

Result : The following hold for any relation R and S

(i) $(\tilde{\tilde{R}}) = R$ *(ii)* $R = S \Leftrightarrow \tilde{R} = \tilde{S}$

(iii) $R \subseteq S \Leftrightarrow \tilde{R} \subseteq \tilde{S}$ *(iv)* $R \tilde{\cap} S = \tilde{R} \cap \tilde{S}$

(v) $R \tilde{\cup} S = \tilde{R} \cup \tilde{S}$

Definition. *The transitive closure of a relation R is the smallest transitive relation containing R. We denote transitive closure of R by R^+. Let X be any finite set containing n elements and R be a relation in X. The relation $R^+ = R \cup R^2 \cup R^3 \cup \ldots \cup R^n$ in X is called the transitive closure of R in X.*

Transitive closures of relations have important applications in certain areas such as networks, syntactic analysis, fault detection and diagnosis in switching circuits, etc.

Example. *Let $X = \{1, 2, 3, 4\}$ and $R = \{(1, 2), (2, 3), (3, 4)\}$ be a relation on X. Find R^+.*

Solution. Given

$$R = \{(1, 2), (2, 3), (3, 4)\}$$
$$R^2 = \{(1, 3), (2, 4)\}$$
$$R^3 = \{(1, 4)\}$$
$$R^4 = \phi.$$
$$\therefore \quad R^+ = R \cup R^2 \cup R^3 \cup R^4 = \{(1, 2), (2, 3), (3, 4), (1, 3), (2, 4), (1, 4)\}$$

Theorem. The transitive closure R^+ of a relation R is a finite set X is transitive. Also for any other transitive relation P in X such that $R \subseteq P$, we have $R^+ \subseteq P$. In this sense, R^+ is the smallest transitive relation containing R.

Proof : To prove R^+ is transitive, let us assume that $a\, R^+\, b$ and $b\, R^+\, c$ for some $a, b, c \in X$.

Since $a\, R^+\, b$, we must have a sequence of elements $d_1, d_2, ..., d_k \in X$ such that $d_1 = a$, $d_k = b$ and $d_1\, R\, d_2$, $d_2\, R\, d_3$, ..., $d_{k-1}\, R\, d_k$.

Here we have assumed that $a\, R^k\, b$ for some k.

Similarly since $b\, R^+\, c$, we must have a sequence of elements, say $e_1, e_2, ..., e_j$ such that $e_i = b$, $e_j = c$ and $e_1\, R\, e_2$, $e_2\, R\, e_3$, ..., $e_{j-1}\, R\, e_j$. Here we have assumed that b $R^j\, c$ for some j. It follows from these assumptions that a $R^{k+j}\, c$.

$$= a\, R^+\, c \quad \Rightarrow \quad R^+ \text{ is transitive.}$$

Let us now assume that $a\, R^+\, b$ for some $a, b \in X$ so that there exists a sequence of elements $c_1, c_2, ..., c_m \in X$ such that $a = c$, $b = c_m$ and $c_i\, R\, c_{i+1}$ for $i = 1, 2, ..., m-1$.

If there is a transitive relation P in X such that $R \subseteq P$, then $c_i\, P\, c_{i+1}$ for $i = 1, 2, ..., m-1$, so that c, P c_m (*i.e.,*) $a\, P\, b$.

Since (a, b) is an arbitrary element of R^+, we see by the same argument that $R^+ \subseteq P$.

Hence R^+ is the smallest transitive relation which includes R.

Example. *Given the relation matrix M_R of a relation R on the set $\{a, b, c\}$ find the relation matrices of $\tilde{R}$, $R^2 = R \circ R$, R^3 $R \circ R \circ R$, and $R \circ \tilde{R}$, where* $M_R = \begin{bmatrix} 1 & 0 & 1 \\ 1 & 1 & 0 \\ 1 & 1 & 1 \end{bmatrix}$.

Solution. Given $M_R = \begin{bmatrix} 1 & 0 & 1 \\ 1 & 1 & 0 \\ 1 & 1 & 1 \end{bmatrix}$

Then (*i*) $M_{\tilde{R}} = \begin{bmatrix} 1 & 1 & 1 \\ 0 & 1 & 1 \\ 1 & 0 & 1 \end{bmatrix}$

(*ii*) $M_{R^2} = M_{R \circ R} = M_R \odot M_R = \begin{bmatrix} 1 & 0 & 1 \\ 1 & 1 & 0 \\ 1 & 1 & 1 \end{bmatrix} \odot \begin{bmatrix} 1 & 0 & 1 \\ 1 & 1 & 0 \\ 1 & 1 & 1 \end{bmatrix} = \begin{bmatrix} 1 & 1 & 1 \\ 1 & 1 & 1 \\ 1 & 1 & 1 \end{bmatrix}$

(*iii*) $M_{R^3} = M_{R \circ R \circ R} = M_{R^2 \circ R} = \begin{bmatrix} 1 & 1 & 1 \\ 1 & 1 & 1 \\ 1 & 1 & 1 \end{bmatrix} \odot \begin{bmatrix} 1 & 0 & 1 \\ 1 & 1 & 0 \\ 1 & 1 & 1 \end{bmatrix} = \begin{bmatrix} 1 & 1 & 1 \\ 1 & 1 & 1 \\ 1 & 1 & 1 \end{bmatrix}$

(*iv*) $M_{R \circ \tilde{R}} = M_R \odot M_{\tilde{R}} = \begin{bmatrix} 1 & 0 & 1 \\ 1 & 1 & 0 \\ 1 & 1 & 1 \end{bmatrix} \odot \begin{bmatrix} 1 & 1 & 1 \\ 0 & 1 & 1 \\ 1 & 0 & 1 \end{bmatrix} = \begin{bmatrix} 1 & 1 & 1 \\ 1 & 1 & 1 \\ 1 & 1 & 1 \end{bmatrix}$.

Definition. *Let M be a square (zero-one) matrix and let r be a positive integer. The rth Boolean power of M is the Boolean product of r factors of M. The rth Boolean product of M is denoted by $M^{(r)}$. Hence $M^{(r)} = M \odot M \odot M \odot \ldots \odot M$ (r times). This is well defined since the Boolean product of matrices is associative.*

Example. *Let* $M = \begin{bmatrix} 0 & 0 & 1 \\ 1 & 0 & 0 \\ 1 & 1 & 0 \end{bmatrix}$, *find $M^{(n)}$ for all positive integers n.*

Solution.
$$M^{(2)} = M \odot M = \begin{bmatrix} 0 & 0 & 1 \\ 1 & 0 & 0 \\ 1 & 1 & 0 \end{bmatrix} \odot \begin{bmatrix} 0 & 0 & 1 \\ 1 & 0 & 0 \\ 1 & 1 & 0 \end{bmatrix} = \begin{bmatrix} 1 & 1 & 0 \\ 0 & 0 & 1 \\ 1 & 0 & 1 \end{bmatrix}$$

$$M^{(3)} = M^{(2)} \odot M = \begin{bmatrix} 1 & 1 & 0 \\ 0 & 0 & 1 \\ 1 & 0 & 1 \end{bmatrix} \odot \begin{bmatrix} 0 & 0 & 1 \\ 1 & 0 & 0 \\ 1 & 1 & 0 \end{bmatrix} = \begin{bmatrix} 1 & 0 & 1 \\ 1 & 1 & 0 \\ 1 & 1 & 1 \end{bmatrix}$$

We find that $M^{(4)} = M^{(3)} \odot M = \begin{bmatrix} 1 & 1 & 1 \\ 1 & 0 & 1 \\ 1 & 1 & 1 \end{bmatrix}$ and $M^{(5)} = \begin{bmatrix} 1 & 1 & 1 \\ 1 & 1 & 1 \\ 1 & 1 & 1 \end{bmatrix}$.

We can see that $M^{(5)} = M^{(6)} = M^{(7)} = \ldots\ldots$

Definition. *Let M_R be the (zero-one) matrix of the relation R on a set X with n elements. Then the matrix of the transitive closure R^+.*

is
$$M_{R^+} = M_R \vee M_R^{(2)} \vee M_R^{(3)} \vee \ldots \vee M_R^{(n)} \text{ where } M_R^{(n)} = M_{R^n}.$$

Example. *Find the matrix of the transitive closure of the relation R, where* $M_R = \begin{bmatrix} 1 & 0 & 1 \\ 0 & 1 & 0 \\ 1 & 1 & 0 \end{bmatrix}$

Solution. Matrix of the transitive closure $M_{R^+} = M_R \vee M_R^{(2)} \vee R_R^{(3)}$.

Here we can see that $M_R^{(2)} = M_R \odot M_R = \begin{bmatrix} 1 & 1 & 1 \\ 0 & 1 & 0 \\ 1 & 1 & 1 \end{bmatrix}$, $M_R^{(3)} = M_R^{(2)} \odot M_R = \begin{bmatrix} 1 & 1 & 1 \\ 0 & 1 & 0 \\ 1 & 1 & 1 \end{bmatrix}$

Then
$$M_{R^+} = \begin{bmatrix} 1 & 0 & 1 \\ 0 & 1 & 0 \\ 1 & 1 & 0 \end{bmatrix} \vee \begin{bmatrix} 1 & 1 & 1 \\ 0 & 1 & 0 \\ 1 & 1 & 1 \end{bmatrix} \vee \begin{bmatrix} 1 & 1 & 1 \\ 0 & 1 & 0 \\ 1 & 1 & 1 \end{bmatrix} = \begin{bmatrix} 1 & 1 & 1 \\ 0 & 1 & 0 \\ 1 & 1 & 1 \end{bmatrix}.$$

3.13 WARSHALL'S ALGORITHM

Warshall's algorithm, is an efficient method for computing the (matrix of) transitive closure of a relation.

Algorithm : Given the relation matrix M_R of a simple digraph, then the following steps produce that matrix of transitive closure of the relation R.

Step 1. $P^{(0)} = M_R$

Step 2. $k = 1$

Step 3. $i = 1$

Step 4. $P_{ij}^{(k)} = P_{ij}^{(k-1)} \vee \left(P_{ik}^{(k-1)} \wedge P_{kj}^{(k-1)}\right) \ \forall\, j = 1$ to n.

Step 5. $i = i + 1$; if $i \leq n$, go to step 4.

Step 6. $k = k + 1$; if $k \leq n$, go to step 3 ; otherwise stop.

Example. *Let R be the relation with directed graph. Find the matrices M_R and M_{R^+} (transitive closure of R) by Warshall's algorithm.*

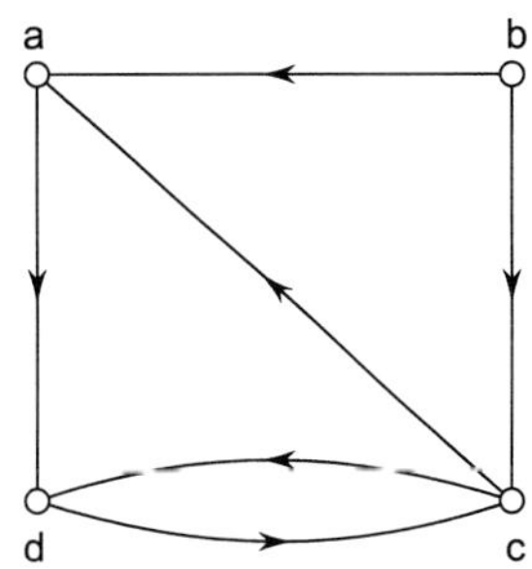

Figure 3.14.

Solution.

$$M_R = \begin{bmatrix} 0 & 0 & 0 & 1 \\ 1 & 0 & 1 & 0 \\ 1 & 0 & 0 & 1 \\ 0 & 0 & 1 & 0 \end{bmatrix} = P^{(0)} \text{ (say)}$$

$$P^{(1)} = \begin{bmatrix} 0 & 0 & 0 & 1 \\ 1 & 0 & 1 & 1 \\ 1 & 0 & 0 & 1 \\ 0 & 0 & 1 & 0 \end{bmatrix} = P^{(2)}.$$

$$P^{(3)} = \begin{bmatrix} 0 & 0 & 0 & 1 \\ 1 & 0 & 1 & 1 \\ 1 & 0 & 0 & 1 \\ 1 & 0 & 1 & 1 \end{bmatrix} \text{ and}$$

$$P^{(4)} = \begin{bmatrix} 1 & 0 & 1 & 1 \\ 1 & 0 & 1 & 1 \\ 1 & 0 & 1 & 1 \\ 1 & 0 & 1 & 1 \end{bmatrix} \text{ is the matrix of the transitive closure.}$$

Warshall's algorithm computes M_{R^+} by efficiently
Computing $P^{(0)} = M_R, P^{(1)}, P^{(2)}, \ldots\ldots, P^{(n)} = M_{R^+}$.

3.14 COMPATIBILITY RELATIONS

Definition. *A relation R in x is said to be a **compatibility** relation if it is reflexive and symmetric.*

Obviously all equivalence relations are compatibility relations, but compatibility relations are need not equivalence relations.

Example. Let X = {ball, bed, dog, hut, egg} and the relation R be given by $R = \{(x, y) \mid x, y \in X \wedge x\,R\,y$ if x and y contains some common letter$\}$.

Then R is the compatibility relation, and x, y are called compatible if $x\,R\,y$. A compatibility relation is sometimes denoted by ≈. Note ball ≈ bed, bed ≈ egg, but ball ≉ egg. Thus ≈ is not transitive.

Denoting ball by x_1, bed by x_2, dog by x_3, hut by x_4 and egg by x_5, the graph of ≈ is as follows :

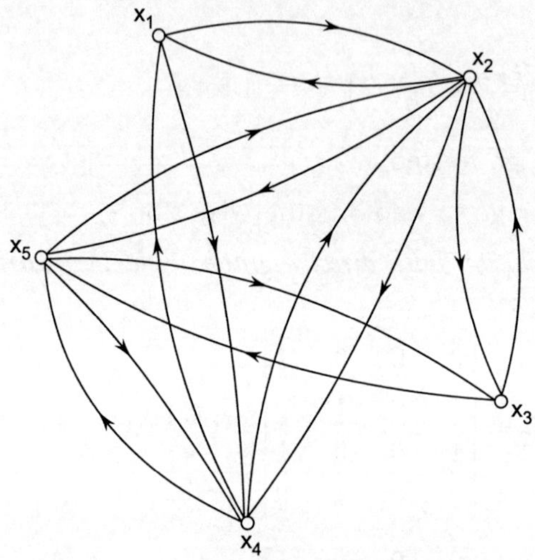

Figure 3.15.

Since ≈ is a compatibility relation, it is not necessary to draw the loops at each element nor it is necessary to draw both $x\,R\,y$ and $y\,R\,x$. Thus, we can simplify the graph of ≈ as shown below :

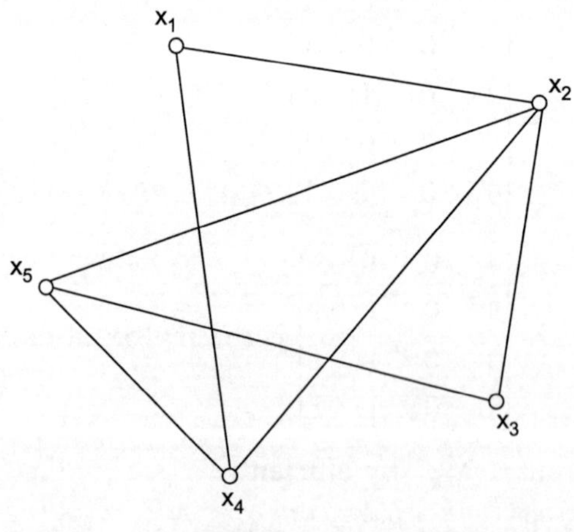

Figure 3.16.

Note that the elements in each of the sets $\{x_1, x_2, x_4\}$ and $\{x_2, x_3, x_5\}$ are related to each other (*i.e.,*) the elements are mutually compatible. Further, these two sets define a covering of X. The set $\{x_2, x_4, x_5\}$ also has elements compatible to each other.

The relation matrix of a compatibility relation is symmetric and has its diagonal elements unity. It is, therefore sufficient to give only the elements of the lower triangular part of the relation matrix in such a case.

The relation matrix for the above compatibility relation is

x_1	1			
x_2	0	1		
x_3	1	1	0	
x_4	0	1	1	1
	x_1	x_2	x_3	x_4

Definition. *Let X be a set and ≈ a compatibility relation on X. A subset $A \subseteq X$ is called a* ***maximal compatibility block*** *if any element of A is compatible to every other element of A and no element of X – A is compatible to all the elements of A.*

Example. The subset $\{x_1, x_2, x_4\}$, $\{x_2, x_3, x_5\}$ and $\{x_2, x_4, x_5\}$ are maximal compatibility blocks in the following figure. These sets are not mutually disjoint, and therefore they only define a covering of X.

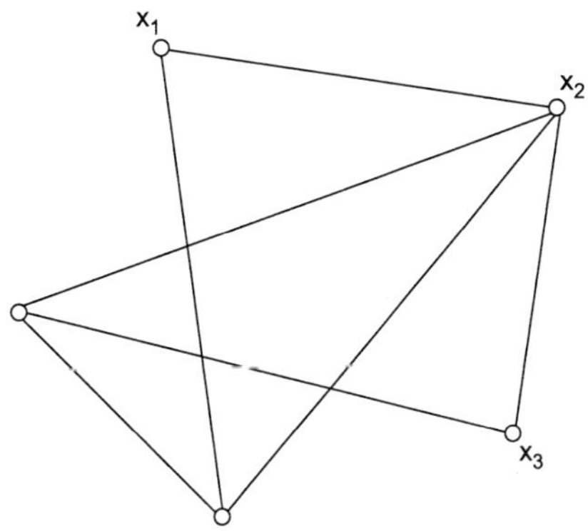

Figure 3.17.

Procedures to find the maximal compatibility blocks corresponding to a compatibility relation.

Method I.

(*a*) First draw a simplified graph of the compatibility relation.

(*b*) Pick from this graph the largest complete polygons. By a largest complete polygon, we mean a polygon in which any vertex is connected to every other vertex. For example, a triangle is always a complete polygon.

In addition to these examples, any element of the set which is related only to itself forms a maximal compatibility block. Similarly, any two elements which are compatible to one another but to no other elements also form a maximal compatibility block.

Method II.

Another procedure for finding the maximal compatibility blocks from the table of the relation matrix can be described in the following manner.

It is assumed that first a simplified table is obtained in which those elements which are only compatible to themselves are deleted, because, they are in a maximal compatibility block by themselves and are in no other compatibility block. Such blocks are included in the list at the end.

(*a*) Start in the rightmost column of the table and proceed to the left until a column containing atleast one non-zero entry is encountered. List all the compatible pairs represented by the entries in that column.

(*b*) Proceed left to the next column that contains atleast one non-zero entry. If any element is compatible to all the members of some previously defined compatibility class, then add this element to that class. If a member is compatible to only some members of a previously defined class, then form a new class which includes all the members that are compatible. Next, list all the compatible pairs not included in any previously defined class.

(*c*) Repeat step (*b*) until all the columns are considered.

The final sets of compatibility classes including those which are isolated elements constitute the maximal compatibility classes.

Example. The maximal compatibility blocks of the relation shown in the following Figure.

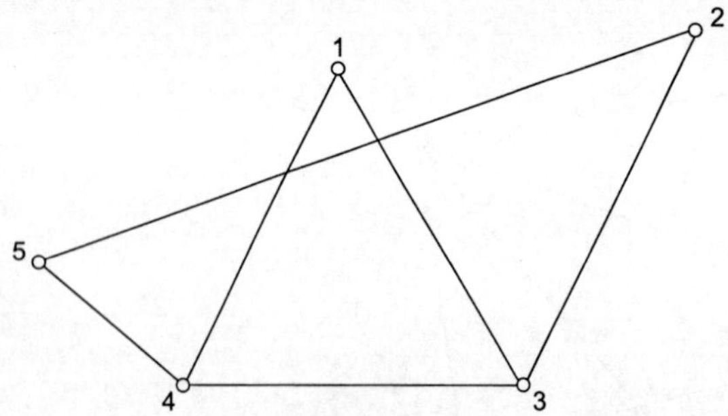

2	0			
3	1	1		
4	1	0	1	
5	0	1	0	1
	1	2	3	4

Figure 3.18.

These are {1, 3, 4}, {2, 3}, {4, 5}, {2, 5}, since 1, 3, 4 are compatible to each others (*i.e.,*) connected to each other, 2 and 3 connected each other, 4 and 5 connected each other, 2 and 5 connected each other, but not 5, 3 ; 2, 4.

Example. In graph shown in Fig. 3.19 and having relation matrix given below, maximal compatibility blocks are {1, 2, 3, 4}, {2, 5}, {3, 6}, {5, 6}.

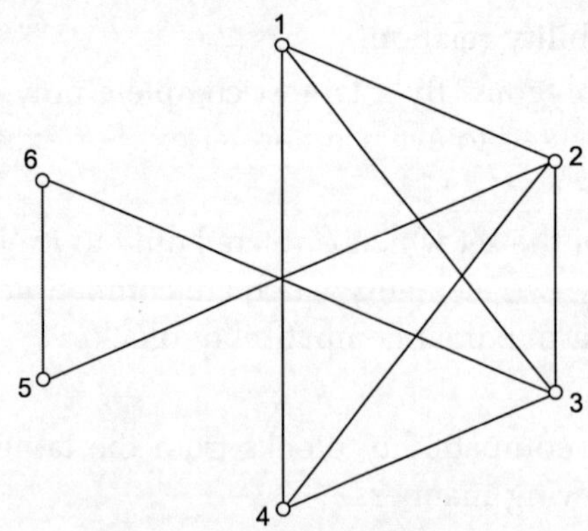

2	1				
3	1	1			
4	1	1	1		
5	0	1	0	0	
6	0	0	1	0	1
	1	2	3	4	5

Figure 3.19.

Example. Let the compatibility relation on a set $\{x_1, x_2, x_3, x_4, x_5, x_6\}$ be given by the matrix given below, then draw the graph and find the maximal compatibility blocks of the relation.

Solution.

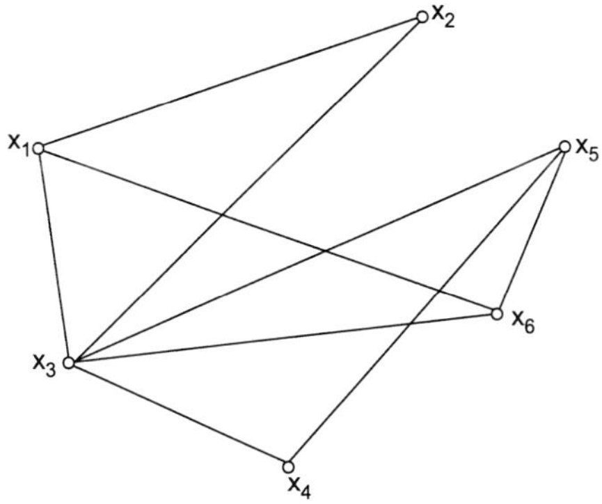

Figure 3.20.

x_2	1				
x_3	1	1			
x_4	0	0	1		
x_5	0	0	1	1	
x_6	1	0	1	0	1
	x_1	x_2	x_3	x_4	x_5

Maximal compatibility blocks are $\{x_1, x_2, x_3\}$, $\{x_1, x_3, x_6\}$, $\{x_3, x_5, x_6\}$, $\{x_3, x_4, x_5\}$.

Example. *Find the relation matrix for the compatibility relation having graph given below and hence find maximal compatibility blocks.*

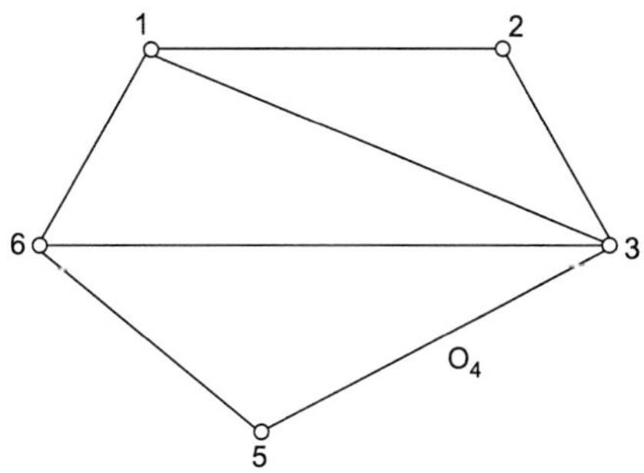

Figure 3.21.

Solution. Relation matrix is

2	0			
3	1	1		
5	0	0	1	
6	1	0	1	1
	1	2	3	5

And maximal compatibility blocks are :

{1, 2, 3}, {1, 3, 6}, {3, 5, 6}, {4}.

EXERCISE 3.2.

1. Find x and y in each of the following cases :

(*a*) $(2x, x + y) = (6, 1)$ (*b*) $(y - 2, 2x + 1) = (x - 1, y + 2)$.

2. If $A = \{x, y\}$, $B = \{1, 2, 3\}$, find $A \times B$, $B \times A$, $A \times A$ and $B \times B$.,

3. Let A = {1, 3, 5}, B = {2, 3}, C = {4, 6}, find the following :
   (*i*) $A \times B$ (*ii*) $B \times C$ (*iii*) $A \times C$
   (*iv*) $(A \cup B) \times C$ (*v*) $A \cap (B \times C)$ (*vi*) $(A \times B) \cup (B \times C)$
   (*vii*) $(A \times B) \cap (B \times C)$
4. If A = {1, 2, 3}, B = {2, 3, 4}, C = {3, 4, 5}, verify that $A \times (B \cup C) = (A \times B) \cup (A \times C)$ and $A \times (B \cap C) = (A \times B) \cap (A \times C)$
5. If A, B, C are any sets, then prove the following :
   (*i*) $A \times (B \cup C) = (A \times B) \cup (A \times C)$
   (*ii*) $A \times (B \cap C) = (A \times B) \cap (A \times C)$
   (*iii*) $A \times (B - C) = (A \times B) - (A \times C)$.
6. Let R be the relation on set A = {1, 2, 3} defined by R = {(1, 2), (2, 3), (1, 3)} then find domain R, range R.
7. Find the number of relations from A = {*a*, *b*, *c*} to B = {*x*, *y*}.
8. Let R and S be the following relations on A = {1, 2, 3}, R = {(1, 1), (1, 2), (2, 3), (3, 1), (3, 3)}, S = {(1, 2), (1, 3), (2, 1), (3, 3)}, then find the following (*i*) $R \cap S$; (*ii*) $R \cup S$; (*iii*) R^c ; (*iv*) R 0 S ; and (*v*) S 0 S.
9. Let A = {1, 2, 3, 4}, consider the following relation R on A, R = {(1, 1), (2, 2), (2, 3), (3, 2), (4, 2), (4, 4)}. Is R
   (*i*) reflexive ? (*ii*) symmetric ?
   (*iii*) transitive (*iv*) anti symmetric ?
10. Give examples of relations R on A = {1, 2, 3} having the stated property.
    (*i*) R is both symmetric and anti symmetric.
    (*ii*) R is neither symmetric nor anti symmetric.
    (*iii*) R is transitive but $R \cup R^{-1}$ is not transitive.
11. Let R, S, T be the relations on A = {1, 2, 3} defined by R = {(1, 1), (2, 2), (3, 3)}, S = {(1, 2), (2, 1), (3, 3)}, T = {(1, 2), (2, 3), (1, 3)}. Determine which of R, S, T are
    (*i*) reflexive ; (*ii*) symmetric ;
    (*iii*) anti symmetric ; (*iv*) transitive.
12. Let A be a set of integers and let ~ be the relation on $A \times A$ defined by $(a, b) \sim (c, d)$ iff $a + d = b + c$. Prove that ~ is an equivalence relation.
13. Let A be a set of non-zero integers and let ~ be the relation on $A \times A$ defined as follows :

    $(a, b) \sim (c, d)$ whenever $ad = bc$.

    Prove that ~ is an equivalence relation.
14. Let P (A) be the power set of any non empty set A, then prove that the relation $\subseteq$ of set inclusion is not an equivalence relation.
15. Let S = {1, 2, 3, 4, 5, 6}. Determine whether each of the following is a partition of S.
    (*i*) {1, 3, 5}, {2, 4}, {3, 6} (*ii*) {1, 5}, {2}, {3, 6}
    (*iii*) {1}, {3, 6}, {2, 4, 5}, {3, 6} (*iv*) {1}, {2}, {3}, {4}, {5}, {6}
16. Let A = {1, 2, 3, 4, 5, 6, 7, 8, 9, 10} and A_1 = {1, 2, 3, 4}, A_2 = {5, 6, 7}, A_3 = {4, 5, 7, 9}, A_4 = {4, 8, 10}, A_5 = {8, 9, 10}, A_6 = {1, 2, 3, 6, 8, 10} which of the following are partitions of A ?
    (*i*) $\{A_1, A_2, A_3\}$ (*ii*) $\{A_1, A_3, A_5\}$

(*iii*) $\{A_3, A_6\}$ (*iv*) $\{A_2, A_3, A_4\}$.

17. Let $A = \{1, 2, 3, 4, 5, 6\}$, consider the partition $P = \{A_1, A_2\}$ where $A_1 = \{1, 3, 5\}$ and $A_2 = \{2, 4, 6\}$, determine the corresponding equivalence relation R.

18. Let $A = \{1, 2, 3, 4, 5, 6, 7\}$, determine a relation R on A by a R b if 3 divides $a - b$. Show that R is an equivalence relation. Also find the partition generated by R.

19. Let $S = \{1, 2, 3, 4\}$ and $A = S \times S$. Define a relation R on A by (a, b) R (a', b') iff $a + b = a' + b'$.

(*i*) Show that R is an equivalence relation.

(*ii*) Compute $\frac{A}{R}$.

20. Let S be the relation on $A = \{1, 2, 3, 4, 5\}$ so that $S = \{(1, 2), (2, 2), (2, 4), (3, 3), (3, 5), (4, 1), (5, 2)\}$

(*a*) Find the following subsets of A :

(*i*) $E = \{x : x \text{ S } 2\}$, $F = \{x : x \text{ S } 3\}$, $G = \{x : 2 \text{ S } x\}$, $H = \{x : 3 \text{ S } x\}$.

(*b*) Find the matrix representation M_S of S.

(*c*) Draw the directed graph of S.

(*d*) Find the composition relation S o S.

21. Let $A = \{1, 2, 3, 4\}$, Relation R on A is

$$R = \{(1, 1), (2, 2), (2, 3), (3, 2), (4, 2), (4, 4)\}$$

Draw its directed graph.

22. Consider the relation $R = \{(1, 3), (1, 4), (3, 2), (3, 3), (3, 4)\}$ on $A = \{1, 2, 3, 4\}$.

(*a*) Find the matrix representation M_R of R.

(*b*) Find R^{-1}.

(*c*) Draw the directed graph of R.

(*d*) Find the composition relation R o R.

23. Find the relation R determined by each of the digraphs given below. Also write the matrix of relation.

(*i*)

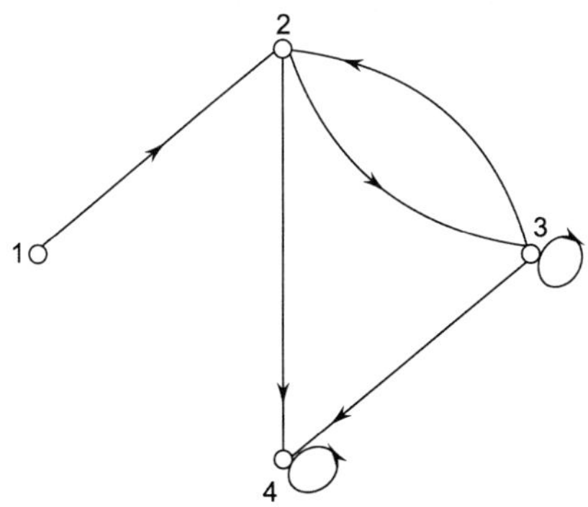

(*ii*)

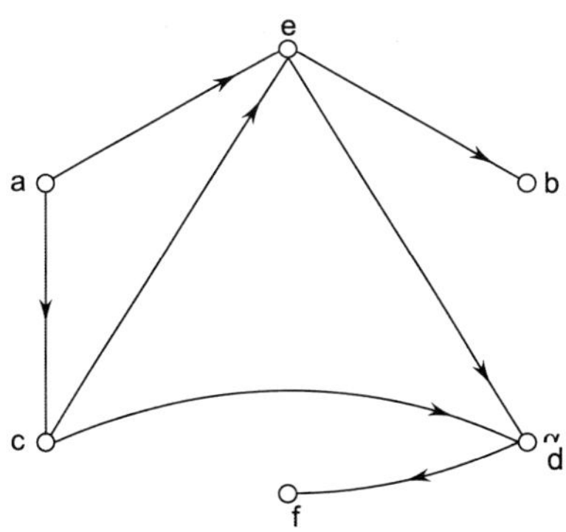

24. Let $A = \{2, 4, 5, 7\}$, and let R be the relation on A having the matrix,

$$M_R = \begin{bmatrix} 1 & 1 & 0 & 1 \\ 0 & 1 & 1 & 0 \\ 0 & 0 & 1 & 1 \\ 1 & 0 & 0 & 0 \end{bmatrix}.$$

Construct the digraph of R.

25. If $M_R = \begin{bmatrix} 1 & 1 & 1 & 0 \\ 0 & 0 & 0 & 1 \\ 0 & 1 & 0 & 0 \\ 0 & 0 & 0 & 0 \end{bmatrix}$ and $M_S = \begin{bmatrix} 0 & 0 & 1 & 1 \\ 0 & 0 & 1 & 0 \\ 1 & 0 & 0 & 0 \\ 1 & 0 & 0 & 0 \end{bmatrix}$ represent the relations R and S defined on A = {1, 2, 3, 4} find $M_R \cdot M_S$ and S o R.

ANSWERS

1. (*a*) $x = 3, y = -2$, (*b*) $x = 2, y = 2$.
2. A × B = {(*x*, 1), (*x*, 2), (*x*, 3), (*y*, 1), (*y*, 2), (*y*, 3)}
   B × A = {(1, *x*), (2, *x*), (3, *x*), (1, *y*), (2, *y*), (3, *y*)}
   A × A = {(*x*, *x*), (*x*, *y*), (*y*, *x*), (*y*, *y*)}
   B × B = {(1, 1), (1, 2), (1, 3), (2, 1), (2, 2), (2, 3), (3, 1), (3, 2), (3, 3)}
3. (*i*) A × B = {(1, 2), (3, 2), (5, 2), (1, 3), (3, 3), (5, 3)}
   (*ii*) B × C = {(2, 4), (2, 6), (3, 4), (3, 6)}
   (*iii*) A × C = {(1, 4), (3, 4), (5, 4), (1, 6), (3, 6), (5, 6)}
   (*iv*) (A ∪ B) × C = {(1, 4), (2, 4), (3, 4), (5, 4), (1, 6), (2, 6), (3, 6), (5, 6)}
   (*v*) A ∩ (B × C) = φ
   (*vi*) (A × B) ∪ (B × C) = {(1, 2), (3, 2), (5, 2), (1, 3), (3, 3), (5, 3), (2, 4), (2, 6), (3, 4), (3, 6)}
   (*vii*) (A × B) ∩ (B × C) = φ.
6. Domain R = {1, 2}, Range R = {2, 3}.
7. $2^6 = 24$.
8. (*i*) R ∩ S = {(1, 2), (3, 3)}
   (*ii*) R ∪ S = {(1, 1), (1, 2), (1, 3), (2, 1), (2, 3), (3, 1), (3, 3)}
   (*iii*) R^c = {(1, 3), (2, 1), (2, 2), (3, 2)}
   (*iv*) R o S = {(1, 2), (1, 3), (1, 1), (2, 3), (3, 3)}
   (*v*) S o S = {(1, 1), (1, 3), (2, 2), (2, 3), (3, 3)}
9. (*i*) No (*ii*) No
   (*iii*) No (*iv*) No.
10. There are several examples for each answer. One possible set of examples follows :
    (*i*) R = {(1, 1), (2, 2)}
    (*ii*) R = {(1, 2), (2, 1), (2, 3)}
    (*iii*) R = {(1, 2)}.
11. (*i*) R (*ii*) R and S (*iii*) R and T (*iv*) R and T.
15. (*i*) No (*ii*) No (*iii*) Yes (*iv*) Yes.
16. (*i*) Yes (*ii*) No (*iii*) Yes (*iv*) No.
17. {(1, 1), (1, 3), (1, 5), (3, 1), (3, 3), (3, 5), (5, 1), (5, 3), (5, 5), (2, 2), (2, 6), (2, 4), (4, 2), (4, 4), (4, 6), (6, 2), (6, 4), (6, 6)}.
18. R = {(1, 1), (1, 4), (1, 7), (2, 2), (2, 5), (3, 3), (3, 6), (4, 1), (4, 4), (4, 7), (5, 2), (6, 3), (6, 6), (7, 1), (7, 4), (7, 7)},
    The partition generated by R = {{1, 4, 7}, {2, 5}, {3, 6}}.

19. $\frac{A}{R} = \{(1, 1), (2, 2), (3, 3), (4, 4)\}, \{(1, 2), (2, 1)\}, \{(1, 3), (3, 1)\}, \{(1, 4), (4, 1)\}, \{(2, 3), (3, 2)\}, \{(2, 4), (4, 2)\}, \{(3, 4), (4, 3)\}\}$.

20. (*a*) $E = \{1, 2, 5\}, F = \{3\}, G = \{2, 4\}, H = \{3, 5\}$.

(*b*) $M_S = \begin{bmatrix} 0 & 1 & 0 & 0 & 0 \\ 0 & 1 & 0 & 1 & 0 \\ 0 & 0 & 1 & 0 & 1 \\ 1 & 0 & 0 & 0 & 0 \\ 0 & 1 & 0 & 0 & 0 \end{bmatrix}$.

(*c*)

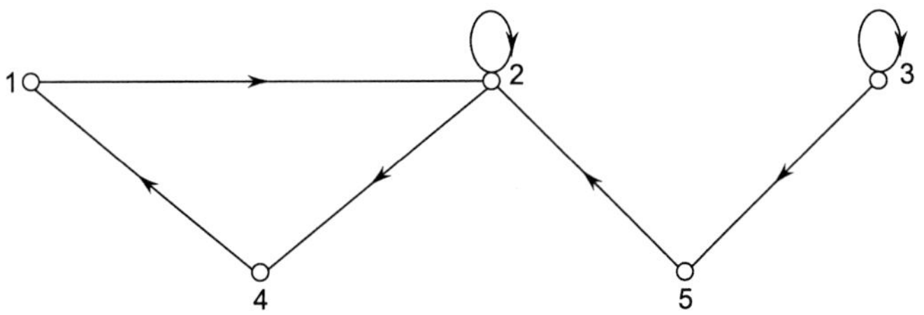

(*d*) $S \circ S = \{(1, 2), (1, 4), (2, 1), (2, 2), (2, 4), (3, 2), (3, 3), (3, 5), (4, 2), (5, 2), (5, 4)\}$.

21.

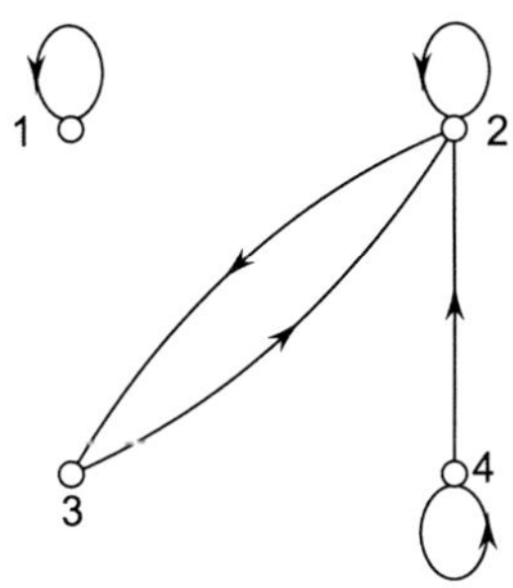

22. (*a*) $M_R = \begin{bmatrix} 0 & 0 & 1 & 1 \\ 0 & 0 & 0 & 0 \\ 0 & 1 & 1 & 1 \\ 0 & 0 & 0 & 0 \end{bmatrix}$

(*b*) $R^{-1} = \{(3, 1), (4, 1), (2, 3), (3, 3), (4, 3)\}$

(*c*)

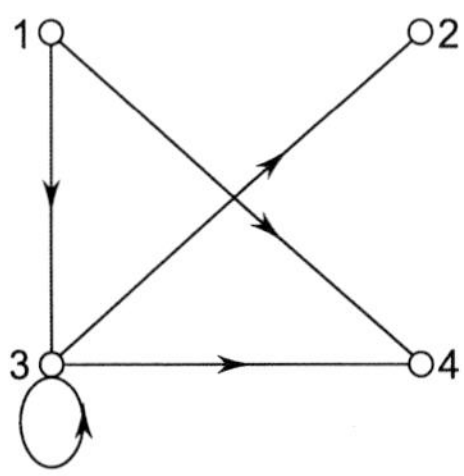

(*d*) $R \circ R = \{(1, 2), (1, 3), (1, 4), (3, 2), (3, 3), (3, 4)\}$

23. (*i*) $R = \{(1, 2), (2, 3), (2, 4), (3, 2), (3, 3), (3, 4), (4, 4)\}$

$$M_R = \begin{bmatrix} 0 & 1 & 0 & 0 \\ 0 & 0 & 1 & 1 \\ 0 & 1 & 1 & 1 \\ 0 & 0 & 0 & 1 \end{bmatrix}.$$

(*ii*) R = {(*a*, *c*), (*a*, *e*), (*c*, *e*), (*c*, *d*), (*d*, *f*), (*e*, *b*), (*e*, *d*)}

$$M_R = \begin{bmatrix} 0 & 0 & 1 & 0 & 1 & 0 \\ 0 & 0 & 0 & 0 & 0 & 0 \\ 0 & 0 & 0 & 1 & 1 & 0 \\ 0 & 0 & 0 & 0 & 0 & 1 \\ 0 & 1 & 0 & 1 & 0 & 0 \\ 0 & 0 & 0 & 0 & 0 & 0 \end{bmatrix}$$

24.

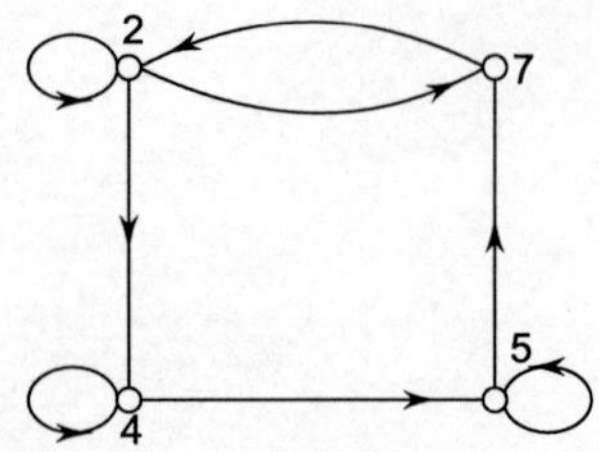

25. $M_R \cdot M_S = \begin{bmatrix} 1 & 0 & 1 & 1 \\ 1 & 0 & 0 & 0 \\ 0 & 0 & 1 & 0 \\ 0 & 0 & 0 & 0 \end{bmatrix}$ and S o R = {(1, 1), (1, 3), (1, 4), (2, 1), (3, 3)}.

3.15 PARTIAL ORDERING RELATIONS

Definition. *A relation R on a set P is called a* ***partial order relation*** *or a* ***Partial Ordering*** *in P iff R is reflexive, anti symmetric and transitive. We denote a partial ordering by the symbol ≤.*

Definition. *A set P on which a partial ordering ≤ is defined is called a* ***partially ordered*** *set (or) a* ***poset*** *and it is denoted by (P, ≤) or [P, ≤].*

Example. Let A be any set and P (A) be the collection of all subsets of A. Then (P (A), ⊆) is a poset.

Example. (**Z**, <) is not a poset because < is not reflexive.

Definition. *Let (P, ≤) be a poset. Elements a, b in P are said to be* ***comparable*** *under ≤ if either a ≤ b or b ≤ a, otherwise they are in* ***comparable***.

Definition. *Let (P, ≤) be a poset. If every pair of elements of A are comparable, then (P, ≤) is called a* ***totally ordered set*** *(or)* ***a chain*** *(or)* ***simply ordered set***. *Here the relation ≤ is called a* ***total order*** *(or)* ***linear order*** *(or)* ***a simple order*** *on P.*

Example 1. Let **Z** be the set of integers and ≤ is the usual ordering on **Z** then (**Z**, ≤) is a poset and also a chain.

Example 2. If *n* is a positive integer, and D_n denote the set of positive divisors of *n*. Then (D_n, 1) is a poset for some values of *n*, (D_n, 1) is a totally ordered but for other values of *n*, (D_n, 1) is not totally ordered. For example (D_8, 1) is totally ordered, but (D_6, 1) is not totally ordered because 2 and 3 are incomparable.

Note. (D_n, ≤) is totally ordered where ≤ is usual ordering relation.

For example, consider D_6 = {1, 2, 3, 6} and its Hasse diagram is in figure. 3.22 which is a totally ordered.

Example 3. **Z**, **N**, **Q**, **R** are chains under partial ordering relation '≤'.

Figure 3.22.

Example. *What is a partial ordering relation ? Let S = {x, y, z} and consider the power set P (s) with relation R given by set inclusion. Is R a partial order ?*

(JNTU November, 2006 - Set 4)

Solution. (See the Definition)

Let $S = \{x, y, z\}$

$\Rightarrow$ $P(S) = \{\phi, \{x\}, \{y\}, \{z\}, \{x, y\}, \{x, z\}, \{y, z\}, \{x, y, z\}\}$

Define a relation R by A R B iff $A \subseteq B\ \forall\ A, B \in P(S)$.

(*i*) We have $A \subseteq A$ for any $A \in P(S)$

$\Rightarrow$ $\subseteq$ is reflexive on P (S).

(*ii*) For any $A, B \in P(S)$, $A \subseteq B$, $B \subseteq A$ then $A = B$.

$\Rightarrow$ R is anti symmetric on P (S).

(*iii*) For any $A, B, C \in P(S)$, $A \subseteq B, B \subseteq C \Rightarrow A \subseteq C$

$\therefore$ R is transitive on P (S).

$\therefore$ R ($\subseteq$) is partial order on P (S).

3.16 HASSE DIAGRAM (OR) POSET DIAGRAM

A partial ordering ≤ on a set P can be represented by means of a diagram known as **Hasse diagram** (or) **poset diagram** of (P, ≤). The **procedure** for drawing Hasse diagram for a poset P as follows :

1. Each element is represented by a small circle.
2. The circle for $x \in P$ is drawn below the circle for $y \in P$ if $x < y$.
3. A line is drawn between *x* and *y* if *y* covers *x* and if $x < y$ but *y* does not cover *x*, then *x* and *y* are not connected directly by a single line.

However, they are connected through one or more elements of P. It is possible to obtain the set of ordered pairs in ≤ from such a diagram.

For a totally ordered set (P, ≤), the Hasse diagram consists of circles one below the other. Thus poset is called a chain.

Example. Consider $P = \{\phi, \{a\}, \{a, b\}, \{a, b, c\}\}$, (P, $\subseteq$) be a poset where $\subseteq$ is set inclusion, then the Hasse diagram of (P, $\subseteq$) is shown below.

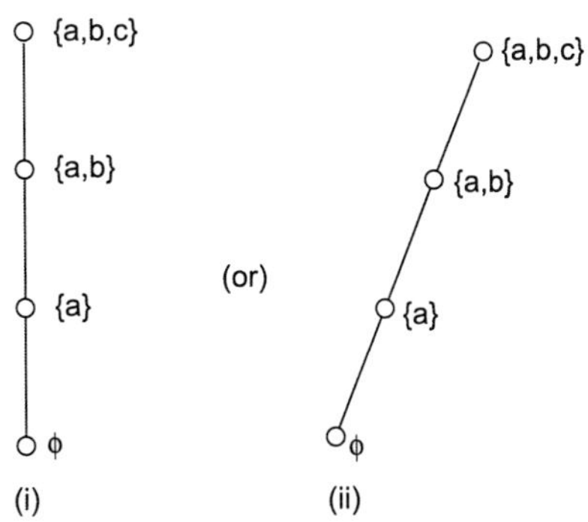

Figure 3.23.

Example. Let $D_6 = \{1, 2, 3, 6\}$. Draw Hasse diagram of $(D_6, |)$. Hasse diagram is

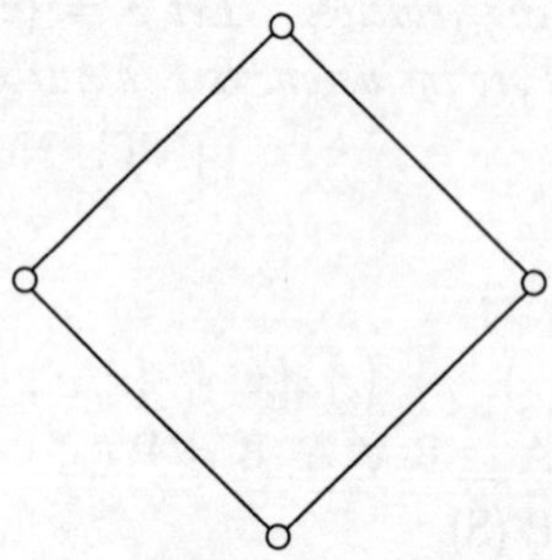

Figure 3.24.

Example. Let $(\{2, 3, 4, 9, 12, 18\}, |)$ be a poset. Draw its Hasse Diagram.

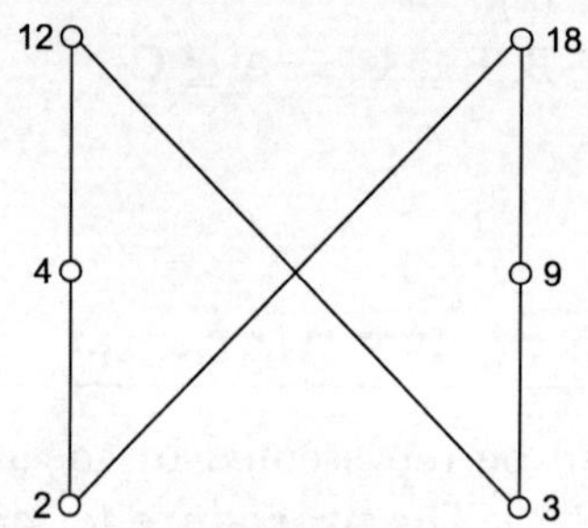

Figure 3.25.

Example. *Let A be any finite set and P (A) be the power set of A ; $\subseteq$ be the inclusion relation on the elements of P (A). Draw Hasse diagrams of (P (A), $\subseteq$) for (i) A = {a} ; (ii) A = {a, b} ; (iii) A = {a, b, c}.*

Solution. (*i*) Given $A = \{a\} \Rightarrow P(A) = \{\phi, \{a\}\}$

Diagram is

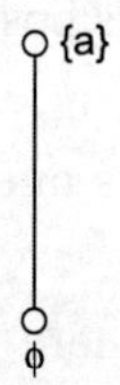

(*ii*) Given $A = \{a, b\} \Rightarrow P(A) = \{\phi, \{a\}, \{b\}, \{a, b\}\}$

Diagram is

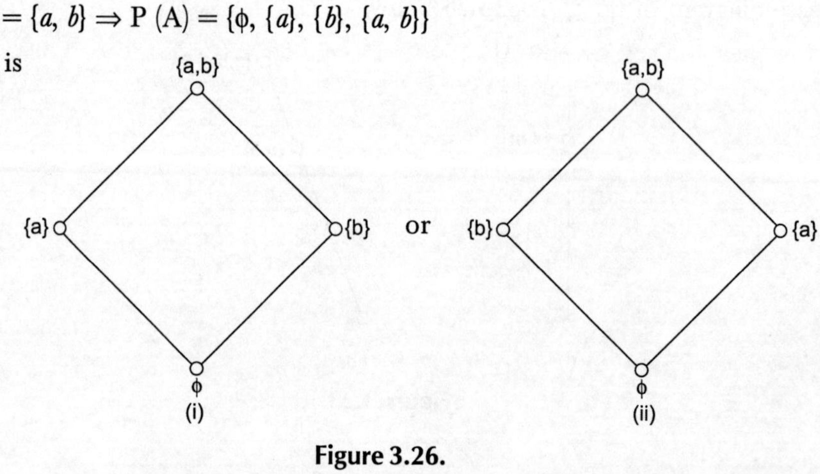

Figure 3.26.

(*iii*) Given A = {*a*, *b*, *c*} ⇒ P (A) = {ϕ, {*a*}, {*b*}, {*c*}, {*a*, *b*}, {*a*, *c*}, {*b*, *c*}, {*a*, *b*, *c*}}

Diagram is

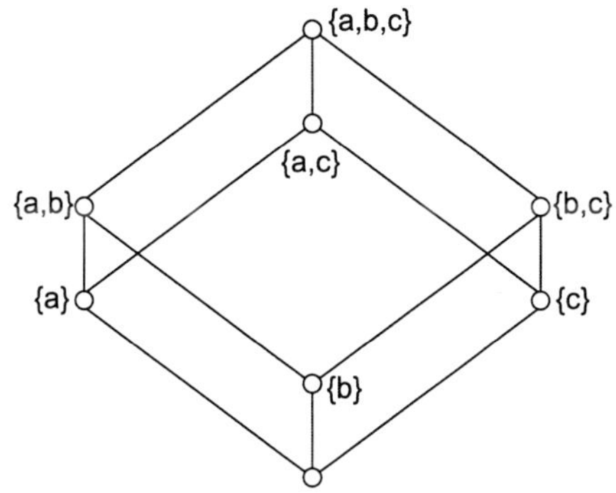

Figure 3.27.

Example. Draw Hasse Diagram of

Poset $(D_{12}, |)$

Solution. D_{12} = {1, 2, 3, 4, 6, 12}

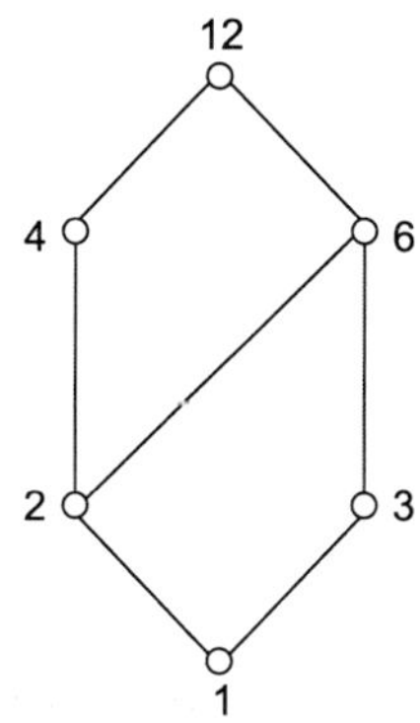

Figure 3.28.

Example. *Draw the Hasse diagram of Poset* $(D_{45}, |)$.

Solution. D_{45} = {1, 3, 5, 9, 15, 45}

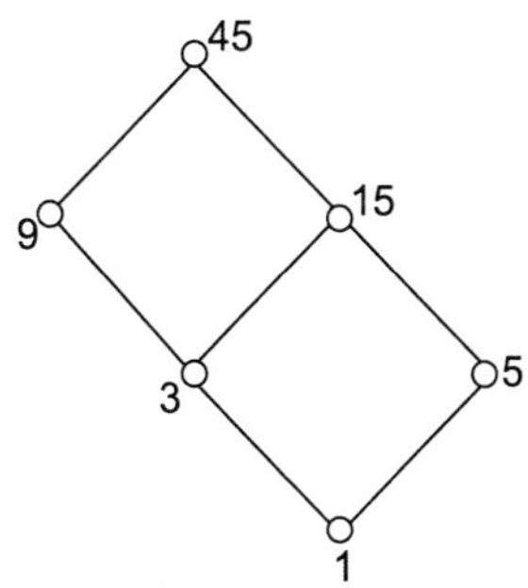

Figure 3.29.

Example. *Let X = {2, 3, 6, 12, 24, 36} then prove (X, |) is a poset and draw its Hasse diagram.*

Solution. Let $a \in X$, $a \mid a \Rightarrow$ '|' is reflexive on X.

Let $a, b \in X$, $a \mid b$, $b \mid a \Rightarrow a = b \Rightarrow$ '|' is anti symmetrion on X.

and let $a, b, c \in X$, $a \mid b$, $b \mid c \Rightarrow a \mid c \Rightarrow$ '|' is transitive on X.

$\therefore$ (X, |) is a poset.

Its Hasse diagram is

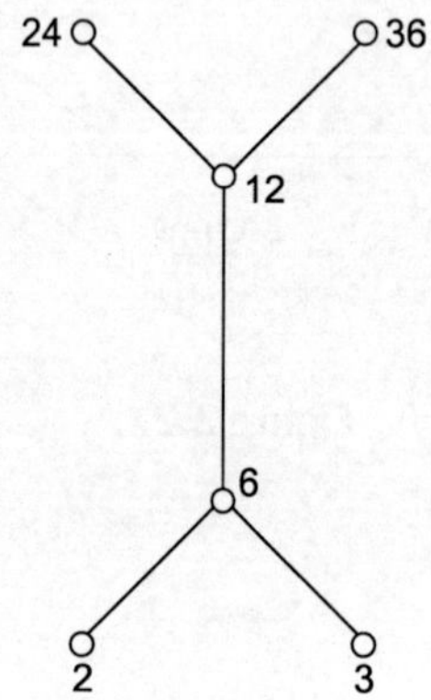

Figure 3.30.

Example. *If A = {1, 2, 3, 4}. Draw Hasse diagram of (P (A), ⊆).*

Solution. Given A = {1, 2, 3, 4}.

then P (A) = {ϕ, {1}, {2}, {3}, {4}, {1, 2}, {1, 3}, {1, 4}, {2, 3}, {2, 4}, {3, 4}, {1, 2, 3}, {1, 2, 4}, {1, 3, 4}, {2, 3, 4}, {1, 2, 3, 4}

Diagram is

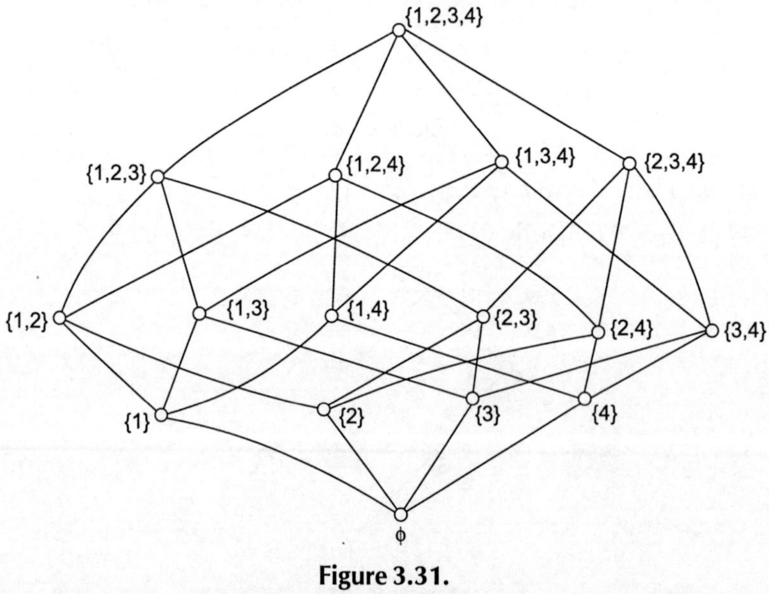

Figure 3.31.

Example. *Let I_{12} is the set of all positive integers which are less than or equal to 12. Draw Hasse diagram of (I_{12}, |).*

Solution. Here $I_{12} = \{1, 2, 3, 4, 5, 6, 7, 8, 9, 10, 11, 12\}$

Diagram is

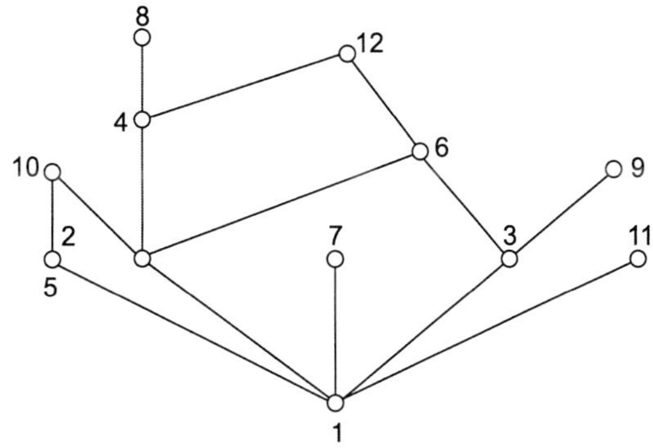

Figure 3.32

Definition. *Let (P, ≤) be a poset, and A ≤ P, then a ∈ A is called a* ***lower bound*** *of A if a ≤ x ∀ x ∈ A and if there are no lower bounds of A which are greater than a, then a is called* ***greatest lower bound*** *(g.l.b) of A (or) infimum (inf) of A.*

Definition. *Let (P, ≤) be a poset, A ≤ P, then a ∈ A is called an* ***upper bound*** *of A if x ≤ a ∀ x ∈ A and if there are no upper bounds of A which are less than a then a is called* ***least upper bound*** *(l.u.b) of A (or) supremum (Sup) of A.*

Example. Consider the poset ({2, 3, 5, 30, 60, 120, 180, 360}, |) having Hasse diagram shown below.

The set {120, 180, 360} has a g.l.b (namely 60), but {2, 3, 5, 30} has no lower bounds and hence no g.l.b and the set {60, 120, 180} has a *l.u.b* (namely 360).

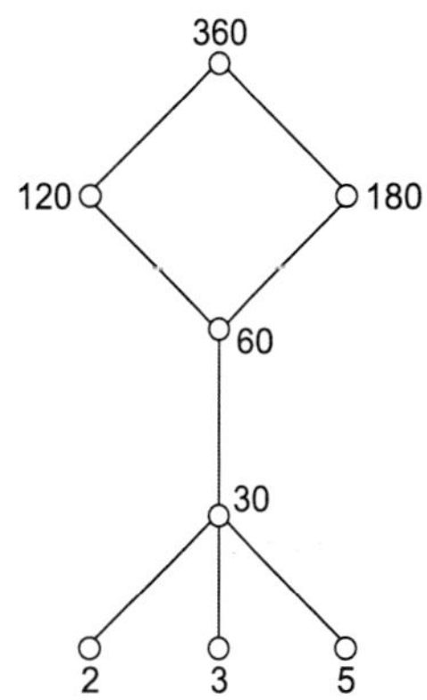

Figure 3.33

3.16.1 Special Elements in Posets

Definition. *Let (P, ≤) be a poset, and A ⊆ P, then*

(*i*) an element $a \in A$ is called least element of A if $a \leq x\ \forall\ x \in A$.

(*ii*) an element $a \in A$ is called the greatest element of A if $x \leq a\ \forall\ x \in A$.

Note 1. If least element of A exists then it is unique.

2. If greatest element of A exists then it is unique.

Definition. *Let (P, ≤) be a poset, A ≤ P. An element a ∈ A is said to be a minimal element of A if ∃ no x in A such that x < a.*

Definition. *Let (P, ≤) be a poset, A ⊆ P. An element a ∈ A is said to be a maximal element of A if ∃ no x in A ∋ a < x.*

Note 1. Minimal (maximal) element is not unique.

2. Least element of A is minimal element, but not converse.

3. Greatest element of A is maximal element, but converse is need not be true.

Example. Consider the poset (D_6, |). 1 is least element in D_6, and 1 is also minimal element in D_6 and in the poset ({2, 3, 4, 9, 12, 18}, |), 2, 3 are minimal elements but these are not least elements. In the poset ({2, 3, 6, 12, 24, 36}, |), 24, 36 are maximal elements but these are not greatest elements.

Theorem. If (P, R) is a poset and P is finite (non-empty), then P has at least one maximal element and at least one minimal element.

Proof : Take an element $a \in$ P.

If a is not maximal, we can find an element $x_1 \in$ P such that $x_1 \neq a$ and a R x_1.

If x_1 is not maximal, we can find an element $x_2 \in$ P such that $x_2 \neq x_1$ and x_1 R x_2.

Proceeding like this, we end up with a finite chain of the form a R x_1, x_1 R x_2, x_2 R x_3, ..., which cannot be extended beyond a certain final stage (because P is finite). Hence, we end up with some $x_k \in$ P which is maximal element of P. Thus, P has at least one maximal element. A similar argument shows that P has at least one minimal element.

Theorem. Every poset has atmost one greatest element and atmost one least element.

Proof : Assume that a and b are greatest elements of (P, R).

Since b is a greatest element, a R b.

Similarly since a is greatest element, b R a.

Since R is antisymmetric relation, $a = b$.

Therefore, two greatest elements of (P, R) cannot be different. That means, if (P, R) has a greatest element, then that element is unique. Thus (P, R) has at most one greatest element.

Similarly, we can prove that (P, R) has atmost one least element.

Theorem : If (P, R) is a poset and A $\subseteq$ P, then A has atmost one *l.u.b.*

Proof : If possible a and b be two *l.u.b.s.* of A. Then a, b are upper bounds of A.

Since b is an upper bound of A, a is *l.u.b* of A, we have a R b.

Since a is an upper bound of A, b is *l.u.b.* of A, we have b R a.

Since R is anti symmetric, $a = b$

∴ A has unique *l.u.b* if exists.

Hence A has atmost one *l.u.b.*

Note. If (P, R) is a poset and A $\subseteq$ P, then A has atmost one *g.l.b.*

Example. *Let D_{24} = {1, 2, 3, 4, 6, 8, 12, 24} and relation divides '1' be a partial ordering on D_{24}. Then draw the Hasse diagram of (D_{24}, 1) and also find the following*

(i) all lower bounds of 8, 12
(ii) all upper bounds of 8, 12
(iii) g.l. b of 8, 12
(iv) l.u.b of 8, 12.
(v) greatest and least elements of this poset if exists.

Solution.

(*i*) The lower bounds of 8 and 12 are 1, 2, 4.

(*ii*) The upper bounds of 8, 12 is 24.

(*iii*) g.l.b of 8, 12 is 4.

(*iv*) l.i.b of 8, 12 is 24.

(*v*) The greatest element is 24 and least element is 1.

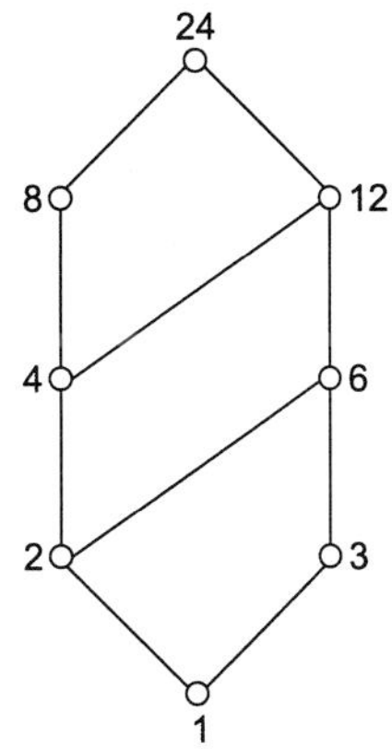

Figure 3.34.

Example. *Determine the greatest and least elements of the following posets if they exists.*

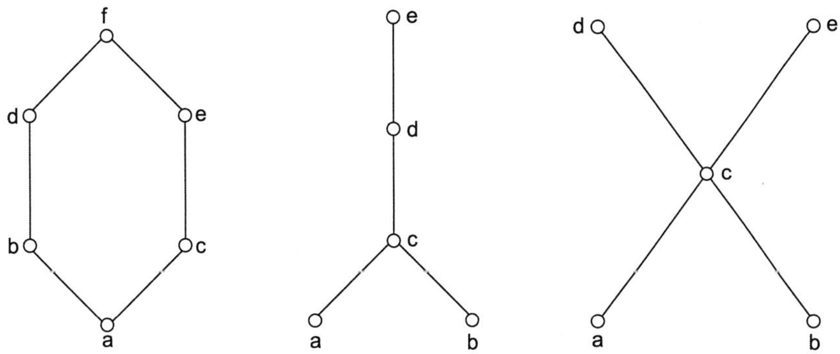

(i) (ii) (iii)

Example 3.35.

Solution.

(*i*) greatest element is *f* and least element is a

(*ii*) greatest element is *e* and no least element.

(*iii*) No greatest element and no least element.

Example. *Determine the maximal and minimal elements of the posets.*

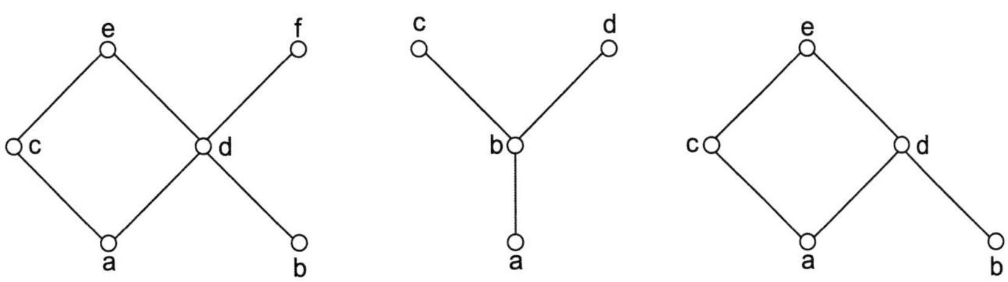

(i) (ii) (iii)

Figure 3.36.

Solution.

(*i*) Maximal elements are *e*, *f* and minimal elements are *a*, *b*.

(*ii*) Maximal elements are *c*, *d* and minimal element is *a*.

(*iii*) Maximal element is *e* and minimal elements are *a*, *b*.

Example. *Draw a poset diagram for each of the following posets and determine all maximal, minimal elements and greatest, least elements if they exist.*

(*i*) $(D_{20}, |)$

(*ii*) $(D_{30}, |)$

(*iii*) $(A, |)$ *where* $A = \{2, 3, 4, 6, 8, 24, 48\}$

(*iv*) $(A, |)$ *where* $A = \{2, 3, 4, 6, 12, 18, 24, 36\}$

Solution. (*i*) $D_{20} = \{1, 2, 4, 5, 10, 20\}$.

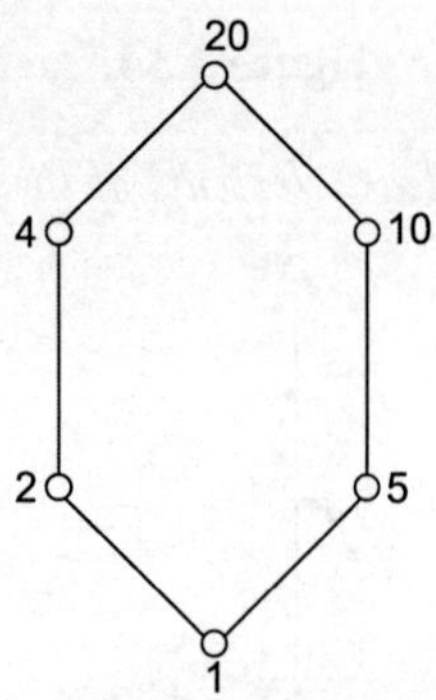

Figure 3.37.

Minimal element = 1 = least element.

Maximal element = 20 = greatest element.

(*ii*) $D_{30} = \{1, 2, 3, 5, 6, 10, 15, 30\}$.

Minimal element = 1

Least element = 1.

Maximal element = 30

Greatest element = 30

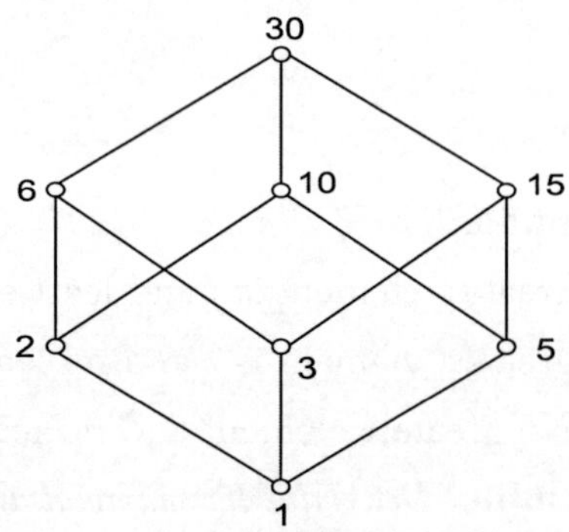

Figure 3.38.

(*iii*) Given A = {2, 3, 4, 6, 8, 24, 48}

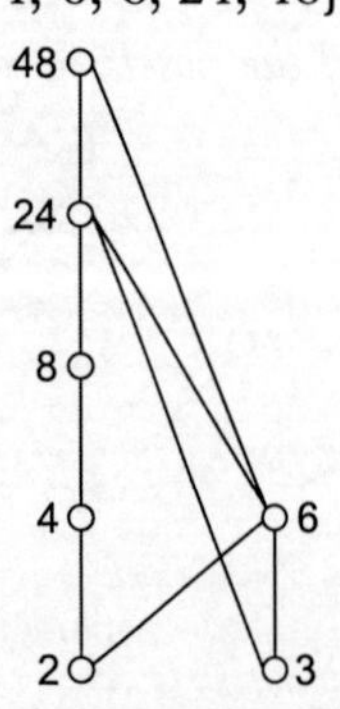

Figure 3.39.

Minimal elements = 2, 3.

Least element = does not exists

Maximal element = 48 = greatest element

(*iv*) Given A = {2, 3, 4, 6, 12, 18, 24, 36}

Minimal elements = 2, 3

Least element = does not exist

Maximal elements = 24, 36.

Greatest element = does not exist.

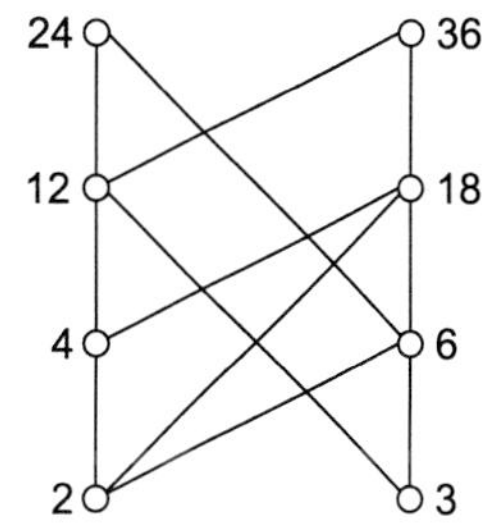

Figure 3.40.

EXERCISE 3.3.

1. Let A = {2, 4, 8, 16, 32}, Draw the Hasse diagram of (A, |)
2. Let A = {1, 2, 3, 4, 6, 12}, Draw the Hasse diagram of (A, |)
3. Draw the Hasse diagram representing the positive divisors of 36.
4. Draw the Hasse diagram for the divisibility relation on the set A in each of the following cases.

   (*a*) A = {1, 2, 3, 5, 6, 10, 15, 30} (*b*) A = {2, 3, 6, 12, 24, 36}

   (*c*) A = {1, 2, 3, 4, 6, 8, 9, 12, 18, 24}
5. Find the matrix of the partial order whose Hasse diagram is

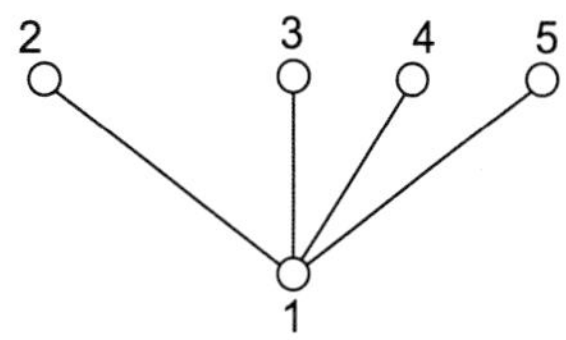

6. Find the maximal and minimal elements of the poset A whose Hasse diagram is given below :

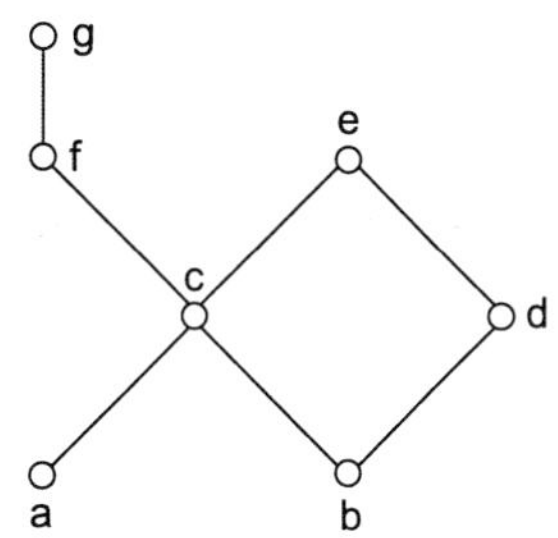

7. Which elements of the poset ({2, 4, 5, 10, 12, 20, 25}, |) are maximal and minimal.
8. Let R be the set of all real numbers with ⊆ as the partial order. Also, let B be the open interval (1, 2). Find

   (*i*) all upper bounds of B (*ii*) all lower bounds of B

   (*iii*) *a l.u.b* of B (*iv*) *a g.l.b* of B.
9. Consider the poset (ρ (S), ⊆), where S = {1, 2, 3, 4}. For each of the following subsets B, determine the infimum and supremum of B.

   (*i*) B = {{1}, {2}} (*ii*) B = {{1}, {2}, {1, 2}}

   (*iii*) B = {{1}, {2}, {3}, {1, 2}} (*iv*) B = {ϕ, {1}, {2}, {1, 2}}

   (*v*) B = {{1}, {1, 2}, {1, 3}, {1, 2, 3}}.

10. For the posets shown in the following Hasse diagrams, find the greatest and least elements, if exists.

(*i*)

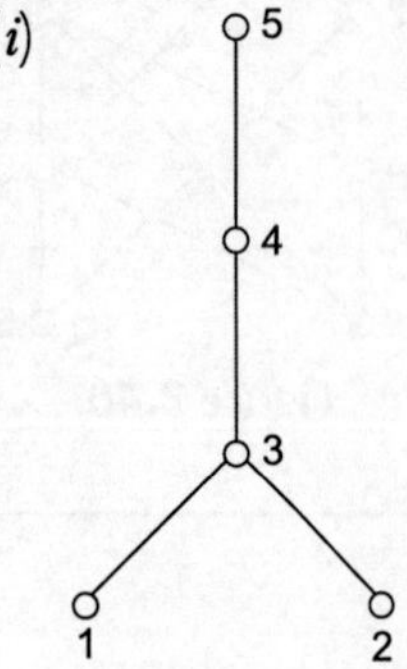

(*ii*)

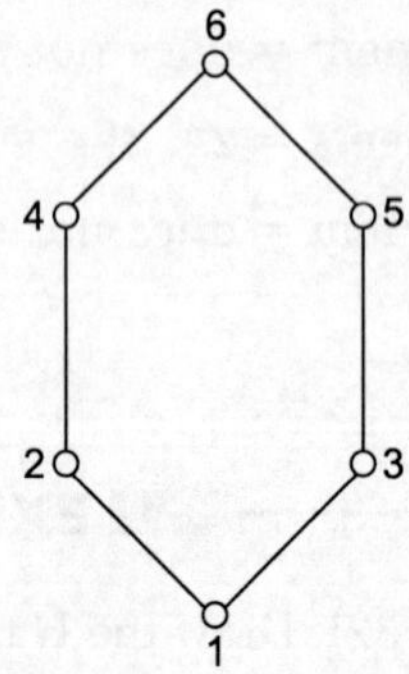

ANSWERS

1.

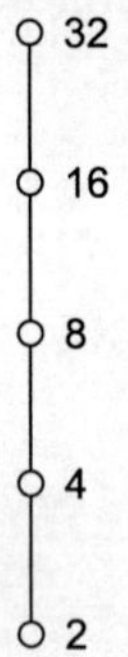

2.

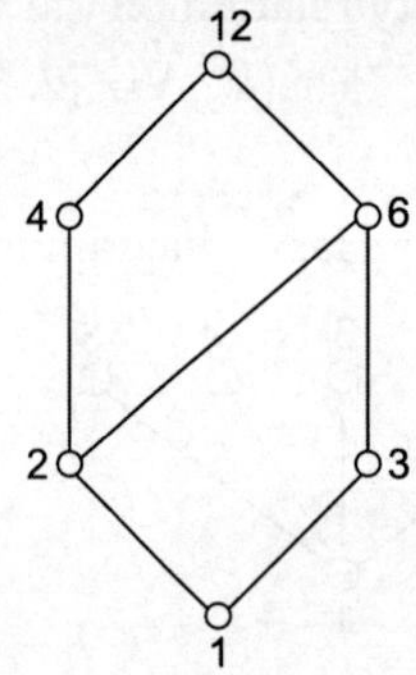

3.

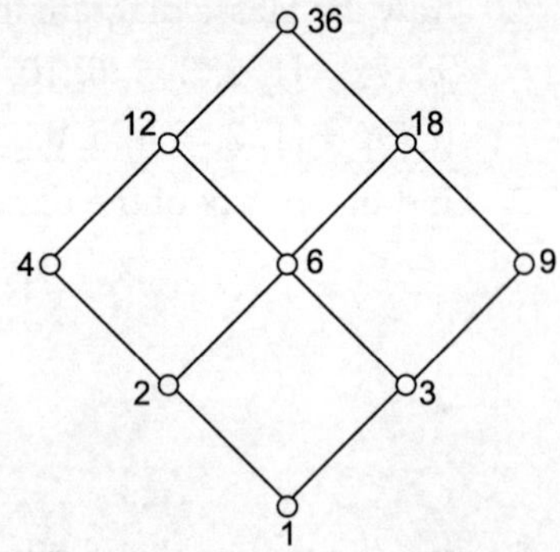

4. (*a*)

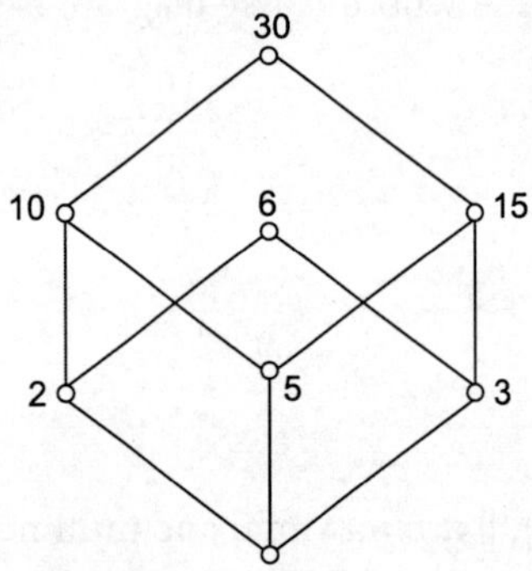

(*b*)

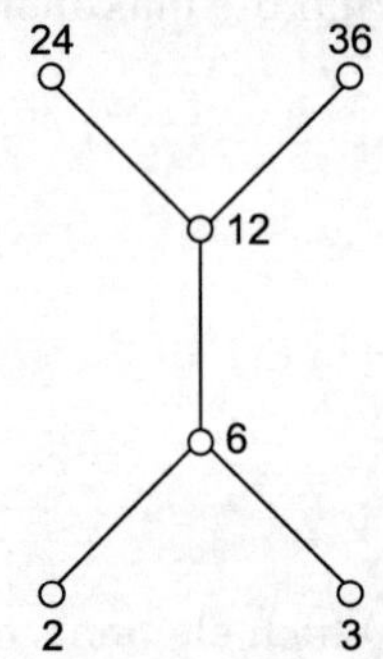

(*c*)

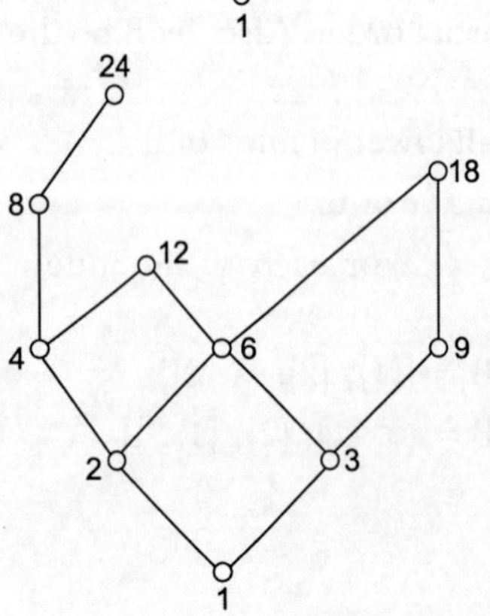

5. $M_R = \begin{bmatrix} 1 & 1 & 1 & 1 & 1 \\ 0 & 1 & 0 & 0 & 0 \\ 0 & 0 & 1 & 0 & 0 \\ 0 & 0 & 0 & 1 & 0 \\ 0 & 0 & 0 & 0 & 1 \end{bmatrix}$

6. Minimal elements are a, b and maximal elements are e and g.
7. Minimal elements are 2, 5 and maximal elements are 12, 20, 25.
8. (*i*) Every real number ≥ 2, is an upper bound of B.
   (*ii*) Every real number ≤ 1, is a lower bound of B.
   (*iii*) 2 is a least upper bound of B.
   (*iv*) 1 is a greatest lower bound of B.
9. (*i*) Infimum : ϕ, supremum : $\{1, 2\}$
   (*ii*) Infimum : ϕ, supremum : $\{1, 2\}$
   (*iii*) Infimum : ϕ, supremum : $\{1, 2, 3\}$
   (*iv*) Infimum : ϕ, supremum : $\{1, 2\}$
   (*v*) Infimum : $\{1\}$, supremum : $\{1, 2, 3\}$
10. (*i*) Greatest element is 5, least element does not exist.
    (*ii*) Greatest element is 6, least element is 1.

3.17 LATTICES

In this section, we introduce lattice as a partially ordered set and we study some properties of lattice.

Definition *of lattice in terms of poset.*

Definition. *A poset (L, ≤) is said to be* ***lattice*** *if every two elements in L have least upper bound (l.u.b) and greatest lower bound (g.l.b).*

The *g.l.b* of a subset $\{a, b\} \subseteq L$ will be denoted by $a \times b$ or $a \wedge b$ (meet or product of a, b) and the *l.u.b* of a subset $\{a, b\} \subseteq L$ will be denoted by $a \oplus b$ (or) $a \vee b$ (Join or sum of a, b).

Note. Every lattice is a poset but all posets are not lattices. For example (X = {2, 3, 6, 12, 24, 36}, |) is a poset but not lattice, because *g.l.b* of 2, 3 is $1 \notin X$.

Example. Let $S = \{a, b\}$, $P(S) = \{\phi, \{a\}, \{b\}, \{a, b\}\}$

Then $(P(S), \subseteq)$ is a lattice because *g.l.b* and *l.u.b* of any pair of elements of P (S) are belongs to P (S) it is clear from Fig. 3.41.

Example. Let $S = \{a, b, c\}$, $P(S) = \{\phi, \{a\}, \{b\}, \{c\}, \{a, b\}, \{a, c\}, \{b, c\}, \{a, b, c\}\}$, then $(P(S), \subseteq)$ is a lattice because *g.l.b* and *l.u.b* of every pair of elements of P (S) are belongs to P (S), is clear from Fig. 3.42.

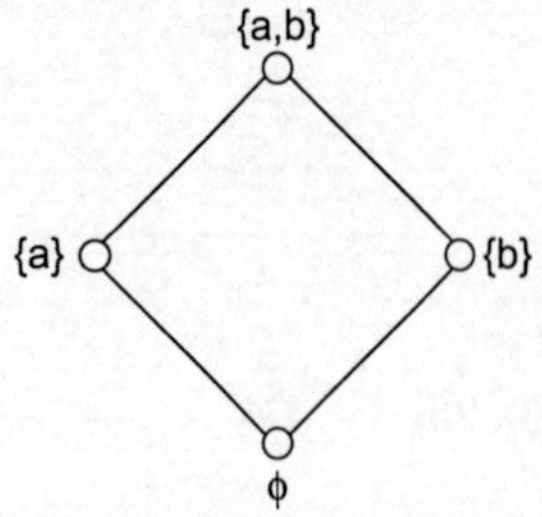

Fig. 3.41

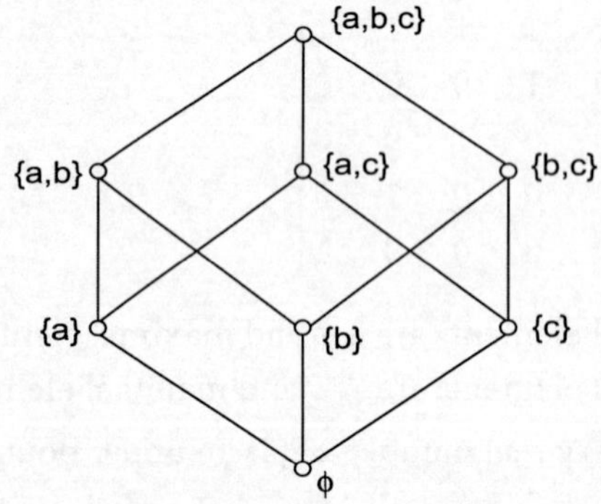

Fig. 3.42

Example. *Show that (D_{105}, |) is a lattice.*

Solution. D_{105} = {1, 3, 5, 7, 15, 21, 35, 105} ⇒ (D_{105}, |) is a poset

∴ (D_{105}, |) is a lattice from the following tables.

Table 3.1

l.u.b	1	3	5	7	15	21	35	105
1	1	3	5	7	15	21	35	105
3	3	3	15	21	15	21	105	105
5	5	15	5	35	15	105	35	105
7	7	21	35	7	105	21	35	105
15	15	15	15	105	15	105	105	105
21	21	21	105	21	105	21	105	105
35	35	105	35	35	105	105	35	105
105	105	105	105	105	105	105	105	105

Table 3.2

g.l.b	1	3	5	7	15	21	35	105
1	1	1	1	1	1	1	1	1
3	1	3	1	1	3	3	1	3
5	1	1	5	1	5	1	5	5
7	1	1	1	7	1	7	7	7
15	1	3	5	1	15	3	5	15
21	1	3	1	7	3	21	7	21
35	1	1	5	7	5	7	35	35
105	1	3	5	7	15	21	35	105

Example. *(D_6, |) is a lattice.*

Solution. *We know that (D_6 = {1, 2, 3, 6}, |) is a poset then (D_6, |) is lattice because every pair of elements of D_6 have g.l.b and l.u.b in D_6.*

Table 3.3

$g.l.b$	1	2	3	6
1	1	1	1	1
2	1	2	1	2
3	1	1	3	3
6	1	2	3	6

Table 3.4

$l.u.b$	1	2	3	6
1	1	2	3	6
2	2	2	6	6
3	3	6	3	6
6	6	6	6	6

Example. *Which of the following posets are lattices on the set {a, b, c, d, e, f, g}.*

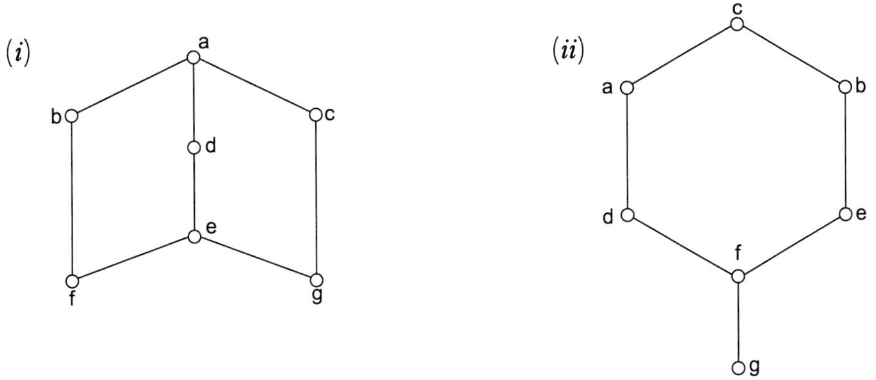

Figure 3.43.

Solution. (*i*) Poset is not a lattice because there is no lower bound for {*f*, *g*}.

(*ii*) Poset is a lattice because for every pair of elements have $l.u.b$ and $g.l.b$ in poset.

Example. *Which of the following posets represents lattices.*

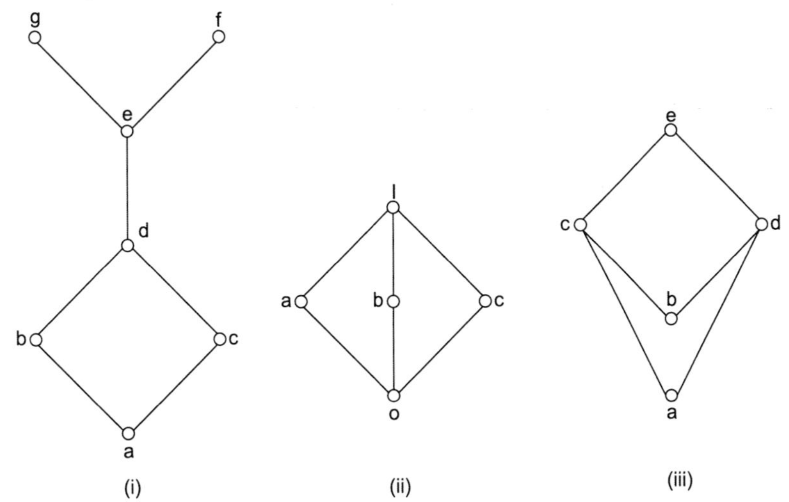

Figure 3.44.

Solution. (*i*) Not a lattice, because $f \vee g$ does not exist.

(*ii*) Lattice.

(*iii*) Not a lattice, because $a \wedge b$ does not exist.

Definition a lattice in terms of algebraic structure

(JNTU November 2006 - Set 4, March 06 - Set 1, May 05 - Set 1, May 04 - Set 2)

Definition. *An algebraic structure (L, ∧, ∨) is said to be* ***lattice*** *if the binary operations ∧, ∨ satisfies the following properties (axioms)*

(*i*) **Idempotent property :**

$$a \wedge a = a,\ a \vee a = \forall\ a \in \text{L}.$$

(*ii*) **Commutative Property :**

$$a \wedge b = b \wedge a,\ a \vee b = b \vee a \forall\ a,\ b \in \text{L}.$$

(*iii*) **Associative Property :**

$$(a \wedge b) \wedge c = a \wedge (b \wedge c),\ (a \vee b) \vee c = a \vee (b \vee c) \ \forall\ a,\ b\, c \in \text{L}.$$

(*iv*) **Absorption Property :**

$$a \wedge (a \vee b) = a,\ a \vee (a \wedge b) = a \ \forall\ a,\ b \in \text{L}.$$

Example. *Show that the set of positive integers N is a lattice with respect to the operations a ∨ b = l.c.m (a, b) and a ∧ b = g.c.d (a, b), l.c.m means least common multiple and g.c.d means greatest common divisor.* (JNTU March 06 - Set 3, Nov 05 - Set 3, May 04 - Set 1)

Solution. Given N is the set of positive integers

$$a \vee b = \text{l.c.m}\ (a, b),\ a \wedge b = \text{g.c.d}\ (a, b)$$

(*i*) If a, b, K ∈ N so that K/a, K/b.

then K is a common divisor of a, b.

If d is the greatest such K, d is called greatest common divisor of a, b and it belong to N.

∴ $$d = \text{g.c.d}\ (a, b) \in \text{N}.$$

(*ii*) If a, b, K ∈ N so that a/K, b/K.

Then K is a common multiple of a, b.

If c is the least such K, c is called least common multiple of a, b and it belongs to N.

∴ $$c = \text{l.c.m}\ (a, b) \in \text{N}.$$

By definition of lattice, (N, ∧, ∨) is a lattice.

Theorem. Let (L, ≤) be a lattice and a, $b\ c \in$ L, then prove that

(*i*) $a \vee (b \wedge c) \leq (a \vee b) \wedge (a \vee c)$

(*ii*) $a \wedge (b \vee c) \geq (a \wedge b) \vee (a \wedge c)$ (Distributive inequalities)

Proof : Let a, b, $c \in$ L where L is lattice.

(*i*) Clearly, $a \leq a \vee b$

$$a \leq a \vee c$$

⇒ a is a lower bound of

$$a \vee b,\ a \vee c$$

⇒ $$a \leq (a \vee b) \wedge (a \vee c) \qquad \text{...(1)}$$

and $$b \wedge c \leq b \leq a \vee b,$$

$$b \wedge c \leq c \leq a \vee c$$

$\Rightarrow$ $b \wedge c$ is a lower bound of $a \vee b$, $a \vee c$

$\Rightarrow$ $b \wedge c \leq (a \vee b) \wedge (a \vee c)$ ...(2)

From (1), (2), $(a \vee b) \wedge (a \vee c)$ is an upper bound of a, $b \wedge$ c.

$\Rightarrow$ $a \vee (b \wedge c) \leq (a \vee b) \wedge (a \vee c)$ (since a l.u.b $\leq$ $u.b.$)

(*ii*) Clearly $a \wedge b \leq a$,

$$a \wedge b \leq b \leq b \vee c$$

$\Rightarrow$ $a \wedge b$ is lower bound of a, $b \vee c$

$\Rightarrow$ $a \wedge b \leq a \wedge (b \vee c)$ ...(3) (since l.b $\leq$ g.l.b)

and $a \wedge c \leq a$

$$a \wedge c \leq c \leq b \vee c$$

$\Rightarrow$ $a \wedge c$ is lower bound of a, $b \vee c$

$\Rightarrow$ $a \wedge c \leq a \wedge (b \vee c)$ ...(4)

From (3), (4), $a \wedge (b \vee c)$ is an upper bound of $a \wedge b$, $a \wedge c$

$\therefore$ $(a \wedge b) \vee (a \wedge c) \leq a \wedge (b \vee c)$ (since l.u.b $\leq$ u.b).

Theorem. Let (L, $\leq$) be *a* lattice, $a, b, c \in$ L if $a \leq c$

Then prove that $a \vee (b \wedge c) \leq (a \vee b) \wedge c$ (Modular inequality)

Proof : We have $a \vee (b \wedge c) \leq (a \vee b) \wedge (a \vee c)$ by distributive inequalities

$$= (a \vee b) \wedge c \qquad (\because \ a \leq c \Rightarrow a \vee c = c)$$

(or) Given $a \leq c$

Clearly $a \leq a \vee b$

$\Rightarrow$ a is lower bound of $a \vee b$, c

$\Rightarrow$ $a \leq (a \vee b) \wedge c$ ...(1)

Clearly, $b \wedge c \leq c$

$$b \wedge c \leq b \leq a \vee b$$

$\Rightarrow$ $b \wedge c$ is lower bound of $a \vee b$, c

$\Rightarrow$ $b \wedge c \leq (a \vee b) \wedge c$ ...(2)

From (1), (2), $(a \vee b) \wedge c$ is upper bound of a, $b \wedge c$

$\therefore$ $a \vee (b \wedge c) \leq (a \vee b) \wedge c$

Theorem. Let (L, N, $\vee$) be *a* lattice, $a, b, c, d \in$ L, prove that (*i*) $(a \wedge b) \vee (b \wedge c) \vee (c \wedge a) \leq (a \vee b) \wedge (b \vee c) \wedge (c \wedge a)$; (*ii*) $(a \wedge b) \vee (c \wedge d) \leq (a \vee c) \wedge (b \wedge d)$

Proof : (*i*) We have $a \wedge b \leq a \leq a \vee b$

$$a \wedge b \leq b \leq b \vee c$$

$$a \wedge b \leq a \leq c \vee a$$

$\Rightarrow$ $a \wedge b$ is lower bound of $a \vee b$, $b \vee c$, $c \vee a$

$\therefore$ $a \wedge b \leq (a \vee b) \wedge (b \vee c) \wedge (c \vee a)$ ...(1)

and $b \wedge c \leq b \leq a \vee b$

$$b \wedge c \leq c \wedge b \vee c$$

$$b \wedge c \leq c \leq c \vee a$$

$\Rightarrow$ $b \wedge c$ is lower bound of $a \vee b$, $b \vee c$, $c \vee a$

$\therefore$ $$b \wedge c \leq (a \vee b) \wedge (b \vee c) \wedge (c \vee a) \quad ...(2)$$

and $$c \wedge a \leq a \leq a \vee b$$
$$c \wedge a \leq c \leq b \vee c$$
$$c \wedge a \leq c \leq c \vee a$$

$\Rightarrow$ c $\wedge$ a is lower bound of $a \vee b$, $b \vee c$, $c \vee a$

$\therefore$ $$c \wedge a \leq (a \vee b) \wedge (b \vee c) \wedge (c \vee a) \quad ...(3)$$

From (1), (2), (3)

$(a \vee b) \wedge (b \vee c) \wedge (c \vee a)$ is an upper bound of $a \wedge b$, $b \wedge c$, $c \wedge a$

$\Rightarrow$ $(a \wedge b) \vee (b \wedge c) \vee (c \wedge a) \leq (a \vee b) \wedge (b \vee c) \wedge (c \vee a)$ sin a l.u.b $\leq$ u.b.

(*ii*) We have $a \wedge b \leq a \leq a \vee c$
$$a \wedge b \leq b \leq b \vee d.$$

$\Rightarrow$ $a \wedge b$ is lower bound of $a \vee c$, $b \vee d$

$\Rightarrow$ $$a \wedge b \leq (a \vee c) \wedge (b \vee d) \quad ...(1)$$

and $$c \wedge d \leq c \leq a \vee c$$
$$c \wedge d \leq d \leq b \vee d$$

$\Rightarrow$ $c \wedge d$ is lower bound of $a \vee c$, $b \vee d$

$\Rightarrow$ $$c \wedge d \leq (a \vee c) \wedge (b \vee d) \quad ...(2)$$

From (1), (2)

$(a \vee c) \wedge (b \vee d)$ is an upper bound of $a \wedge b$, $c \wedge d$.

$\Rightarrow$ $(a \wedge b) \vee (c \wedge d) \leq (a \vee c) \wedge (b \vee d)$ since l.u.b $\leq$ u.b.

Theorem. Let (L, $\leq$) be a lattice a, b, $c \in$ L, prove that that the following are equivalent :

(*i*) $a \vee (b \wedge c) = (a \vee b) \wedge (a \vee c)$

(*ii*) $a \wedge (b \vee c) = (a \wedge b) \vee (a \wedge c)$

Proof : Assume $a \vee (b \wedge c) = (a \vee b) \wedge (a \vee c)$ ...(1)

Let a, b, $c \in$ L,

$$\begin{aligned}(a \wedge b) \vee (a \wedge c) &= [(a \wedge b) \vee a] \wedge [(a \wedge b) \vee c] && \text{(by (1))}\\ &= a \wedge [(a \vee c) \wedge (b \vee c)] && \text{(by absorption, (1))}\\ &= [a \wedge (a \vee c)] \wedge (b \vee c) && \text{(by associative)}\\ &= a \wedge (b \vee c) && \text{(by absorption)}\end{aligned}$$

and on other hand assume
$$a \wedge (b \vee c) = (a \wedge b) \vee (a \wedge c) \quad ...(2)$$

Let a, b, $c \in$ L,

$$\begin{aligned}(a \vee b) \wedge (a \vee c) &= [(a \vee b) \wedge a] \vee [(a \vee b) \wedge c] && \text{(by (2))}\\ &= a \vee [(a \wedge c) \vee (b \wedge c)] && \text{(by absorption, (2))}\\ &= [a \vee (a \wedge c)] \vee (b \wedge c) && \text{(by associative)}\\ &= a \vee (b \wedge c) && \text{(by absorption)}\end{aligned}$$

$\therefore$ $$a \vee (b \wedge c) = (a \vee b) \wedge (a \vee c).$$

Theorem. Let (L, ≤) be a lattice and $a, b \in$ L. Then prove the following :

(*i*) $a \vee b = b$ iff $a \leq b$

(*ii*) $a \wedge b = a$ iff $a \leq b$

Proof : (1) suppose $a \leq b$

we have $a \leq b$

$\Rightarrow$ b is upper bound of a, b.

$\Rightarrow$ $a \vee b \leq b$ ...(1)

Since $a \vee b$ is upper bound of a, b, $b \leq a \vee b$ ...(2)

From (1), (2) $a \vee b = b$

Conversely, suppose $a \vee b = b$

Since $a \leq a \vee b$, $a \leq b$.

(*ii*) Suppose $a \leq b$

We have $a \leq a$

$\Rightarrow$ a is lower bound of a, b.

$\Rightarrow$ $a \leq a \wedge b$ ...(3)

Since l.b ≤ g.l.b.

Since $a \wedge b$ is lower bound of a, b, $a \wedge b \leq a$ ...(4)

From (3), (4), $a \wedge b = a$

Conversely suppose $a \wedge b = a$

Since $a \wedge b \leq b$, $a \leq b$.

Theorem. Let (L, ≤) be a lattice $a, b, c, d \in$ L. Then prove the following :

(*i*) $b \leq c \Rightarrow a \wedge b \leq a \wedge c$, $a \vee b \leq a \vee c$. (Isotonicity laws)

(*ii*) $a \leq b$, $c \leq d \Rightarrow a \vee c \leq b \vee d$, $a \wedge c \leq b \wedge d$.

Proof : (*i*) We have $b \leq c$ iff $b \wedge c = b$.

To show that $a \wedge b \leq a \wedge c$, it is enough to show that

$$(a \wedge b) \wedge (a \wedge c) = a \wedge b$$

Now $(a \wedge b) \wedge (a \wedge c) = (a \wedge a) \wedge (b \wedge c)$ (?)

$= a \wedge (b \wedge c) = a \wedge b$ since $b \wedge c = b$.

$\Rightarrow$ $a \wedge b \leq a \wedge c$.

and
$$\begin{aligned}(a \vee b) \vee (a \vee c) &= [(a \vee b) \vee a] \vee c \\ &= [a \vee (a \vee b)] \vee c \\ &= [(a \vee a) \vee b] \vee c \\ &= [(a \vee a) \vee (b \vee c) \\ &= (a \vee a) \vee (b \vee c) \\ &= a \vee (b \vee c) \qquad (b \leq c \Rightarrow b \vee c = c) \\ &= a \vee c\end{aligned}$$

$\Rightarrow$ $a \vee b \leq a \vee c$ (Since $x \leq y \Rightarrow x \vee y = y$)

(*ii*) We have $a \leq b \Rightarrow a \vee b = b$, $a \wedge b = a$

$c \leq d \Rightarrow c \vee d = d$, $c \vee d = c$.

Now $(a \vee c) \vee (b \vee d) = [(a \vee c) \vee b] \vee d$

$= [b \vee (a \vee c)] \vee d$

$= [(b \vee a) \vee c] \vee d$

$= (b \vee a) \vee (c \vee d)$

$= b \vee d$

$\therefore$ $a \vee c \leq b \vee d$;

since $x \vee y = y \Rightarrow x \leq y$

and $(a \wedge c) \wedge (b \wedge d) = [(a \wedge c) \wedge b] \wedge d$

$= [b \wedge (a \wedge c)] \wedge d$

$= [(b \wedge a) \wedge c] \wedge d$

$= (b \wedge a) \wedge (c \wedge d) = a \wedge c$

$\therefore$ $a \wedge c \leq b \wedge d$

Theorem. A totally ordered set (chain) is a lattice. Is converse true ? Justify your answer.

Proof : Let (P, ≤) be a chain

Let $x, y \in P \Rightarrow x \leq y$ (or) $y \leq x$

if $x \leq y \Rightarrow$ l.u.b $\{x, y\} = y \in P$, g.l.b $\{x, y\} = x \in P$

and if $y \leq x \Rightarrow$ l.u.b $\{x, y\} = x$, g.l.b $\{x, y\} = y \in P$

⇒ (P, ≤) be a lattice.

Converse is need not true.

For example, $(P(A) = \{\phi, \{a\}, \{b\}, \{a, b\}\}, \subseteq)$ is a lattice

but it is not a chain because $\{a\}$ $\{b\}$ are not comparable.

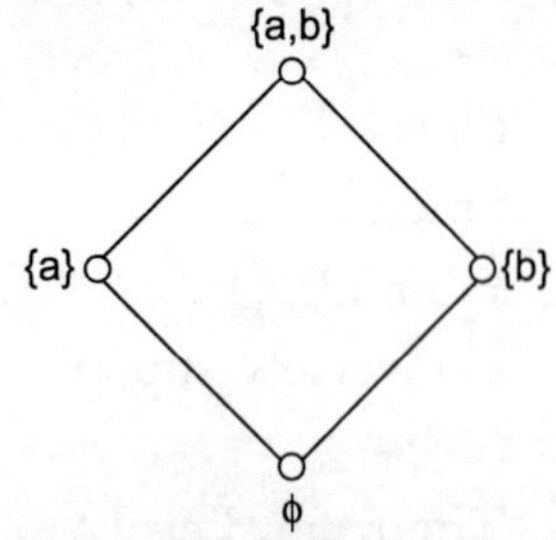

Figure 3.45.

Example. *If $A = \{a, b\}$, then prove that $(P(A), \subseteq)$ is a lattice.*

Solution. Given $A = \{a, b\}$

$\Rightarrow$ $P(A) = \{\phi, \{a\}, \{b\}, \{a, b\}\}$

Clearly $(P(A), \subseteq)$ is a poset (verify)

Table 3.5

g.l.b	ϕ	{a}	{b}	A
ϕ	ϕ	ϕ	ϕ	ϕ
{a}	ϕ	{a}	ϕ	{a}
{b}	ϕ	ϕ	{b}	{b}
A	ϕ	{a}	{b}	A

Table 3.6

l.u.b	ϕ	{a}	{b}	A
ϕ	ϕ	{a}	{b}	A
{a}	{a}	{a}	A	A
{b}	{b}	A	{b}	A
A	A	A	A	A

Since all entries in above two tables are in P (A), $(P(A), \subseteq)$ is a lattice.

Example. *Let C be a collection of sets which are closed under intersection and union. Verify whether (C ∩, ∪) is a lattice.*(JNTU March, 2006, Set 1, May 05 - Set 1, May 04 - Set 2, May 04 - Set 2)

Solution. Let C be a collection of sets and given that C is closed under union and intersection.

i.e., for any A, B ∈ C ⇒ A ∪ B ∈ C and A ∩ B ∈ C

Define a relation ⊆ on C.

First we have to show that (C, ⊆) is a poset.

(*i*) Let A ∈ C ⇒ A ⊆ A ⇒ ⊆ is a reflexive on C.

(*ii*) Let A, B ∈ C, A ⊆ B, B ⊆ A ⇒ A = B

⇒ ⊆ is symmetric on C.

(*iii*) Let A, B, D ∈ C, A ⊆ B, B ⊆ D ⇒ A ⊆ D

⇒ ⊆ is transitive on C.

∴ (C, ⊆) is a poset.

Note that A ⊆ A ∪ B, B ⊆ A ∪ B.

Define ⊆ on ⊂ by A ∨ B = A ∪ B, A ∨ B = A ∩ B.

l.u.b of A, B = A ∨ B = A ∪ B ∈ C and

g.l.b. of A, B = A ∧ B = A ∩ B ∈ C for all A, B ∈ C.

Hence (C, ∩, ∪) is a lattice.

Definition. *Let (L, ∧, ∨) be a lattice, S ⊆ L, then S is said to be sublattice of lattice L if ∧, ∨ are binary operations on S. (i.e.,) S is closed under ∧, ∨*

(*i.e.,*) $a \wedge b \in S$, $a \vee b \in S\ \forall\ a, b \in S$.

Example. Let (L, ≤) be a lattice, where L = $\{a_1, a_2, a_3, a_4, a_5, a_6, a_7, a_8\}$ and its hasse diagram is as follows :

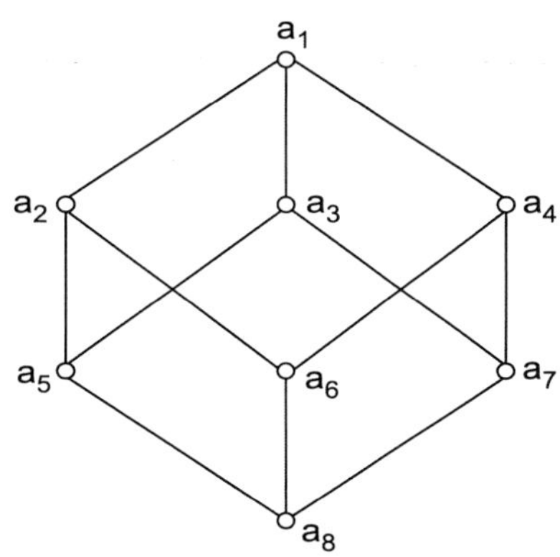

Figure 3.46.

Now consider the subsets of L given by

$$S_1 = \{a_1, a_2, a_4, a_6\}$$
$$S_2 = \{a_3, a_5, a_7, a_8\} \text{ and}$$
$$S_3 = \{a_1, a_2, a_4, a_8\}$$

We see that $(S_1, \leq)$, $(S_2, \leq)$ are sublattices of $(L, \leq)$, but $(S_3, \leq)$ is not a sublattice, because $a_2, a_4 \in S_3$, but $a_2 \wedge a_4 = a_6 \notin S_3$

Table 3.7

$\wedge$	a_1	a_2	a_4	a_6
a_1	a_1	a_2	a_4	a_6
a_2	a_2	a_2	a_6	a_6
a_4	a_4	a_6	a_4	a_6
a_6	a_6	a_6	a_6	a_6

Table 3.8

$\vee$	a_1	a_2	a_4	a_6
a_1	a_1	a_1	a_1	a_1
a_2	a_1	a_2	a_1	a_2
a_4	a_1	a_1	a_4	a_4
a_6	a_1	a_2	a_4	a_6

From these tables, we observed that S_1 is closed under $\wedge$, $\vee$

$\therefore$ S_1 is a sublattice of L.

And similarly we can easily verify that S_2 is a sublattice of L.

Example. *Consider the lattice L shown in Fig. 3.47. Find four sublattices with 5 elements.*

Solution. The sublattices of L are

$$S_1 = \{a_1, a_2, a_4, a_5, a_7\}$$
$$S_2 = \{a_1, a_2, a_3, a_4, a_7\}$$
$$S_3 = \{a_1, a_3, a_5, a_6, a_7\}$$
$$S_4 = \{a_1, a_2, a_3, a_6, a_7\}.$$

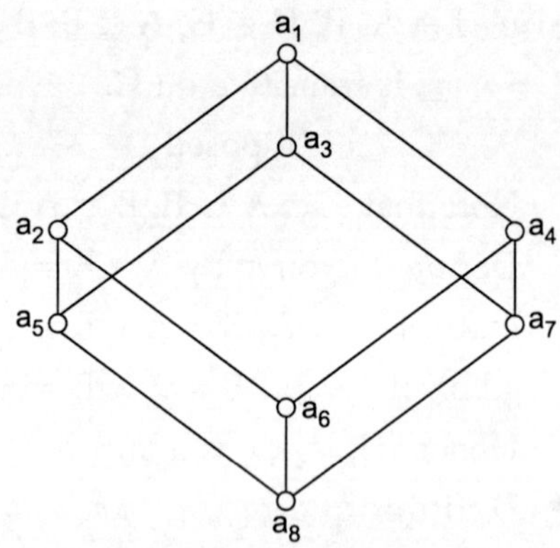

Figure 3.47.

Example. *Consider the lattice L shown in Fig. 3.48, find posets which are not sublattices of L.*

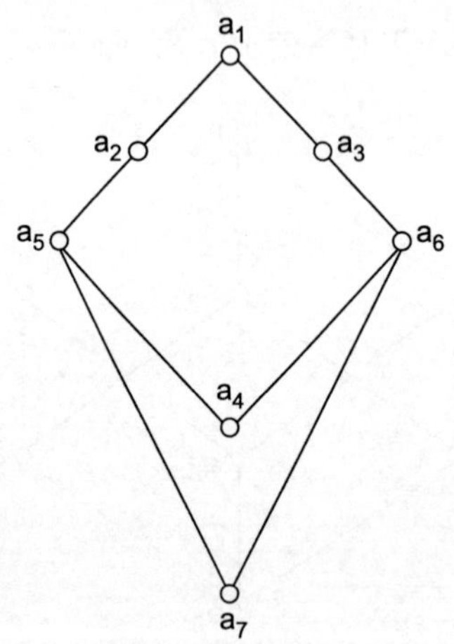

Figure 3.48.

Solution. The Hasse diagram S_1 is not a sublattice of L because $a_5 \wedge a_6 \notin$ S, $a_5 \vee a_6 \notin S_1$. and S_2 is not a sublattice of L because $a_5 \vee a_6 \notin S_2$.

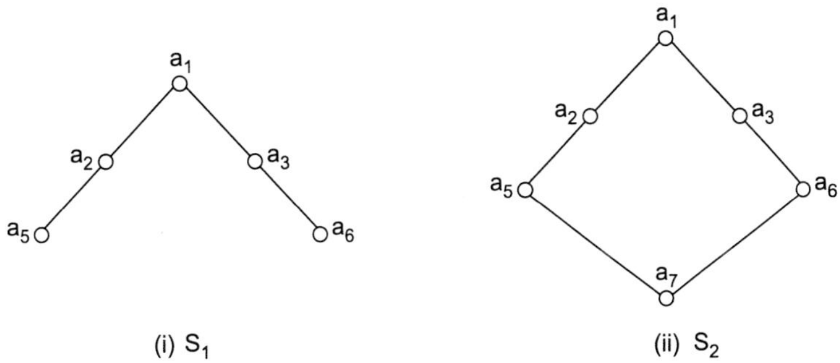

Fig. 3.49.

3.17.1 Types of Lattices

1. **Modular Lattice :** A lattice L is said to be modular if $a \le c \Rightarrow a \vee (b \wedge c) = (a \vee b) \wedge c$.

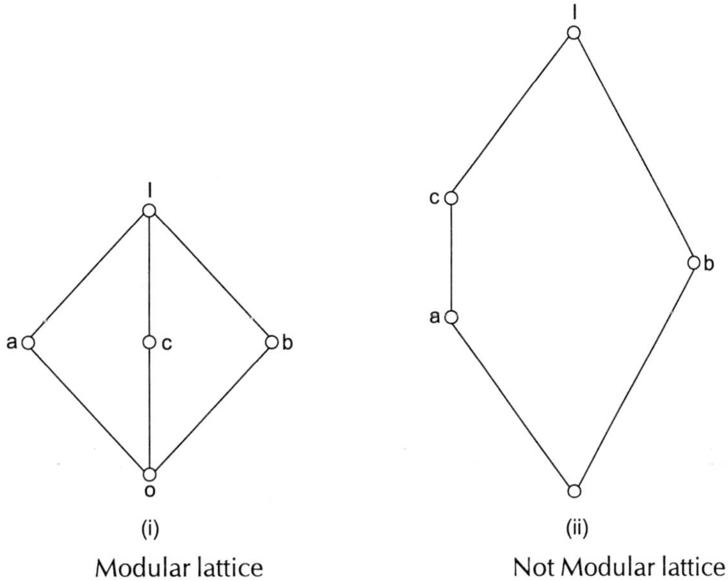

Fig. 3.50.

because $\quad a \vee (b \wedge c) = a \vee o = a$

$(a \vee b) \wedge c = 1 \wedge c = c.$

2. **Complemented Lattice :** An element a in a bounded lattice L is said to be complement of b if $a \wedge b = 0$, $a \vee b = 1$.

   A lattice L is said to be complemented lattice if it is bounded and every element in L has atleast one complement.

   Note. The complement of a in L is denoted by $\bar{a}$.

   and $\quad \bar{0} = 1, \bar{1} = 0.$

   Note. Every chain is not complementary lattice.

Examples. *Consider the lattices given by following Hasse diagrams, which of them are complemented.*

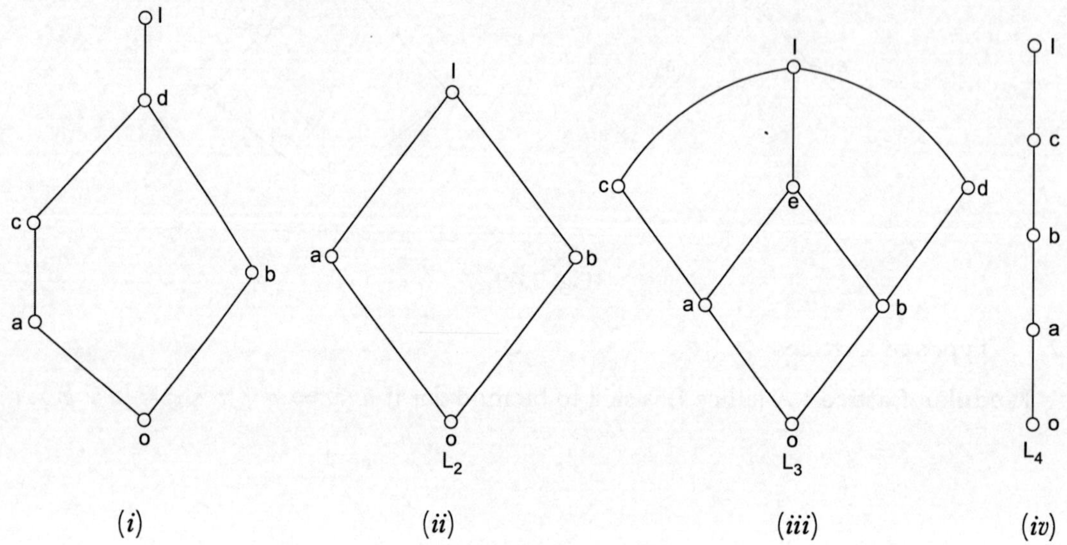

Fig. 3.51.

Solution.

(*i*) Lattice L_1 is not complemented because b has no complement.

(*ii*) Lattice L_2 is complemented.

(*iii*) Lattice L_3 is not complemented because *e* has no complement.

(*iv*) Lattice L_4 is not complemented because each element does not have complements.

3. **Bounded Lattices :** A lattice L is said to be bounded if it has both a least element and a greatest element.

Note 1. In bounded lattice, least element is denoted by 0 and greatest element is denoted by 1.

Note 2. If L is a bounded lattice, then for any element $a \in L$, we have the following identities.

(*i*) $0 \le a \le 1$. (*ii*) $a \vee 0 = a,\ a \wedge 0 = 0$ (*iii*) $a \wedge | = a,\ a \vee | = |$.

Examples 1. The set of non-negative integers with usual ordering $0 < 1 < 2 < 3 < ......$ has a least element zero but there is no greatest element.

2. The power set P (U) of all subsets of a universal set U is a bounded lattice since it has least element ϕ, greatest element U.

4. **Distributive Lattices :** A lattice L is said to be distributive if for any elements $a, b, c \in L$, the following distributive laws hold.

(*i*) $a \wedge (b \vee c) = (a \wedge b) \vee (a \wedge c)$

(*ii*) $a \vee (b \wedge c) = (a \vee b) \wedge (a \vee c)$

Example.

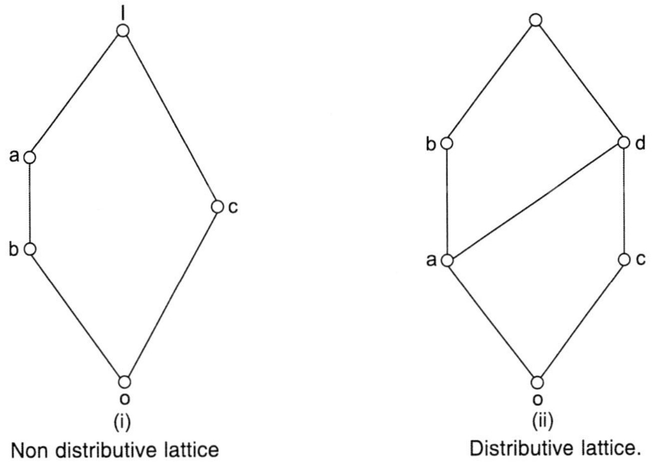

Figure 3.52.

Theorem. Show that every distributive lattice is modular. Is converse true ? Justify your answer.

Proof : Suppose L is a distributive lattice.

Let $x, y, z \in L \ni x \leq z$

$$x \vee (y \wedge z) = (x \vee y) \wedge (x \vee z) = (x \vee y) \wedge z$$

$\therefore$ L is modular lattice.

Converse is need not to be true. For example, lattice shown in Fig. 3.53.

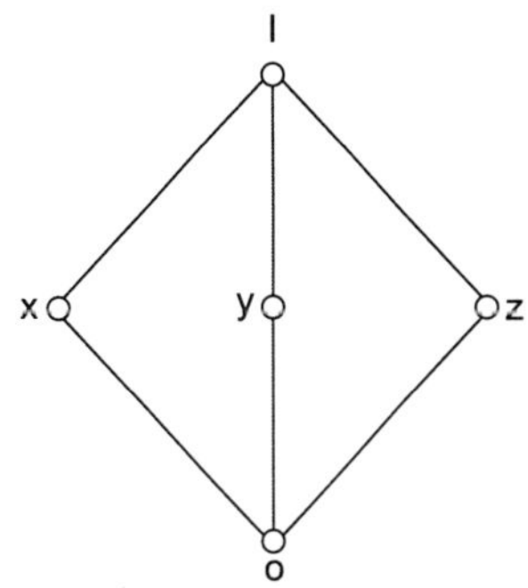

Figure 3.53.

For any $x, y, z \in L, x \vee (y \wedge z) = x \vee 0 = x,$

$$(x \vee y) \wedge (x \vee z) = 1 \wedge 1 = 1,$$

So $x \vee (y \wedge z) = x \neq 1 = (x \vee y) \wedge (x \vee z)$

$\therefore$ L is not distributive, and we know that L is modular lattice.

Theorem. Let L be a bounded distributive lattice and $a \in$ L. If a has a complement in L, then the complement is unique.

Proof : Suppose b and c are two complements of a in L.

Then $a \vee b = 1, a \wedge b = 0$

$$a \vee c = 1, a \wedge c = 0.$$

Since L is distributive, the distributive laws hold. Using these laws, we find that

$$\begin{aligned} b = b \vee 0 = b \vee (a \wedge c) &= (b \vee a) \wedge (b \vee c) \\ &= 1 \wedge (b \vee c) = (a \vee c) \wedge (b \vee c) \\ &= (a \wedge b) \vee c = 0 \vee c = c \end{aligned}$$

$\therefore$ The complement of a is unique.

Theorem. In a distributive lattice L, $a \wedge x = a \wedge y$, $a \vee x = a \vee y$, then prove $x = y$.

Proof : Suppose L is distributive lattice, let $a, x, y \in$ L.

Suppose $\quad a \wedge x = a \wedge y,\ a \vee x = a \vee y,$

Then
$$x = (a \vee x) \wedge x = (a \vee y) \wedge x = x \wedge (a \vee y)$$
$$= (x \wedge a) \vee y = (a \wedge x) \vee y = (a \wedge y) \vee y = y.$$

$\therefore \quad x = y.$

Theorem. Prove that every chain is a distributive lattice.

Proof : 3.1 theorem, every chain is lattice.

Let $a, b, c \in$ chain (L, $\leq$). Then there exist the following two possible cases :

(*i*) $a \leq b$ (or) $a \leq c$

(*ii*) $b \leq a$ (or) $c \leq a$.

In case (*i*), we have $a \wedge (b \vee c) = \mathrm{a}$

($\because \quad a \leq b$, or $a \leq c \Rightarrow a \leq b \vee c \Rightarrow a \wedge (b \vee c) = a$)

and $\quad (a \wedge b) \vee (a \wedge c) = a.$

In case (*ii*), we have
$$a \wedge (b \vee c) = b \vee c \text{ and}$$
$$(a \wedge b) \vee (a \wedge c) = b \vee c.$$

Thus, in both cases, distributive law holds.

$\therefore$ A chain is a distributive lattice.

Example. *Consider the lattices given by following Hasse diagrams, which of them are complemented ?*

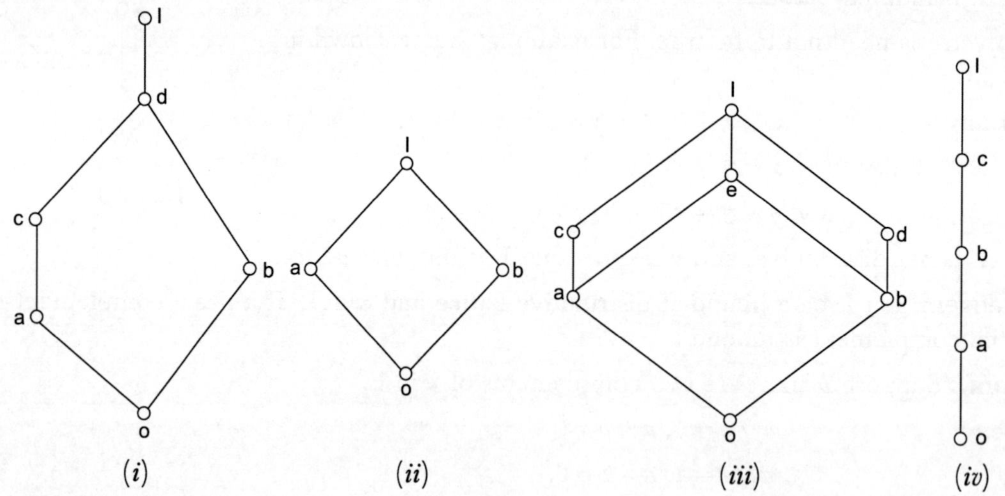

Figure 3.54.

Solution.

(*i*) Lattice is not complemented because b has no complement.

(*ii*) Lattice is complemented.

(*iii*) It is not complemented, because e has no complement.

(*iv*) It is not complemented, because a, b, c have no complements.

Example. *Check whether the following lattice distributive (or) not ?*

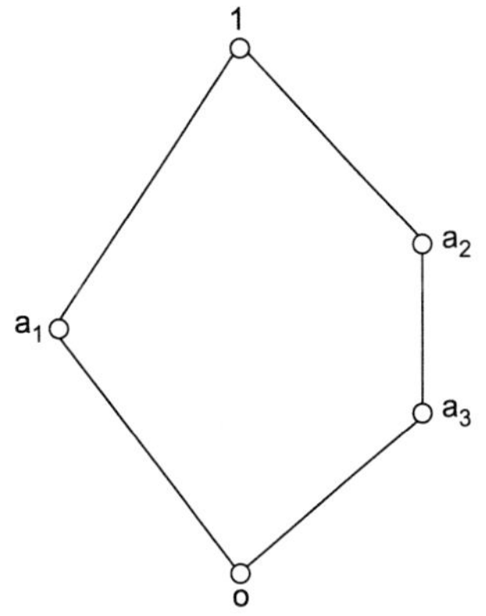

Figure 3.55.

Solution. $a_1 \wedge (a_2 \vee a_3) = a_1 \wedge a_2 = 0,$

$(a_1 \wedge a_2) \vee (a_1 \wedge a_3) = 0 \vee 0 = 0$

$\therefore$ $a_1 \wedge (a_2 \vee a_3) = (a_1 \wedge a_2) \vee (a_1 \wedge a_3)$

but $a_2 \wedge (a_1 \vee a_3) = a_2 \wedge 1 = a_2,$

$(a_2 \wedge a_1) \vee (a_2 \wedge a_3) = 0 \wedge a_3 = a_3$

$\therefore$ $a_2 \wedge (a_1 \vee a_3) = a_2 \neq a_3 = (a_2 \wedge a_1) \vee (a_2 \wedge a_3)$

Hence lattice is not distributive.

Example. If a, b, c are elements of a distributive lattice $(L, \wedge, \vee)$ and $a \vee b = a \vee c$, $a \wedge b = a \wedge c$, then $b = c$.

Proof : Let $(L, \wedge, \vee)$ be a distributive lattice, $a, b, c, \in L$

$\therefore$

$$
\begin{aligned}
b &= b \vee (a \wedge b) && \text{(Absorption)} \\
&= b \vee (a \wedge c) && (\because\ a \wedge b = a \wedge c) \\
&= (b \vee a) \wedge (b \vee c) && \text{(Distributive)} \\
&= (a \vee b) \wedge (c \vee b) && \text{(commutative)} \\
&= (a \vee c) \wedge (c \vee b) && (\because\ a \vee b = a \vee c) \\
&= (c \vee a) \wedge (c \vee b) = c \vee (a \wedge b) \\
&= c \vee (a \wedge c) = c.
\end{aligned}
$$

Example. *Determine which of the following lattices are modular :*

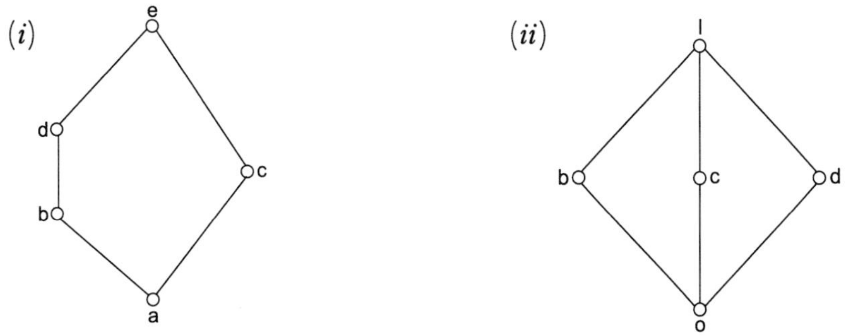

Figure 3.56.

Solution. (*i*) Here $b \leq d$, $b \vee (c \wedge d) = b \vee a = b$

and $(b \vee c) \wedge d = e \wedge d = d$

$\therefore$ $b \vee (c \wedge d) \neq (b \vee c) \wedge d$

$\therefore$ (*i*) is not a modular lattice.

(*ii*) Here $0 \leq b$, $0 \vee (c \wedge b) = 0 \vee 0 = 0$, $(0 \vee c) \wedge b = c \wedge b = 0$

$\therefore$ $0 \vee (c \wedge b) = (0 \vee c) \wedge b$

and $0 \leq d$, $0 \vee (c \wedge d) = 0 \vee 0 = 0$, $(0 \vee c) \wedge d = c \wedge d = 0$

$\therefore$ $0 \vee (c \wedge d) = (0 \vee c) \wedge d$

Thus for any x, y, z in the given lattice, we have

$$x \leq y \Rightarrow x \vee (z \wedge y) = (x \vee z) \wedge y$$

$\therefore$ Lattice is modular.

Example. *Let A = {a, b, c}, then prove the following :*

(i) (P (A), ∪, ∩) is a bounded lattice.

(ii) (P (A), ∪, ∩) is a complemented lattice.

Solution. Given A = {*a*, *b*, *c*}, then

P (A) = {ϕ, {*a*}, {*b*}, {*c*}, {*a*, *b*}, {*a*, *c*}, {*b*, *c*}, A} and its Hasse diagram is given below and clearly (P (A), ∪, ∩) is a lattice.

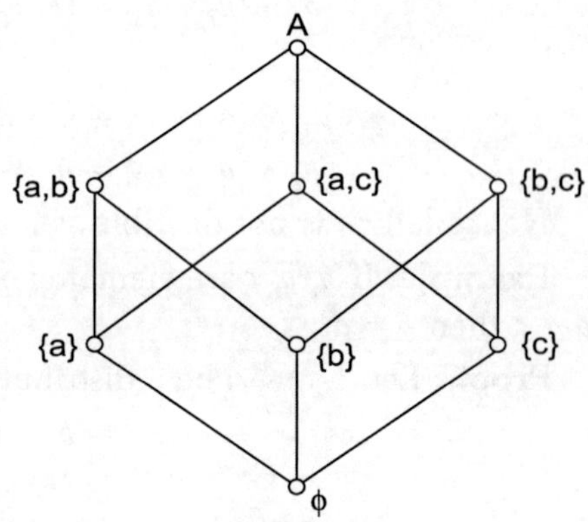

Figure 3.57.

(*i*) P (A) contains least member, say ϕ, P (A) contains greatest member, say A,

Then (P (A), ∪, ∩) is a bounded lattice.

(*ii*) Complement of {*a*} is {*b*, *c*},

Complement of {*b*} is {*a*, *c*}

Complement of {*c*} is {*a*, *b*},

then (P (A), ∪, ∩) is a Complemented lattice.

Duality Principle of Lattices

Any statement about lattices involving the operations $\wedge$ and $\vee$ and the relations $\leq$ and $\geq$ remains true if $\wedge$ is replaced by $\vee$, $\vee$ by $\wedge$, $\leq$ by $\geq$ and $\geq$ by $\leq$.

The operations $\wedge$, $\vee$ are called duals of each other as the relations $\leq$ and $\geq$, similarly lattices (L, $\leq$) and (L, $\geq$) are called duals of each other.

Example. *Find the dual of the following statements :*

(*i*) $(a \wedge b) \vee c = (b \vee c) \wedge (c \vee a)$

(*ii*) $(a \wedge b) \vee a = a \wedge (b \vee a)$

Solution. (*i*) $(a \vee b) \wedge c = (b \wedge c) \vee (c \wedge a)$

(*ii*) $(a \vee b) \wedge a = a \vee (b \wedge a)$.

Lattice Homomorphism, Isomorphism

Definition. *Let (L, *, ⊕), (S, ∧, ∨) are two lattices. A mapping f : L → S is said to be* ***lattice homomorphism*** *if*

(i) $f(a * b) = f(a) \wedge f(b)$

(ii) $f(a \oplus b) = f(a) \vee f(b)\ \forall\ a, b \in L$.

and if f is one-one, onto from L to S, then f is called **isomorphism**. Hence lattices L, S are said to be **isomorphic**. The function f from L to S is said to be **order preserving** relative to the ordering $\leq_1$ in L and $\leq_2$ in S iff $a \leq_1 b \Rightarrow f(a) \leq_2 f(b)$. $\forall\ a, b \in L$.

Example. Let D_6 denote lattice and power set P (A) of the set $A = \{a, b\}$ denotes the another lattice. Since posets D_6 and P (A) are isomorphic, therefore both lattices D_6 and P (A) are isomorphic.

EXERCISE 3.4.

1. A lattice on the set $A = \{a, b, c, d, e, f, g\}$ is given in figure shown below.

Find

(i) $f \wedge g$;

(ii) $d \wedge e$;

(iii) $a \wedge (b \wedge c)$;

(iv) $a \wedge (b \vee c)$;

(v) $f \vee g$;

(vi) $d \vee e$;

(vii) $d \vee b$; and

(viii) $d \wedge b$.

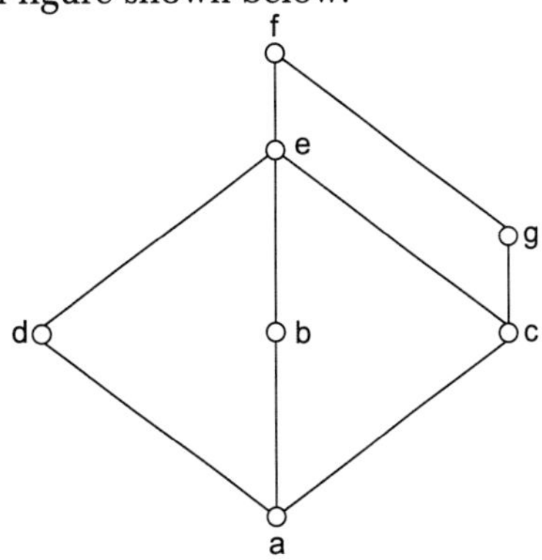

Figure 3.58.

2. Which of the following posets are lattices.

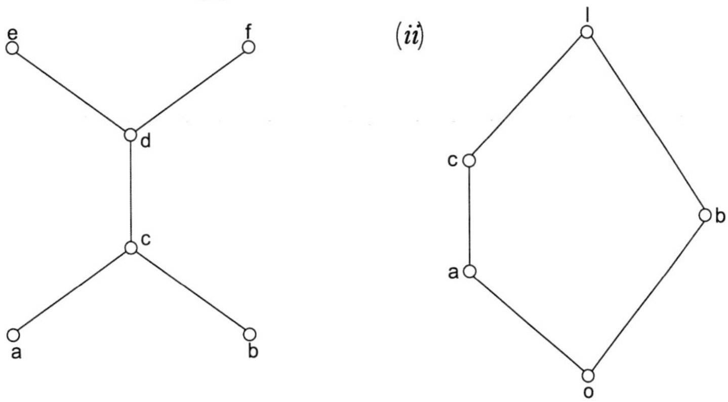

Figure 3.59.

3. Show that in a lattice if $a \leq b$ and $c \leq d$, then $a \wedge c \leq b \wedge d$.

4. Show that a sub lattice of a distributive lattice is distributive.

5. Show that in a complemented, distributive lattice,

(i) $(a \wedge b)' = a' \vee b'$ (ii) $(a \vee b)' = a' \wedge b'$.

(iii) $a \wedge b' = 0$ iff $a' \wedge b = 1$.

6. Show that, if a bounded lattice has two or more elements, then prove $0 \neq 1$.
7. Find the complement of each element of $(D_{12}, |)$.
8. Show that a lattice is distributive if and only if $(a \wedge b) \vee (b \wedge c) \vee (c \wedge a) = (a \vee b) \wedge (b \vee c) \wedge (c \vee a)$.
9. Show that every distributive lattice is modular, but not conversely.
10. Which of the following lattices are modular.

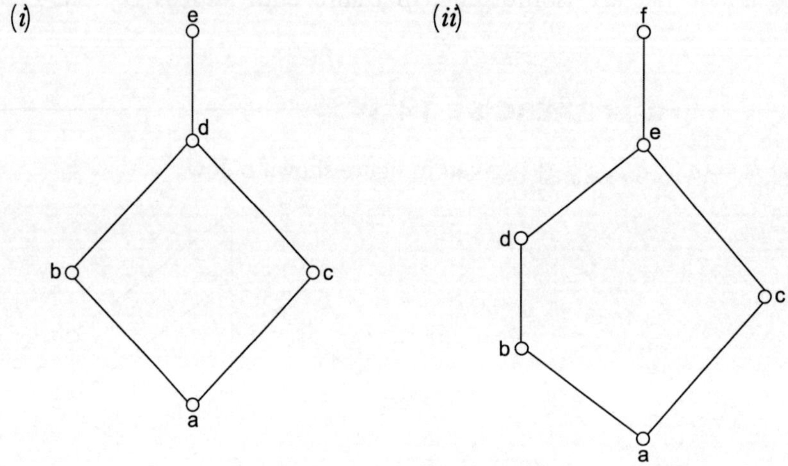

Figure 3.60.

ANSWERS

1. (*i*) *g* (*ii*) *d* (*iii*) *a* (*iv*) *a*
   (*v*) *f* (*vi*) *e* (*vii*) *e* (*viii*) *a*
2. (*i*) Not a lattice, because g.l.b of *a*, *b* does not exist.
   (*ii*) A lattice.
7. $1' = 42$, $2' = 21$, $3' = 14$, $6' = 7$, $7' = 6$, $14' = 3$, $21' = 2$, $42' = 1$.
10. (*i*) It is modular.
    (*ii*) It is not modular, because $b \leq d$, $b \vee (c \wedge d) \neq (b \vee c) \wedge d$.

3.18 FUNCTIONS

In the study of mathematics, functions provide an important unifying concept. Functions are also familiar in computer science as components (or) programs that formalize the relationship between the input and output for a computation.

Definition. *Let A and B be any two sets. A relation f from A to B is said to be a **function** (mapping) from A to B if for every $a \in A$, there is a unique $b \in B \ni (a, b) \in f$. A function f from A to B is also can be written as $f : A \rightarrow B$ which is read as "a function of A to B" (or) "f is mapping from A to B" (or) "f maps A to B". The pictorial representation of f is as shown below :*

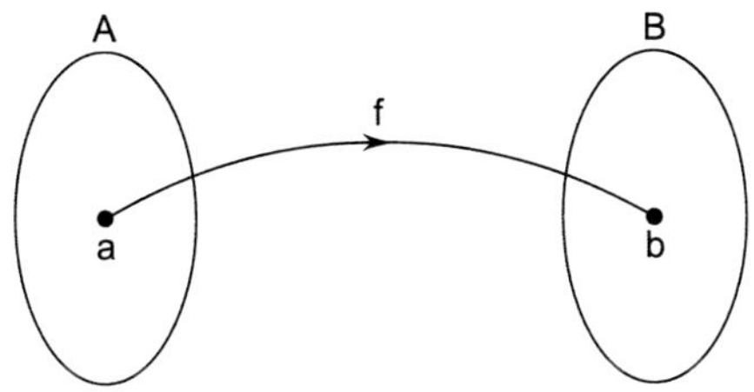

Figure 3.61.

Note 1. Functions are called **mappings** or **transformations** or **correspondence**.

2. If $f: A \to B$ be a function then A is called **domain** of f, B is called **codomain** of f.

3. Every function from A to B is a relation from A to B. But every relation from A to B need not be a function from A to B.

4. One-one and many-one relations are functions but one-many and many-many relations are not functions.

Definition. *If domain and codomain of a function f are both the same set. Then function f is called* ***operator*** *or* ***transformation*** *on that set.*

i.e., $f: A \to A$ is called operator or transformation on A.

Example 1. *Let N be the set of all natural numbers and R be the set of all real numbers. Then* $f(x) = \sqrt{x}\ \forall x \in N$ *is a function from N to R.*

Example 2. Let $A = \{1, 2, 3\}$, $B = \{p, q, r, s\}$ and

$f = \{(1, p), (2, r), (3, s)\}$ then f is function from A to B and this function may be represented by a diagram as shown below.

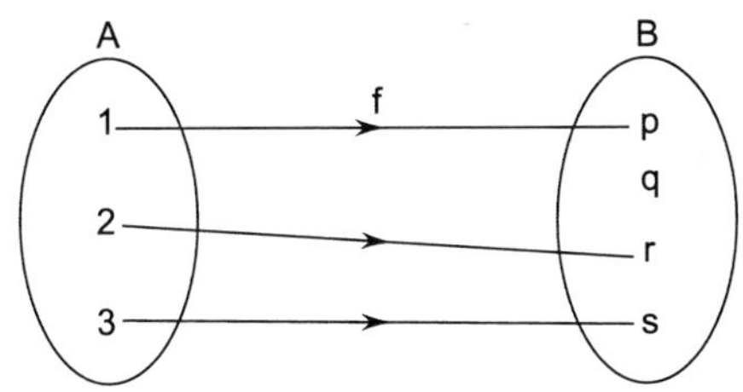

Fig. 3.62.

Example 3. Let R be the set of real numbers, A function $f: R \to R$ defined by $f(x) = x^3\ \forall x \in R$, then f is called operator or transformation on R.

Example 4. Let $f = \{(1, 2), (2, 3), (3, 1), (4, 4)\}$ be a function of $\{1, 2, 3, 4\}$ into itself and f is called operator or transformation on $\{1, 2, 3, 4\}$.

Definition. *Let* $f: A \to B$ *be a function and an element* $a \in A$ *is associated with an element* $b \in B$, *then b is called f-image of a or b is called* ***the image*** *of a under the mapping f and is written as* $f(a) = b$. *Also a is called the* ***pre-image*** *(or)* ***inverse image*** *of b.*

Definition. *The set of those elements in B which appear as the f-image of atleast one element in A is called the range of f. The range of* $f: A \to B$ *is denoted by* $f(A)$ *and* $f(A) = f(a) \mid a \in A\}$.

Note 1. $f(A) \subseteq B$

2. $f \subseteq A \times B$

3. Every element of A has a unique f-image in B.

4. Two or more elements of A can have the same f-image in B.

5. There may be some elements in B which are not f-images of elements of A.

Example. *If $A = \{-1, 0, 1\}$ and $f : A \to R$ defined by $f(x) = x^2 \; \forall x \in A$. Then find range of f and write domain and codomain of f.*

Solution. Range of $f = \{0, 1\}$ and domain of f is $A = \{-1, 0, 1\}$ codomain is R.

Example. *State whether or not each of the relations given below defines a function of $A = \{a, b, c\}$ into $B = \{1, 2, 3\}$.*

(i) $f = \{(a, 2), (a, 3), (b, 3), (c, 1)\}$ (ii) $f = \{(a, 2), (b, 3)\}$

(iii) $f = \{(a, 1), (b, 3), (c, 1)\}$ (iv) $f = \{(a, 1), (b, 2), (c, 3)\}$

(v) $f = \{(a, 1), (b, 1), (c, 1)\}$ (vi) $f = \{(a, 1), (a, 2), (b, 3), (c, 3)\}$

(vii) $f = \{(a, 1), (a, 2), (b, 3), (c, 2)\}$.

Solution.

(i)

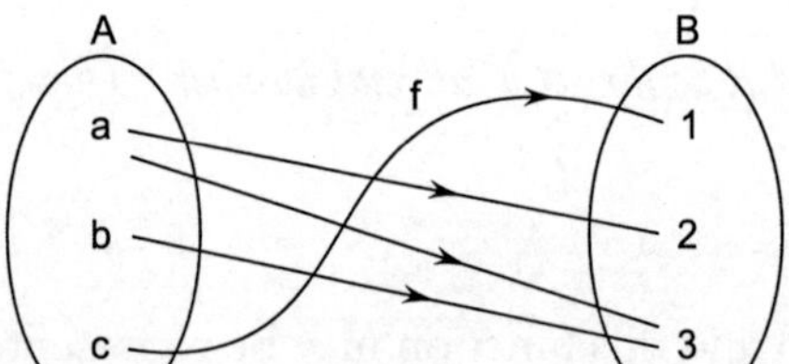

∴ f is not function because a has not unique image in B.

(ii)

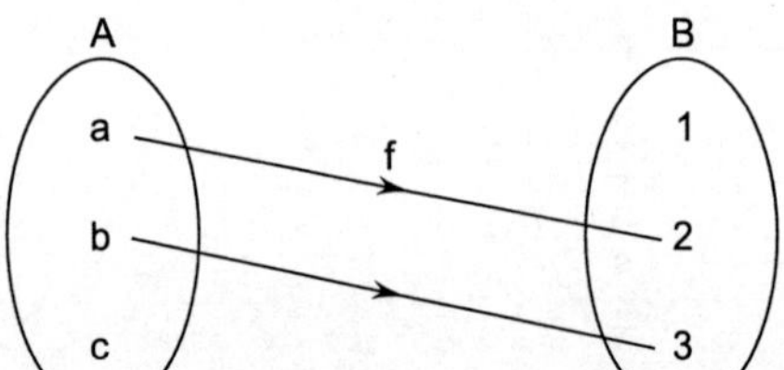

∴ f is not a function because c has no f-image in B.

(iii)

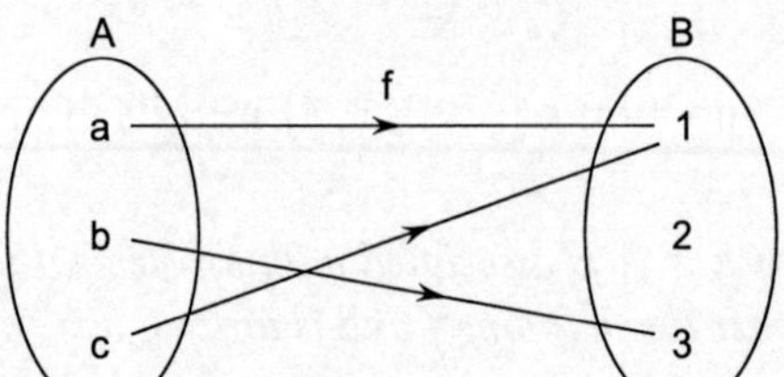

∴ f is a function.

Figure 3.63... continued...

(iv)

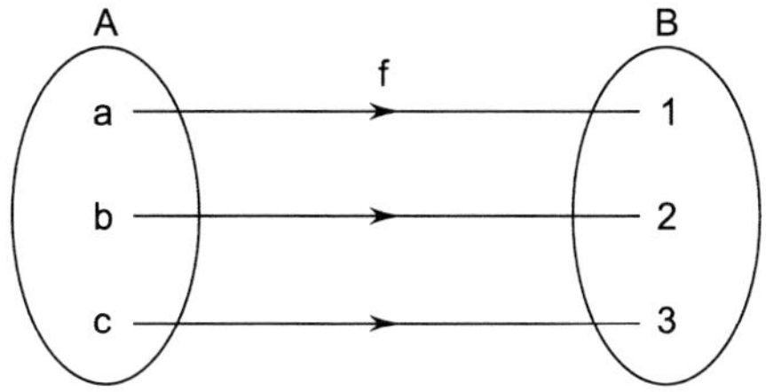

∴ f is a function.

(v)

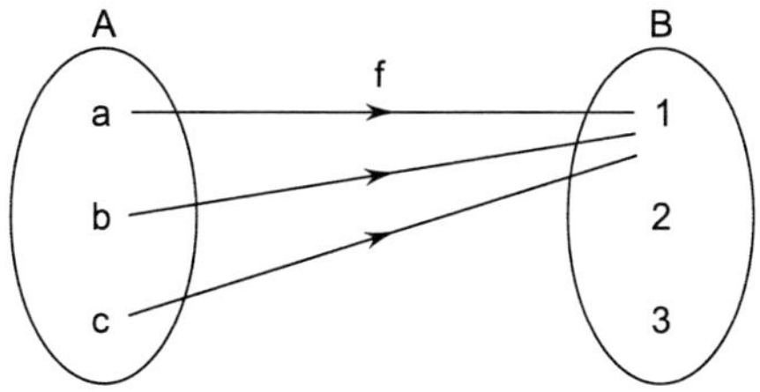

∴ f is a functioin.

(vi)

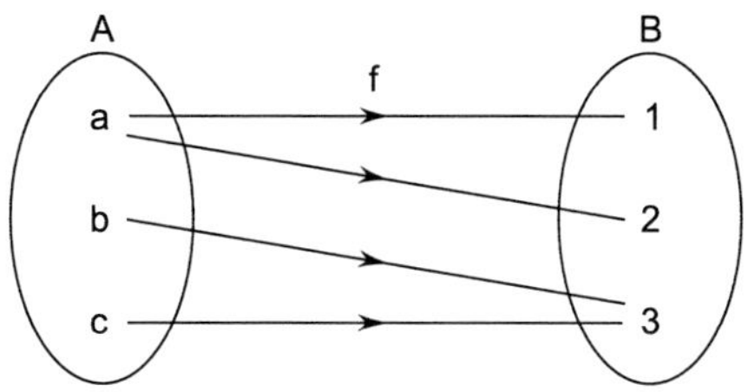

∴ f is not a function because $a \in$ A have not unique image in B.

(vii)

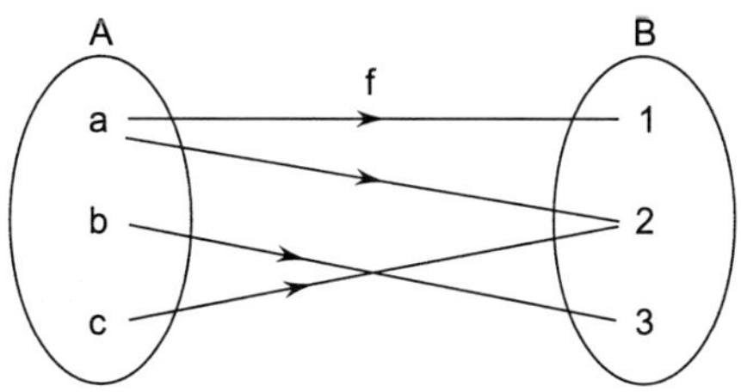

∴ f is not a function because f is many many relation.

Figure 3.63.

Definition. *A mapping $f : A \to B$ is called a one-one (Injective or 1 – 1) function if distinct elements of A have distinct images in B ... (i.e.,), if $a_1, a_2 \in A \ni a_1 \neq a_2 \Rightarrow f(a_1) \neq f(a_2)$. Thus $f : A \to B$ is one-one if and only if for any $a_1, a_2 \in A \ni f(a_1) = f(a_2) \Rightarrow a_1 = a_2$.*

Note 1. If A and B are finite sets and f: A → B is injective then n (A) ≤ n (B).

2. If n (A) = m, n (B) = n, then the number of possible injective functions from A to B is ${}^{n}p_{m}$.

Example 1. If A = {1, 2, 3} and B = {p, q, r, s} and if f: A → B such that f = {(1, p), (2, r), (3, s)}, then f is one-one.

Example 2. If R is a set of real numbers and f: R → R is defined by $f(x) = x^3$, then f is one-one function.

Example 3. f: R → R where R is a set of real number defines by $f(x) = x^2$ is not one-one because $f(-2) = 4$, $f(2) = 4$.

Definition. *A function $f : A \to B$ is said to be into function if there is atleast one element of B which is not the f-image of any element of A.*

Note. f: A $\to$ B into function if f(A) $\subset$ B.

Example. f: R $\to$ R where R better set of real numbers, such that $f(x) = x^2$ is an into function because – 1 of codomain R has no preimage in the domain R.

Definition. *A function $f : A \to B$ is said to be onto (or) surjective function if for every element $b \in B$ there an element $a \in A \ni f(a) = b$.*

Note. f: A $\to$ B onto if f (A) = B

(*ii*) Range of f = codomain of f.

Note 1. If A, B are finite sets and f: A $\to$ B is surjective, then n (B) $\leq n$ (A).

2. If n (A) = m and n (B) = m, then the number of possible surjective functions from A to B is m !.

3. If n (A) = m, n (B) = 2, then the number of possible surjective mappings from A to B is $2^m - 2$.

Example 1. If **Z** is the set of all integers, f: **Z** $\to$ **Z** defined by $f(x) = x + 1$, then f is an onto function.

2. If A = {– 1, 1, – 2, 2, – 3, 3}, B = {1, 4, 9} and f: A $\to$ B is defined by

$$f(x) = x^2 \ \forall \ x \in A. \text{ Then } f \text{ is onto.}$$

3. If **N** is the set of natural numbers, f: N $\to$ N is defined by $f(x) = 3x + 2$ is not an onto function, because there is no integer a in domain **N** $\ni f(a) = 6$.

Definition. *If the function f is both one-one and onto then f is called a one-one onto or bijective function.*

Note 1. If A and B are finite sets and f: A $\to$ B is bijective, then n (A) $\leq n$ (B) and n (B) $\leq n$ (A). Hence n (A) = n (B).

Note 2. If n (A) = m = n (B), then the number of possible bijections from A to B is m !.

Example 1. Let A = {a, b, c, d}, B = {1, 2, 3, 4}, then f = {(a, 4), (b, 1), (c, 2), (d, 3)} is a bijective functioin.

Example 2. Let A = {a, b, c, d, e} and f: A $\to$ A defined by $f(a) = d$, $f(b) = c$, $f(c) = e$, $f(d) = a$, $f(e) = b$, then f is a bijection from A to A.

Example 3. $f(x) = 5x - 2$ defined on R is bijection.

Definition. ***One-one into function :*** *A function $f : A \to B$ is said to be one-one into function if to each element of A there corresponds one element of B but there are some elements of B, which do not correspond to any of the elements of A as shown in the following figure.*

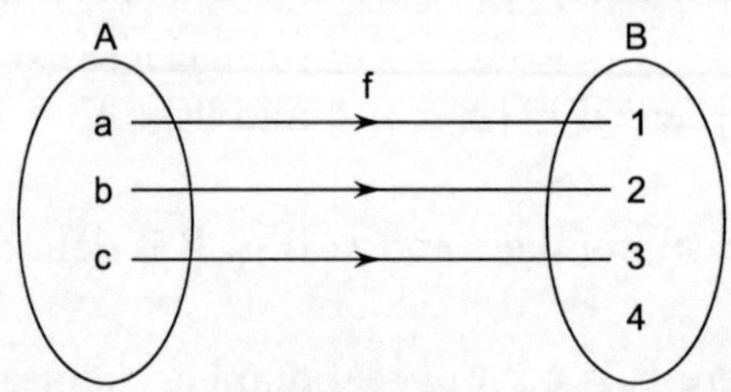

Fig. 3.64.

Definition. ***Many-one into function :*** *A function $f : A \to B$ is said to be many-one into function if two or more elements of a set A corresponds to the same element of B and there are some elements in B which do not correspond to any of the elements of A as shown in the following figure.*

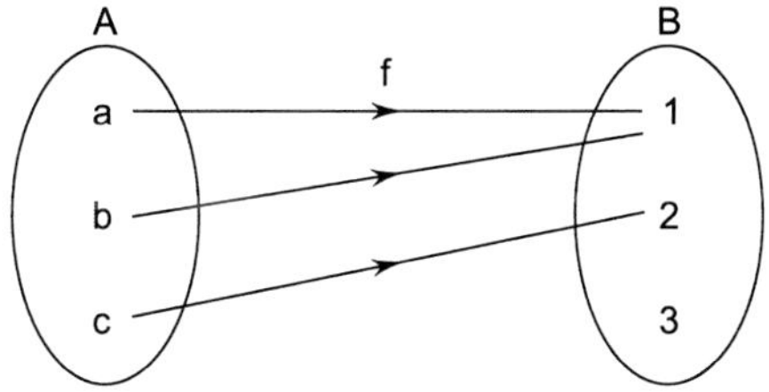

Fig. 3.65.

Definition. ***Many-one onto function :*** *A function $f : A \to B$ is said to be many-one onto function if each element in B is joined to at least one element in A and two or more elements in A are joined to the same element in B as shown in the following figure.*

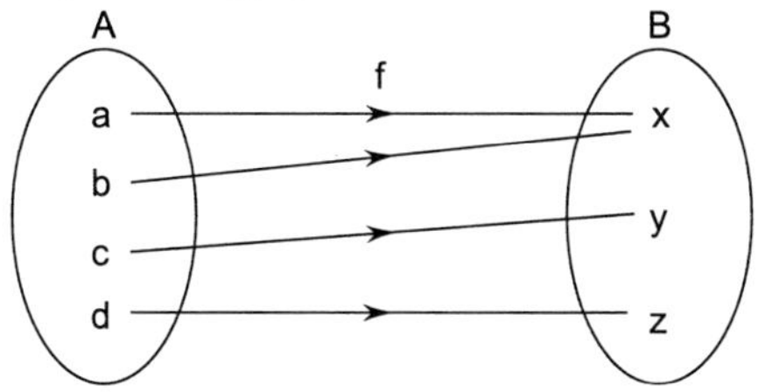

Fig. 3.66.

Definition. *A function $f : A \to A$ defined by $f(x) = x$ is called identity function on A and it is denoted by I_A or I.*

Example. If A = {1, 2, 3}, then $f = \{(1, 1), (2, 2), (3, 3)\}$ is an identity function on A.

Note. Every identity function is a bijection.

Inverse Functions :

Theorem. If f is a bijection from A to B, then the relation f^{-1} from B to A is a function from B to A.

Proof : Let f: A → B is a bijection. Let $f = \{(a, b) \mid a \in A, b \in B\}$

$\Rightarrow \qquad f^{-1} = \{(b, a) \mid (a, b) \in f\}$

Let $y \in B$. Since f is onto $\exists\, x \in A$ such that $(x, y) \in f \Rightarrow (y, x) \in f^{-1}$

$\therefore \qquad \forall\, y \in B\ \exists\, x \in A$ such that $(y, x) \in f^{-1}$

Further $\qquad (y, x_1) \in f^{-1}, (y, x_2) \in f^{-1} \Rightarrow (x_1, y), (x_2, y) \in f$

$\Rightarrow \qquad f(x_1) = y = f(x_2) \Rightarrow x_1 = x_2 \qquad (\because f \text{ is one-one})$

$\therefore$ The relation f^{-1} is a function from B to A.

Definition. *If $f : A \to B$ is a one-one, onto function (i.e., bijection), then the mapping f^{-1} from B to A is called inverse mapping of the function f from A to B.*

Note. Let $a \in A\ \exists\ b \in B \ni f(a) = b$

Thus $\qquad f(a) = b \Rightarrow f^{-1}(b) = a.$

Theorem. If $f: A \to B$ is bijection, then $f^{-1}: B \to A$ is unique.

Proof : Given $f: A \to B$ is bijection.

If possible $g: B \to A$, $h: B \to A$ be two inverse mappings of f.

Let b be arbitrary element of B $\exists\, a_1 \in A \ni g(b) = a_1$

$\Rightarrow \quad f(a_1) = b.$

Also $b \in B \;\exists\, a_2 \in A \ni h(b) = a_2 \Rightarrow f(a_2) = b$

$\therefore \quad f(a_1) = f(a_2) \Rightarrow a_1 = a_2 \qquad (\because f \text{ is one-one})$

$\Rightarrow \quad g(b) = h(b) \;\forall\, b \in B \Rightarrow g = h$

$\therefore$ Inverse of f is unique *i.e.*, f^{-1} is unique.

Theorem. Let A and B be two non-empty sets. If $f: A \to B$ is a bijection, then $f^{-1}: B \to A$ is also a bijection.

Proof : Given $f: A \to B$ is a bijection

$\therefore \quad f^{-1}: B \to A$ is a unique function.

(*i*) **To prove that f^{-1} is one-one.**

Let b_1 and b_2 be any two different elements of B

i.e., $\quad b_1 \neq b_2.$

Let $\quad f^{-1}(b_1) = a_1$ and $f^{-1}(b_2) = a_2$ such that $a_1, a_2 \in A$, then $b_1 = f(a_1)$, $b_2 = f(a_2)$.

Now $\quad b_1 \neq b_2 \Rightarrow f(a_1) \neq f(a_2)$

$\Rightarrow \quad a_1 \neq a_2 \qquad (\because f \text{ is a bijection})$

$\Rightarrow \quad f^{-1}(b_1) \neq f^{-1}(b_2)$

$\Rightarrow \quad f^{-1}$ is one-one.

(*ii*) **To prove that f^{-1} is onto.**

Let a be an element of A.

Then $\exists$ an element $b \in B \ni b = f(a)$ (or) $a = f^{-1}(b)$.

Thus, a is the f^{-1} image of the element $b \in B$.

Hence f^{-1} is onto.

$\therefore \quad f^{-1}: B \to A$ is a bijection.

Example. *Iff $f: Q \to Q$ be defined by $f(x) = 5x + 4 \;\forall\, x \in Q$, where Q is the set of rational numbers. Then prove that f is one-one onto and find f^{-1}.*

Solution. Given $f: Q \to Q$ be function defined by $f(x) = 5x + 4 \;\forall\, x \in Q$, where Q is the set of rational numbers.

Let $\quad x_1, x_2 \in Q \text{ (domain)} \Rightarrow f(x_1), f(x_2) \in Q \text{ (codomain)}$

$$f(x_1) = f(x_2) \Rightarrow 5x_1 + 4 = 5x_2 + 4 \Rightarrow x_1 = x_2$$

$\Rightarrow \quad f$ is one-one.

and let $y \in Q$ (codomain). If possible, let $x \in Q$ (domain) $\ni 5x + 4 = y$

i.e., $\quad x = \dfrac{y-4}{5} \in Q \text{ (domain)}$

Now, $$f\left(\frac{y-4}{5}\right) = 5\left(\frac{y-4}{5}\right) + 4 = y$$

$\therefore$ For any $y \in Q$ (domain) $\exists \frac{y-4}{5} \in Q$ (domain) $\ni f\left(\frac{y-4}{5}\right) = y$

$\therefore$ f is onto $\Rightarrow f$ is one-one onto and hence f^{-1} exists.

We have $f(x) = y \Leftrightarrow f^{-1}(y) = x$

$$f(x) = 5x + 4 = y \Rightarrow x = \frac{y-4}{5}$$

$\therefore$ $$f^{-1}(y) = \frac{y-4}{5} \Rightarrow f^{-1}(x) = \frac{x-4}{5} \quad \forall\, x \in Q.$$

Example. *If $f : R \to R \ni f(x) = \frac{2x+3}{5}$, find $f^{-1}(x)$.*

Solution. f is bijection (verify)

We have $$f(x) = y \Leftrightarrow f^{-1}(y) = x$$

$$f(x) = \frac{2x+3}{5} = y \Rightarrow x = \frac{5y-3}{2} \Rightarrow f^{-1}(y) = \frac{5y-3}{2}$$

$\Rightarrow$ $$f^{-1}(x) = \frac{5x-3}{2}.$$

Example. *If $f : R \to R^+$ such that $f(x) = \left(\frac{1}{2}\right)^x$ find f^{-1}.*

Solution. f is a bijection (verify).

We have $$f(x) = y \Leftrightarrow f^{-1}(y) = x.$$

$$f(x) = \left(\frac{1}{2}\right)^x = y \Rightarrow x = \log_{\frac{1}{2}} y \Rightarrow f^{-1}(y) = \log_{\frac{1}{2}} y$$

$\therefore$ $f^{-1} : R^+ \to R$ defined by $f^{-1}(x) = \log_{\frac{1}{2}} x$.

Example. *If $f : R^+ \to R$ such that $f(x) = \log_8 x$ find $f^{-1}(x)$.*

Solution. f is a bijection (verify)

We have $$f(x) = y \Leftrightarrow f^{-1}(y) = x$$

$$y = f(x) = \log_8^x \Rightarrow x = 8^y \Rightarrow f^{-1}(y) = 8^y.$$

$\therefore$ $f^{-1} : R \to R^+$ such that $f^{-1}(x) = 8^x$.

Example. *Let $A = \{x \mid -1 \le x \le 1\} = B$, for each of the following functions from A to B, state whether it is injective, surjective or bijective (i) $f(x) = \frac{x}{2}$; (ii) $g(x) = |x|$; (iii) $h(x) = x^2$; (iv) $l(x) = \sin \pi x$.*

Solution. Given $A = \{x \mid -1 \le x \le 1\}$, $B = \{x \mid -1 \le x \le 1\}$.

(*i*) $f: A \to B$ such that $f(x) = \frac{x}{2}$.

Let $a_1, a_2 \in A_x \Rightarrow f(a_1), f(a_2) \in B$.

Now $f(a_1) = f(a_2) \Rightarrow \frac{a_1}{2} = \frac{a_2}{2} \Rightarrow a_1 = a_2 \Rightarrow f$ is injective and

$$f(x) = y \Rightarrow \frac{x}{2} = y \Rightarrow x = 2y \Rightarrow f(2y) = y$$

If $y = -1 \Rightarrow f(-2) = -1$, but $-2 \notin A$.

$\therefore$ $y = -1$ has no preimage in A in B,

$\therefore$ f is not surjective.

$\Rightarrow$ f is not bijective.

(*ii*) Given $g: A \to B$, $g(x) = |x|$.

Let $a_1, a_2 \in A \Rightarrow g(a_1), g(a_2) \in B$.

Now $g(a_1) = g(a_2) \Rightarrow |a_1| = |a_2| \Rightarrow a_1 = \pm a_2$

$\Rightarrow$ g is not injective.

and g is not surjective because $-1 \in B$ has no preimage in A.

Hence g is not bijective.

(*iii*) Given $h: A \to B$, $h(x) = x^2$

Let $a_1, a_2 \in A \Rightarrow h(a_1) = h(a_2)$

Now $h(a_1) = h(a_2) \Rightarrow a_1^2 = a_2^2 \Rightarrow a_1 = \pm a_2$

$\Rightarrow$ h is not injective.

and $h(x) = y \Rightarrow x^2 = y \Rightarrow x = \sqrt{y}$.

$\therefore$ $h(\sqrt{y}) = y$.

Let $y = -B \;\exists$ no $i \in A$, such that $f(i) = i^2 = -1$.

$\therefore$ h is not surjective.

Hence h is not bijective.

(*iv*) Given $1: A \to B$, $1(x) = \sin \pi x$.

Let $a_1, a_2 \in A \Rightarrow 1(a_1), 1(a_2) \in B$.

Now $1(a_1) = 1(a_2) \Rightarrow \sin \pi a_1 = \sin \pi a_2$

$\Rightarrow$ $\pi a_1 = \pi a_2$ or $\pi - \pi a_2$

$\Rightarrow$ $a_1 = a_2$ or $1 - a_2$

$\Rightarrow$ 1 is not injective (Note that $1(1) = 1(0) = 1(-1) = 0$)

and let $1(x) = y \Rightarrow \sin \pi x = y$

$$\Rightarrow \qquad \pi x = \sin^{-1} y$$

$$\Rightarrow \qquad x = \frac{1}{\pi}\sin^{-1} y$$

For any $\quad y \in B \; \exists \; x = \frac{1}{\pi}\sin^{-1} y \in A$ such that $l\,(x) = l\left(\frac{1}{\pi}\sin^{-1} y\right) = y$

$\therefore$ l is surjective.

Hence l is bijective.

Example. *Show that the function $f : R \to R$ given by $f(x) = \sin x$ is neither one-one nor onto.*

Solution. Since $f\left(\frac{\pi}{4}\right) = \sin\frac{\pi}{4} = \frac{1}{\sqrt{2}}, f\left(\frac{3\pi}{4}\right) = \sin\frac{3\pi}{4} = \frac{1}{\sqrt{2}}$, f is not one-one and for any $x \in R$, $-1 \le \sin x \le 1$, range of $f(x) = \sin x$ is $[-1, 1] \ne R$ = codomain of f.

$\Rightarrow$ f is not onto.

Example. *Let $f : R \to R$ such that $f(x) = \cos(5x + 2)$. Is f one-one ? Is f invertible ?*

Solution. Given f: R $\to$ R such that $f(x) = \cos(5x + 2)$.

Since $\quad \cos(5x + 2) = \cos(2\pi + 5x + 2)$, where $5x + 2, 2\pi + 5x + 2 \in R$

we have $\quad f(5x + 2) = f(2\pi + 5x + 2)$.

$\therefore$ f is not one-one.

For every $\quad x \in R, \cos(5x + 2) \in [-1, 1]$

$\Rightarrow \quad$ Range of $f(x) = [-1, 1] \ne R$

$\therefore$ f is not onto.

Hence f is not invertible.

Example. *Show that function $f(x) = x^3$ and $g(x) = x^{\frac{1}{3}}$ for $x \in R$ are inverse of each other.*

(J.N.T.U May 05 - Set 3, Nov-04 - Set 3).

Solution. Given $\quad f(x) = x^3$

We have $\quad f(x) = y \Leftrightarrow f^{-1}(y) = x$

Take $\quad f(x) = y \Rightarrow x^3 = y \Rightarrow x = y^{\frac{1}{3}}$

$\therefore \quad f^{-1}(y) = y^{\frac{1}{3}} \Rightarrow f^{-1}(x) = x^{\frac{1}{3}}$

Consider $\quad f^{-1} = g \Rightarrow g(x) = x^{\frac{1}{3}} \quad$ ...(1)

and $\quad g(x) = y \Rightarrow x^{\frac{1}{3}} = y \Rightarrow x = y^3$.

$\therefore \quad g^{-1}(y) = y^3$

Consider $\quad g^{-1} = f \Rightarrow f(y) = y^3 \Rightarrow f(x) = x^3 \quad$ ...(2)

From (1), (2), f, g are inverse of each other.

Composite Function

Let $\quad f: A \to B$ and $g: B \to C$

Let $\quad a \in A$, since $f: A \to B$ there exists an element $b \in B \ni f(a) = b$.

Since $\quad g: B \to C$ there exists an element $c \in C \ni g(b) = c \Rightarrow g(f(a)) = c$

$\Rightarrow \quad (g \circ f)(a) = c$

$g \circ f$ is a function from A to C $\Rightarrow g \circ f: A \to C$

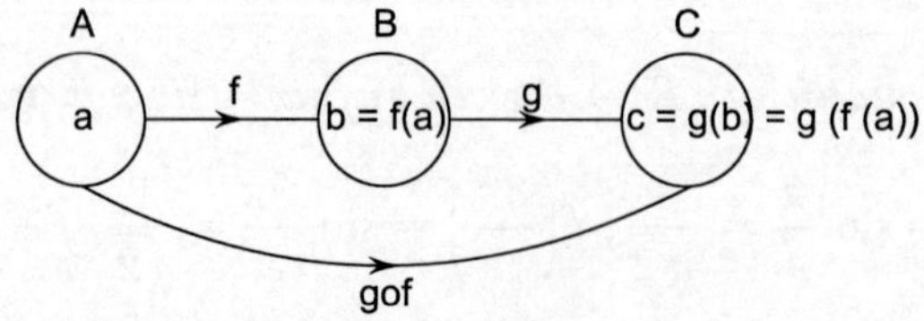

Fig. 3.67.

Thus if $f: A \to B$ and $g: B \to C$ are two functions, then the function $g \circ f$ from A to C such that $(g \circ f)(a) = g(f(a))$ for every $a \in A$, is called the composite function of f and g.

Note 1. Range of $f \subseteq$ domain of g.

2. In $g \circ f$, perform first f and then perform g.

3. Even if $g \circ f$ is possible, $f \circ g$ may not be possible.

4. Even if $f \circ g$ and $g \circ f$ are possible, $f \circ g$ may or may not be equal to $g \circ f$.

5. Composition of functions satisfies associative law *i.e.*, $h \circ (g \circ f) = (h \circ g) \circ f$.

Example. *Let* $X = \{1, 2, 3\}$, $Y = \{p, q\}$, $Z = \{a, b\}$, $f: X \to Y$ *and* $g: Y \to Z$, *so that* $f = \{(1, p), (2, p), (3, q)\}$, $g = \{(p, b), (q, b)\}$, *then find* $g \circ f$.

Solution. $g \circ f = \{(1, b), (2, b), (3, b)\}$

Example. *Let* $X = \{1, 2, 3\}$ *and* f, g, h *and* s *are functions from* X *to* X *given by* $f = \{(1, 2), (2, 3), (3, 1)\}$, $g = \{(1, 2), (2, 1), (3, 3)\}$, $h = \{(1, 1), (2, 2), (3, 1)\}$, $s = \{(1, 1), (2, 2), (3, 3)\}$.

Find $f \circ g$, $g \circ f$, $f \circ h \circ g$, $s \circ g$, $g \circ s$, $s \circ s$ *and* $f \circ s$.

Solution.

$$f \circ g = \{(1, 3), (2, 2), (3, 1)\}$$
$$g \circ f = \{(1, 1), (2, 3), (3, 2)\}$$
$$f \circ h \circ g = \{(1, 3), (2, 2), (3, 2)\}$$
$$s \circ g = \{(1, 2), (2, 1), (3, 3)\}$$
$$g \circ s = \{(1, 2), (2, 1), (3, 3)\}$$
$$s \circ s = \{(1, 1), (2, 2), (3, 3)\}$$
$$f \circ s = \{(1, 2), (2, 3), (3, 1)\}$$

And we observed that $g \circ f \neq f \circ g$, $s \circ g = g = g \circ s$, $s \circ s = s$, $f \circ s = s \circ f = f$, $g \circ s = g = s \circ g$.

Example. *Let* $f(x) = x + 2$, $g(x) = x - 2$, *and* $h(x) = 3x$ *for* $x \in R$, *where R is the set of real numbers. Find (i)* $g \circ f$, *(ii)* $f \circ g$, *(iii)* $f \circ f$, *(iv)* $g \circ g$, *(v)* $f \circ h$, *(vi)* $h \circ g$, *(vii)* $h \circ f$, *and* $(f \circ h) \circ g$, *(viii)* $f \circ (h \circ g)$.

Solution. (*i*) $(g \circ f)(x) = g(f(x)) = g(x+2) = (x+2) - 2 = x.$

(*ii*) $(f \circ g)(x) = f(g(x)) = f(x-2) = (x-2) + 2 = x$

(*iii*) $(g \circ g)(x) = g(g(x)) = g(x-2) = (x-2) - 2 = x - 4$

(*iv*) $(f \circ h)(x) = f(h(x)) = f(3x) = 3x + 2.$

(*v*) $(h \circ g)(x) = h(g(x)) = h(x-2) = 3(x-2) = 3x - 6$

(*vi*) $(h \circ f)(x) = h(f(x)) = h(x+2) = 3(x+2) = 3x + 6.$

(*vii*) $((f \circ h) \circ g)(x) = (f \circ h)(g(x)) = (f \circ h)(x-2)$

$$= f(h(x-2)) = f(3(x-2)) = f(3x-6)$$

$$= 3x - 6 + 2 = 3x - 4.$$

(*viii*) $(f \circ (h \circ g))(x) = f(h(g(x))) = f(h(x-2))$

$$= f(3(x-2)) = f(3x-6) = 3x - 6 + 2 = 3x - 4.$$

Note that $(f \circ h) \circ g = f \circ (h \circ g).$

Example. *Let $f : R \to R$ be given by $f(x) = -x^2$ and*

$g : R_+ \to R_+$ be given by $g(x) = \sqrt{x}$, where R_+ is the set of non-negative real numbers and R is the set of real numbers. Find $f \circ g$ and $g \circ f$, if possible.

Solution. Codomain of g is $R_+ \subseteq R$ domain of f

$\Rightarrow$ $f \circ g$ is possible.

$$\therefore \quad (f \circ g)(x) = f(g(x)) = f\left(\sqrt{x}\right) = -\left(\sqrt{x}\right)^2 = -x$$

and since codomain of f is $R \not\subseteq R_+ =$ domain of g, $g \circ f$ is not possible.

Example. *If $f(x) = \dfrac{x+1}{x-1}$, $x \neq 1$, show that (i) $(f \circ f \circ f \circ f)(x) = x$*

(ii) $(f \circ f \circ f \circ f \circ f)(x) = f(x)$

Solution. Given $f(x) = \dfrac{x+1}{x-1}$, $x \neq 1$

$$\Rightarrow \quad (f \circ f)(x) = f(f(x)) = \frac{\dfrac{x+1}{x-1} + 1}{\dfrac{x+1}{x-1} - 1} = \frac{\dfrac{x+1+x-1}{x-1}}{\dfrac{x+1-x+1}{x-1}} = x.$$

$$\Rightarrow \quad (f \circ f \circ f)(x) = f((f \circ f)(x)) = f(x)$$

Then, $(f \circ f \circ f \circ f)(x) = f((f \circ f \circ f)(x)) = f(f(x)) = x.$

(*ii*) $(f \circ f \circ f \circ f \circ f)(x) = f((f \circ f \circ f \circ f)(x)) = f(x).$

Important Results on Composition of Functions

Theorem. If f: A $\to$ B and g : B $\to$ C are two one-one (injective) functions then prove $g \circ f$: A $\to$ C is one-one. Is converse true ? Justify your answer.

Proof : $f: A \to B$, $g: B \to C$ are one-one $\Rightarrow g \circ f: A \to C$.

Let $a_1, a_2 \in A \Rightarrow f(a_1), f(a_2) \in B$

$\Rightarrow g(f(a_1)), g(f(a_2)) \in C.$

$\Rightarrow (g \circ f)(a_1), (g \circ f)(a_2) \in C \Rightarrow g \circ f: A \to C.$

Now $(g \circ f)(a_1) = (g \circ f)(a_2) \Rightarrow g(f(a_1)) = g(f(a_2))$

$\Rightarrow f(a_1) = f(a_2)$ ($\because$ g is one-one)

$\Rightarrow a_1 = a_2$ ($\because$ f is one-one)

$\Rightarrow g \circ f: A \to C$ is one-one (injective)

Converse of the above theorem is not true because if $f: A \to B$, $g: B \to C$ are functions and $g \circ f$ is one-one.

Consider $A = \{1, 2\}$, $B = \{p, q, r\}$, $C = \{s, t\}$

Let $f = \{(1, p), (2, q)\}$, and $g = \{(p, s), (q, t), (r, t)\}$

Now $g \circ f = \{(1, s), (2, t)\} \Rightarrow g \circ f: A \to C$ is one-one.

But $g: B \to C$ is not one-one.

Theorem. If $f: A \to B$, $g: B \to C$ are functions and $g \circ f$ is an injective, then f is necessarily an injective.

Proof : Let $x, y \in A$.

Since $f: A \to B$, $f(x), f(y) \in B$.

Now $f(x) = f(y) \Rightarrow g(f(x)) = g(f(y))$

$\Rightarrow (g \circ f)(x) = (g \circ f)(y)$

$\Rightarrow x = y$ since $g \circ f$ is one-one.

$\Rightarrow f$ is an injective.

Theorem. If $f: A \to B$, $g: B \to C$ are two onto (surjective) functions, then the function $g \circ f: A \to C$ is onto. Is converse true ? Justify your answer.

Proof : Given $f: A \to B$, $g: B \to C$ are onto $\Rightarrow g \circ f: A \to C$.

Let $c \in C$.

Since $g: B \to C$ is an onto function $\exists\ b \in B \ni g(b) = c$

Since $f: A \to B$ is an onto function $\exists\ a \in A \ni f(a) = b$

Now $g(b) = c \Rightarrow g(f(a)) = c \Rightarrow (g \circ f)(a) = c$.

Hence for any element $c \in C\ \exists\ a \in A \ni (g \circ f)(a) = c$.

$\therefore$ $g \circ f: A \to C$ is an onto function.

The converse of the theorem is not true.

If $f: A \to B$, $g: B \to C$ are functions, and $g \circ f$ is a surjection.

Then f not surjection.

For, consider $A = \{a, b\}$, $B = \{3, 4, 5\}$ and $C = \{1, 2\}$

Let $f = \{(a, 3), (b, 4)\}$ and $g = \{(3, 1), (4, 2), (5, 2)\}$

Now $g \circ f = \{(a, 1), (b, 2)\} \Rightarrow (g \circ f)(A) = \{1, 2\} = C$ and hence $g \circ f$ is a surjection from A to C. Also $f(A) = \{3, 4\} \neq B$

$\therefore$ $f : A \to B$ is not a surjection.

Theorem. If $f : A \to B$ and $g : B \to C$ such that $g \circ f$ is a surjection, then g is necessarily a surjection.

Proof : Let $c \in C$.

since $g \circ f : A \to C, \exists\, a \in A \ni (g \circ f)(a) = c$ (*i.e.*, $g(f(a)) = c$

since $g : B \to C$, $f(a) \in B$, for any $c \in C$, there exists an element belonging to B.

Hence g is a surjection.

Theorem : If $f : A \to B$, $g : B \to C$ are two bijective functions, then $g \circ f : A \to C$ is a bijection. Is converse true ? Justify your answer.

Proof : Given $f : A \to B$, $g : B \to C$ are bijective functions.

$\Rightarrow$ $g \circ f : A \to C$

Since f, g are injective, $g \circ f : A \to C$ is injective

Since f, g are surjective, $g \circ f : A \to C$ is surjective

$\Rightarrow$ $g \circ f : A \to C$ is bijective.

Converse of this theorem is not true.

Because if $f : A \to B$, $g : B \to C$ are functions, $g \circ f : A \to C$.

Consider $A = \{1, 2\}$, $B = \{p, q, r\}$, $C = \{s, t\}$, and let $f = \{(1, p), (2, q)\}$, $g = \{(p, s), (q, t), (r, t)\} \Rightarrow$ $g \circ f = \{(1, s), (2, t)\} \Rightarrow g \circ f : A \to C$ is one-one, onto

$\Rightarrow$ $g \circ f$ is bijective, but $g : B \to C$ is not bijective ($\because$ g is not one-one)

Theorem. If $f : A \to B$, $g : B \to C$ and $h : C \to D$ are functions, then $h \circ (g \circ f) = (h \circ g) \circ f$ (Associative law). (JNTU, November, 2004 - Set 1)

Proof. Given $f : A \to B, g : B \to C \Rightarrow g \circ f : A \to C$.

Now $g \circ f : A \to C, h : C \to D \Rightarrow h \circ (g \circ f) : A \to D$

and $g : B \to C, h : C \to D \Rightarrow h \circ g : B \to D$

Now, $f : A \to B, h \circ g : B \to D \Rightarrow (h \circ g) \circ f : A \to D$.

Thus $h \circ (g \circ f)$, $(h \circ g) \circ f$ both exist and have same domain and codomain D.

Let $a \in A$,

$$[h \circ (g \circ f)](a) = h[(g \circ f)(a)] = h[g(f(a))] = (h \circ g)(f(a))$$
$$= [(h \circ g) \circ f](a)$$

$$\therefore \quad h \circ (g \circ f) = (h \circ g) \circ f.$$

Theorem. If $f : A \to B$, $g : B \to C$ and both f, g are onto, show that $g \circ f$ is also onto. Is $g \circ f$ one-one if both g and f are one to one ? Justify. [JNTU March-06 Set 2,]

Proof : $g\ o\ f$ is clearly onto

and $g\ o\ f$ is clearly one-one

Theorem. Show that the composite of any function with the identity function is the function it self (or) show that $f\ o\ I = I\ o\ f = f$.

Proof : Let $f: A \to B,\ I_A : A \to A,\ I_B : B \to B$

$\Rightarrow$ $f\ o\ I_A : A \to B.$

and $(f\ o\ I_A)\ (a) = f(I_A\ (a)) = f(a)$ $(\because\ I_A : A \to A \text{ identity})$

$\Rightarrow$ $f\ o\ I_A = f$ ...(i)

and $I_B\ o\ f : A \to B$

$\Rightarrow$ $(I_B\ o\ f)\ (a) = I_B\ (f(a)) = f(a)$ $(\because\ I_B \text{ is identity on } B)$

$\Rightarrow$ $I_B\ o\ f = f$ ...(ii)

From (i), (ii), $f\ o\ I_A = f = I_B\ o\ f.$

Theorem. If $f: A \to B$ is a bijection, then $f^{-1}\ o\ f = I_A$ and $f\ o\ f^{-1} = I_B$.

Proof : Given $f: A \to B$ is a bijection $\Rightarrow f^{-1} : B \to A$ is a bijection.

By definition, $f\ o\ f^{-1} : B \to B$ and $f^{-1}\ o\ f : A \to A$.

To prove $f^{-1}\ o\ f = I_A$, let $a \in A$.

Since $f: A \to B$ is a bijection, $\exists$ unique $b \in B \ni f(a) = b$.

Since f is a bijection, $a = f^{-1}\ (b)$.

$\therefore$ $(f^{-1}\ o\ f)\ (a) = f^{-1}\ (f(a)) = f^{-1}\ (b) = a = I_A\ (a).$

$\Rightarrow$ $f^{-1}\ o\ f = I_A.$

Similarly, we can easily prove $f\ o\ f^{-1} = I_B$.

Theorem. If $f: A \to B$ and $g: B \to A$ are two functions such that $g\ o\ f = I_A$ and $f\ o\ g = I_B$. Then prove that $g = f^{-1}$. [JNTU November 05 - Set 4, November 2004 - Set 4)

Proof : Given $f: A \to B,\ g: B \to A.$

Then $g\ o\ f: A \to A,\ f\ o\ g: B \to B \Rightarrow g\ o\ f = I_A,\ f\ o\ g = I_B.$

Let $a_1, a_2 \in A \Rightarrow f(a_1), f(a_2) \in B.$

Now $f(a_1) = f(a_2) \Rightarrow g\ (f(a_1)) = g\ (f(a_2)) \Rightarrow (g\ o\ f)\ (a_1) = (g\ o\ f)\ (a_2)$

$\Rightarrow$ $I_A\ (a_1) = I_A\ (a_2) \Rightarrow a_1 = a_2$

$\Rightarrow$ f is one-one.

And let b any element in B,

$I_B\ (b) = (f\ o\ g)\ (b)$

$\Rightarrow$ $b = f(g\ (b)) \Rightarrow f(g\ (b)) = b$

$\exists$ pre-image $g\ (b) \in A$ for b, under the mapping f

$\therefore$ f is onto.

Thus f is one-one onto and hence $f^{-1} : B \to A$ exists and is also one-one onto.

Next we shall show that $g = f^{-1}$.

Now $g : B \rightarrow A$ and $f^{-1} : B \rightarrow A$

Let $a \in A$ and b be the f-image of a where $b \in B$

$\therefore \qquad f(a) = b \Rightarrow a = f^{-1}(b)$

Now $\qquad g(b) = g(f(a)) = (g \, o \, f)(a) = I_A(a) = a$

$\Rightarrow \qquad g(b) = f^{-1}(b) \Rightarrow g = f^{-1}.$

Theorem. If $f : A \rightarrow B$, $g : B \rightarrow C$ are bijective functions, then $(g \, o \, f)^{-1} = f^{-1} \, o \, g^{-1}$.

Proof : Let $f : A \rightarrow B$, $g : B \rightarrow C$ are bijectioins $\Rightarrow g \, o \, f : A \rightarrow C$ is a bijection.

$\Rightarrow \qquad (g \, o \, f)^{-1} : C \rightarrow A$ is a bijection.

Also $\qquad g^{-1} : C \rightarrow B$ and $f^{-1} : B \rightarrow A$ are bijections $\Rightarrow f^{-1} \, o \, g^{-1} : C \rightarrow A$ is a bijection.

Let c be any element of C. Then $\exists$ an element $b \in B \ni g(b) = c \Rightarrow b = g^{-1}(c)$.

Also $\exists$ an element $a \in A \ni f(a) = b \Rightarrow a = f^{-1}(b)$

Now $\qquad (g \, o \, f)(a) = g(f(a)) = g(b) = c \Rightarrow a = (g \, o \, f)^{-1}(c)$

$\Rightarrow \qquad (g \, o \, f)^{-1}(c) = a \qquad ...(1)$

Also $\qquad (f^{-1} \, o \, g^{-1})(c) = f^{-1}(g^{-1}(c)) = f^{-1}(b) = a \qquad ...(2)$

From (1), (2), $(g \, o \, f)^{-1}(c) = (f^{-1} \, o \, g^{-1})(c) \Rightarrow (g \, o \, f)^{-1} = f \, o \, g^{-1}$.

Example. *If R is a set of real numbers and* $f : R \rightarrow \mathbf{R}^{-1}$, $g : R \rightarrow R$ *are two functions defined by* $f(x) = 3x - 2$, $g(x) = x^2 + 1$ *find the values of (i)* $(f \, o \, g)(x)$ *; (ii)* $(f \, o \, g)(x + 1)$*, (iii)* $(f \, o \, g)(2)$ *(iv)* $(g \, o \, f)(x)$ *(v)* $(g \, o \, f)(o)$ *(vi)* $(f \, o \, f)(x)$ *(vii)* $(f \, o \, f)(-1)$ *(viii)* $(g \, o \, g)(x)$ *(ix)* $(g \, o \, f)(1)$

Solution. Given $f(x) = 3x - 2$, $g(x) - x^2 + 1$.

(*i*) $(f \, o \, g)(x) = f(g(x)) = f(x^2 + 1) = 3(x^2 + 1) - 2 = 3x^2 + 1.$

(*ii*) $(f \, o \, g)(x + 1) = f(g(x + 1) = f((x + 1)^2 + 1) = f(x^2 + 2x + 2)$

$= 3(x^2 + 2x + 2) - 2 = 3x^2 + 6x + 4$

(or) we have $\qquad (f \, o \, g)(x) = 3x^2 + 1 \qquad$ from (*i*)

$\Rightarrow \qquad (f \, o \, g)(x + 1) = 3(x + 1)^2 + 1 = 3(x^2 + 2x + 1) + 1$

$= 3x^2 + 6x + 4.$

(*iii*) $(f \, o \, g)(2) = f(g(2)) = f(2^2 + 1) = f(5) = 3(5) - 2 = 13$

(or) we have $\qquad (f \, o \, g)(x) = 3x^2 + 1 \qquad$ from (*i*)

$(f \, o \, g)(2) = 3(2)^2 + 1 = 13$

(*iv*) $(g \, o \, f)(x) = g(f(x)) = g(3x - 2) = (3x - 2)^2 + 1 = 9x^2 - 12x + 5$

(*v*) $(g \, o \, f)(0) = 9(0)^2 - 12(0) + 5 = 5 \qquad$ (From (*iv*))

(*vi*) $(f \, o \, f)(x) = f(f(x)) = f(3x - 2) = 3(3x - 2) - 2 = 9x - 8$

(*vii*) $(f \, o \, f)(-1) = 9(-1) - 8 = -17 \qquad$ (From (*vi*))

(*viii*) $(g \, o \, g)(x) = g(g(x)) = g(x^2 + 1) = ((x^2 + 1)^2 + 1)$

$= x^4 + 2x^2 + 2.$

(*ix*) $(g \, o \, g)(1) = g(g(1)) = g(1^2 + 1) = g(2) = 2^2 + 1 = 5$

(or) we have $(g \circ g)(x) = x^4 + 2x^2 + 2$ (from (*viii*))

$\Rightarrow$ $(g \circ g)(1) = 1^4 + 2(1)^2 + 2 = 5$

Example. Let $f(x) = x + 3$ and $g(x) = x^3$ be two functions defined on the set of real numbers. Show that $g \circ f \neq f \circ g$.

Solution. $(f \circ g)(x) = f(g(x)) = f(x^3) = x^3 + 3$

and $(g \circ f)(x) = g(f(x)) = g(x + 3) = (x + 3)^3$.

But $(f \circ g)(1) = 4$, $(g \circ f)(1) = 4^3 = 64$.

Hence $f \circ g \neq g \circ f$.

(or) $(f \circ g)(1) = f(g(1)) = f(1) = 1 + 3 = 4$,

$(g \circ f)(1) = g(f(1)) = g(1 + 3) = g(4) = 4^3 = 64$

$\therefore$ $(f \circ g)(1) \neq (g \circ f)(1)$

$\Rightarrow$ $f \circ g \neq g \circ f$.

Definition. *If $A \subseteq R$ then the function $f : A \to B$ is called a* ***real variable*** *function.*

Definition. *If $B \subseteq R$, then the function, $f : A \to B$ is called a* ***real valued*** *function.*

Definition. *If $A \subseteq R$, $B \subseteq R$, then the function $f : A \to B$ is called a real function.*

Example. a^x $(a > 0)$, e^x, $\sin x$, $\log x$ are all real functions (*i.e.,*) these are all real, real valued functions.

Note. If f and g are real functions with domain A, B sets respectively, then both f and g defined on $A \cap B$. We define functions $f + g$, $f - g$, fg, $\frac{f}{g}$, $f + k$, kf, f^n, $\sqrt{f}$, and $|f|$ as follows :

(*i*) $(f + g)(x) = f(x) + g(x)$

(*ii*) $(f - g)(x) = f(x) - g(x)$

(*iii*) $(fg)(x) = f(x)\, g(x)$

(*iv*) $\left(\frac{f}{g}\right)(x) = \frac{f(x)}{g(x)}$, if $g(x) \neq 0$

(*v*) $(f + k)(x) = f(x) + k$, where k is constant.

(*vi*) $(kf)(x) = kf(x)$, where k is constant.

(*vii*) $f^n(x) = (f(x))^n$

(*viii*) $\sqrt{f}(x) = \sqrt{f(x)}$

(*ix*) $|f| = |f(x)|$

Example. *If* $f(x) = x^2$ *for* $x < 0$

$= x$ *for* $0 < x < 1$

$= \frac{1}{x}$ *for* $x \geq 1$.

Then (i) $(3f - 6)(x)$ *(ii)* $f^2(3)$ *(iii)* $f(-6) + f\left(\frac{1}{2}\right) - f(2)$

Solution. (*i*) $(3f - 6)(x) = 3f(x) - 6 = \begin{cases} 3x^2 - 6 & \text{for} \quad x < 0 \\ 3x - 6 & \text{for} \quad 0 < x < 1 \\ \frac{3}{x} - 6 & \text{for} \quad x \geq 1 \end{cases}$

(*ii*) $f^2(3) = (f(3))^2 = \left(\frac{1}{3}\right)^2 = \frac{1}{9}$

(*iii*) $f(-6) + f\left(\frac{1}{2}\right) - f(2)$

$$= (-6)^2 + \frac{1}{2} - \frac{1}{2} = 36$$

Example. *If* $f = \{(1, 2), (2, -3), (3, -1)\}$ *is a function, find (i)* $2f$ *(ii)* $2 + f^2$ *(iii)* f^2 *and also write range of each function.*

Solution. Given $\quad f(1) = 2, f(2) = -3, f(3) = -1.$

(*i*) $(2f)(x)$ $\quad = 2f(x) \Rightarrow 2f = \{(1, 2f(1), (2, 2f(2)), (3, 2f(3))\}$

$= \{(1, 4), (2, -6), (3, -2)\}$

Range of $\quad 2f = \{4, -6, -2\}$

(*ii*) $(2 + f)(x)$ $\quad = 2 + f(x)$

$\Rightarrow \quad 2 + f = \{(1, 2 + f(1)), (2, 2 + f(2)), (3 + 2 + f(3))\}$

$= \{(1, 2 + 2), (2, 2 - 3), (3, 2 - 1)\} = \{(1, 4), (2, -1), (3, 1)\}$

Range of $\quad 2 + f = \{4, \; 1, 1\}$

(*iii*) $f^2(x)$ $\quad = (f(x))^2 \Rightarrow f^2 = \{(1, (f(1)^2), (2, (f(2))^2), (3, (f(3))^2)\}$

$= \{(1, 2^2), (2, (-3)^2), (3, (-1)^2)\}$

$= \{(1, 4), (2, 9), (3, 1)\}$

Range of $\quad f^2 = \{4, 9, 1\}$

Example. *Consider* $f : \mathbf{Z}^+ \to \mathbf{Z}^+$ *defined by* $f(a) = a^2$. *Check if f is one-to-one and/or onto and/or into using suitable explanation.* [JNTU Nov-06, Set 4]

Solution. Given $\quad f : \mathbf{Z}^+ \to \mathbf{Z}^+$ defined by $f(a) = a^2$

Let $\quad a, b \in \mathbf{Z}^+$ (domain)

Take $\quad f(a) = f(b) \Rightarrow a^2 = b^2 \Rightarrow a = \pm b$

$\therefore$ f is not one-one. $\quad [\because \; f(2) = 4 = f(-2)]$

and f is not onto because $2 \in \mathbf{Z}^+$ (codomain) has no preimage in $\mathbf{Z}^+$ (domain).

hence f is into function.

Example. *Define a bijective function. Explain with reasons whether the following functions are bijective or not. Find also the inverse of each of the functions.*

(i) $f(x) = 4x + 2$, *A = set of real numbers.*

(ii) $f(x) = 3 + \frac{1}{x}$, *A = set of non-zero real numbers.*

(iii) $f(x) = (2x + 3) \bmod 7, A = N_7$. [JNTU Nov 06, - Set 1. Feb 07 - Set 4]

Solution. (*i*) Given $f: A \to A$ by $f(x) = 4x + 2$, where $A = R$.

Let $x, y \in A$ (domain) and $f(x) = f(y)$

$\Rightarrow$ $4x + 2 = 4y + 2 \Rightarrow x = y \Rightarrow f$ is one-one.

Let $y \in A$ (codomain) $\exists\ \frac{y-2}{4} \in A$ (domain) $\ni f\left(\frac{y-2}{4}\right) = y$

$\therefore$ f is onto.

Hence f is bijection.

Let $f(x) = y \Rightarrow 4x + 2 = y \Rightarrow x = \frac{y-2}{4}$

$\therefore$ $f^{-1}(y) = x = \frac{y-2}{4} \Rightarrow f^{-1}(y) = \frac{y-2}{4}\ \forall\ y \in A$.

(*ii*) Given $f: A \to A$ defined by $f(x) = 3 + \frac{1}{x}\ \forall\ x \in A$, where $A = R_0$.

Let $x, y \in A$ (domain) and $f(x) = f(y)$

$\Rightarrow$ $3 + \frac{1}{x} = 3 + \frac{1}{y} \Rightarrow x = y \Rightarrow f$ is one-one.

f is not onto, because $3 \in A$ (codomain) has no preimage in A (domain)

$\therefore$ f is not bijective.

Then we can not find f^{-1}.

(*iii*) Given $f: A \to A$ defined by $f(x) = (2x + 3) \pmod 7$,

where $A = N_7 = \{0, 1, 2, 3, 4, 5, 6\}$.

We have $f(0) = 3, f(1) = 5, f(2) = 0, f(3) = 2, f(4) = 4, f(5) = 6, f(6) = 1$.

Then f is one-one and onto.

$\Rightarrow$ f is bijection.

Example. *Let f and g be functions from the positive real numbers to positive real numbers defined by $f(x) = [2x]$, $g(x) = x^2$, calculate f o g and g o f.* [JNTU - Nov. 06 - Set 1]

Solution. Given $f(x) = [2x], g(x) = x^2$

$$f \circ g(x) = f(g(x)) = f(x^2) = [2x^2]$$

$$g \circ f(x) = g(f(x)) = g([2x]) = [2x]^2.$$

Example. *Check whether the following functions are even, odd.*

(i) $f(x) = \sin^4 x + 5$

(ii) $f(x) = x^3 + 4 \tan x$

(iii) $x + \cos x$

Solution. (*i*) Given : $f(x) = \sin^4 x + 5$

$\Rightarrow$ $f(-x) = [\sin(-x)]^4 + 5 = (-\sin x)^4 + 5 = \sin^4 x + 5 = f(x)$

$\therefore$ f is even function.

(*ii*) Given $f(x) = x^3 + 4 \tan x.$

$$f(-x) = (-x)^3 + 4 \tan(-x) = -x^3 - 4 \tan x = -f(x)$$

$\Rightarrow$ f is odd function.

(*iii*) Given $f(x) = x + \cos x$

$$f(-x) = -x + \cos(-x) = -x + \cos x \neq f(x) \text{ and } \neq -f(x)$$

$\therefore$ f is neither even nor odd functioin.

Example. *If $f(x) = px^4 + qx^2 + r$ is an even function, find r.*

Solution. Given $f(x)$ is even function $\Rightarrow f(-x) = f(x)$

$\Rightarrow$ $p(-x)^4 + q(-x)^2 + r = px^4 + qx^2 + r$

$\Rightarrow$ $r \in R.$

Example. *If $f(x) = ax^3 + bx + c$ is an odd function find c.*

Solution. $f(x)$ is odd $\Rightarrow f(-x) = -f(x)$

$\Rightarrow$ $a(-x)^3 + b(-x) + c = -(ax^3 + bx + c)$

$\Rightarrow$ $-ax^3 - bx + c = -ax^3 - bx - c$

$\Rightarrow$ $c = -c \Rightarrow 2c = 0 \Rightarrow c = 0.$

3.19 SPECIAL TYPES OF FUNCTIONS

1. **Implicit function :** If the relation between two variables x and y is $f(x, y) = 0$ and $y = g(x)$ be a function of x, such that for every x for which $g(x)$ is defined, then $f(x, g(x))$ is called an **implicit** function of x given by $f(x, y) = 0$.

   Example. $x^2 - xy + y^2 + 2 = 0.$
2. **Even Function :** A function $f(x)$ is said to be even function if $f(-x) = f(x)$ $\forall$ $x \in$ domain of f.

   Example. 4, $x^2 + 5$, x^4, $\sin^2 x$, $\cos x$, $|x|$ are even functions.
3. **Odd Function :** A function $f(x)$ is said to be odd function if $f(-x) = -f(x)$. $\forall x \in R$

   Example. x, $\sin x$, $x^3 - 7x$ are odd functions.

   Note. There are functions, which are neither even nor odd.

   Example. $x^2 - 5x + 2$, $\sin x + \sec x$, $|x - 2|$, $\sin x + \cos x$.
4. **Polynomial Function :** A function of the form

   $a_0 x^n + a_1 x^{n-1} + a_2 x^{n-2} + \ldots\ldots + a_n$ where $a_0 \neq 0$, a_1, a_2, ..., a_n are real constants and n is a positive integer (or) 0 is called a **polynomial** function of degree n in x.

   Example. $2x^2 + 4x + 3$ is a polynomial function of degree 2 in x.
5. **Rational Function :** A function of the form $\dfrac{\phi(x)}{\psi(x)}$, where $\phi(x)$, $\psi(x)$ are polynomial functions and $\psi(x) \neq 0$ is called a **rational** function.

   Example. $f(x) = \dfrac{5x^2 - 6}{2x^2 + 4x + 3}$ is a rational function.

6. **Exponential Function :** A function $f(x) = a^x$ $(a > 0)$ is called an exponential function.
   Example. $f: R \to R^+$ such that $f(x) = 3^x$.
7. **Logarithmic function :** A function $f(x) = \log_a x$ where $a \neq 1$ and x are positive numbers is called a logarithmic function.
   Example. $f: R^+ \to R$ such that $f(x) = \log_2 x$.
8. **Step function :** Let $x \in R$, A function $f(x) = [x]$ where $f(x) = n$ (an integer) such that $n \leq x < n + 1$ is called a **step** function.

   Note. $f(x) = [x]$ is the unique integer in $[x - 1, x]$, $x - [x]$ is called fractional part and sometimes this fractional part $x - [x]$ is denoted by $]\ x\ [$.

   Note. $[x + k] = [x] + k$, where k is an integer.
9. **Floor and Ceiling functions :** Let x be any real number. Then
   (*i*) The floor function $f(x)$ defined for x is the largest integer less than or equal to x and it is denoted by $f(x) = \lfloor x \rfloor$.

Example.

(*i*) $f(4.25) = \lfloor 4.25 \rfloor = 4$, (*ii*) $f(-3.56) = \lfloor -3.56 \rfloor = -4$; (*iii*) $f(\sqrt{8}) = \lfloor \sqrt{8} \rfloor = 2$,

(*iv*) $f(-2) = \lfloor -2 \rfloor = -2$.

(*ii*) A ceiling function $f(x)$ defined for x is the smallest (least) integer greater than or equal to x and it is denoted by $f(x) = \lceil x \rceil$

Example. (*i*) $f(4.25) = \lceil 4.25 \rceil = 5$ (*ii*) $f(-3.56) = \lceil -3.56 \rceil = -3$

(*iii*) $f(\sqrt{8}) = \lceil \sqrt{8} \rceil = 3$ (*iv*) $f(-2) = \lceil -2 \rceil = -2$.

10. **Peano's successor function :** Let P be the set of all positive integers and $\sigma : P \to P$ be such that $\sigma(n) = n + 1$, where $n \in P$, then σ is called Peano's successor function.
    Example. $\sigma(1) = 2$, $\sigma(2) = 3$, $\sigma(3) = 4$, ...
11. **Characteristic Function :** Let A be a non-empty set and $U = \{u_1, u_2, ..., u_n\}$ be the universal set, then the characteristic function of A is defined as $f_A : U \to \{0, 1\}$ such that

$$f_A(u_i) = \begin{cases} 1 & ; \text{ when } u_i \in A \\ 0 & ; \text{ when } u_i \notin A \end{cases}.$$

Example. If $A = \{3, 6, 7\}$ and $U = \{1, 2, 3, 4, ..., 9, 10\}$, then $f_A(1) = 0$, $f_A(2) = 0$, $f_A(3) = 1$, $f_A(4) = 0$, $f_A(5) = 0$, $f_A(6) = 1$ and so on.

Note. f_A is onto but it is not one-one.

Theorem. If f_A is characteristic function of A

Then (*i*) $f_A(x) = 0 \Leftrightarrow A = \phi$

(*ii*) $f_A(x) = 1 \Leftrightarrow A = U$.

Theorem. The characteristic functions of subsets satisfy the following properties :

(*i*) $f_{A \cap B}(x) = f_A(x) \cdot f_B(x)\ \forall\ x$

(*ii*) $f_{A \cup B}(x) = f_A(x) + f_B(x) - f_A(x) f_B(x)\ \forall\ x$.

Proof : (*i*) By definitions of f_A, f_B, we have

$$f_A(x)\, f_B(x) = \begin{cases} 1 & ; \quad \text{when } x \in A \cap B \\ 0 & ; \quad \text{otherwise} \end{cases}$$

(*ii*) If $x \in A$, then $f_A(x) = 1 \Rightarrow f_A(x) + f_B(x) - f_A(x)\, f_B(x)$

$$= 1 + f_B(x) - 1\, f_B(x) = 1$$

If $x \in B$, then $f_B(x) = 1 \Rightarrow f_A(x) + f_B(x) - f_A(x)\, f_B(x) = 1$ and if $x \notin A \cup B$, then both $f_A(x) = 0, f_B(x) = 0,$

then $\quad f_A(x) + f_B(x) - f_A(x)\, f_B(x) = 0$

Hence $\quad f_A(x) + f_B(x) - f_A(x)\, f_B(x) = \begin{cases} 1 & ; \quad \text{if } x \in A \cup B \\ 0 & ; \quad \text{otherwise} \end{cases}$

$$= f_{A \cup B}.$$

$$\therefore \qquad f_{A \cup B} = f_A + f_B - f_A\, f_B.$$

12. **Mod functions :** For each positive integer number, m the mod-m function is defined as $f_m(a) = b$, $0 \le b < m$, a is non-negative integer.

The term mod is used for mathematical congruence relation and is defined as $a \equiv b \pmod m$ iff m divides $b - a$.

Here m is called the modulus and $a \equiv b \pmod m$ is read as a is congruent to b modulus m. If $m \neq 0$ and a are non-negative integers, then we can write $a = \text{km} + r$, $0 \le r < m$. In other words, when a is divided by m to obtain remainder r, we may also write it as a (mod m).

Example. (*i*) $f_6(37) = 1$ because $37 = 6 \times 6 + 1$ and $37 = 1 \pmod 6$

(*ii*) $f_6(4) =$ because $4 = 6 \times 0 + 4$ and $4 \equiv 4 \pmod 6$

13. **Restriction function :** If $f: X \to Y$ and $A \subseteq X$, then $f \cap (AXY)$ is a function from $A \to Y$ called the restriction of f to A and is sometimes written as $\frac{f}{A}$. If g is restriction of f, then f is called the extension of g.

Note 1. $\frac{f}{A} : A \to Y$ is such that $\left(\frac{f}{A}\right)(a) = f(a)\ \forall\ a \in A$. The domain of $\frac{f}{A}$ is A and domain of f is X.

2. If g is a restriction of f. Then $D(g) \subseteq D(f)$ and $g(x) = f(x)$ for $x \in D(g)$ and $g \subseteq f$.

Example. Let $f: R \to R$ be given by $f(x) = x^2$ when R is the set of real numbers. If W is the set of natural numbers with 0. Then $W \subseteq R$ and

$$\frac{f}{N} = \{(0, 0), (1, 1), (2, 4), (3, 9), \ldots\ldots\}.$$

Example. Let $f: R \to R$ be given by $f(x) = |x|$ where $|x|$ denote the absolute value of x. Let R_+ be the set of positive real numbers and $g: R_+ \to R$ be given by $g(x) = x$, then g is a restriction of f (*i.e.,*) $g = \frac{f}{R_+}$ and f is the extension of g in R.

14. **Hashing Function :** The central computer (server) at an university maintains records for each student. How many memory locations can be alloted so that the student records can

be retrieved fastly ? The solution to this problem is to use suitably chosen hashing function. Records are identified using a key, which uniquely identifies each students record. For example, student records are identified using the Register number | Enrolment number of the student as the key. A hashing function h assigns memory location $h(p)$ to the record that has p as its key.

The most common hashing function is the function $h(p) = p \bmod q$, where q is the number of available memory locations. (The hashing function should be onto so that all memory allocations are possible). For example, $q = 11$, the record of the student with register number 15 is assigned to memory location 4, since $h(15) = 4 \bmod 11$. Similarly, since $h(122) = 122$ mode $11 = 2$, the record of the student with register number 121 is assigned to memory location 2.

Since a hashing function is not one-one, more than one file may be assigned to a memory location. When this happens, we say that a collision occurs. To resolve a collision to assign the first free location following the occupied memory location assigned by the hashing function. For instance, after making two earlier assignments we assign location 5 to the record of the student with register number 256.

3.20 RECURSION

Recursive Functions : In this section we first define a class of functions inductively and show that any such function can be evaluated in a purely mechanical manner.

Definition. *Any function $f : \mathbf{N}^n \to \mathbf{N}$ is called total function.*

Example. $f(x, y) = x - y \ \forall \ x, y \in \text{N}$.

Definition. *If $f : D \to N$ where $D \subseteq N^n$, then f is called partial function.*

Example. $g(x, y) = x + y$, $x, y \in \mathbf{N}$ which satisfy $x \geq y$.

Now we give a set of three functions called the **initial functions**, which are used in defining other functions by induction.

1. **Zero function :** A function $\text{Z} : \text{N} \to$ by $\text{Z}(x) = 0 \ \forall \ x \in \text{N}$ is called zero function.
2. **Successor function :** A function $\text{S} : \text{N} \to \text{N}$ by $\text{S}(x) = x + 1 \ \forall \ x \in \text{N}$ is called successor function.
3. **Projection function :** A function $\text{U}_i^n : \text{N}^n \to \text{N}$ by $\text{U}_i^n(x_1, x_2, ..., x_i, ..., x_n) = x_i \ \forall \ x_i \in \text{N}$, $i = 1, 2, ..., n$, is called projection function. It is also called the generalized identity function.

Example. $\text{U}_1^2(x, y) = x$, $\text{U}_2^3(3, 4, 7) = 4$.

Now we extend the definition of composition of functions for more than one variable.

Definition. *Let $f_1(x, y)$, $f_2(x, y)$ and $g(x, y)$ be any three functions. The composition of g with f_1, f_2 is a function is given by*

$$h(x, y) = g(f_1(x, y), f_2(x, y))$$

Here domain of g is $\text{R}_{f_1} \times \text{R}_{f_2}$ where $\text{R}_{f_1}, \text{R}_{f_2}$ are ranges of f_1, f_2, and domain of h is $\text{D}_{f_1} \wedge \text{D}_{f_2}$, where $\text{D}_{f_1}, \text{D}_{f_2}$ are domains of f_1, f_2.

Note. If f_1, f_2, g are total, then h is also total function.

Definition. *The composition of g with* $f_1, f_2, \ldots\ldots, f_n$ *is*

$$h(x_1, x_2, \ldots\ldots, x_m) = g(f_1(x_1, x_2, \ldots\ldots, x_m), \ldots\ldots, f_n(x_1, x_2, \ldots\ldots, x_m))$$

Note. If $f_1, f_2, \ldots\ldots, f_n$ and g are total then h is total function.

Example. $f_1(x, y) = x + y$, $f_2(x, y) = xy + y^2$, $g(x, y) = xy$

Then
$$\begin{aligned} h(x, y) &= g(f_1(x, y), f_2(x, y)) \\ &= g(x + y, xy + y^2) = (x + y)(xy + y^2) \end{aligned}$$

Here h is total because f_1, f_2 and g are total.

Given a function $f(x_1, x_2, \ldots\ldots, x_n)$ of n variables, it is convenient to consider $n - 1$ of the variables as fixed and vary only the remaining variable over N (or) over a subset of N. For example, we may treat x as fixed and vary y in $f(x, y)$ to obtain $f(x, 0), f(x, 1), f(x, 2), \ldots\ldots$ In a mechanical process of computing the value of a function, this procedure can be used conveniently although it appears too cumbersome for hand computation. Thus, to determine $f(2, 3)$ where $f(x, y) = x + y$, we assume that $f(2, 0) = 2$ is given and then proceed to evaluate $f(2, 1)$, $f(2, 2)$ and finally $f(2, 3)$. Each functional value (except $f(2, 0)$ is computed by adding 1 to the previous value until the desired result is obtained. The computation for $f(2, 3)$ is

$$\begin{aligned} f(2, 1) &= f(2, 0) + 1 = 2 + 1 = 3 \\ f(2, 2) &= f(2, 1) + 1 = 3 + 1 = 4 \\ f(2, 3) &= f(2, 2) + 1 = 4 + 1 = 5 \end{aligned}$$

In general, given a function $f(x_1, x_2, \ldots\ldots, x_n)$ of variables, consider $n - 1$ of the variables as fixed and vary only the remaining variable over N. Here fixed $n - 1$ variables are called parameters, while the one which is assumed to vary is considered a variable for this discussion.

Definition. *Recursive function :* *Let* $g(x_1, x_2, \ldots\ldots, x_n)$, $h(x_1, x_2, \ldots\ldots, x_n, y, z)$ *be functions of n and n + 2 variables respectively. Define a function* $f(x_1, x_2, \ldots\ldots, x_n, y)$ *of n + 1 variables by* $f(x_1, x_2, \ldots\ldots, x_n, 0) = g(x_1, x_2, \ldots\ldots, x_n)$ *and* $f(x_1, x_2, \ldots\ldots, x_n, y + 1) = h(x_1, x_2, \ldots\ldots, x_n, y, f(x_1, x_2, \ldots\ldots, x_n, y))$, *then f is called a recursive function (or) recursion. Here the value of f at y + 1 is expressed in terms of the value of f at y. The variables* $x_1, x_2, \ldots\ldots, x_n$ *are treated as parameters.*

Definition. *Primitive recursive :* *A function f is called primitive recursive if it can be obtained from the initial functions by a finite number of operations of composition and recursion.*

Example. *Show that the function f (x, y) = x + y is primitive recursive, hence find f (2, 3).*

Solution. Consider $x + (y + 1) = (x + y) + 1$ so that $f(x, y + 1) = x + (y + 1)$

$$= (x + y) + 1 = f(x, y) + 1 = S(f(x, y))$$

and
$$f(x, 0) = x$$

Now we define $f(x, y)$ as $f(x, 0) = x = U_1'(x)$,

$$\begin{aligned} f(x, y + 1) &= S(f(x, y)) \\ &= S(U_3^3(x, y, f(x, y))) \end{aligned}$$

$\therefore$ f is primitive recursive function.

And by above definition of

$$f, f(2, 0) = U_1'(2) = 2.$$
$$f(2, 1) = S(U_3^3(2, 0, f(2, 0)))$$
$$= S(f(2, 0)) = S(2) = 3.$$
$$f(2, 2) = S(U_3^3(2, 1, f(2, 1))) = S(f(2, 1)) = S(3) = 4$$
and $$f(2, 3) = S(U_3^3(2, 2, f(2, 2))) = S(f(2, 2)) = S(4) = 5.$$
(or) $$f(2, 3) = S(f(2, 2))$$
$$= S(5(f(2, 1)))$$
$$= S(S(S(f(2, 0))))$$
$$= S(S(S(2))) = S(S(3)) = S(4) = 5.$$

Example. *Show that the function $f(x) = K$, where K is constant, is primitive recursive.*

Solution. $$f(x + 1) = K$$
$$= f(x)$$
$$= U_2^2(x_1 f(x))$$

$\therefore$ f is primitive recursive.

Example. *Using recursion, define the multiplication function * given by $g(x, y) = x * y$*

Solution. Since $g(x, 0) = x * 0 = 0$ and $g(x, y + 1) = g(x, y) + x$
$$= x * y + x$$

We write, $$g(x, 0) = 0 = Z(x),$$
$$g(x, y + 1) = (x * y) + x$$
$$= f(U_3^3(x, y, g(x, y)), U_1^3(x, y, g(x, y))).$$
where $$f(x, y) = x + y$$
which is primitive recursive by previous example 1.

$\therefore$ g is primitive recursive function.

Example. *Show that the function $x!$ is primitive recursive, where $0! = 1$ and $n! = n * (n - 1)!$*

Solution. Let $f(x) = x!$ then $f(0) = 0! = 1 = S(0)$

and $$f(x + 1) = (x + 1)!$$
$$= (x + 1) * x!$$
$$= (x + 1) * f(x)$$
$$= x * f(x) + f(x)$$
$$= U_1^2(x, f(x)) * U_2^2(x, f(x)) + U_2^2(x, f(x))$$

since $g(x, y) = x * y$, $h(x, y) = x + y$ are primitive recursive, f is primitive recursive.

The following are some of the primitive recursive functions which are used frequently.

1. Sign function (or) non-zero test function (Sg) :

$$Sg(0) = 0, Sg(y + 1) = 1.$$

Since $$Sg(0) = Z(0), Sg(y + 1) = S(Z(U_2^2(y, Sg(y))))$$

2. Zero-test function $(\overline{Sg})$

$$\overline{Sg}\ (0) = 1,\ \overline{Sg}\ (y + 1) = 0$$

Since $\quad \overline{Sg}\ (0) = S\ (0)\ \overline{Sg}\ (y + 1) = Z\ (U_2^2\ (y, Sg\ (y)))$

3. Predecessor function (P) :

$$P\ (0) = 0,\ P\ (y + 1) = y = U_1^2\ (y, P\ (y))$$

Note that $\quad P\ (0) = 0,\ P\ (1) = 0,\ P\ (2) = 1,\ P\ (3) = 2,\ \ldots\ldots$

4. Odd and even parity function (Pr) :

$$Pr\ (0) = 0,\ Pr\ (y + 1) = \overline{Sg}\ (U_2^2\ (y, Pr\ (y)))$$

Note that $\quad Pr\ (0) = 0,\ Pr\ (1) = 1,\ Pr\ (2) = 0,\ Pr\ (3) = 1,\ Pr\ (4) = 0,\ \ldots\ldots$

5. Proper subtraction function ($\dot{-}$) :

$$x \dot{-} 0 = x,\ x \dot{-} (y + 1) = p\ (x \dot{-} y)$$

Note that $\quad x \dot{-} y = 0$ for $x < y$ and $x \dot{-} y = x - y$ for $x \geq y$.

6. Absolute value function (| |) :

$$|\ x - y\ | = (x \dot{-} y) + (y \dot{-} x)$$

7. Minimum of x and y (min (x, y))

$$\min\ (x, y) = x \dot{-} (x \dot{-} y).$$

8. Maximum of x and y (max (x, y))

$$\max\ (x, y) = y + (x \dot{-} y)$$

9. The square function $(f(y) = y^2)$:

$$f(y) = y^2 = U_1'\ (y) * U_1'\ (y)$$

Example. *Show that $f\ (x, y) = x^y$ is a primitive recursive function.*

Solution. Note that $f(x, 0) = x^0 = \begin{cases} 1 & \text{if} \quad x \neq 0 \\ 0 & \text{if} \quad x = 0 \end{cases}$

$$f(x, y + 1) = x^{y+1} = x^y * x = x * x^y$$

Hence define $\quad f(x, y) = x^y$ as $f(x, 0) = x^0 = Sg\ (x)$

$$f(x, y + 1) = x * x^y = x * f(x, y)$$

$$= U_1^3\ (x, y, f(x, y)) * U_3^3\ (x, y, f(x, y)).$$

$\therefore \quad f(x, y)$ is primitive recursive function.

Example. *Show that the function $f\ (x) = \left[\dfrac{x}{2}\right]$ which is equal to the greatest integer, which is $\leq \dfrac{x}{2}$ is primitive recursive.*

Solution. Note that $\quad \dfrac{x}{2} = \begin{cases} \frac{x}{2} & \text{if} \quad x \text{ is even} \\ \frac{x-1}{2} & \text{if} \quad x \text{ is odd} \end{cases}.$

$\therefore$

$$f(0) = \left[\frac{0}{2}\right] = 0 = Z\ (x)$$

and $f(y+1) = \left[\frac{y+1}{2}\right] = \left[\frac{y}{2}\right] + Pr(y)$, where $Pr(y)$ is parity of y which is primitive recursive.

$\Rightarrow \quad f(y+1) = f(y) + Pr(y)$

$\therefore$ f is primitive recursive function.

Theorem. Show that $f(x, y) = x^y + y^x$ is primitive recursive.

Solution. We know that x^y is primitive recursive by *ex* ...

Similarly y^x is also primitive recursive.

Since $g(x, y) = x + y$ is primitive, recursive, $f(x, y) = x^y + y^x$ is primitive recursive function.

Example. *Show that if (x, y) defines the remainder upon division of y by x, then it is a primitive recursive function.* [JNTU - Nov - 05, Set 1, May 05, Set 3]

Solution. Consider the following case for $f(x, y)$

$$f(3, 8) = \text{Remainder of } \frac{8}{3} = 2,$$

$$f(3, 7) = 1, f(3, 6) = 0, \ldots$$

For $y = 0$, $f(x, 0) = 0$. Also the value of $f(x, y)$ increased by 1, when y is increased by 1, until the value becomes equal to x, in which case it is put equal to 0, and the process continuous. We therefore, build a function which increases by 1 each time y increases by 1. (*i.e.,*) $S(f(x, y))$. Now we multiply this function by another primitive recursive function which becomes too whenever $S(f(x, y)) = x$.

Also this other function must be 1 when ever $S(f(x, y)) \neq x$, but $S(f(x, y))$ is always $\leq x$ and hence such a function is $Sg(x \dot{-} S(f(x, y)))$. Thus the required definition of $f(x, y)$ is

$$f(x, o) = o$$

$$f(x, y+1) = S(f(x, y) * Sg(x \dot{-} S(f(x, y)))$$

Hence f is primitive recursive function.

Definition. *If the n-triples are defined over the set of natural numbers only, then such n-any relation is called number-theoretic.*

Definition. ***Characteristic function :*** *The characteristic function of a relation R is defined as* ψ_R

$$(x_1, x_2, \ldots\ldots, x_n) = \begin{cases} 1 & \text{if} \quad (x_1, x_2, \ldots\ldots, x_n) \in R \\ 0 & \text{if} \quad (x_1, x_2, \ldots\ldots, x_n) \notin R \end{cases}$$

Hence $\psi_A : R \to N$ and $R \subseteq N^n$, an n-ary relation.

Definition. *A relation R is said to be primitive recursive if its characteristic function is primitive recursive.*

Example. *Show that* $\{(x, x) \mid x \in N\}$ *which defines the relation of equality is primitive recursive.*

[JNTU May 05 - Set 3, Nov 04 - Set 3]

Solution. Let $\quad R = \{(x, x) \mid x \in N\}$

Characteristic function

$$\psi_R(x, y) = \begin{cases} 1 & \text{if} \quad (x, y) \in R \ (=) \\ 0 & \text{if} \quad (x, y) \notin R \ (=) \end{cases}$$

$$= \begin{cases} 1 & \text{if} \quad x = y \\ 0 & \text{if} \quad x \neq y \end{cases}$$

$$= \overline{Sg}\ (|\ x - y\ |) \text{ which is primitive recursive function.}$$

Since for $x = y$, $\overline{Sg}\ (|\ x - x\ |) = \overline{Sg}\ (0) = 1$, for $x \neq y$, $\overline{Sg}\ (|\ x - y\ |) = 0$.

$\Rightarrow$ ψ_R is primitive recursive.

$\Rightarrow$ Relation R is primitive recursive.

Example. *Show that for any fixed K the relation given by* $\{(K, y) \mid y > K\}$ *is primitive recursive.* [JNTU Nov 05 - Set 1 and 2]

Solution. Let $R = \{(K, y) \mid y > K\}$

Here we have to find a function such that it is equal to 1 if $y > K$ and 0 if $y \leq K$.

Such function is $Sg\ (y \dot{-} K)$ which is defined by

$$Sg\ (y \dot{-} K) = \begin{cases} 1 & \text{if} \quad y - K > 0 \\ 0 & \text{if} \quad y - K < 0 \end{cases}$$

$$= \begin{cases} 1 & \text{if} \quad y > K \\ 0 & \text{if} \quad y \leq K \end{cases}$$

Take characteristic function $\psi_R\ (y, K) = Sg\ (y \dot{-} K)$ which is primitive recursive.

$\therefore$ ψ_R is primitive recursive.

Hence relation R is primitive recursive.

Example. *Show that the function* $f(x_1, x_2, y)$ *defined as*

$$f(x_1, x_2, y) = \begin{cases} x_2 & \text{if} \quad x_1 > y \\ (x_1 * y) + x_2 & \text{if} \quad x_1 \leq y \end{cases}$$

is primitive recursive.

Solution. We have $$x \dot{-} y = \begin{cases} 0 & \text{for} \quad x_1 \leq y \\ x_1 - y & \text{for} \quad x_1 > y \end{cases}$$

$$\overline{Sg}\ (x_1 \dot{-} y) = \begin{cases} 1 & \text{for} \quad x_1 \leq y \\ 0 & \text{for} \quad x_1 > y \end{cases}$$

$$(x_1 * y) * \overline{Sg}\ (x_1 \dot{-} y) = \begin{cases} x_1 * y & \text{for} \quad x_1 \leq y \\ 0 & \text{for} \quad x_1 > y \end{cases}$$

$$x_2 + (x_1 * y) * \overline{Sg}\ (x_1 \dot{-} y) = \begin{cases} x_2 + (x_1 * y) & \text{for} \quad x_1 \leq y \\ x_2 & \text{for} \quad x_1 > y \end{cases}$$

$$= \begin{cases} x_2 & \text{if} \quad x_1 > y \\ (x_1 * y) + x_2 & \text{if} \quad x_1 \le y \end{cases}$$

$$= f(x_1, x_2, y)$$

$\therefore$ $\quad f(x_1, x_2, y) = x_2 + (x_1 \times y) * \overline{Sg}\ (x_1 \dot{-} y)$ which is primitive recursive.

Regular function : Let $g\ (x_1, x_2, \ldots\ldots, x_n, y)$ be a total function. If there exists at least one value of y, say $\bar{y} \in N$, such that the function $g\ (x_1, x_2, \ldots\ldots, x_n, \bar{y}) = 0$ for all n-triples $(x_1, x_2, \ldots\ldots, x_n) \in N^n$, then g is called a regular function.

Example. Consider the function $g\ (x, y) = y \dot{-} x$.

Since $\quad g\ (x, 0) = 0 \dot{-} x = 0\ \forall\ x$ (*i.e.*, $g\ (x, y) = 0$ for some $y = 0$), $g\ (x, y)$ is regular.

Example. Consider the total function $g\ (x, y) = |\ y^2 - x\ |$. Here $g\ (x, y) = 0$ for only those values of x which are perfect squares. Therefore, no values of y is N is such that $|\ y^2 - x\ | = 0$ for all x. Hence $g\ (x, y)$ is not regular.

Total Function : A function $f\ (x_1, x_2, \ldots\ldots, x_n)$ is said to be defined from a total function $g\ (x_1, x_2, \ldots\ldots, x_n, y)$ by minimization (or) μ operation if $f\ (x_1, x_2, \ldots\ldots, x_n)$

$$= \begin{cases} \mu_y \left(g\left(x_1, x_2, \ldots\ldots, x_n, y\right) = 0\right) & \text{if there is such a } y \\ \text{undefined} & \text{otherwise} \end{cases}$$

where μ_y means the least y greater than or equal to zero.

Recursive Function : A function is said to be recursive if it can be obtained from the initial functions by a finite number of applications of the operations of composition, recursion and minimization over regular functions.

Partial Recursive Function : A function is said to be partial recursive if it can be obtained from the initial functions by a finite number of applications of the operations of composition, recursion and minimization.

As was done in the case of primitive recursive functions, it is not necessary to start always from the initial functions in order to construct other functions in the class. In fact, if a set of recursive functions is known, they can be used along with the admissible operations to generate other recursive functions.

Example. *Show that the function* $f\ (x) = \dfrac{x}{2}$ *is a partial recursive function.*

Solution. Let $g\ (x, y) = |\ 2y - x\ |$ and g is not regular because $|\ 2y - x\ | = 0$ only for even values of x.

Define $\quad f(x) = \mu_y\ (|\ 2y - x\ | = 0)$

$$= \frac{x}{2} \text{ for } x \text{ is even.}$$

Here, the smallest y for which $|\ 2y - x\ | = 0$ is $\dfrac{x}{2}$

$\therefore\quad g$ is partial recursive.

Example. *Let $\lfloor\sqrt{x}\rfloor$ be the greatest integer $\le \sqrt{x}$. Show that $\lfloor\sqrt{x}\rfloor$ is primitive recursive.*

Solution. We have $(y+1)^2 \dot{-} x = \begin{cases} 0 & \text{for} \quad (y+1)^2 < x \\ \text{non-zero} & \text{for} \quad (y+1)^2 \ge x \end{cases}$

$$\therefore \qquad \overline{Sg}\,((y+1)^2 \dot{-} x) = \begin{cases} 1 & \text{if} \quad (y+1)^2 < x \\ 0 & \text{if} \quad (y+1)^2 \ge x \end{cases}$$

The smallest value of y for which $(y+1)^2 \ge x$ is $\lfloor\sqrt{x}\rfloor$.

Hence $$\lfloor x \rfloor = \mu_y\left(\overline{Sg}\left((y+1)^2 \dot{-} x\right) = 0\right)$$

and since $\lfloor\sqrt{x}\rfloor$ is defined for all x, $\lfloor\sqrt{x}\rfloor$ is a recursive function.

Characteristic function : Let U be an universal set and A be a subset of U. The function ψ_A : $U \to \{0, 1|$ is defined by

$$\psi_A(x) = \begin{cases} 1 & \text{if} \quad x \in A \\ 0 & \text{if} \quad x \notin A \end{cases}$$ is called the characteristic function of the set A

Recursive Set : A set 'A' is said to be recursive if its characteristic function ψ_A is recursive.

Partial Recursive Set : A set 'A' is said to be partial recursive if its characteristic function ψ_A is partial recursive.

Note that if A is recursive, that is ψ_A is recursive, then ψ A is also recursive because $\psi_{\sim A} = 1 \dot{-} \psi_A = \overline{Sg}\,(\psi_A)$ and if A and B are recursive, then $A \cap B$ and $A \cup B$ are also recursive since $\psi_{A \cap B} = \psi_A \times \psi_B$ and $\psi_{A \cup B} = (\psi_A + \psi_B) \dot{-} \psi_{A \cap B}$.

Example. *Show that the sets of even and odd natural numbers are both recursive.*

[JNTU - November 2005 - Set 1 and 2]

Solution. Let A be the set of even natural numbers and ~ A be the set of odd natural numbers.

Characteristic function of A is $\psi_A = \begin{cases} 1 & \text{if} \quad x \in A \\ 0 & \text{if} \quad x \notin A \end{cases}$ = parity function, which is primitive recursive.

$\therefore$ ψ_A is primitive recursive.

$\Rightarrow$ ψ_A is recursive $\Rightarrow$ A is recursive

$\Rightarrow$ ~ A is also recursive.

Example. *Show that the set of divisors of a positive integer n is recursive.*

Solution. If x divides n, then $\frac{n}{x} = i$, $1 \le i \le n$.

(*i.e.*,) $x \times i = n$, so $x \le n$ is a division of n if $|\, x \times i - n \,| = 0$ for one fixed value of i, and if x is not a divisor, $|\, x \times i - n \,| \ne 0$ for all $1 \le i \le n$.

Let B be the set of divisors of n.

$$\text{Consider } \sum_{i=1}^{n} \overline{Sg}\,|\,x * i - n\,| = \begin{cases} \overline{Sg}\,(0) = 1 & \text{if} \quad x \text{ is a divisor} \\ \overline{Sg}\,(\text{a non-zero}) = 0 & \text{if} \quad x \text{ is not a divisor} \end{cases}$$

$$= \begin{cases} 1 & \text{if} \quad x \in B \\ 0 & \text{if} \quad x \notin B \end{cases}$$

$$= \psi_B(x).$$

Hence B is recursive.

Example. *Let D (x) denote "number of divisors of x". Show that D (x) is primitive recursive.*

[JNTU March 06 - Set 2].

Solution. We know that, the function which defines the remainder upon division of y by x is primitive recursive. We shall denote such a function by $rm\,(x, y)$.

If a number x divides y, then $rm\,(x, y) = 0$ and $\overline{Sg}\ rm\,(x, y) = 1$, so that the number of divisors for y is given by

$$D\,(y) = \sum_{x=1}^{y} \overline{Sg}\,(rm\,(x, y))$$

$\therefore$ D (y) is primitive recursive.

Example. *Show that the predicate "x is a prime" is primitive recursive.*

Solution. A number x ($\neq 0, \neq 1$) is a prime if it has only two divisors 1 and x. Let $\overline{Pr}$ be the set containing non primes. Consider $Sg\,(D\,(x) \dot{-} 2) + \overline{Sg}\,(|\,x - 1|) + \overline{Sg}\,(|\,x - 0\,|)$

$$= \begin{cases} 1 & \text{if} \quad x \in \overline{Pr} \\ 0 & \text{if} \quad x \notin \overline{Pr} \end{cases}$$

$$= \psi_{\overline{Pr}}\,(x)$$

$\therefore$ $\psi_{\overline{Pr}}$ is primitive recursive.

Since $\psi_{\overline{Pr}}\,(x) = 1 \dot{-} \psi_{\overline{Pr}}\,(x)$, $\psi_{Pr}\,(x)$ is also primitive recursive.

$\Rightarrow$ Pr is primitive recursive.

Ackerm ann's function : The function A (x, y) defined by

$$A\,(0, y) = y + 1$$

$$A\,(x + 1, 0) = A\,(x, 1)$$

$A\,(x + 1, y + 1) = A\,(x, A\,(x + 1, y))$ is known as Ackermann's function. Observe that one can construct the value of A (x, y) for fixed values of x and y by using the above definition. Therefore, A (x, y) is well defined and total.

Now, we demonstrate, how the above definition can be used in finding the value of A (3, 3).

$$A\,(3, 3) = A\,(2, A\,(3, 2))$$

	A (3, 2) = A (2, A (3, 1))
	A (3, 1) = A (2, A (3, 0)) = A (2, A (2, 1)) = A (2, A (0, 4)) = A (2, 5).
Since	A (2, 1) = A (1, A (2, 0)) = A (1, A (1, 1)),
	A (1, 1) = A (0, A (1, 0)) = A (0, A (0, 1)) = A (0, 2) = 3.
Now	A (2, 1) = A (1, 3) = A (0, A (1, 2))
	A (1, 2) = A (0, A (1, 1)) = A (0, 3) = 4
So	A (2, 1) = A (0, 4) = 5
Again	A (2, 2) = A (1, 5) = A (0, A (1, 4))
	A (1, 4) = A (0, A (1, 3))
	A (1, 3) = A (0, A (1, 2)) = A (0, 4) = 5
	A (1, 4) = A (0, 5) = 6
	A (2, 2) = A (0, 6) = 7
Again,	A (2, 5) = A (1, A (2, 4))
	A (2, 4) = A (1, A (2, 3))
	A (2, 3) = A (1, A (2, 2)) = A (1, 7)
	A (1, 7) = A (0, A (1, 6))
	A (1, 6) = A (0, A (1, 5))
	A (1, 5) = A (0, A (1, 4)) = A (0, 6) = 7.
	A (1, 6) = A (0, 7) = 8
	A (1, 7) = A (0, 8) = 9, so A (2, 3) = 9.
Similarly	A (2, 4) = 11, A (2, 5) = 13 so A (3, 1) = 13.
Now	A (3, 2) = A (2, 13) = 29 by similar method.
Finally	A (3, 3) = A (2, 29) = 61.
Since	A (2, n) = $2n + 3$
Note that	A (0, n) = $n + 1$, A (1, n) = $n + 2$, A (2, n) = $2n + 3$.

EXERCISE 3.5.

1. Define the following and give examples :

(*a*) Function (*b*) one-one function

(*c*) onto function (*d*) one-one onto function.

2. Let A = {1, 2, 3, 4}. Determine whether or not the following relations on A are functions :

(*i*) $f = \{(2, 3), (1, 4), (2, 1), (3, 2), (4, 4)$

(*ii*) $g = \{(1, 1), (3, 1), (4, 2)\}$

(*iii*) $h = \{(2, 1), (3, 4), (1, 4), (2, 1), (4, 4)\}$

3. Determine whether or not the following relations are functions :

(*i*) $\{(x, y) \mid x, y \in \mathbf{Z}, y = 3x + 1\}$ (*ii*) $\{(x, y) \mid x, y \in \mathbf{Z}, y = x^2 + 3\}$

(*iii*) $\{(x, y) \mid x, y \in \mathrm{Z}, y^2 = x\}$ (*iv*) $\{(x, y) \mid x, y \in \mathrm{Q}, x^2 + y^2 = 1\}$.

4. If A = {1, 2, 3, 4} and B = {*a*, *b*, *c*, *d*}, determine if the following functions are one-one or onto or both.
   (*i*) $f = \{(1, a), (2, a), (3, b), (4, d)\}$ (*ii*) $g = \{(1, c), (2, b), (3, a), (4, a)\}$
   (*iii*) $h = \{(1, d), (2, b), (3, a), (4, c)\}$.
5. Let $A = \{x \in R \mid x \neq 2\}$ and define $f(x) = \frac{x}{x-2}$ for all $x \in A$, show that *f* is one-one.
6. Check whether the function $f: R \to R$ defined by $f(x) = x^2 - 8$ is one-one, onto or both, justify ?
7. Let $f(x) = 2x - 3$, $g(x) = x^2 + 3x + 5$, find a formula for $g \circ f$ and $f \circ g$. (JNTU Nov 04, Set 2)
8. Let A = {1, 2, 3} and *f*, *g*, *h* be functions from A to A given by *f* = {(1, 2), (2, 3), (3, 1)}, *g* = {(1, 2), (2, 1), (3, 3)}, *h* = {(1, 1), (2, 2), (3, 1)}. Find $f \circ g$, $g \circ f$, $f \circ h \circ g$.
9. Let $f: A \to R$ be defined by $f(x) = \frac{x-2}{x-3}$, where $A = R - \{3\}$. Find f^{-1}.
10. Let $f: R \to R$ be defined by $f(x) = 2x + 5$. Show that *f* is bijection and find f^{-1}.
11. Let A = {1, 2, 3, ..., *n* ($n \geq 2$) and B = {*a*, *b*}, find the number of surjections from A to B.
12. If A = {1, 2, 3, 4, 5} and there are 6720 one-one functions from A to B, then find *n* (B).
13. If a set A has 5 elements, how many bijective functions are there from A to A.
14. Let A = {1, 2, 3, 4, 5, 6} and B = {6, 7, 8, 9, 10}. If a function $f: A \to B$ is defined by
    f = {(1, 7), (2, 7), (3, 8), (4, 6), (5, 9), (6, 9)}, determine $f^{-1}(6)$, $f^{-1}(9)$ and if $B_1 = \{7, 8\}$, $B_2 = \{8, 9, 10\}$ find $f^{-1}(B_1)$ and $f^{-1}(B_2)$.
15. In each of the following cases, a function $f: R \to R$ is given. Determine whether *f* is one-one (or) onto (or) both and if *f* is not onto, find its range.
    (*i*) $f(x) = 2x - 3$ (*ii*) $f(x) = x^3$ (*iii*) $f(x) = x^2$
    (*iv*) $f(x) = x^2 + x$ (*v*) $f(x) = \sin x$.
16. Show that $f(x) = K$, where K is constant, is primitive recursive.
17. Show that $x!$ is primitive recursive, where $0! = 1$ and $n! = n \times (n-1)!$
18. Show that the set of divisors of a positive integer *n* is recursive.
19. If $f(x) = x^2 - 3x + 2$, find (*i*) $f(x^2)$; (*ii*) $f(y - x)$; (*iii*) $f(x + 3)$. (Nov 04 JNTU - Set 1)
20. If $f: R \to R$, $g: R \to R$, where R is the set of real numbers. Find $f \circ g$, $g \circ f$ where $f(x) = x^2 - 2$, $g(x) = x + 4$. State whether these functions are injective, surjective or bijective.
    (JNTU Nov 05 - Set 4, Nov 04 - Set 4).

ANSWERS

2. (*i*) Not a function (*ii*) Not a function (*iii*) Function
3. (*i*) Function (*ii*) Function
   (*iii*) Not a function (*iv*) Not a function.
4. (*i*) Not one-one, not onto (*ii*) Not one-one, not onto (*iii*) one-one, onto.
6. Onto but not one-one.
7. $g \circ f(x) = 4x^2 - 6x + 5$, $f \circ g(x) = 2x^2 + 6x + 7$.
8. $f \circ g$ = {(1, 3), (2, 2), (3, 1)}, $g \circ f$ = {(1, 1), (2, 3), (3, 2)}, $f \circ h \circ g$ = {(1, 3), (2, 2), (3, 2)}

9. $f^{-1}(y) = \dfrac{2-3y}{1-y}$

10. $f^{-1}(y) = \dfrac{y-5}{2}$

11. $2^n - 2$

12. 8

13. $5\,! = 120$.

14. $f^{-1}(6) = \{4\}, f^{-1}(9) = \{5, 6\}, f^{-1}(B_1) = \{1, 2, 3\}, f^{-1}(B_2) = \{3, 5, 6\}$.

15. (*i*) One-one, onto
 (*ii*) One-one, onto
 (*iii*) Neither one-one nor onto, range is $(0, \infty)$
 (*iv*) Neither one-one nor onto, range is $\left(-\dfrac{1}{4}, \infty\right)$
 (*v*) Neither one-one nor onto, range is $(-1, 1)$

19. (*i*) $f(x^2) = x^4 - 3x^2 + 4$ (*ii*) $f(y-x) = x^2 + y^2 - 2xy - 3y + 3x + 2$
 (*iii*) $f(x+3) = x^2 + 3x + 2$

20. $f \circ g = x^2 + 8x + 14$, $g \circ f = x^2 + 2$.
 f is not injective, g is injective, and
 f is surjective, g is surjective, f is not bijective, g is bijective.

3.21 PIGEONHOLE PRINCIPLE AND APPLICATIONS

If $n + 1$ pigeons (objects) are distributed among n pigeonholes (categories of the desired characteristics), then some hole contains atleast 2 pigeons. If $2n + 1$ pigeons are distributed among n pigeonholes, then some hole contains atleast 3 pigeons. In general, if K is an integer and $Kn + 1$ pigeons are distributed among n pigeonholes. Then some hole contains at least K + 1 pigeons. This follows since the average number of pigeons perhole is $K + \dfrac{1}{n}$ is $\left\lceil K + \dfrac{1}{n} \right\rceil =$ K + 1.

Generalization : If K pigeons are assigned to n Pigeonholes then one of the pigeonhole must contains at least $\left\lfloor \dfrac{K-1}{n} \right\rfloor + 1$ pigeons.

Example 1. *If 401 letters were delivered to 50 apartments, then prove that some apartment received atleast 8 letters.*

Solution. Average number of letters delivered to each apartment is $\dfrac{401}{50}$, then some apartment received atmost $\left\lfloor \dfrac{401}{50} \right\rfloor = 8$ letters.

Example 2. *Show that, in a group of 61 people at least 6 people were born in the same month.*

Solution. Average number of people were born in a month $= \left\lfloor \frac{61}{12} \right\rfloor + 1 = 6.$

Example 3. *Suppose there are 26 students and 7 cars to transport them. Then atleast one car must have 4 passengers.*

Solution. Here $K = 26$, $n = 7$.

By generalized pigeonhole principle, at least $\left\lfloor \frac{26-1}{7} \right\rfloor + 1 = 4$ passengers can transport by atleast one car.

3.21.1 The Pigeonhole Principle

Theorem. If A is the average number of pigeons per hole, then some pigeonhole contains atleast A pigeons and some pigeonhole contains atmost A pigeons.

Proof : If n is the number of pigeonholes and m_i is the number of pigeons in the ith hole, then we prove that either $m_1 \geq A$ or $m_2 \geq A$ or ... or $m_n \geq A$. If possible assume that $m_1 < A$, $m_2 < A$, ... and $m_n < A$.

Then $m_1 + m_2 + ... + m_n < nA =$ the total number of pigeons. This is a contradiction since $m_1 + m_2 + ... + m_n$ is equal to the total number of pigeons.

$\therefore \quad m_1 \geq A$ or $m_2 \geq A$ or ...... or $m_n \geq A$

and if possible assume that $m_1 > A$, $m_2 > A$, ... and $m_n > A$.

Then $m_1 + m_2 + ... + m_n > nA =$ The total number of pigeons. This is a contradiction since $m_1 + m_2 + ... + m_n =$ the total number of pigeons.

$\therefore \quad m_1 \leq A$ or $m_2 \leq A$ or ...... or $m_n \leq A$.

Note. Now the number of pigeons in a pigeonhole is necessarily an integer, but the average A need not be an integer. The pigeonhole principle also can be stated as follows :

If A is the average number of pigeons per hole, then some pigeonhole contains at least $\lceil A \rceil$ pigeons and some pigeonhole contains atmost $\lfloor A \rfloor$ pigeons.

Example. *Use the pigeonhole principle to prove the following :*

(i) Given 37 positive integers, then there must be at least 4 of them that have the same remainder when divided by 12.

(ii) Given 8 distinct integers, then some pair of them have the same remainder when divided by 7.

Solution.

(*i*) The average number of positive integers so that the remainder fixed number r when divided by 12 is $\frac{37}{12}$. Then there are at least $\left\lfloor \frac{37}{12} \right\rfloor + 1 = 4$ positive integers that have the same remainder when divided by 12.

(*ii*) Let $x_1, x_2, ..., x_8$ are 8 distinct integers. If each integer is divided by 7 and their remainders are recordsed, the only possibilities for these remainders are 0, 1, 2, 3, 4, 5 and 6. Thus, we have 8 remainders to be distributed among 7 possibilities from 0 to 6. In other words, it is like distributing 8 pigeons among 7 pigeonholes. We conclude that there are atleast 2 pigeons is some hole or in this framework, there are atleast 2 of the remainders which are equal.

CHAPTER

4

Algebraic Systems (Algebraic Structures)

4.1 INTRODUCTION

In this unit we shall first explain what is meant by an algebraic system and then give several examples of familiar algebraic systems and discuss some of their properties. Here we will discuss about semigroups, monoids, groups and abelian groups.

Semi groups are very important in the theory of sequential machines, formal languages and in certain applications relating to computer arithmetic such as multiplication monoids are used in a number of applications but most particularly in the area of syntactic analysis and formal languages. Groups are useful in the design of fast adders and error-correcting codes.

4.2 BINARY OPERATION AND PROPERTIES

Definition. *Let A be any set. A mapping $f : A \times A \to A$ is called a binary operation on A.*

i.e., $\exists$ unique $f(a, b) \in$ A for every $a \in$ A, $b \in$ A.

We denote a binary operation by a symbol such as

$+, -, \bullet, \times, \div, 0, *, \square, \sqsubset$, etc.

Note. The binary operation on A is sometimes known as binary composition on A.

Example 1. + is a binary operation on natural numbers set N.

Example 2. $\times$ is a binary operation on rational numbers set Q.

We can observe that whether the following operations binary on corresponding sets which are given below as tabular form.

Table 4.1.

operation \ set	N	W	Z	Q	Q′	R	C	Qo	Ro	Co
+	✓	✓	✓	✓	✓	✓	✓	✓	✓	✓
–	×	×	✓	✓	×	✓	✓	×	×	×
×	✓	✓	✓	✓	×	✓	✓	✓	✓	✓
÷	×	×	×	×	×	×	×	✓	×	×

Definition. *A set together with a number of binary operations on the set is called an algebraic structure or algebraic system.*

Example 1. (N, +) is an algebraic structure.

2. (Z, +, –, ×) is an algebraic structure.
3. (N, –) is not an algebraic structure.
4. (N, +, –) is not an algebraic structure since – is not binary operation even though + is binary operation on N.
5. (Q, +, –, ×) is an algebraic structure.

Properties of Binary Operation :

1. **Closure Property :**

   For any $a, b \in G, a \circ b \in G$.

2. **Associativity :**

   For any $a, b, c \in G, (a \circ b) \circ c = a \circ (b \circ c)$

3. **Existence of identity :**

   To each $a \in G \ \exists \ e \in G \ni \ a \circ e = a = e \circ a$.

   Here e is called identity in G.

4. **Existence of inverse :**

   To each $a \in G \ \exists \ b \in G \ni \ a \circ b = e = b \circ a$.

   Here b is called inverse of a in G.

5. **Commutative :**

   For any $a, b \in G, a \circ b = b \circ a$.

6. **Distributive Properties :**

   For any $a, b, c \in G, a * (b \circ c) = (a * b) \circ (a * c)$

   $(b \circ c) * a = (b * a) \circ (c * a)$

7. **Cancellation Properties :**

   For any $a, b, c \in G$,

   $a \circ b = a \circ c \ \Rightarrow \ b = c$ (Left Cancellation Law)

   $b \circ a = c \circ a \ \Rightarrow \ b = c$ (Right Cancellation Law)

Definition. ***Groupoid or Quasi Group***

A set G with a binary operation O defined on it is called a groupoid or a quasi group. Thus (G, O) is a groupoid or a quasi group.

Example. (N, +) is a groupoid but (N, –) is not a groupoid, because $2 - 3 = -1 \notin$ N.

4.2.1 Composition Table for Finite Sets

A binary composition in a finite set can be shown in a tabular form known as **Composition Table**.

Suppose S = $\{a_1, a_2, ..., a_n\}$ a finite set having n elements. Suppose there is a composition in S denoted '0'. We write the elements of the set S in a horizontal row as well as in a vertical column. The element $a_i \; o \; a_j$ associated to the ordered pair (a_i, a_j) is placed at the intersection of the row headed by a_i and the column headed by a_j.

The composition table for a finite group contains each element exactly once in each of its rows and columns and the composition table is shown below.

Composition Table 4.2.

0	a_1	a_2	......	a_j	......	a_n
a_1	$a_1 \circ a_1$	$a_1 \circ a_2$	......	$a_1 \circ a_j$	......	$a_1 \circ a_n$
a_2	$a_2 \circ a_1$	$a_2 \circ a_2$	......	$a_2 \circ a_j$	......	$a_2 \circ a_n$
::	::	::	......	::	......	::
a_i	$a_i \circ a_1$	$a_i \circ a_2$	......	$a_i \circ a_j$	......	$a_i \circ a_n$
::	::	::	......	::	......	::
a_n	$a_n \circ a_1$	$a_n \circ a_2$	......	$a_n \circ a_j$	......	$a_n \circ a_n$

Example 1. Let G = {1, – 1} and multiplication x be a binary operation defined on G. Then the composition table can be constructed as shown below :

Table 4.3.

x	1	– 1
1	1	– 1
– 1	– 1	1

Example 2. Let G = {1, – 1, i, – i}, where i^2 = – 1 and multiplication x be a binary operation defined on G. Then the composition table can be constructed as shown below :

Table 4.4.

x	1	– 1	i	– i
1	1	– 1	i	– i
– 1	– 1	1	– i	i
i	i	– i	– 1	1
– i	– i	i	1	– 1

4.3 SEMI GROUPS AND MONOIDS

Definition. *An algebraic structure (S, o) is said to be a* ***semi group*** *if o is associative in S.*

In other words, a non empty set S, together with a binary operation o is a semi group if for any, a, b, $c \in$ S, $(a \; o \; b) \; o \; c = a \; o \; (b \; o \; c)$

Example 1. Let N be the set of natural numbers set. (N, +) is a semi group because $a + b \in$ N $\forall\ a, b \in$ N and $(a + b) + c = a + (b + c)\ \forall\ a, b, c \in$ N.

Example 2. Let E = {2, 4, 6, 8, ...} be the set of even numbers, (E, +), (E, ×) are semi groups.

Example 3. Let A = {1, 3, 5, 7, ...} be the set of odd numbers, then (A, +) is not a semi group because 1 + 3 = 4 ∉ A.

Example 4. Let P (A) be the power set of a non empty set A.

Then (P (A), ∪) and (P (A), ∩) are semi goups.

Definition. *An algebraic structure (M, o) is said to be a* ***Monoid*** *if*

(i) o is associative in M.

i.e., $(a \text{ o } b) \text{ o } c = a \text{ o } (b \text{ o } c)\ \forall\ a, b, c \in M.$

(ii) Identity exists in M.

i.e., To each $a \in M\ \exists\ e \in M \ni a \text{ o } e = a = e \text{ o } a.$

Monoid also can be denoted by (M, o, e) where e is the identity in M.

Example 1. Let Z be the set of integers, then (Z, +) is a monoid. Here O is the identity.

Example 2. (Z, .) is a monoid. Here 1 is the identity.

Example 3. Let S is the set of all odd integers and . is usual multiplication. Then (S, .) is a monoid. 1 is the identity.

Example 4. Let N be the set of natural numbers, then (N, +) is not a monoid, because identity O is not in N.

4.3.1 Cyclic Semi Groups and Cyclic Monoids

Definition. *A semi group (monoid) (A, ×) is said to be cyclic if for some* $a \in A$*, every element* $x \in A$ *is of the form* a^n*, where n is an integer.*

i.e., A = $\{a^n / n \in Z\}$. Thus the cyclic semi group (Monoid) (A, ×) is said to be generated by the element a and a is called generator of the cyclic semi group (Monoid).

Note. In case of semi group (Monoid) (A, +), if every element $x \in$ A is of the form $n \cdot a$ for some a ∈ A, then A is called cyclic semi group (Monoid).

Example. Let W = {0, 1, 2, 3, ...} of all non-negative integers. Then (W, +) is an infinite cyclic semi group and cyclic monoid. Here 1 ∈ W is the generator because 2 = 2 (1), 3 = 3 (1), 4 = 4 (1), etc.

Theorem. Every cyclic monoid, is commutative.

Proof : Let (M, •) be a cyclic monoid, and is generated by an element $a \in$ M.

Let $x, y \in \text{M} \Rightarrow x = a^r, y = a^s$ where $r, s \in$ Z.

Then $x \cdot y = a^r \cdot a^s = a^{r+s} = a^{s+r} = a^s \cdot a^r = y \cdot x$

∴ (M, .) is commutative.

Corallary : Every cyclic semi group is commutative.

Theorem. Any semi group can be extended to a monoid by adjoining an identity element.

Proof : We shall indicate the method of proof by an example.

For, let N = {1, 2, 3, ...} be the set of natural numbers. Clearly (N, +) is a semigroup and it will not be a monoid as it does not contain the additive identity element O. If we include O to the set N that is A = {0, 1, 2, 3, ...} then (A, +) is a monoid with the additive identity element O ∈ A. Thus any semi group can be extended to a monoid by adjoining an identity element.

Example 1. Let Z be the set of integers, the operation o is defined as $a \text{ o } b = a + b + ab \; \forall \; a, b \in Z$. Then (Z, *o*) is a semi group.

Solution. Closure Property :

Let $\qquad a, b \in Z \Rightarrow a + b, ab \in Z$

$\Rightarrow \qquad a + b + ab \in Z \Rightarrow a \text{ o } b \in Z.$

Associativity :

Let $\qquad a, b, c \in Z,$

$$(a \text{ o } b) \text{ o } c = (a + b + ab) \text{ o } c = (a + b + ab) + c + (a + b + ab)\, c$$
$$= a + b + c + ab + bc + ca + abc$$

and
$$a \text{ o } (b \text{ o } c) = a \text{ o } (b + c + bc) = a + (b + c + bc) + a\,(b + c + bc)$$
$$= a + b + c + ab + bc + ca + abc$$

$\therefore \qquad (a \text{ o } b) \text{ o } c = a \text{ o } (b \text{ o } c)$

Hence (Z, *o*) is a semi group.

Example 2. *Show that the set Q under the binary operation o defined by* $a \text{ o } b = \dfrac{a+b}{2}$ *is not a semi group.*

Solution. Closure Property : Let $a, b \in Q \Rightarrow \dfrac{a+b}{2} \in Q$

$\Rightarrow \qquad a \text{ o } b \in Q$

Associativity :

Let $\qquad a, b, c \in Q$

$$\therefore \qquad (a \text{ o } b) \text{ o } c = \frac{a+b}{2} \text{ o } c = \frac{\dfrac{a+b}{2} + c}{2} = \frac{a+b+2c}{4}$$

and
$$a \text{ o } (b \text{ o } c) = a \text{ o } \left(\frac{b+c}{2}\right) = \frac{a + \dfrac{b+c}{2}}{2} = \frac{2a+b+c}{4}$$

$\therefore \qquad (a \text{ o } b) \text{ o } c \neq a \text{ o } (b \text{ o } c).$

Hence (Q, *o*) is not a semi group.

Example 3. *The set Z with binary operation o such that* $a \text{ o } b = a^b$. *Then (Z, o) is not a semi group.*

Solution. Consider $(2 \text{ o } 2) \text{ o } 3 = 2^2 \text{ o } 3 = 4 \text{ o } 3 = 4^3 = 64$

and $\qquad 2 \text{ o } (2 \text{ o } 3) = 2 \text{ o } (2^3) = 2 \text{ o } 8 = 2^8 = 256$

$\therefore \qquad (2 \text{ o } 2) \text{ o } 3 \neq 2 \text{ o } (2 \text{ o } 3)$ for $2, 3 \in Z$.

Hence (Z, *o*) is not a semi group.

Example 4. *Show that the set N of natural numbers is a semi group under the operation * where x * y = max {x, y}. Is it a monoid ?*

Solution. Closure Property :

Let $x, y \in N$

$$x * y = \max\{x, y\} = \text{either } x \text{ or } y \in N$$

Associativity :

Let $x, y, z \in N$.

$$(x * y) * z = \max\{x, y\} * z = \max\{\max\{x, y\}, z\}$$
$$= \max\{x, y, z\}$$

and $$x * (y * z) = x * \max\{y, z\} = \max\{x, \max\{y, z\}\}$$
$$= \max\{x, y, z\}$$

$\therefore$ $$(x * y) * z = x * (y * z)$$

$\therefore$ (N, *) is a semi group.

And Let $x \in N$, $x * o = \max\{x, o\} = x$, $o * x = \max\{o, x\} = x$

$\therefore$ o is the identity, but it is not belongs to N.

Hence (N, *) is not a monoid.

Example 5. *Let N be the set of all natural numbers. The operation * on N defined by $x * y = \text{Min}\{x, y + 2\}$. Then prove that (N, *) is not a semi group.*

Solution. Closure Property :

Let $x, y \in N$

Then $x * y = \text{Min}\{x, y + z\} \in N$

Associativity does not satisfies because for $5, 2, 1 \in N$,

$$5 * 2 = \text{Min}\{5, 2 + 2) = 4$$

$\Rightarrow$ $$(5 * 2) * 1 = \text{Min}\{4, 1 + 2\} = 3$$

and $$2 * 1 = \text{Min}(2, 1 + 2\} = 2$$

$\Rightarrow$ $$5 * (2 * 1) = 5 * 2 = \text{Min}\{5, 2 + 2\} = 4$$

$\therefore$ $$(5 * 2) * 1 = 3 \neq 4 = 5 * (2 * 1)$$

Example 6. *Prove that the set $S = \{a, b, c\}$ with the following composition table is not a semi group.*

Solution. Since all the entries in the composition table are in S, S is closed w.r.t. * And for any $a, b, c \in S$, $(c * a) * b = b * b = b$ and $c * (a * b) = c * c = c$

$\therefore$ $$(c * a) * b \neq c * (a * b)$$

and $$(a * b) * c = c \neq a = a * (b * c)$$

$\Rightarrow$ * is not associative.

Table 4.5

*	a	b	c
a	a	c	b
b	c	b	a
c	b	a	c

Hence (S, *) is not a semi group.

Example 7. *Complete the composition table so that 0 is associative in $S = \{a, b, c\}$*

Table 4.6 (*i*).

0	*a*	*b*	*c*
a	*c*	*a*	*b*
b	*a*	*b*	*c*
c	.	.	*a*

Solution. Check that $(c \text{ o } b) \text{ o } c = c \text{ o } (b \text{ o } c)$

$\therefore$ $\quad c \text{ o } (b \text{ o } c) = c \text{ o } c = a$

Then $\quad (c \text{ o } b) \text{ o } c = a \Rightarrow c \text{ o } b$ must be c.

Again $\quad c \text{ o } (a \text{ o } c) = c \text{ o } b = c$

$\Rightarrow$ $\quad (c \text{ o } a) \text{ o } c = c \Rightarrow c \text{ o } a$ must be b.

Hence composition table is

Table 4.6 (*ii*).

0	*a*	*b*	*c*
a	*c*	*a*	*b*
b	*a*	*b*	*c*
c	*b*	*c*	*a*

Example 8. *In each of the following cases a binary operation * on a set $A = \{a, b\}$ is defined through a composition table. Determine (A, *) is a semi group or a monoid or neither.*

Table 4.7.

(*i*)

*	*a*	*b*
a	*b*	*a*
b	*a*	*b*

(*ii*)

*	*a*	*b*
a	*a*	*a*
b	*b*	*b*

(*iii*)

*	*a*	*b*
a	*a*	*b*
b	*a*	*a*

Solution. Given * is binary operation on A in each case (*i*), (*ii*), (*iii*).

Next we have to verify only the associative law and the existence of identity, if any

(*i*) We find that

$$a * (a * a) = a * b = a = b * a = (a * a) * a$$
$$a * (a * b) = a * a = b = b * b = (a * a) * b$$
$$a * (b * a) = a * a = (a * b) * a$$
$$a * (b * b) = a * b = (a * b) * b$$
$$b * (a * a) = b * b = b = a * a = (b * a) * a$$
$$b * (a * b) = b * a = a = a * b = (b * a) * b$$
$$b * (b * a) = b * a = (b * b) * a$$
$$b * (b * b) = b * b = (b * b) * b.$$

Then * is associative $\Rightarrow$ (A, *) is a semi group.

And $\quad a * b = a,\ b * a = a,\ b * b = b$

Then b is the identity element in A.

$\therefore$ (A, *) is a monoid.

(*ii*) As in the above case, we check that the associative property holds. Then (A, *) is a semi-group. But neither a nor b is the identity, because $a * b = a$, $b * a = b$. Then (A, *) is not a monoid.

(*iii*) We find that $b * (a * b) = b * b = a$

$$(b * a) * b = a * b = b$$

$\Rightarrow$ * is not associative.

Then (A, *) is not a semi group.

Hence (A, *) is not a monoid.

Example 9. *Let S be a non empty set and o be an operation on S defined by a o b = a for a, b ∈ S. Determine whether o is commutative and associative in S.* [February 07 - JNTU - Set 2]

Solution.

Let $a, b \in S, a \, o \, b = a, b \, o \, a = b$

$\therefore$ $a \, o \, b \neq b \, o \, a$

$\Rightarrow$ '*o*' is not commutative in S.

and let $a, b, c \in S, (a \, o \, b) \, o \, c = a \, o \, c = a$

$$a \, o \, (b \, o \, c) = a \, o \, b = a$$

$\therefore$ $(a \, o \, b) \, o \, c = a \, o \, (b \, o \, c)$

$\Rightarrow$ '*o*' is associative in S.

Example 10. Let (S, *) be a semi group where S = $\{a, b\}$, $a * a = b$. Show that.

(*i*) $a * b = b * a$

(*ii*) $b * b = b$ [JNTU - MCA - February, 2007)

Solution. (*i*) Given $a * a = b$

$\therefore$ $a * b = a * (a * a) = (a * a) * a = b * a$

$\therefore$ $a * b = b * a$

(*ii*) Here $a * b = a$ or b

If $a * b = a$, then $b * b = (a * a) * b = a * (a * b) = a * a = b$

If $a * b = b$, then $b * b = (a * a) * b = a * (a * b) = a * b = b$

In either case, $b * b = b$.

Example 11. *Let (S, *) be a semi group. If a * c = c * a and b * c = c * b for a, b, c in S. Show that (a * b) * c = c * (a * b).*

Solution. Let $a, b \in S$, $(a * b) * c = a * (b * c) = a * (c * b)$

$$= (a * c) * b = (c * a) * b$$

$$= c * (a * b)$$

Example 12. *Let (S, *) be a semi group and a, b ∈ S. If a * b = b and b * a = a. Show that a, b, a * b, b * a are idempotent.*

Solution.
$$a * a = a * (b * a) = (a * b) * a = b * a = a,$$
$$b * b = b * (a * b) = (b * a) * b = a * b = b.$$
$$(a * b) * (a * b) = (a * b * a) * b = (a * (b * a)) * b$$
$$= (a * a) * b = a * (a * b) = a * b$$
and
$$(b * a) * (b * a) = (b * a * b) * a = (b * (a * b)) * a$$
$$= (b * b) * a = b * (b * a) = b * a.$$

EXERCISE 4.1

1. Are the following operations binary on the indicated sets.
   (*i*) The usual addition on R.
   (*ii*) The usual multiplication on Q.
   (*iii*) o is an operation on Z - {0} defined by
   $$a \; o \; b = \text{l.c.m. of } a, b.$$
   (*iv*) o is an operation in R defined by $a \; o \; b = 3a - 4b$ for $a, b \in$ R.
   (*v*) U is an operation in S and S = {A, B, C, D} where A = $\{a_1, a_2\}$, B = $\{a_2, a_4\}$, C = $\{a_3, a_4, a_5\}$, D = $\{a_5\}$.
2. Let Z be the set of integers. o is an operation in Z defined by $a \; o \; b = a + b + 1$. Prove that (Z, o) is a semi group.
3. Let S be a non-empty set. P (S) is the set of all power sets of S. Prove that (P (S), ∩) is a semi group.
4. Show that the set of all rational numbers forms a monoid w.r.t. usual multiplication.
5. Let S is the set of all 2 × 2 matrices such that each element in S are rational numbers.
   If matrix multiplication is the binary operation on S, then (S, .) is a monoid where . is usual multiplication.
6. Show that the set G = {......, $2^{-2}, 2^{-1}, 1, 2, 2^2$, ...} is a monoid under multiplication.
7. Show that the set of all positive rational numbers forms a monoid under the composition o defined by $a \; o \; b = \dfrac{ab}{4}$.
8. The operation o defined on Z such that $a \; o \; b = a + b - ab$ for $a, b \in$ Z. Show that (Z, o) is a monoid.
9. Let a binary operation o be defined on the set A = {a, b, c, d} as follows in the tabular form.

0	a	b	c	d
a	a	b	c	d
b	b	a	b	c
c	c	b	d	a
d	d	b	a	c

   Compute (*i*) $a \; o \; d$ and $d \; o \; a$; (*ii*) $b \; o \; d$ and $d \; o \; b$; (*iii*) $(a \; o \; b) \; o \; c$ and $a \; o \; (b \; o \; c)$.
10. Which of the following tables defines a semi group, a monoid.

0	a	b
a	b	b
b	a	a

*	a	b	c
a	c	b	a
b	b	c	b
c	a	b	c

11. Prove that the natural numbers set w.r.t. * operation is not a semi group where $x * y = \text{Max}\{x + 2, y\}$.

12. Let (S, *) be a semi group let $a \in S$. Define a binary operation o on S by $x \, o \, y = x * a * y$. Show that o is an associative in S.

4.3.2 Homomorphism of Semi Groups and Monoids

Definition. *Let (S, *), (T, o) be two semi groups. A mapping $f : S \to T$ is said to be semi group homomorphism, iff $f(a * b) = f(a) \, o \, f(b) \; \forall \, a, b \in S$.*

Definition. *A semi group homomorphism f is said to be a semi group monomorphism if f is one-one.*

Definition. *A semi group homomorphism f is said to be a semi group epimorphism if f is onto.*

Definition. *A semi group homomorphism f is said to be a semi group isomorphism if f is one-one, onto.*

Definition. *Two semi groups (S, *), (T, o) are said to be isomorphic if there exists a semi group isomorphic mapping from S to T.*

Definition. *A semi group homomorphism $f : S \to S$ is said to be a semi group automorphism if f is one-one, onto.*

Example. Let (N, +), and $(Z_m, +_m)$ be any two semi groups where $Z_m = \{\bar{0}, \bar{1}, \bar{2}, \ldots\ldots, \overline{m-1}\}$.

Define $f : N \to Z_m$ by $f(a) = \bar{a} \; \forall \; a \in N$.

Then $\qquad f(a + b) = \overline{a +_m b} = \bar{a} +_m \bar{b} = f(a) +_m f(b)$

$\therefore$ f is a semi group homomorphism.

Example 2. Let (Z, +), (R, ×) are two semi groups.

Define $\; f : Z \to R$ by $f(x) = 3^x \; \forall \; x \in Z$.

Let $\qquad x, y \in Z, f(x + y) = 3^{x+y} = 3^x \,.\, 3^y = f(x) \times f(y)$

$\therefore$ f is homomorphism from Z to R.

Theorem. Let (S, *), (T, □), (V, ⊕) be semi groups and $g : S \to T$ and $h : T \to V$ be semi group homomorphisms. Then $h \, o \, g : S \to V$ is a semi group homomorphism from (S, *) to (V, ⊕).

Proof : Let $a, b \in S$. Then $(h \, o \, g)(a * b) = h(g(a * b))$

$$= h(g(a) \,\square\, g(b))$$
$$= h(g(a)) \oplus h(g(b))$$
$$= (h \, o \, g)(a) \oplus (h \, o \, g)(b)$$

and $\qquad h \, o \, g : S \to V$

Then $h \, o \, g$ is a semi group homomorphism from (S, *) to (V, ⊕),

Theorem. The set of all semi group automorphisms of a semi group is a semi group under the composition of mappings.

Proof : Let A (S) = {f : f is an semi group automorphism on S}. Where S is a semi group.

Let $f, g \in A(S) \Rightarrow f, g$ are one-one, onto and automorphisms on S.

$\Rightarrow \; g \, o \, f$ is one-one, onto on S.

Let $\quad a, b \in S, (g \, o \, f)(ab) = g(f(ab)) = g(f(a) \, f(b))$

$$= g(f(a)) \,.\, g(f(b)) = (g \, o \, f)(a) \,.\, (g \, o \, f)(b).$$

$\Rightarrow$ $g \, o \, f$ is semigroup homomorphism from S to S.

$\Rightarrow$ $g \, o \, f$ is an semi group automorphism from S to S.

$\Rightarrow$ $g \, o \, f \in A\,(S)$

$\therefore$ A (S) is closed w.r.t. composition of mappings '*o*'.

And clearly composition of mappings '*o*' in A (*s*) is associative

Hence (A (S), o) is a semi group.

Note : (A (S), o) is not commutative semi group

because $f \, o \, g \neq g \, 0 \, f.$

Theorem. The set of all semi group automorphisms of a semi group is a monoid.

Proof : By above theorem,

(A (S), o) is a semi group.

where $A\,(S) = \{f : f \text{ is an semi group automorphism on } S\}$

Let $f \in A\,(S)$, $I : S \to S$ be a function $\ni I\,(a) = a \; \forall \; a \in S$.

Let $a \in S$, $(f \, o \, I)\,(a) = f(I\,(a)) = f(a)$,

and $(I \, o \, f)\,(a) = I\,(f(a)) = f(a)$

$\Rightarrow$ $f \, o \, I = f = I \, o \, f$

and $I\,(ab) = ab = I\,(a)\,I\,(b) \; \forall \; a_1 \; b \in S.$

$\Rightarrow$ I is homomorphism on S.

$\Rightarrow$ I is identity in A (S).

Hence (A (S), *o*) is a monoid.

Theorem. Let (S, .) be a semi group and (S*, *o*) is a semi group of functions from S to S under the composition of mappings. Then $g : S \to S^*$ is a homomorphism.

Proof : Let $a \in S$, define $g : S \to S^*$ by $g\,(a) = f_a$

where $f_a \in S^*$, $f_a : S \to S$ defined by $f_a\,(b) = ab$ for any $b \in S$.

Let $a, b \in S \Rightarrow ab \in S \Rightarrow g\,(a\,.\,b) = f_{ab}$

$= f_a \, o \, f_b$

Since $f_{ab}\,(c) = (ab)\;c = a\,(bc) = f_a\,(bc) = f_a\,(f_b\,(c))$

$= (f_a \, o \, f_b)\;(c)$

$\therefore$ $g\,(a\,.\,b) = f_a \, o \, f_b = g\,(a) \; o \; g\,(b)$

$\Rightarrow$ $g : S \to S^*$ is *a* homomorphism.

Definition. *Let (M, *, e_M), (T, o, e_T) be any two monoids with the identity elements e_M, e_T respectively. A mapping $f : M \to T$ is said to be a monoid homomorphism if (i) $f\,(a * b) = f\,(a) \; o \; f\,(b) \; \forall \; a, b \in M$.*

(*ii*) $f(e_M) = e_T$.

Definition. *Let (M, *, e_M), (T, o, e_T) are two monoids and $f : M \to T$ is homomorphism. Then*

(i) f is said to be a monoid monomorphism if f is one-one.

(ii) f is said to be a monoid epimorphism if f is onto.

(iii) f is said to be a monoid isomorphism iff f is one-one, onto.

Theorem. For any commutative monoid (M, 0), the set of idempotent elements of M forms a submonoid.

Proof : Let (M, 0) be a commutative monoid.

Let $S = \{a \in M \mid a \text{ o } a = a\}$ be the set of idempotent elements of M.

Clearly $e \in S$ because $e \text{ o } e = e$.

Let $a, b \in S$ with $a \text{ o } a = a$ and $b \text{ o } b = b$.

Now
$$\begin{aligned}(a \text{ o } b) \text{ o } (a \text{ o } b) &= (a \text{ o } b) \text{ o } (b \text{ o } a) && (\because \text{ (M, 0) is an abelian})\\ &= a \text{ o } (b \text{ o } b) \text{ o } a\\ &= a \text{ o } b \text{ o } a && (\because b \text{ o } b = b)\\ &= a \text{ o } (b \text{ o } a)\\ &= a \text{ o } (a \text{ o } b)\\ &= (a \text{ o } a) \text{ o } b\\ &= a \text{ o } b\end{aligned}$$

$\therefore$ $a \text{ o } b \in S$.

Hence (S, 0) is a submonoid.

4.3.3 Sub Semi groups and Submonoids

Definition. *Let (S, O) be a semi group. A subset T of S is said to be a sub semigroup w.r.t. O if (T, O) is a semi group.*

Example. Let N = {1, 2, 3, ...} be the set of natural numbers, then (N, +) and (N, ×) are semi groups. Consider E = {2, 4, 6, ...} of all even natural numbers. Then (E, +) and (E, ×) are also semi groups. Hence (E, +) is a sub semi group of (N, +) and (E, ×) is a sub semigroup of (N, ×).

Definition. *Let (M, O, e) be a monoid. A subset T of M is said to be a submonoid of monoid M if (T, O, e) is a monoid.*

Example. Let M = {1, – 1, *i*, – *i*} then (M, ×, 1) is a monoid.

Consider T = {1, – 1} then (T, ×, 1) is a monoid.

Since $T \subseteq M$, T is a submonoid of monoid M.

4.4 GROUPS

A **group** is the algebraic structure consisting of a set together with a binary operation defined on it such that the elements of the given set have some relationships among themselves. The applications of group theory in studying the structure of human genes, theory of relativity, and in quantum mechanics have been appreciated.

Definition. *An algebraic structure (G, O) is said to be* **group** *if 'O' satisfies the following conditions (also called group axioms).*

1. **Closure :** $a \text{ o } b \in G \ \forall \ a, b \in G$.
2. **Associativity :** $(a \text{ o } b) \text{ o } c = a \text{ o } (b \text{ o } c) \ \forall \ a, b, c \in G$.

3. **Existence of identity :** To each $a \in G \exists e \in G \ni a\ o\ e = a = e\ o\ a$. Here e is called identity element in G.
4. **Existence of inverse :** To each $a \in G \exists b \in G \ni$.
   $a\ o\ b = e = b\ o\ a$. Here b is called inverse of a in G and it is denoted by $b = a^{-1}$.

Note 1. If 0 is binary operation then no need to apply closure property.

Note 2. Inverse of identity element in a group is itself.

Definition. *A group (G, 0) is said to be abelian group (commutative group) if $a\ o\ b = b\ o\ a$ (Commutative property)*

$$\forall\ a, b \in G.$$

Definition. *A group (G, 0) is said to be* ***finite*** *group if G contains a finite number of distinct elements. Otherwise the group (G, 0) is called* ***infinite*** *group.*

The number of elements in a finite group is called the order of the group, is denoted by 0 (G) or | G |.

Example. *The set of all integers forms infinite abelian group with respect to usual addition.*

Solution. Let Z be the set of integers.

Closure Property : Let $a, b \in Z \Rightarrow a + b \in Z$.

Associativity : Let $a, b, c \in Z \Rightarrow (a + b) + c = a + (b + c)$

Existence of identity : Let $a \in Z \Rightarrow a + 0 = a = 0 + a$

To each $a \in Z \exists 0 \in Z \ni a + 0 = a = 0 + a$.

Here 0 is the additive identity in Z.

Existence of Inverse : Let $a \in Z \Rightarrow a + (-a) = 0 = (-a) + a$

To each $a \in Z \exists -a \in Z \ni a + (-a) = 0 = (-a) + a$.

Here $-a$ is the additive inverse of a in Z.

Commutative : Let $a, b \in Z \Rightarrow a + b = b + a$.

Hence (Z, +) is an infinite abelian group.

Example. (N, +) is not a group because additive identity does not exists in N. *i.e.,* $0 \notin$ IN.

Example. *The set of all non-zero rational numbers form an infinte abelian group w.r.t. usual multiplication.*

Solution. Let Q_0 be the set of all non-zero rational numbers.

Closure Property : Let $a, b \in Q_0 \Rightarrow a\ .\ b \in Q_0$

Associativity : Let $a, b, c \in Q_0 \Rightarrow (a\ .\ b)\ .\ c = a\ .\ (b\ .\ c)$

Existence of identity : Let $a \in Q_0 \Rightarrow a\ .\ 1 = a = 1\ .\ a$.

To each $a \in Q_0 \exists 1 \in Q_0 \ni a\ .\ 1 = a = 1\ .\ a$.

Here 1 is the multiplicative identity in Q_0.

Existence of inverse : Let $a \in Q_0 \Rightarrow a\ .\ \frac{1}{a} = 1 = \frac{1}{a}\ .\ a$

To each $a \in Q_0 \ \exists \ \frac{1}{a} \in Q_0 \ni a . \frac{1}{a} = 1 = \frac{1}{a} . a.$

Here $\frac{1}{a}$ is the inverse of a in Q_0.

Commutative : Let $a, b \in Q_0 \Rightarrow a . b = b . a.$

Hence $(Q_0, .)$ is an infinite abelian group.

Example. (Q, .) is not a group because 0 has no inverse in Q.

Example. *Show that the set* $G = \{..., -3m, -2m, -m, 0, m, 2m, 3m, ...\}$ *of multiples of integers by a fixed integer m is a group w.r.t. +.*

Solution. Closure Property : Let $a, b \in G$ then $a = rm$, $b = sm$, where r, s are some integers.

$$\therefore \quad a + b = rm + sm = (r + s)\, m \in G \text{ since } r + s \in Z.$$

Associativity : The elements of G are all integers and we know that the addition of integers is an associative composition.

Existence of identity : Let $a \in G \Rightarrow a + 0 = a = 0 + a.$

Here 0 is the identity in G.

Existence of Inverse : Let $a \in G \Rightarrow a = rm \Rightarrow -a = -rm.$

$$\therefore \quad a + (-a) = rm - rm = 0 \text{ and } (-a) + a = -rm + rm = 0.$$

To each $a \in G \ \exists -a \in G \ni a + (-a) = 0 = (-a) + a.$

Here $-a$ is the inverse of a in G.

Hence (G, +) is a group.

Example. *Show that the four fourth roots of unity form a group w.r.t. multiplication.*

(i.e.,) $(\{1, -1, i, -i\}, .)$ *is a group.*

Solution. Let $G = \{1, -1, i, -i\}$. First we form the composition table. Now we make the following conclusions :

1. **Closure Property :**

   Since all the entries in the composition table are elements of the set G, G is closed w.r.t. '.'.

Table 4.8.

.	1	−1	i	$-i$
1	1	−1	i	$-i$
−1	−1	1	$-i$	i
i	i	$-i$	−1	1
$-i$	$-i$	i	1	−1

2. **Associativity :** The elements of G are all complex numbers and the multiplication of complex numbers is associative. Then associative property satisfies in G w.r.t. '.'
3. **Existence of Identity :**

   From the composition table we see that the row headed by the element. 1 Just coincides with the top row of the composition table. Thus we have $1\,(1) = 1$, $(1\,(-1) = -1$, $1\,(i) = i$, $1\,(-i) = -i.$

In other words, we have $1 \in G$ and $1 \,.\, a = a \;\forall\; a \in G$.

Similarly, we can see that $1 \in G$ and $a \,.\, 1 = a \;\forall\; a \in G$.

$\therefore \qquad 1 \,.\, a = a = a \,.\, 1 \Rightarrow 1$ is the identity in G.

4. Existence of Inverse :

We know that the identity element is its own inverse. Therefore, inverse of 1 is 1. From the composition table, we see that $(-1)(-1) = 1$, $i(-i) = 1$, $-i(i) = 1$.

Hence inverses $1, -1, i, -i$ are $1, -1, -i, i$ respectively, which are in G. Hence, (G, .) is a group.

Note 1. To see that the composition is commutative we observe from the composition table that the entries in the first, second, third and fourth rows of the table just coincide with the corresponding entries in the first, second, third and fourth columns. Therefore, we have $ab = ba \;\forall\; a, b \in G$ and the composition is commutative.

Note 2. $G = \{1, -1, i, -i\}$ is an abelian group w.r.t., multiplication.

Example. *Show that ({1, – 1}, .) is finite abelian group.*

Solution. Let $G = \{1, -1\}$. First we form composition table. Now we make the following conclusion.

1. Closure Property :

Since all the entries in the composition table are elements of G, G is closed w.r.t. '.'.

Composition table

Table 4.9.

.	1	–1
1	1	–1
–1	–1	1

2. Associativity : The elements of G are all complex numbers and the multiplication of complex numbers is associative. Then '.' is associative in G.

3. Existence of Identity :

From table, we have $\qquad 1 \,.\, 1 = 1, -1 \,.\, 1 = -1.$

Then 1 is the identity in G.

4. Existence of Inverse :

We have $\qquad 1 \,.\, 1 = 1, -1 \,.\, -1 = 1$

Then inverses of 1, – 1 are 1, – 1 respectively which are in G.

5. Commutative : We have $1 \,.\, -1 = -1 = -1 \,.\, 1$.

Then '.' is commutative in G.

$\therefore \qquad (G, .) = (\{1, -1\}, .)$ is an abelian group.

Since G contains 2 elements, G is finite abelian group.

Example. *Show that ({1, ω, ω²}, .) is finite abelian group. (or) show that the set of all cube roots of unity form finite abelian group.*

Solution. Let G = $\{1, \omega, \omega^2\}$. First we form the composition table. Now we make the following conclusions :

1. **Closure Property :** Since all the entries in the composition table are elements of G, G is closed w.r.t. '.'.

   Composition Table

Table 4.10.

.	1	ω	ω^2
1	1	ω	ω^2
ω	ω	ω^2	1
ω^2	ω^2	1	ω

2. **Associativity :** The elements of G are complex numbers and multiplication of complex numbers is associative. Then '.' is associative in G.
3. **Existence of identity :** We have $1 \,.\, 1 = 1$, $1 \,.\, \omega = \omega$, $1 \,.\, \omega^2 = \omega^2$. Then 1 is the identity in G.
4. **Existence of Inverse :** We have $1 \,.\, 1 = 1$, $\omega \,.\, \omega^2 = 1$, $\omega^2 \,.\, \omega = 1$, then inverses of $1, \omega, \omega^2$ are $1, \omega^2, \omega$ respectively which are in G.
5. **Commutative :** Since corresponding rows and columns in the composition table are identical, '.' is commutative in G.

Since G contains 3 elements, (G, .) is finite abelian group.

Example. *Prove that the set of all n n^{th} roots of unity forms a finite abelian group of order n w.r.t. multiplication.*

Solution. We have $1 = 1^{\frac{1}{n}} = (i + i0)^{\frac{1}{n}} = (\cos 0 + i \sin 0)^{\frac{1}{n}}$

$$= (\cos 2r\pi + i \sin 2r\pi)^{\frac{1}{n}} \text{ where } r \text{ is any integer.}$$

$$= \cos\frac{2r\pi}{n} + i \sin\frac{2r\pi}{n} \quad \text{(by De-Moivre's theorem)}$$

$$= e^{i\frac{2r\pi}{n}},\ r = 0, 1, 2, ..., n - 1.$$

Consider
$$G = \left\{1, e^{\frac{i2\pi}{n}}, e^{\frac{i4\pi}{n}}, ..., e^{\frac{i2(n-1)\pi}{n}}\right\}$$

which is the set of all n n^{th} roots of unity.

$$\Rightarrow \qquad G = \{1, \omega, \omega^2, \omega^3, ..., \omega^{n-1}\}, \text{ where } \omega = e^{\frac{2\pi i}{n}}$$

Closure Property : Let $a, b \in$ G. Then both a and b are n^{th} roots of unity. Therefore $a^n = 1$ and $b^n = 1$.

$\therefore \quad (ab)^n = a^n b^n = (1)(1) = 1 \Rightarrow ab$ is also an nth root of unity

$\Rightarrow \quad ab \in G.$

Associative : The elements of G are all complex numbers and the multiplication of complex numbers is associative.

Then '.' is associative in G.

Existence of identity : We have $1 \in G$ and $1 \,.\, a = a = a \,.\, 1 \;\forall\; a \in G$. Therefore, 1 is the identity in G.

Existence of inverse : Clearly the inverse of identity element 1 is itself. If ω^r, $1 \le r \le n-1$ is any other element of G.

Then $\quad \omega^r \,.\, \omega^{n-r} = \omega^n = 1$ and

$$\omega^{n-r} \,.\, \omega^r = \omega^n = 1$$

Here ω^{n-r} is also an element of G and ω^{n-r} is inverse of ω^r.

Commutative : All elements of G are complex numbers. Since the multiplication of complex numbers is commutative, the multiplication is commutative in G.

Hence (G, .) is an abelian group.

Since G is finite, (G, .) is finite abelian group.

Example. *Show that Q_1 of rational numbers other than 1 is an infinite abelian group w.r.t., the operation * defined by $a * b = a + b - ab$. $\forall\, a, b \in Q_1$.*

Solution. Closure Property : Let $a, b \in Q_1$. Then a and b are rational numbers such that $a \neq 1$, $b \neq 1$.

Now, $a * b = a + b - ab \in Q_1$ because $a + b,\, ab \in Q \Rightarrow a + b - ab \in Q$ and $a + b - ab \neq 1$ (if $a + b - ab = 1 \Rightarrow a + b - ab - 1 = 0 \Rightarrow (a-1)(b-1) = 0 \Rightarrow a = 1$ or $b = 1$).

Associativity : If $a, b, c \in Q_1$, then

$$(a * b) * c = (a + b - ab) * c = (a + b - ab) + c - (a + b - ab)\, c$$
$$= a + b + c - ab - bc - ca + abc.$$

Also $$a * (b * c) = a * (b + c - bc) = a + (b + c - bc) - a\,(b + c - bc)$$
$$= a + b + c - ab - bc - ca + abc$$

$\therefore \quad (a * b) * c = a * (b * c) \;\forall\; a, b, c \in Q_1.$

Existence of Identity : Let $a \in Q_1 \Rightarrow a \neq | \in Q.$

$$a * e = a \Rightarrow a + e - ae = a$$

$\Rightarrow \quad e\,(1 - a) = 0 \Rightarrow e = 0$ since $a \neq 1$

and $\quad a * 0 = a + 0 - a0 = a,\; 0 * a = 0 + a - 0a = a$

$\therefore \quad a * 0 = a = 0 * a \Rightarrow 0$ is the identity in Q_1.

Existence of Inverse : Let $a \in Q_1 \Rightarrow 1 \neq a \in Q.$

$$a * b = e \Rightarrow a + b - ab = 0 \Rightarrow b(1 - a) = -a \Rightarrow b = \frac{a}{a-1}.$$

Since $\quad a \neq 1, b = \dfrac{a}{a-1} \in Q_1.$

$$a * \frac{a}{a-1} = a + \frac{a}{a-1} - a.\frac{a}{a-1} = \frac{a(a-1) + a - a^2}{a-1} = 0$$

and $\quad \dfrac{a}{a-1} * a = 0 \Rightarrow a * \dfrac{a}{a-1} = 0 = \dfrac{a}{a-1} * a$

$\Rightarrow \quad \dfrac{a}{a-1}$ is the inverse of a in Q_1.

Commutative : $\quad a * b = a + b - ab = b + a - ba = b * a.$

Hence $(Q_1, .)$ is an infinite abelian group.

Example. *Let A be any non-empty set. S is the set of all bijections (one-one and onto mappings) from A to A. Then show that S is a group w.r.t. composition of mapping 'O' as binaary operation.*

Solution. Let S be the set of all bijections from A to A. Let $g, f \in$ S.

By definition of composition of mappings, for $x \in$ A, $(g \, o \, f)(x) = g(f(x))$.

Closure : Let $f, g \in S \Rightarrow f, g$ are bijections on A.

$\Rightarrow \quad g \, o \, f : A \to A.$

Let $\quad a, b \in A, (g \, o \, f)(a) = (g \, o \, f)(b) \Rightarrow g(f(a)) = g(f(b))$

$\Rightarrow \quad f(a) = f(b) \qquad (\because \ g$ is one-one)

$\Rightarrow \quad a = b \qquad (\because \ f$ is one-one).

$\Rightarrow \quad g \, o \, f : A \to A$ is one-one.

Since g is onto on A, $\exists \ b \in A \ni g(b) = c$ for $c \in$ A.

Since f is onto on A, $\exists \ a \in A \ni f(a) = b$ for $b \in$ A.

$$g(b) = c \Rightarrow g(f(a)) = c \Rightarrow (g \, o \, f)(a) = c$$

$\Rightarrow \quad g \, o \, f : A \to A$ is onto.

$\Rightarrow \quad g \, o \, f : A \to A$ is bijection.

$\Rightarrow \quad g \, o \, f \in S.$

Associativity : Let $h, g, f \in S$

For $\quad x \in A, (h \, o \, (g \, o \, f))(x) = h[(g \, o \, f)(x)] = h[g(f(x))]$

$$= [(h \, o \, g)(f(x)] = [(h \, o \, g) \, o \, f](x)$$

$\Rightarrow \quad h \, o \, (g \, o \, f) = (h \, o \, g) \, o \, f.$

Existence of Identity : Let I be the identity mapping from A to A. For $x \in$ A, I $(x) = x$, and we know that I is bijection.

Hence inverses of f_1, f_2, f_3, f_4, f_5 and f_6 are f_1, f_2, f_3, f_4, f_6 and f_5 respectively which are in G. Hence $(g, 0)$ is a group.

Example. *Show that the sets of all ordered pairs (a, b) of real numbers for which $a \neq 0$ w.r.t., the operation X defined by $(a, b) \times (c, d) = (ac, bc + d)$ is a group. Is the group commutative ?*

[JNTU – February, 2007 - Set 4].

Solution. Closure : Let $(a, b), (c, d) \in S$

$\Rightarrow \quad (ac, bc + d) \in S$ since $a \neq 0,\ c \neq 0 \Rightarrow ac \neq 0$.

Associativity : Let $(a, b), (c, d), (f, g) \in S$

$$\Rightarrow \quad ((a, b) \times (c, d)) \times (f, g) = (ac, bc + d) + (f, g)$$

$$= (acf, (bc + d) f + g) = (acf, bcf + df + g)$$

$$= (a, b) \times (cf, df + g) = (a, b) \times ((c, d) \times (f, g))$$

Existence of identity : Let $(a, b) \in S$.

Let $\quad (x, y) \in S$ such that $(x, y) \times (a, b) = (a, b)$

$\therefore \quad (xa, ya + b) = (a, b) \Rightarrow xa = a,\ ya + b = b$

$\Rightarrow \quad x = 1 \qquad (\because\ a \neq 0)$ and $ya = 0$

$\Rightarrow \quad x = 1$ and $y = 0 \qquad (\because\ a \neq 0)$

$\therefore \quad (a, b) \times (1, 0) = (a, b) = (1, 0) \times (a, b)$

$\Rightarrow$ (1, 0) is the identity in S.

Existence of inverse : Let $(a, b) \in S$.

Let $\quad (x, y) \in S$ such that $(x, y) \times (a, b) = (1, 0)$

$\Rightarrow \quad (xa, ya + b) = (1, 0) \Rightarrow xa = 1,\ ya + b = 0$

$\Rightarrow \quad x = \dfrac{1}{a},\ y = \dfrac{-b}{a}$ since $a \neq 0$.

$$\therefore \quad (a, b) \times \left(\frac{1}{a}, \frac{-b}{a}\right) = (1, 0) = \left(\frac{1}{a}, \frac{-b}{a}\right) \times (a, b)$$

$\Rightarrow \quad \left(\dfrac{1}{a}, \dfrac{-b}{a}\right)$ is the inverse of (a, b) in S.

Commutative : Let $(a, b), (c, d) \in S$.

$\therefore \quad (a, b) \times (c, d) = (ac, bc + d)$

and $\quad (c, d) \times (a, b) = (ca, da + b)$

$\therefore \quad (a, b) \times (c, d) \neq (c, d) \times (a, b)$.

Hence S is a group but not a commutative group w.r.t. X.

Example. *If G is the set of all positive rational numbers, then prove that G is an abelian group under the composition defined by 0 such that $a \circ b = \dfrac{ab}{3}$ for $a, b \in G$.* [JNTU, November 2006 - Set 1]

Solution. Given G is the set of all positive rational numbers, and for $a, b \in$ G, we have $a \text{ o } b = \dfrac{ab}{3}$.

Closure : Let $a, b \in G \Rightarrow a \text{ o } b = \dfrac{ab}{3} \in G$

Associativity : Let $a, b, c \in G \Rightarrow (a \text{ o } b) \text{ o } c = \dfrac{ab}{3} \text{ o } c$

$$= \frac{\frac{ab}{3}.c}{3} = \frac{abc}{9} \text{ and}$$

$$a \text{ o } (b \text{ o } c) = a \text{ o } \frac{bc}{3} = \frac{a.\frac{bc}{3}}{3} = \frac{abc}{9}$$

$\therefore$ $(a \text{ o } b) \text{ o } c = a \text{ o } (b \text{ o } c)$

Existence of Identity : Let $a \in$ G.

Let $e \in G \ni a \text{ o } e = a \Rightarrow \dfrac{ae}{3} = a \Rightarrow e = 3$

$\therefore$ $a \text{ o } e = a \text{ o } 3 = \dfrac{a.3}{3} = a$ and $e \text{ o } a = 3 \text{ o } a = \dfrac{3.a}{3} = a.$

$\therefore$ $a \text{ o } 3 = 3 \text{ o } a = a.$

$\Rightarrow$ $e = 3$ is the identity in G.

Existence Inverse : Let $a \in$ G.

Let $b \in G \ni a \text{ o } b = e \Rightarrow \dfrac{a.b}{3} = 3 \Rightarrow b = \dfrac{9}{a}$

$\therefore$ $a \text{ o } b = a \text{ o } \dfrac{9}{a} = \dfrac{a.\frac{9}{a}}{3} = 3,\ b \text{ o } a = \dfrac{9}{a} \text{ o } a = \dfrac{\frac{9}{a}.a}{3} = 3.$

$\therefore$ $a \text{ o } \dfrac{9}{a} = 3 = \dfrac{9}{a} \text{ o } a \Rightarrow b = \dfrac{9}{a}$ is the inverse of a in G.

Commutativity : Let $a, b \in$ G

$$a \text{ o } b = \frac{a.b}{3} = \frac{b.a}{3} = b \text{ o } a$$

Hence (G, 0) is an abelian group.

Theorem : In a group (G,0), the identity element is unique.

Proof : If possible, let e_1, e_2 be two identity elements in the group (G, 0).

Since e_1 is an identity in G, $e_1 \ o \ e_2 = e_2 = e_2 \ o \ e_1$ ...(1)

Since e_2 is an identity in G, $e_2 \ o \ e_1 = e_1 = e_1 \ o \ e_2$ ...(2)

From (1), (2), $e_1 = e_2$.

$\therefore$ The identity element in a group (G, 0) is unique.

Theorem : In a group (G, o), inverse of any element is unique.

Proof : Let e be the identity element in the group (G, 0).

If $a \in$ G, then a will have an inverse.

If possible, let $b \in$ G, $c \in$ G are two inverses of a in G.

$\Rightarrow$ $a \ o \ b = b \ o \ a = e$ and $a \ o \ c = c \ o \ a = e$

$\therefore$ $c \ o \ (a \ o \ b) = c \ o \ e = c$ ...(1)

$(c \ o \ a) \ o \ b = e \ o \ b = b$ ...(2)

Since $c \ o \ (a \ o \ b) = (c \ o \ a) \ o \ b$, $b = c$ from (1), (2).

$\therefore$ Inverse of any element in a group (G, 0) is unique.

Theorem : (Cancellation Laws)

Let (G, .) be a group. Then for $a, b, c \in$ G, $ab = ac \Rightarrow b = c$ (Left cancellation law)

and $ba = ca \Rightarrow b = c$ (Right cancellation law)

Proof : Let (G, .) be a group e be the identity in G.

For any $a, b, c \in$ G, $ab = ac \Rightarrow a^{-1}(ab) = a^{-1}(ac)$

$\Rightarrow$ $(a^{-1} a) \ b = (a^{-1} a) \ c$

$\Rightarrow$ $eb = ec$

$\Rightarrow$ $b = c$.

And $ba = ca \Rightarrow (ba) \ a^{-1} = (ca) \ a^{-1}$

$\Rightarrow$ $b \ (aa^{-1}) = c \ (aa^{-1}) \Rightarrow be = ce \Rightarrow b = c$.

Note 1. If G is an additive group, $a + b = a + c \Rightarrow b = c$ and $b + a = c + a \Rightarrow b = c$.

2. (N, +) is a semi group. For $a, b, c \in$ N,

$$a + b = a + c \Rightarrow b = c \quad \text{and} \quad b + a = c + a \Rightarrow b = c,$$

But (N, +) is not a group.

$\therefore$ In a semi group even if cancellation law holds the semi group is not a group.

Theorem : Let (G, .) be a group, for any $a \in$ G, then $(a^{-1})^{-1} = a$.

Proof : Let e be the identity in G.

By definition of inverse, $aa^{-1} = e = a^{-1} a$.

Then $(a^{-1})^{-1} = a$.

Theorem : Let (G, .) be a group. For any $a, b \in$ G, $(ab)^{-1} = b^{-1} a^{-1}$.

Proof : Given (G, .) be a group. Let e be the identity in G.

$$a, b \in G \Rightarrow ab \in G, a^{-1} \in G, b^{-1} \in G.$$

$\Rightarrow \qquad ab \in G, b^{-1} a^{-1} \in G$

Now $\qquad (ab)(b^{-1} a^{-1}) = a[b(b^{-1} a^{-1})]$ (by associativity)

$= a[(bb^{-1}) a^{-1}]$

$= a[e a^{-1}]$

$= aa^{-1} = e$

and $\qquad (b^{-1} a^{-1})(ab) = [(b^{-1} a^{-1}) a] b = [b^{-1} (a^{-1} a)] b$ (by associativity)

$= (b^{-1} e) b = b^{-1} b = e.$

$\therefore \qquad (ab)(b^{-1} a^{-1}) = e = (b^{-1} a^{-1})(ab)$

Hence $\qquad (ab)^{-1} = b^{-1} a^{-1}.$

Note 1. If (G, +) is a group, then $-(a + b) = (-b) + (-a)$

2. $(b^{-1} a^{-1})^{-1} = ((ab)^{-1})^{-1} = ab.$

Theorem. If a, b are any two elements of a group (G, .) which commute. Show that

(*i*) a^{-1} and b commute

(*ii*) b^{-1} and a commute and

(*iii*) a^{-1} and b^{-1} commute.

Proof : (*i*) Given $ab = ba$

$\Rightarrow \qquad (ab) a^{-1} = (ba) a^{-1}$

$\Rightarrow \qquad a(ba^{-1}) = b(aa^{-1}) = be = b$

$\Rightarrow \qquad a^{-1} a(ba^{-1}) = a^{-1} b$

$\Rightarrow \qquad e\, ba^{-1} = a^{-1} b$

$\Rightarrow \qquad ba^{-1} = a^{-1} b.$

$\therefore$ a^{-1} and b commute.

(*ii*) We have $ab = ba \Rightarrow b^{-1}(ab) = b^{-1}(ba)$

$\Rightarrow \qquad (b^{-1} a) b = (b^{-1} b) a = ea = a$

$\Rightarrow \qquad (b^{-1} a)(bb^{-1}) = ab^{-1}$

$\Rightarrow \qquad b^{-1} a\, e = ab^{-1} \;\Rightarrow\; b^{-1} a = ab^{-1}$

$\therefore$ b^{-1} and a commute.

(*iii*) We have $ab = ba.$

$\Rightarrow \qquad (ab)^{-1} = (ba)^{-1}$

$\Rightarrow \qquad b^{-1} a^{-1} = a^{-1} b^{-1}$

$\therefore$ a^{-1} and b^{-1} commute.

Example. *All groups of order 4 and less are commutative.*

Solution. Let (G, .) be a group.

1. Let 0 (G) = 1. Then G = {*e*} where *e* is the identity element.

Here $\qquad e \,.\, e = e$

$\Rightarrow$ G is commutative.

2. Let 0 (G) = 2. Then G = {e, a} where e is the identity element. Then $e \cdot e = e$, $e \cdot a = a = a \cdot e$ and $a \cdot a = e$.

The composition table is shown in table.

Table 4.12

•	e	a
e	e	a
a	a	e

3. Let 0 (G) = 3. Then G = {e, a, b} where e is the identity element. We have (i) $ee = e$, $ea = a$ and $eb = b$

   (ii) $ae = a$, $ab = e$ ($\because$ $ab = b \Rightarrow ab = eb \Rightarrow a = e$, which is absurd) and $aa = b$.

   (iii) $be = b$, $ba = e$ ($\because$ $ba = a \Rightarrow ba = ea \Rightarrow b = e$ which is absured) and $bb = a$.

The composition table is

Table 4.13

•	e	a	b
e	e	a	b
a	a	b	e
b	b	e	a

From table G is abelian ($\because$ rows, columns of table are equal)

4. Let 0 (G) = 4, then G = {e, a, b, c} where e is the identity. Here we have two cases.

   (i) $e^{-1} = e$, $a^{-1} = a$, $b^{-1} = b$, $c^{-1} = c$

   (ii) $e^{-1} = e$, $a^{-1} = a$, $b^{-1} = c$, $c^{-1} = b$.

Case (i), we have

Composition table of case (i)

Table 4.14.

•	e	a	b	c
e	e	a	b	c
a	a	e	c	b
b	b	c	e	a
c	c	b	a	e

(i) $ee = e$, $ea = a$, $eb = b$, $ec = c$

(ii) $ae = a$, $aa = e$, $ab = c$ ($\because$ $ab = b \Rightarrow ab = eb \Rightarrow a = e$, which is absurd) and $ac = b$

(iii) $be = b$, $bb = e$, $ba = c$ ($\because$ $ba = a \Rightarrow ba = ea \Rightarrow b = e$ which is absurd) and $bc = a$.

(iv) $ce = c$, $cc = e$, $cb = a$ ($\because$ $cb = a \Rightarrow cb = eb \Rightarrow c = e$ which is absurd) and $ca = b$.

Case (ii), we have (i) $ee = e$, $ea = a$, $eb = b$, and $ec = c$

(ii) $ae = a$, $aa = e$, $ab = c$ ($\because$ $ab = b \Rightarrow ab = eb \Rightarrow a = e$ which is absurd) and $ac = b$.

Composition table for case (ii)

Table 4.15.

•	e	a	b	c
e	e	a	b	c
a	a	e	c	b
b	b	c	a	e
c	c	b	e	a

(*iii*) $be = b$, $bc = e$, $ba = c$ $\quad$ ($\because$ $ba = a \Rightarrow ba = ea \Rightarrow b = e$ which is absurd)

(*iv*) $ce = c$, $cb = e$, $ca = b$ $\quad$ ($\because$ $ca = a \Rightarrow ca = ea \Rightarrow c = e$, which is absurd)

Clearly G is commutative in either case.

Example. *Let S be non-empty set. Verify whether the following are groups or not.*

[JNTU - November, 2006 - Set 2]

(*i*) (*P (s)*, $\cup$),

(*ii*) (*P (S)*, $\cap$) *where P (S) is the power set of S.*

Solution. (*i*) Let A, B $\in$ P (S) $\Rightarrow$ A $\cup$ B $\in$ P (S)

$\therefore$ P (S) is closed w.r.t. $\cup$

Let A, B, C $\in$ P (S) $\Rightarrow$ (A $\cup$ B) $\cup$ C = A $\cup$ (B $\cup$ C)

$\therefore$ $\cup$ is associative in P (S).

Let A $\in$ P (S) $\Rightarrow$ A $\cup$ ϕ = A = ϕ $\cup$ A

$\therefore$ ϕ is identity in P (S).

Let A $\in$ P (S), $\exists$ no B in P (S) $\ni$ A $\cup$ B = ϕ = B $\cup$ A

$\therefore$ A has no inverse in P (S)

Hence (P (S), $\cup$) is not a group, but it is a monoid.

(*ii*) Let A, B, $\in$ P (S) $\Rightarrow$ A $\cap$ B $\in$ P (S)

P (S) is closed w.r.t. $\cap$.

Let A, B, C $\in$ P (S) $\Rightarrow$ (A $\cap$ B) $\cap$ C = A $\cap$ (B $\cap$ C)

$\therefore$ $\cap$ is associative in P (S).

Let A $\in$ P (S) $\Rightarrow$ A $\cap$ S = A = S $\cap$ A

$\therefore$ S is identity in P (S).

Let A $\in$ P (S) $\exists$ no B in S $\ni$ A $\cap$ B = S = B $\cap$ A

$\therefore$ A has no inverse in P (S)

Hence (P (S), $\cap$) is not a group, but it is a monoid.

4.4.1 Addition modulo *m*, congruence, multiplication modulo *p*.

Definition. *Let a, b be any two distinct integers, m be a fixed positive integer. Then the addition modulo m denoted by* $a +_m b$, *is defined as* $a +_m b = r$, $0 \le r < m$, *where r is the least non-negative integer remainder when the sum of a and b is divided by m.*

Example 1. $5 +_8 7 = 4$; since $5 + 7 = 12 = 1\ (8) + 4$

2. $4 +_6 2 = 0$; since $4 + 2 = 6 = 1\ (6) + 0$

3. $-9 +_7 1 = 6$; since $-9 + 1 = -8 = -2\ (7) + 6$

Definition. *If a, b are two integers such that a – b is divisible by a fixed positive integer m. Then we say that a is congruent to b modulo m. We write it as* $a \equiv b \pmod{m}$

Note 1. $a \equiv b \pmod{m}$ iff $\dfrac{m}{a-b}$.

2. $a +_m b \equiv (a + b) \pmod{m}$

3. $a +_m b = b +_m a.$

Example 1. $15 \equiv (-1) \pmod{4}$, $15 \equiv 3 \pmod{4}$, etc.

2. $16 \equiv (-2) \pmod{9}$, $16 \equiv 7 \pmod{9}$, etc.

Definition. *Let p be any fixed positive integer. The multiplication modulo p of any two integers a and b is denoted by* $a \times_p b$ *and is defined as* $a \times_p b = r$, $0 \le r < p$ *where r is the least non-negative remainder, when the ordinary product of a and b divided by p.*

Note 1. $a \times_p b \equiv ab \pmod{p}$

2. $a \times_p b = b \times_p a$

Example 1. $5 \times_8 7 = 3$; since $5 \times 7 = 35 = 4\ (8) + 3$

2. $4 \times_6 2 = 2$; since $4 \times 2 = 8 = 1\ (6) + 2$

Theorem. The set $G = \{0, 1, 2, ..., m-1\}$ of first m non-negative integers is a finite abelian group under the operation of addition modulo m.

Example 1. *Prove that the set G = {0, 1, 2, 3, 4} is an abelian group of order 5 w.r.t. addition modulo 5.*

Solution. The composition table for G w.r.t. $+_5$ is given below :

Given $G = \{0, 1, 2, 3, 4\}$

Composition Table

Table 4.16.

$+_5$	0	1	2	3	4
0	0	1	2	3	4
1	1	2	3	4	0
2	2	3	4	0	1
3	3	4	0	1	2
4	4	0	1	2	3

Closure Property :

Since all the entries in the composition table are in G, G is closed w.r.t. $+_5$.

Associativity : Since elements of G are integers, $+_5$ is associative in G.

Identity : From first row, first column, we have 0 is the identity.

Inverse : From table, inverses of 0, 1, 2, 3, 4 are 0, 4, 3, 2, 1 respectively, which are in G.

Commutative : Since corresponding rows, columns are identical, $+_5$ is commutative in G.

Hence (G, $+_5$) is an abelian group.

Theorem. The set G = {1, 2, 3, ..., $p-1$} of $p-1$ integers, p being prime, is a finite abelian group of order $p-1$, under the multiplication modulo p.

Example 2. *Prove that ({1, 2, 3, 4}, $\times_5$) is an abelian group.*

Solution. Let G = {1, 2, 3, 4}

The composition table for G w.r.t. X_5 is given below.

Table 4.17.

$\times_5$	1	2	3	4
1	1	2	3	4
2	2	4	1	3
3	3	1	4	2
4	4	3	2	1

Closure Property :

Since all the entries in the composition table are in G, G is closed w.r.t. $\times_5$.

Associative Property :

Since elements of G are non-zero real numbers, $\times_5$ is associative in G.

Identity :

Let $a \in G \Rightarrow a \times_5 1 = a = 1 \times_5 a$

∴ 1 is the identity in G.

Inverse :

From table, we have

$$1 \times_5 1 = 1, 2 \times_5 3 = 1,$$
$$3 \times_5 2 = 1, 4 \times_5 4 = 1.$$

Hence inverses of 1, 2, 3, 4 are 1, 3, 2, 4 respectively, which are in G.

Commutative : Let $a, b \in G \Rightarrow a \times_5 b$ = The remainder when $a \times b$ is divided by 5.

= The remainder when $b \times a$ is divided by 5.

$= b \times_5 a.$

Hence (G, $\times_5$) is an abelian group.

Definition. *The operation congruence modulo 'm' is an equivalence relation in the set of integers. So the operation 'congruence modulo m' partitions Z into disjoint equivalence classes called* ***residue classes modulo m*** *(or)* ***congruence classes modulo m.***

If $a \in Z$, then the residue class of a is $\bar{a}$ or $[a]$ or $\{a\}$

where $\bar{a} = \{x \mid x \in Z$ and $x - a$ is divisible by $m\}$.

Similarly, if $b \in Z$, then $\bar{b} = \{x \mid x \in Z$ and $x - b$ is divisible by $m\}$

Note : $\bar{a} = \bar{b} \Rightarrow a \equiv b \pmod{m}$ (or) $m \mid a - b$.

Definition. *{0, 1, 2, 3, ..., m − 1} is called the complete set of least positive residues modulo m.*

Definition. Let $m \in N$ and $r \in Z$.

Let $\bar{r} = \{x \mid x \in Z, x \equiv r \pmod{m}\}$, then the set

$Z_m = \{\bar{0}, \bar{1}, \bar{2}, \bar{3}, ..., \overline{m-1}\}$ is called the complete set of least positive residues classes modulo m (or) simply set of residue classes modulo m.

Here $\bar{r} = \{......, -2m + r, -m + r, r, m + r, 2m + r, ...\}$

Example. If $m = 6$, the set of least positive residues modulo m

$= \{0, 1, 2, 3, 4, 5\}$ and the set of residue classes modulo

$m = \{\bar{0}, \bar{1}, \bar{2}, \bar{3}, \bar{4}, \bar{5}\}$

where

$$\bar{0} = \{......, -12, -6, 0, 6, 12, ...\},$$
$$\bar{1} = \{......, -11, -5, 1, 7, 13, ...\}$$
$$\bar{2} = \{......, -10, -4, 2, 8, 14, ...\},$$
$$\bar{3} = \{......, -9, -3, 3, 9, 15, ...\}$$
$$\bar{4} = \{......, -8, -2, 4, 10, 16, ...\},$$
$$\bar{5} = \{......, -7, -1, 5, 11, 17, ...\}$$

We observe that $\bar{a} = \overline{a+m} = \overline{a+2m} = ...$

Thus, for the above examples,

$$\bar{0} = \bar{6} = \overline{12} = ...,$$
$$\bar{1} = \bar{7} = \overline{13} = ..., \quad \bar{2} = \bar{8} = \overline{14} = ...$$

Note 1. The elements $\bar{0}, \bar{1}, \bar{2}, ..., \overline{m-1}$ of Z_m are all distinct.

Note 2. If $a \in Z$ and if r is the remainder when a is divided by m, then $\bar{a} = \bar{r}$. Because by division algorithm, $a = qm + r$, $q \in Z$ and r is the remainder such that $0 \le r \le m - 1$.

∴ $a - r = qm \Rightarrow a - r$ is divisible by m.

⇒ $a \equiv r \pmod{m}$

⇒ $\bar{a} = \bar{r}$.

Note 3. The residue class $\bar{0}$ is called the zero residue class.

Addition of residue classes.

Definition. *Let m be fixed positive integer, for $\bar{a}, \bar{b} \in Z_m$, we define addition of residue classes, denoted by $+_m$ as $\bar{a} +_m \bar{b} = \overline{a +_m b}$.*

Note. If r is the remainder $(0 \le r < m)$ when $a + b$ is divided by m, then $\overline{a +_m b} = \bar{r}$

Example 1. $\bar{4} +_5 \bar{3} = \overline{4 +_5 3} = \bar{2}$, Example 2: $\bar{4} +_7 \bar{3} = \bar{0}$

Theorem. The set of residue classes modulo m is an abelian group of order m w.r.t., addition of residue classes.

Definition. *Let m be fixed positive integer. For* $\bar{a}, \bar{b} \in G$, *we define multiplication of residue classes denoted by* $\times_p$ *as* $\bar{a} \times_p \bar{b} = \overline{a \times_p b}$.

Note : If r is the remainder $(0 \le r < p)$ when $a \times b$ is divided by p, then $\bar{a} \times_p \bar{b} = \bar{r}$.

Theorem. The set of non-zero residue classes modulo a prime integer p forms an abelian group of order $(p - 1)$ w.r.t., multiplication of residue classes.

Example 3. *Show that the non-zero residue classes modulo 5 form an abelian group w.r.t., multiplication of residue classes.*

Solution. Let $G = \{\bar{1}, \bar{2}, \bar{3}, \bar{4}\}$ be the non-zero residue classes modulo 5.

Table 4.18

$\times_5$	$\bar{1}$	$\bar{2}$	$\bar{3}$	$\bar{4}$
$\bar{1}$	$\bar{1}$	$\bar{2}$	$\bar{3}$	$\bar{4}$
$\bar{2}$	$\bar{2}$	$\bar{4}$	$\bar{1}$	$\bar{3}$
$\bar{3}$	$\bar{3}$	$\bar{1}$	$\bar{4}$	$\bar{2}$
$\bar{4}$	$\bar{4}$	$\bar{3}$	$\bar{2}$	$\bar{1}$

Closure Property :

Since all the entries in the composition table are in G, G is closed w.r.t. $\times_5$.

Associativity :

Let $\bar{a}, \bar{b}, \bar{c} \in G$,

$$(\bar{a} \times_5 \bar{b}) \times_5 \bar{c} = \overline{a \times_5 b} \times_5 \bar{c}$$

$$= \overline{(a \times_5 b) \times_5 c} = \overline{a \times_5 (b \times_5 c)} = \bar{a} \times_5 \left(\overline{b \times_5 c}\right)$$

$$= \bar{a} \times_5 (\bar{b} \times_5 \bar{c}).$$

Identity : From table, we have

$$\bar{1} \times_5 \bar{1} = \bar{1},\ \bar{2} \times_5 \bar{1} = \bar{2},\ \bar{3} \times_5 \bar{1} = \bar{3},\ \bar{4} \times_5 \bar{1} = \bar{4}$$

Then $\bar{1}$ is the identity in G.

Inverse : From table, we have

$$\bar{1} \times_5 \bar{1} = \bar{1},\ \bar{2} \times_5 \bar{3} = \bar{1},\ \bar{3} \times_5 \bar{2} = \bar{1},\ \bar{4} \times_5 \bar{4} = \bar{1}$$

Then inverses of $\bar{1}, \bar{2}, \bar{3}, \bar{4}$ are $\bar{1}, \bar{3}, \bar{2}, \bar{4}$ respectively which are in G.

Commutative : Since corresponding rows, columns of the composition table are identical, $\times_5$ is commutative in G.

Hence (G, $\times_5$) is a finite abelian group.

Example 4. *Show that the set of residue classes modulo 5. Form an abelian group w.r.t., the addition of residue classes.*

Solution. Let G = $\{\bar{0}, \bar{1}, \bar{2}, \bar{3}, \bar{4}\}$ be the set of residue classes modulo 5.

Table 4.19.

$+_5$	$\bar{0}$	$\bar{1}$	$\bar{2}$	$\bar{3}$	$\bar{4}$
$\bar{0}$	$\bar{0}$	$\bar{1}$	$\bar{2}$	$\bar{3}$	$\bar{4}$
$\bar{1}$	$\bar{1}$	$\bar{2}$	$\bar{3}$	$\bar{4}$	$\bar{0}$
$\bar{2}$	$\bar{2}$	$\bar{3}$	$\bar{4}$	$\bar{0}$	$\bar{1}$
$\bar{3}$	$\bar{3}$	$\bar{4}$	$\bar{0}$	$\bar{1}$	$\bar{2}$
$\bar{4}$	$\bar{4}$	$\bar{0}$	$\bar{1}$	$\bar{2}$	$\bar{3}$

Closure Property :

Since all the entries in the composition table are in G, G is closed w.r.t. $+_5$.

Associativity : Let $\bar{a}, \bar{b}, \bar{c} \in$ G

Then, $$\left(\bar{a} +_5 \bar{b}\right) +_5 \bar{c}$$

$$= \left(\overline{a +_5 b}\right) +_5 \bar{c} = \overline{(a +_5 b) +_5 c}$$

$$= \overline{a +_5 (b +_5 c)} = \bar{a} +_5 \left(\overline{b +_5 c}\right) - \bar{a} +_5 \left(\bar{b} +_5 \bar{c}\right)$$

Identity : From first row, first column of composition table, $\bar{0}$ is the identity in G.

Inverse : From table, we have $\bar{0} +_5 \bar{0} = \bar{0}$

$$\bar{1} +_5 \bar{4} = \bar{0},\ \bar{2} +_5 \bar{3} = \bar{0},\ \bar{3} +_5 \bar{2} = \bar{0},\ \bar{4} +_5 \bar{1} = \bar{0}.$$

Therefore, inverses of $\bar{0}, \bar{1}, \bar{2}, \bar{3}, \bar{4}$ are $\bar{0}, \bar{4}, \bar{3}, \bar{2}, \bar{1}$ respectively, which are in G.

Commutative : Let $\bar{a}, \bar{b} \in$ G.

$$\bar{a} +_5 \bar{b} = \overline{a +_5 b} = \overline{b +_5 a} = \bar{b} +_5 \bar{a}.$$

Hence (G, $+_5$) is an abelian group.

Example 5. *In a group (G, .) for every $a \in G$, $a^2 = e$, prove that G is an abelian group.*

Solution. Let $a, b \in$ G $\Rightarrow ab \in$ G.

Given that for every $a \in$ G, $a^2 = e \Rightarrow a \,.\, a = e \Rightarrow a^{-1} = a$

$$ab \in G \Rightarrow (ab)^2 = e \Rightarrow (ab)\,(ab) = e$$

$$\Rightarrow \quad (ab)^{-1} = ab \Rightarrow b^{-1}\, a^{-1} = ab$$

$$\Rightarrow \quad ba = ab \text{ since } b \in G \Rightarrow b^{-1} = b.$$

∴ (G, .) is an abelian group.

Example 6. *Show that in a group G for* $a, b \in G$,
$(ab)^2 = a^2 b^2$ *iff (G, .) is an abelian.*

Solution. Let $a, b \in G$ and $(ab)^2 = a^2 b^2$.

$$\Rightarrow \quad (ab)(ab) = (aa)(bb)$$

$$\Rightarrow \quad a(ba)b = a(ab)b$$

$$\Rightarrow \quad ba = ab \text{ by left, right cancellation laws.}$$

$\Rightarrow$ (G, .) is an abelian group.

Conversely suppose G is an abelian group.

$$\therefore \quad (ab)^2 = (ab)(ab) = a(ba)b = a(ab)b \qquad (\because \; ab = ba)$$

$$= (aa)(bb) = a^2 b^2.$$

Example 7. *If (G, .) is a group such that* $(ab)^m = a^m b^m$ *for three consecutive integers of m for all* $a, b \in G$, *show that G is abelian.*

Solution. Let $a, b \in G$ let m, $m + 1$, $m + 2$ be three consecutive integers.

By hypothesis $\quad (ab)^m = a^m b^m, (ab)^{m+1} = a^{m+1} b^{m+1},$

$$(ab)^{m+2} = a^{m+2} \, . \, b^{m+2}$$

$$\therefore \quad (ab)^{m+2} = (ab)^{m+1} (ab).$$

$$\Rightarrow \quad a^{m+2} b^{m+2} = a^{m+1} b^{m+1} \, . \, ab.$$

$$\Rightarrow \quad a \, . \, a^{m+1} \, . \, b^{m+1} b = a \, a^m b^m b \, ab.$$

$$\Rightarrow \quad a^{m+1} b^{m+1} = a^m b^m ba$$

$$\Rightarrow \quad (ab)^{m+1} = (ab)^m (ba)$$

$$\Rightarrow \quad (ab)^m (ab) = (ab)^m (ba)$$

$$\Rightarrow \quad ab = ba.$$

Example 8. *If (G, .) is a group of even order, prove that G must have an element* $a \neq e$ *satisfying* $a^2 = e$.

Solution. Let 0 (G) = $2n$ where $n \in N$.

We shall prove that G must have an element $a \neq e$ such that $a^{-1} = a$. Suppose G has no element, other than the identity element e, which is its own inverse. We know that in a group every element possesses a unique inverse and the inverse of the identity element is itself.

Further, if $b, c \in G$ and $b^{-1} = c$, then $c^{-1} = b$.

So by excluding e, the remaining $(2n - 1)$ elements of G must be divided into pairs such that each pair consists of an element and its inverse. It is not possible, since 0 (G) = $2n - 1$ which is odd. Our assumption is wrong.

Hence in the group G of even order there is an element ($\neq e$) which is its own inverse.

If a is taken as that element in G, then $a = a^{-1} \Rightarrow aa = e$

$$\Rightarrow \quad a^2 = e.$$

EXERCISE 4.2.

1. Define a group, abelian group. Give one example to each.
2. Show that the following are groups :
   (*i*) The set C of all complex numbers w.r.t. +.
   (*ii*) The set R of all real numbers w.r.t. +.
   (*iii*) The st C_0 of all non-zero complex numbers w.r.t. ×.
   (*iv*) The set R_0 of all non-zero real numbers w.r.t. ×.
   (*v*) The set {1, – 1} w.r.t. ×.
3. Does the set of all odd integers form a group w.r.t. addition.
4. Show that the set of vectors defined as directed line segments does not form a group w.r.t. scalar (dot) product.
5. Show that the set of all positive rational numbers forms an abelian group under the composition defined by $a * b = \frac{ab}{4}$.
6. Let R be the set of all real numbers and 0 a binary operation on R defined by $a o\, b = a + b + ab$. Determine the identity element in R and inverse of a when $a \neq -1$.
7. Show that the R_{-1} of real numbers other than – 1 is an abelian group w.r.t. the operation * defined by $a * b = a + b + ab$.
8. If $g = \{x \mid x$ is a rational number and $0 < x \leq 1\}$. Show that (G, .) is not a group.
9. Prove that the set of all $m \times n$ matrices whose elements are numbers (integers, real or complex) form an infinite abelian group w.r.t. matrix addition.
10. Prove that the set of all $n \times n$ non-singular matrices having their elements as rational (real or complex) numbers is an infinite non-abelian group w.r.t. matrix multiplication.

ANSWERS

1. (R, +), (R_0, .) are groups and also abelian groups.
3. Not a group.
6. Identity is 0 and inverse of a is $\frac{-a}{1+a}$.

4.4.2 Order of an Element of a Group.

Definition. *Let (G, 0) be a group, a ∈ G, then the order of the element a is defined as the least positive integer n such that a o a o a o ...... o a (n times) = e, where e is the identity in G. Order of element a is denoted by 0 (a).*

Note 1. If (G, .) be a group, $a \in$ G, then 0 $(a) = n$, if ∃ least positive integer $n \ni a^n = e$, where e is the identity in G.

If there exists no positive integer $n \ni a^n = e$. Then, we say that a is of infinite order or zero order.

Note 2. If $a^m = e$ in a group (G, .) where m is a positive integer. Then 0 (a) is finite, $\leq m$.

Note 3. Order of identity element in any group is 1.

Example 1. If G = {1, – 1}, then (G, .) is a group and 0 (1) = 1, 0 (– 1) = 2 ($\because$ $(-1)^2 = 1, (-1)^4 = 1, (-1)^6 = 1, \ldots\ldots$)

Example 2. If G = {1, ω, ω²} then (G, .) is a group and 0 (1) = 1, 0 (ω) = 3, 0 (ω²) = 3 ($\because$ $(\omega^2)^3 = 1, (\omega^2)^6 = 1, \ldots\ldots$)

Example 3. In a group ({1, – 1, *i*, – *i*}, .), 0 (1) = 1, 0 (– 1) = 2, 0 (*i*) = 4, 0 (– *i*) = 4.

Example 4. In a group (Q_0, .), 0 (1) = 1 ($\because$ 1 is the identity in Q_0)

0 (– 1) = 2 and other element order in Q_0 is infinite, since $2 \in Q_0$, $2^n \neq 1$ for any positive integer *n*.

Definition. *If (G, +) is a group, $a \in G$, then 0 (a) = m, if there exists least positive integer m ∋ ma = e = 0 and if ma = 0 in (G, +), then 0 (a) ≤ m.*

Example. In the group (Z, +) order of 0 is 1 and for other element order in Z is infinite or zero. Since $0 \neq a \in Z$, there is no $m \in N \ni ma = 0$.

Example 1. *Find the order of each element of the group {1, – 1, i, – i} w.r.t., usual multiplication.*

Solution. Identity element of group G is 1.

Now $\quad 0 (1) = 1$, since $(-1)^2 = 1$, $0(-1) = 2$,

Since $\quad (i)^1 = i, (i)^2 = -1, (i)^3 = -i, (i)^4 = 1,$

we have $\quad 0 (i) = 4.$

and since $(-i)^1 = -i, (-i)^2 = -1, (-i)^3 = i, (-i)^4 = 1$, we have $0(-i) = 4$.

Example 2. *Find the order of each element of the group*

G = {0, 1, 2, 3, 4, 5}, the composition being addition modulo 6.

Solution. In additive notation, 0 (*a*) = *m* if *ma* = 0 where *m* is least positive integer.

Since 0 is the additive identity,

$$0 (a) = 1.$$

Now $\quad 1 +_6 1 +_6 1 +_6 1 +_6 1 +_6 1 = 0 \Rightarrow 6 (1) = 0 \Rightarrow 0 (1) = 6$

$$2 +_6 2 +_6 2 = 0 \Rightarrow 3 (2) = 0 \Rightarrow 0 (2) = 3$$

$$3 +_6 3 = 0 \Rightarrow 2 (3) = 0 \Rightarrow 0 (3) = 2$$

$$4 +_6 4 +_6 4 = 0 \Rightarrow 3 (4) = 0 \Rightarrow 0 (4) = 3.$$

and $\quad 5 +_6 5 +_6 5 +_6 5 +_6 5 +_6 5 = 0 \Rightarrow 6 (5) = 0 \Rightarrow 0 (5) = 6.$

Theorem 1. If a is an element of order *n* of a group G and *p* is prime to *n*, then a^p is also of order *n*.

Proof : Let 0 (*a*) = *n*

$\therefore \quad a^n = e \Rightarrow (a^n)^p = e^p \Rightarrow (a^p)^n = e$

$\Rightarrow \quad 0 (a^p) \leq n \Rightarrow m \leq n$, where $0 (a^p) = m$.

Since *n*, *p* are relatively prime, $px + ny = 1$.

$\therefore \quad a = a^1 = a^{px+ny} = a^{px} . a^{ny} = a^{px} (a^n)^y = (a^p)^x . e^y$

$$= (a^p)^x e = (a^p)^x$$

Now $\quad a^m = [(a^p)^x]^m = (a^p)^{mx} = [(a^p)^m]^x = e^x = e \quad (\because\ 0\ (a^p) = m \Rightarrow (a^p)^m = e)$

$\therefore \quad a^m = e \Rightarrow 0\ (a) \le m \Rightarrow n \le m.$

Hence $\quad m = n \Rightarrow 0\ (a^p) = 0\ (a).$

Theorem 2. Let G is an abelian group. If $a, b \in$ G such that $0\ (a) = m$, $0\ (b) = n$ and $(m, n) = 1$, then $0\ (ab) = m\ .\ n.$

Proof : Let (G, .) is an abelian group, and e be the identity in G.

Since $a, b \in$ G such that $0\ (a) = m$, $0\ (b) = n$

We have $\quad a^m = e,\ b^n = e.$

Also $\quad ab \in$ G, let $0\ (ab) = p.$

Now $\quad (ab)^{mn} = a^{mn}\ b^{mn} = (a^m)^n\ (b^n)^m = e^n\ e^m = ee = e.$

$\Rightarrow \quad p \mid mn$ since $0(a) = n$ then $a^m = e$ iff n/m ...(1)

Also $\quad (ab)^{pn} = [(ab)^p]^n = e^n = e$

and $\quad (ab)^{pn} = a^{pn}\ b^{pn} = a^{pn}\ (b^n)^p = a^{pn}\ e^p = a^{pn}\ e = a^{pn}.$

$\Rightarrow \quad a^{pn} = e \Rightarrow m \mid pn.$

Since $\quad (m, n) = 1,\ m \mid p$ ...(2)

Similarly, we can prove that $n \mid p$ ...(3)

From (2), (3) and from $(m, n) = 1$, we have mn $\mid p$ ...(4)

From $\quad$ (1), (4), $mn = p$

Hence $\quad 0\ (ab) = p = mn = 0\ (a)\ 0\ (b).$

Theorem 3. The order of every element of a finite group is finite and is less than or equal to the order of the group.

Proof : Let (G, .) be a finite group. Let $a \in$ G.

By closure, we have $a^2, a^3, ..., \in$ G.

Since G is finite, all the positive integral powers of a

i.e., $a, a^2, a^3, ...$ cannot be distinct elements of G.

Let $\quad a^r = a^s$ where $r, s \in$ N and $r > s.$

$\therefore \quad a^r = a^s \Rightarrow a^{r-s} = a^0 = e \Rightarrow a^m = e$ where $r - s = m.$

Since $r > s$, m is a positive integer.

$\therefore \quad \exists$ a positive integer m such that $a^m = e.$

Hence if n is the least positive integral value of m such that $a^n = e$, then $0\ (a) = n$

$\therefore \quad 0\ (a)$ is finite.

To prove that $0\ (a) \le 0$ (G).

If possible, let $0\ (a) > 0$ (G).

Let $0\ (a) = n$; by closure, we have, $a, a^2, a^3, ..., a^n \in$ G

No two of these elements are equal.

(For, if possible, let $a^r = a^s$, $1 \leq s < r \leq n$. Then $a^{r-s} = e$. Since $0 < r - s < n$, $0\,(a) < n$, which is a contradiction).

Hence the n elements $a, a^2, \ldots\ldots, a^n$ are distinct elements of G. Since $0\,(a) > 0\,(G)$, this is not possible. Therefore, $0\,(a) \leq 0\,(G)$.

Theorem 4. In a group (G, .), if $a \in G$, then $0\,(a) = 0\,(a^{-1})$.

Proof : Let $0\,(a) = p$ and $0\,(a^{-1}) = q$.

Now $\quad 0\,(a) = p \Rightarrow a^p = e \Rightarrow (a^p)^{-1} = e^{-1} \Rightarrow a^{-p} = e.$

$\Rightarrow \quad (a^{-1})^p = e \Rightarrow 0\,(a^{-1}) \leq p \Rightarrow q \leq p \quad \ldots(1)$

Further, $\quad 0\,(a^{-1}) = q \Rightarrow (a^{-1})^q = e \Rightarrow a^{-q} = e$

$\Rightarrow \quad (a^{-q})^{-1} = e^{-1} \Rightarrow a^q = e$

$\Rightarrow \quad 0\,(a) \leq q \Rightarrow p \leq q \quad \ldots(2)$

From (1), (2), $\quad 0\,(a) = 0\,(a^{-1})$.

Theorem 5. The order of any positive integral power of an element a in a group G cannot exceed the order of a.

Proof : Let $0\,(a) = n$.

If m is a positive integer, then $a^m \in G$.

$$0\,(a) = n \Rightarrow a^n = e \Rightarrow (a^n)^m = e^m \Rightarrow a^{mn} = e$$

$\Rightarrow \quad (a^m)^n = e \Rightarrow 0\,(a^m) \leq n \Rightarrow 0\,(a^m) \leq 0\,(a).$

Theorem 6. If a is an element of a group G such that $0\,(a) = n$, then $a^m = e$ iff $n \mid m$.

Proof : Let $n \mid m$. $\exists$ a positive integer $q \ni m = nq$.

Also $\quad 0\,(a) = n \Rightarrow a^n = e$

$\therefore \quad a^m = a^{nq} = (a^n)^q = e^q = e.$

Conversely suppose that $a^m = e$.

Since n is a positive integer, by division algorithm there exists integers q and r such that $m = nq + r$, $0 \leq r < n$.

Now, $\quad a^m = a^{nq+r} = a^{nq} \,.\, a^r = (a^n)^q \,.\, a^r = e^q\, a^r = ea^r = a^r.$

$$a^m = e \Rightarrow a^r = e\text{; which is absurd since } 0\,(a) = n$$

$\therefore \quad r = 0$

$\Rightarrow \quad m = nq \Rightarrow n \mid m.$

Example 3. *For any two elements a, b $\in$ G where G is a group, then $0\,(a) = 0\,(b^{-1}\,ab)$.*

Solution. Let $a, b \in G \Rightarrow b^{-1}, a, b \in G \Rightarrow b^{-1}\,ab \in G$.

Now $\quad (b^{-1}\,ab)^2 = (b^{-1}\,ab)\,(b^{-1}\,ab) = (b^{-1}\,a)\,(bb^{-1})\,(ab)$

$$= (b^{-1}\,a)\,e\,(ab) = (b^{-1}\,a)\,(ab) = b^{-1}\,(aa)\,b = b^{-1}\,a^2\,b.$$

We can prove by induction, $(b^{-1}\,ab)^k = b^{-1}\,a^k\,b$, where k is a positive integer.

Let $\quad 0\,(a) = n$ and $0\,(b^{-1}\,ab) = m$

$\Rightarrow \quad a^n = e$ and $(b^{-1}\,ab)^m = e$

Now $a^n = e \Rightarrow (b^{-1} ab)^n = b^{-1} a^n b = b^{-1} eb = b^{-1} b = e.$

$\Rightarrow$ $0 (b^{-1} ab) \leq n \Rightarrow m \leq n$ ...(1)

and $(b^{-1} ab)^m = e \Rightarrow b^{-1} a^m b = e \Rightarrow a^m b = b$

$\Rightarrow$ $a^m = bb^{-1} = e \Rightarrow 0 (a) \leq m \Rightarrow n \leq m$ ...(2)

From (1), (2), $m = n \Rightarrow 0 (a) = 0 (b^{-1} ab).$

Example 4. *For any two elements a, b ∈ G where (G, .) is a group, 0 (ab) = 0 (ba).*

Solution. Let $a, b \in G \Rightarrow ab, ba \in G$.

Now $a^{-1} (ab) a = (a^{-1} a) (ba) = e (ba) = ba$

$\Rightarrow$ $ba = a^{-1} (ab) a \Rightarrow 0 (ba) = 0 (a^{-1} (ab) a)$

$\Rightarrow$ $0 (ba) = 0 (ab)$ $(\because \; a, x \in G \Rightarrow 0 (a) = 0 (x^{-1} ax))$

Example 5. *Let G is a group, a ∈ G. If 0 (a) = n and m | n, then prove that* $0 (a^m) = \dfrac{n}{m}$.

Solution. In a group G, $a \in G \ni 0 (a) = n$.

Also given that $m \mid n$. We have to prove that $0 (a^m) = \dfrac{n}{m}$

$a \in G \Rightarrow a^m \in G$ where m is a positive integer.

$m \mid n \Rightarrow n = mq$ where q is some positive integer.

$\therefore$ $0 (a) = n \Rightarrow a^n = e \Rightarrow a^{mq} = e$ $(\because \; m \mid n \, \exists \, q \in Z \ni n = mq)$

$\Rightarrow$ $(a^m)^q = e.$

n is least so that $a^n = e \Rightarrow mq$ is least so that $a^{mq} = e$

$\Rightarrow$ q is least so that $(a^m)^q = e \Rightarrow 0 (a^m) = q \Rightarrow 0 (a^m) = \dfrac{n}{m}$.

Example 6. *Let S is a semi group. If for all x, y ∈ S,* $x^2y = yx^2$, *prove that S is an abelian group.*

Solution. Given S is a semi group let $x \in S$.

If $y \in S$ such that $x^2y = yx^2 = y$, then $x^2 \in S$ is an identity in S

$\Rightarrow$ x^2 is the identity element in S $\forall \; x \in S$.

Let $x^2 = e \; \forall \; x \in S.$

Further, $\forall \; x \in S, x^2 = e \Rightarrow xx = e \Rightarrow x^{-1} = x \in S.$

$\therefore$ S is a group.

Also $x, y \in S \Rightarrow x, y \in S \Rightarrow xy = (xy)^{-1} = y^{-1} x^{-1} = yx$

$\Rightarrow$ S is an abelian group.

Example 7. *Let G is a group with identity element e. If x, y ∈ G ∋* $xy^2 = y^3x$ *and* $yx^3 = x^2y$, *then prove that x = y = e.*

Solution. Given G is a group with identity element e.

If $\quad x, y \in G$

such that $\quad xy^2 = y^3 x \quad$ ...(1)

$yx^3 = x^2 y \quad$ ...(2)

then we have to prove that $x = y = e$.

From (1), $\quad xy^2 = y^3 x \Rightarrow x^2\, y^2 = xy^3\, x$

$\Rightarrow \quad x^2 y = xy^3\, xy^{-1} = (xy^2)\, yxy^{-1}$

$\Rightarrow \quad yx^3 = y^3 xy\, xy^{-1}$. (by (2), (1)

$\Rightarrow \quad x^3 y = y^2\, xyx$

$\Rightarrow \quad x\,(x^2 y) = y^2 x\, yx \Rightarrow x\,(yx^3) = y^2\, xyx$

$\Rightarrow \quad xyx^2 = y^2 xy \Rightarrow xy\, x^2 y = y^2\, xy^2$

$\Rightarrow \quad xy\,(yx^3) = y^2\,(y^3 x)$ (by (2), (1))

$\Rightarrow \quad xy^2\, x^3 = y^5\, x$

$\Rightarrow \quad xy^2\, x^2 = y^5 \Rightarrow (y^3 x)\, x^2 = y^5$ (by (1))

$\Rightarrow \quad x^3 = y^2 \quad$ ...(3)

$\Rightarrow \quad x^4 = xy^2 = y^3 x \Rightarrow x^3 = y^3 \Rightarrow y^2 = y^3$ (by (3))

$\Rightarrow \quad y^2 y = y^2 e \Rightarrow y = e$

From (2), $\quad ex^3 = x^2 e \Rightarrow x^2 x = x^2 e \Rightarrow x = e$

$\therefore \quad x = y = e.$

Example 8. *If (G, .) is a group and* $ab\,a^{-1} = b^i$. *Prove that* $a^r b\, a^{-r} = b^{i^r}$.

Solution. We prove this result by induction.

For $\quad r = 2,\ a^2\, b\, a^{-2} = aa\, ba^{-1}\, a^{-1} = a\,(aba^{-1})\, a^{-1}$

$= a\, b^i\, a^{-1} = (ab\, a^{-1})^i$

$= (b^i)^i = b^{i^2}$

Given result is true for $\quad r = 2$.

Suppose that the given result is true for $r = n$.

Then $\quad a^n b\, a^{-n} = b^{i^n}$.

$\therefore \quad a^{n+1}\, b\, a^{-(n+1)} = a\,(a^n\, b\, a^{-n})\, a^{-1} = a\, b^{i^n}\, a^{-1}$

$= \left(ab\,a^{-1}\right)^{i^n} = \left(b^i\right)^{i^n} = b^{i\,i^n}$

$= b^{i^{n+1}}$.

Given result is true for $\quad r = n + 1$.

Hence finite mathematical induction, $a^r\, b\, a^{-r} = b^{i^r}$ is true for all natural numbers of r.

Example 9. *If* $x^i = gx\, g^{-1}$, *prove that* $O\,(x) = i^{O\,(g)} - 1$.

Solution. We have $x = g^{O\,(g)}\, x\, g^{O\,(g)} = x^{i^{O(g)}}$

by previous example 8, hence the result follows

Example 10. *Let G be a group and let a, b ∈ G. If* $a^{-1}\,ba = b^{-1}$ *and* $b^{-1}\,ab = a^{-1}$. *Prove that* $a^2 = b^{-2}$, $a^2 = b^2$, *and* $a^4 = e = b^4$.

Solution. Let $a, b \in$ group (G, .)

We have $\quad a = ae = a\,(b^{-1}\,b) = a\,(b^{-1})\,b = a\,(a^{-1}\,ba)\,b = bab$

Again $\quad a^{-2} = a^{-1}\,a^{-1} = a^{-1}\,(b^{-1}\,ab) = a^{-1}\,b^{-1}\,(a)\,b = a^{-1}\,b^{-1}\,(bab)\,b$

$a^{-2} = a^{-1}\,ab^2 = b^2 \Rightarrow a^2 = b^{-2}$ ...(1) ($\because\ a = bab$)

and $\quad a^{-2} = (a^{-1})^2 = (b^{-1}\,ab)^2 = b^{-1}\,a^2b = b^{-1}\,b^{-1}$ ($\because\ a^{-2} = b^2 \Rightarrow b^{-1} = a^2 b$)

$\Rightarrow \quad a^{-2} = b^{-2} \Rightarrow a^2 = b^2$ ...(2)

Hence $\quad a^2\,.\,a^2 = b^{-2}\,b^2 = e$

$\Rightarrow \quad a^4 = e$

and we have $\quad a^{-2} = b^2$, from (1)

$a^2 = b^2$

$\Rightarrow \quad a^{-2}\,a^2 = b^2\,b^2$

$\Rightarrow \quad e = b^4$

Finally, $\quad a^4 = e = b^4$.

Example 11. *In a group G for a, b ∈ G,* $0\,(a) = 5$, $b \neq e$ *and* $aba^{-1} = b^2$. *Show that* $0\,(b) = 31$.

[JNTU - November, 2006 - Set 4]

Solution. By example 9. $0\,(b) = i^{0\,(G)} - 1$

since $\quad aba^{-1} = b^2$

$\Rightarrow \quad 0\,(b) = 2^5 - 1 = 31$.

4.5 SUB GROUPS

Definition. *Let (G, 0) be a group. Let H be a non empty subset of G such that (H, 0) be a group. Then H is called a subgroup of G.*

Example 1. (Z, +) is a subgroup of group (Q, +).

Example 2. ({1, – 1}, .) is a subgroup of group ({1, – 1, *i*, – *i*}, .)

Example 3. (N, +) is not a subgroup of group (Z, +) because identity 0 does not exist in N.

Note. Suppose e is the identity element in a group G. Then $\{e\} \subseteq$ G and we have $e\,.\,e = e$, $e^{-1} = e$, so $\{e\}$ is a subgroup of G. Also G ⊆ G. So G is also a subgroup of G. These two subgroups $\{e\}$, G of G are called trivial or improper subgroups of G. All other subgroups, if exists, are called non-trivial (or) proper subgroups of G.

Theorem. The identity of a subgroup H of a group (G, .) is same as the identity of group G.

Proof : Let $a \in$ H and e' be the identity of H.

Since H is a subgroup, (H, .) is a group $\Rightarrow a.\ e' = a$ ...(1)

Let e be the identity in G.

$$a \in H \Rightarrow a \in G \Rightarrow a \, . \, e = a \qquad ...(2)$$

Also $\qquad e' \in H \Rightarrow e' \in G.$

From (1), (2), $\qquad ae' = ae \Rightarrow e' = e$ (by Left Cancellation Law).

Theorem. The inverse of any element of a subgroup H of a group (G, .) is same as the inverse of that element regarded as an element of the group G.

Proof : Let e be the identity in (G, .)

Since H is a subgroup of G, e is also the identity in H.

Let $\qquad a \in H \Rightarrow a \in G.$

Let b be the inverse of a in H and c be the inverse of a in G, then $ab = e$ and $ac = e$

$\Rightarrow \qquad ab = ac \Rightarrow b = c$ (by Left Cancellation Law)

Theorem. If H is any subgroup of a group G, then $H^{-1} = H$. Is converse true ? Justify your answer.

Proof : Let H be a subgroup of a group (G, .).

Let $\qquad h^{-1} \in H^{-1}$

By definition, of $\quad H^{-1}, h \in H.$

Since H is a subgroup of G,

$$h^{-1} \in H$$

$\therefore$ Therefore, $\qquad H^{-1} \subseteq H.$

and let $\qquad h \in H \Rightarrow h^{-1} \in H$, since H is a group.

$\Rightarrow \qquad (h^{-1})^{-1} \in H^{-1}$

$\Rightarrow \qquad h \in H^{-1}.$

Therefore, $\qquad H \subseteq H^{-1}$

Hence $\qquad H^{-1} = H.$

Converse is need not be true. For example, $H = \{-1\}$ such that $H^{-1} = H$, but $H = \{-1\}$ is not a subgroup of (G, .).

where $\qquad G = \{1, -1\}$, because $(-1)(-1) = 1 \notin H$ (closure is not true).

Theorem : If H is any subgroup of a group (G, .), then HH = H.

[JNTU, February, 2007 - Set I]

Proof : Let $x \in HH$ so that $x = h_1 \, . \, h_2$, where $h_1 \in H$ and $h_2 \in H$.

Since H is a subgroup, $h_1 \, h_2 \in H \Rightarrow x \in H$

$$\therefore \qquad HH \subseteq H \qquad ...(1)$$

and let $h_3 \in H$ and e be the identity in H.

$$\text{Then} \qquad h_3 = h_3 \, e \in HH \Rightarrow H \subseteq HH \qquad ...(2)$$

From (1), (2) results, $\quad HH = H.$

Theorem : A non-empty subset H of a group (G, .) is a subgroup of G iff (*i*) $ab \in H \; \forall \; a, b \in H$. (*ii*) $a^{-1} \in H \; \forall \; a \in H$.

Proof : Let H be a subgroup of a group (G, .) ⇒ (H, .) is a group.

by closure, $ab \in H \; \forall \; a, b \in H$

by inverse property, $a^{-1} \in H \; \forall \; a \in H$.

Conversely suppose (*i*), (*ii*) are satisfies.

Next we have to prove other two properties

Let $\quad a, b, c \in H \Rightarrow a, b, c \in G$

$\Rightarrow \quad (a \,.\, b) \,.\, c = a \,.\, (b \,.\, c) \Rightarrow$ '.' is associative in H.

and let $\quad a \in H, a^{-1} \in H$ (by (*ii*))

$\Rightarrow \quad a \,.\, a^{-1} \in H$ (by (*i*)) $\Rightarrow e \in H$, is the identity

⇒ (H, .) is a group ⇒ H is a subgroup of group G.

Note. If the operation in G is +, then the conditions in the above theorem can be stated as follows :

A non-empty subset H of a group $(a, +)$ is a subgroup of G iff (*i*) $a, b \in H \Rightarrow a + b \in H$; (*ii*) $a \in H \Rightarrow -a \in H$.

Theorem. Let H be the non-empty subset of a group (G, .). The necessary and sufficient condition for H to be a subgroup of G is $ab^{-1} \in H \; \forall \; a, b \in H$ where b^{-1} is the inverse of b in G.

[JNTU - Feb-07 - Set 3]

Proof : The condition is necessary.

Since H is a group by itself, $b \in H \Rightarrow b^{-1} \in H$.

Let $a \in H, b^{-1} \in H \Rightarrow a \,.\, b^{-1} \in H$ (by closure property)

The condition is sufficient.

Suppose $ab^{-1} \in H \; \forall \; a, b \in H$.

1. Since $H \neq \phi$, let $a \in H$.
   By hypothesis $a \in H, b \in H \Rightarrow ab^{-1} \in H$.
   For any $a \in H, a \in H \Rightarrow a \,.\, a^{-1} \in H \Rightarrow a \,.\, a^{-1} \in G \Rightarrow e \in G$. Since the identity in G and H is same, $e \in H$.
2. We have $e \in H, a \in H \Rightarrow e \,.\, a^{-1} \in H$ (by $ab^{-1} \in H \; \forall \; a, b \in H$)
   $\Rightarrow \quad a^{-1} \in H$ is inverse of a.
3. Let $b \in H \Rightarrow b^{-1} \in H$
   and let $a \in H, b^{-1} \in H \Rightarrow a\,(b^{-1})^{-1} \in H$
   $\Rightarrow \quad ab \in H \Rightarrow$ H is closed w.r.t. '.'.
4. Let $a, b, c \in H \Rightarrow a, b, c \in G$
   $\Rightarrow \quad (a \,.\, b) \,.\, c = a \,.\, (b \,.\, c) \Rightarrow$ '.' is associative in H.
   Hence (H, .) is a group.
   Since $H \neq \phi$, subset of G, H is a subgroup of group G.

Note 1. If the operation in G is +, then the condition in the above theorem can be stated as follows :

$$a, b \in H \Rightarrow a - b \in H.$$

Note 2. The above theorem can be used to prove that a certain non-empty subset of a given group is a subgroup of the group.

Criterion for the product of two subgroups to be a subgroup.

Theorem : If H and K are two subgroups of a group (G, .), then HK is a subgroup of G iff HK = KH.

Proof : Let H, K be any two subgroups of (G, .)

Suppose $\quad HK = KH$

To prove that HK is a subgroup of G, it is enough to prove that $(HK)(HK)^{-1} = HK$.

$$\therefore \quad (HK)(HK)^{-1} = (HK)(K^{-1}H^{-1}) \text{ since } (HK)^{-1} = K^{-1}H^{-1}$$
$$= H(KK^{-1})H^{-1}$$
$$= H K H^{-1} \text{ since } KK^{-1} = K$$
$$= (HK)H^{-1} = (KH)H^{-1} = K(HH^{-1}) = KH = HK.$$
$$\therefore \quad (HK)(HK)^{-1} = HK.$$

$\therefore$ HK is a subgroup of (G, .).

Conversely suppose HK is a subgroup of group (G, .)

$$\Rightarrow \quad (HK)^{-1} = HK \text{ since K is a subgroup, } K^{-1} = K.$$
$$\Rightarrow \quad K^{-1}H^{-1} = HK$$
$$\Rightarrow \quad KH = HK.$$

Theorem. If H_1 and H_2 are two subgroups of a group (G, .), then $H_1 \cap H_2$ is also a subgroup of G.

Proof : Let H_1, H_2 be two subgroups of G.

Let e be the identity in G $\Rightarrow e \in H_1\ e \in H_2$

$$\Rightarrow \quad e \in H_1 \cap H_2 \Rightarrow H_1 \cap H_2 \neq \phi$$

Let $a \in H_1 \cap H_2$, $b \in H_1 \cap H_2 \Rightarrow a \in H_1$, $a \in H_2$ and $b \in H_1$, $b \in H_2$

Since H_1 is a subgroup, $ab^{-1} \in H_1$

Since H_2 is a subgroup, $ab^{-1} \in H_2$

$\Rightarrow \quad ab^{-1} \in H_1 \cap H_2$

$\therefore \quad H_1 \cap H_2$ is a subgroup of group (G, .).

Corallary : Intersection of an arbitrary family of subgroups of a group is a subgroup of the group *i.e.,* if $\{H_i \mid i \in \Delta\}$ is any set of subgroups of a group (G, .), then $\bigcap_{i \in \Delta} H_i$ is a subgroup of group (G, .).

Note. The union of two subgroups of a group need not be a subgroup of the group. For example, Let (Z, +) be the group of all integers, and

$$H_1 = 2Z = \{......, -6, -4, -2, 0, 2, 4, 6, ...\}$$
$$H_2 = 3Z = \{......, -12, -9, -6, -3, 0, 3, 6, 9, ...\} \text{ be two subgroups of Z.}$$

We have $H_1 \cup H_2 = \{......, -12, -9, -6, -4, -3, -2, 0, 2, 3, 4, 6, 9, ...\}$

Since $4 \in H_1 \cup H_2$, $3 \in H_1 \cup H_2$ does not imply $4 + 3 \in H_1 \cup H_2$

$\Rightarrow$ $H_1 \cup H_2$ is not closed under +.

$\therefore$ $H_1 \cup H_2$ is not a subgroup of (Z, +).

Theorem. The union of two subgroups of a group is a subgroup iff one is contained in the other.

Proof : Let H_1 and H_2 be two subgroups of a group (G, .). To prove that $H_1 \cup H_2$ is a subgroup iff $H_1 \subseteq H_2$ or $H_2 \subseteq H_1$. Suppose $H_1 \subseteq H_2$ or $H_2 \subseteq H_1$.

Let $H_1 \subseteq H_2 \Rightarrow H_1 \cup H_2 = H_2$.

Since H_2 is a subgroup of group (G, .), $H_1 \cup H_2$ is a subgroup of G.

Similarly, $H_2 \subseteq H_1$, $H_1 \cup H_2 = H_1$.

Since H_1 is a subgroup of group (G, .), $H_1 \cup H_2$ is a subgroup of G. Conversely suppose $H_1 \cup H_2$ is a subgroup of G.

If possible, $H_1 \not\subseteq H_2$ and $H_2 \not\subseteq H_1$.

Since $H_1 \not\subseteq H_2$, $\exists\ a \in H_1$ and $a \notin H_2$ ...(1)

Again $H_2 \not\subseteq H_1$, $\exists\ b \in H_2$ and $b \notin H_1$ ...(2)

From (1), (2), we have that $a \in H_1 \cup H_2$, $b \in H_1 \cup H_2$.

Since $H_1 \cup H_2$ is a subgroup of G, $a \,.\, b \in H_1 \cup H_2$.

$\therefore$ $ab \in H_1$ or $ab \in H_2$ or $ab \in H, \cap H_2$.

Suppose $ab \in H_1$.

Since H_1 is a subgroup of G, $a \in H_1$ and $ab \in H_1$.

$\Rightarrow$ $a^{-1} \in H_1$ and $ab \in H_1 \Rightarrow a^{-1}(ab) \in H_1 \Rightarrow (a^{-1}a)\, b \in H_1$

$\Rightarrow$ $e\, b \in H_1 \Rightarrow b \in H_1$ which is absurd by (2).

$\therefore$ $ab \notin H_1$.

Similarly, we can prove that $ab \notin H_2$.

$\therefore$ $ab \notin H_1 \cap H_2$

$\Rightarrow$ $ab \notin H_1 \cup H_2$ which is a contradiction that $H_1 \cup H_2$ is a group.

We must have $H_1 \subseteq H_2$ or $H_2 \subseteq H_1$.

Definition. *Let (G, 0) be a group. Then centre of the group G is the set of those elements of G which commute with every element of G. The centre of G is denoted by Z (G), and is defined as*

$$Z(G) = \{a \in G \mid a \circ x = x \circ a \ \forall\ x \in G\}.$$

Definition. *If a is an element of a group G, then the normalizer on a in G is the set of all those elements of G which commute with a. The normalizer of a in G is denoted by N (a) and N (a) =* $\{x \in G \mid ax = xa\}$

Note 1. If e is the identity in group G, $ex = xe = x\ \forall\ x \in G$, then N (e) = G.

Note 2. The normalizer N (a) is a subgroup of G.

EXERCISE 4.3.

1. Find the order of each element of the multiplicative group $\{1, \omega, \omega^2\}$.
2. Define order of an element of a group (G, .) and find order of each element in a group $(R_0, .)$ where R_0 is non-zero real numbers set.
3. If every element of a group G except the identity element is of order 2, then prove that the group G is abelian.
4. Show that the equation $x^2\, ax = a^{-1}$ is solvable in a group G iff a is the cube of some element in G.
5. Define a subgroup of a group G and give one subgroup of the multiplicative group $\{1, -1, i, -1\}$.
6. Show that all the subgroups of an abelian group are abelian.
7. Show that the set of all elements a of an abelian group G which satisfy $a^2 = e$ forms *a* subgroup of G.
8. Let (G, .) be an abelian group, and H, K be two subgroups of G. Show that HK is *a* subgroup of G.
9. If G is a group and $N(a) = \{x \in G \mid ax = xa\}$ for $a \in G$, then prove that $N(a)$ is a subgroup of G.

ANSWERS

1. $0(1) = 1$, $0(\omega) = 3$, $0(\omega^2) = 3$
2. $0(1) = 1$, $0(-1) = 2$, order of every element except $1, -1$ is 0 or ∞.
5. $(\{1, -1\}, .)$

4.6 COSETS AND LAGRANGE'S THEOREM

Definition. *Let (H,) be a subgroup of the group (G, .), let $a \in G$, then the set $aH = \{ah \mid h \in H\}$ is called a left coset of H in G generated by a and the set $Ha = \{ha \mid h \in H\}$ is called a right coset of H in G, generated by a.*

Note :

1. aH, Ha are called cosets of H generated by a in G. These are also called residue classes modulo the subgroup of the group.
2. If e is the identity in G, then $eH = \{eh \mid h \in H\} = \{h \mid h \in H\} = H$ and $He = \{he / h \in H\} = \{h \mid h \in H\} = H$.
3. If the operation in G is denoted by +, then the left coset of H in G generated by a, denoted by $a + H$ and $a + H = \{a + h \mid h \in H\}$. Similarly, the right coset of H in G generated by $a = H + a = \{h + a \mid h \in H\}$.
4. Let H be a subgroup of the group (G, .) and $a, b \in G$, then
   (*i*) $a(bH) = (ab)H$ and $(Hb)a = H(ba)$
   (*ii*) $x \in aH \Rightarrow yx \in y(aH)$ for $y \in G \Rightarrow yx \in (ya)H$.

Theorem. If H is any subgroup of a group (G, .), $h \in G$, then $h \in H$ iff $hH = H = Hh$.

Proof : Let $h \in H$.

Let h' be an arbitrary element of H. Then $hh' \in h$H.

Since H is a subgroup of G, $h, h' \in H \Rightarrow hh' \in H$.

Thus every element of hH is also an element of H.

$\therefore \quad hH \subseteq H \quad ...(1)$

Again $\quad h' = eh' = (hh^{-1})\, h' = h\,(h^{-1}\, h') \in hH.$

$(\because \; e$ is the identity in H, $h \in H \Rightarrow h^{-1} \in H$ and $h^{-1} \in H, h' \in H \Rightarrow h^{-1}\, h' \in H)$

$\therefore \quad H \subseteq hH \quad ...(2)$

From (1), (2), $\quad hH = H.$

Similarly, we can prove that

$$H = hH.$$

$\therefore \quad h \in H \Rightarrow hH = H = Hh.$

Conversely suppose $\quad hH = H = Hh.$

Let $\quad h \in G$, since $h = he$, $h \in hH$,

But $\quad hH = H, h \in H$

Similarly, $\quad Hh = H \Rightarrow h \in H.$

Hence $\quad hH = H = Hh \Rightarrow h \in H.$

Theorem. If a, b are any two elements of a group (G, .) and H any subgroup of G, then

(*i*) $Ha = Hb$ iff $ab^{-1} \in H$

(*ii*) $aH = bH$ iff $a^{-1}\, b \in H$.

Proof : (*i*) Suppose, $Ha = Hb$

Since $\quad a = ea \in Ha, a \in Hb \Rightarrow ab^{-1} \in (Hb)\, b^{-1} = H\,(bb^{-1})$

$\Rightarrow \quad ab^{-1} \in He = H.$

Now $\quad ab^{-1} \in H \Rightarrow H\, ab^{-1} = H \Rightarrow H\, ab^{-1}\, b = Hb$

$\Rightarrow \quad Ha\, e = Hb \Rightarrow Ha = Hb.$

(*ii*) Suppose $aH = bH$.

Since $\quad b = be \in bH, b \in aH$

$\Rightarrow \quad a^{-1}\, b \in a^{-1}\, aH = (a^{-1}\, a)\, H = eH = H.$

On other hand, $\quad a^{-1}\, b \in H \Rightarrow a^{-1}\, bH = H$

$\Rightarrow \quad aa^{-1}\, bH = aH$

$\Rightarrow \quad e\, bH = aH \Rightarrow bH = aH.$

$\therefore \quad aH = bH \Leftrightarrow a^{-1}\, b \in H.$

Theorem. *If a, b are any two elements of a group (G, .) and H any subgroup of G, then (i) $a \in bH \Leftrightarrow aH = bH$*

(*ii*) $a \in Hb \Leftrightarrow Ha = Hb.$

Proof : (*i*) $a \in bH \Rightarrow b^{-1}\, a \in b^{-1}\, bH = eH = H.$

$\Rightarrow \quad b^{-1} a \in H$

$\Rightarrow \quad b^{-1} aH = H$

$\Rightarrow \quad bb^{-1} aH = bH \Rightarrow e\,aH = bH \Rightarrow aH = bH$

Conversely suppose $aH = bH$

$\therefore \quad a \in aH \Rightarrow a \in bH.$

(*ii*) Left to the reader.

Theorem. Any two left (right) cosets of a subgroup are either disjoint or identical.

Proof : Let H be a subgroup of a group G.

Let aH and bH be two left cosets of H in G.

If aH and bH are disjoint, there is nothing to prove.

If $aH \cap bH \neq \phi$, then there exists at least one element c such that $c \in aH$ and $c \in bH$.

Let $\quad c = ah_1$, and $c = bh_2$ where $h_1, h_2 \in H$.

$\therefore \quad ah_1 = bh_2 \Rightarrow ah_1\, h_1^{-1} = bh_2\, h_1^{-1}$

$\Rightarrow \quad ae = b\,(h_2\, h_1^{-1}) \Rightarrow a = b\,(h_2\, h_1^{-1})$

Since H is a subgroup, $h_2\, h_1^{-1} \in H$

Let $\quad h_3 = h_2\, h_1^{-1}$

$\therefore \quad h_3 \in H.$

Now $\quad a = bh_3 \Rightarrow aH = bh_3\, H = bH \qquad (\because\ h_3 \in H \Rightarrow h_3\, H = H).$

Two left cosets are identical if they are not disjoint.

Hence $\quad aH \cap bH = \phi$ or $aH = bH$.

Similarly, we can prove that $Ha \cap Hb = \phi$ or $Ha = Hb$.

Corallary : H is any subgroup of a group G. If the cosets aH, bH, cH, ..., are all disjoint, then $G = H \cup aH \cup bH \cup \ldots\ldots$, where H is the coset corresponding to the identity element in G and also $G = H \cup Ha \cup Hb \cup Hc \cup \ldots\ldots$

Theorem. If H is a subgroup of a group G. Then $G = H \cup Ha \cup Hb \cup Hc \cup \ldots\ldots$ where a, b, c, ... are elements of G.

Proof : Let G be a group. Therefore, each element of any right coset of H in G is an element of G. Hence the union of all right cosets of H in G is a subset of G.

$$\therefore \quad \bigcup_{x \in G} Hx \subseteq G \qquad \ldots(1)$$

Also if x is any element of G, then $x \in Hx$

$$\text{Therefore,} \quad x \in \bigcup_{x \in G} Hx \Rightarrow G \subseteq \bigcup_{x \in G} Hx \qquad \ldots(2)$$

$$\text{From (1), (2),} \quad \bigcup_{x \in G} Hx = G.$$

Note. If H is a subgroup of a group G, then $G = H \cup aH \cup bH \cup cH \cup \ldots\ldots$, where a, b, c, ... are elements of G.

Coset Decomposition :

From the definition of left and right cosets and their properties, we note that

(*i*) Left (right) cosets are subsets of a group G.

(*ii*) No left (right) coset of H in G is empty.

(*iii*) Any two left (right) cosets of H in G are either disjoint or identical.

(*iv*) The union of all left (right) cosets of H in G are either disjoint or identical.

(*v*) The union of all left (right) cosets of H in G is equal to G. Therefore, the set of all left (right) cosets of H in G gives us apartition of G. This partition is called the left (right) coset decomposition of G with respect to the subgroup H.

Definition. *If H is a subgroup of a group G, the number of distinct left (right) cosets of H in G is called the index of H in G and is denoted by [G : H] or I_G (H).*

Theorem : (Lagrange's theorem)

The order of each subgroup of a finite group is a divisor of the order of the group.

Proof : Let (G, .) be a group of finite order *n*.

Let H be a subgroup of G and let 0 (H) = *m*.

Suppose $h_1, h_2, \ldots\ldots, h_m$ are *m* members of H.

Let $a \in$ G. Then $Ha = \{h_1a, h_2a, \ldots\ldots, h_ma\}$

H*a* has *m* distinct members, since $h_ia = h_ja \Rightarrow h_i = h_j$.

Therefore, each right coset of H in G has m distinct members. Any two distinct right cosets of H in G are disjoint *i.e.,* they have no common element.

since G is a finite group, the number of distinct right cosets of H in G will be finite, say, equal to K.

By theorem, in 246 page,

$$G = Ha_1 \cup Ha_2 \cup \ldots\ldots, \cup Ha_k$$

$$\Rightarrow \quad 0\,(G) = 0\,(Ha_1) + 0\,(Ha_2) + \ldots\ldots + 0\,(Ha_k)$$

$$\Rightarrow \quad n = m + m + \ldots\ldots + m\ (k \text{ times})$$

$$\Rightarrow \quad n = mk.$$

$$\Rightarrow \quad m/n \Rightarrow 0\,(H) \mid 0\,(G).$$

Note. Here K is index of H in G $\Rightarrow k = \frac{n}{m} = \frac{0\,(G)}{0\,(H)}$.

Example. Let G = {1, – 1, *i*, – *i*} be a group and H = {1, – 1} be a subgroup of G.

$$\Rightarrow \quad 0\,(G) = 4,\ 0\,(H) = 2.$$

$\therefore$ The index of H in G $= \frac{0\,(G)}{0\,(H)} = \frac{4}{2} = 2.$

$\Rightarrow$ The number of distinct right (left) cosets of H in G is 2.

and H1 = {– 1, 1} = H (– 1)

H*i* = {– *i*, *i*} = H (– *i*)

Then distinct right cosets of H in G are H1, H*i*.

4.7 NORMAL SUBGROUPS

Definition. *A subgroup H of a group (G, .) is said to be a normal subgroup of G if* $xh\,x^{-1} \in H\ \forall\, x \in G, h \in H$.

Note 1. H = {e} and H = G are normal subgroups of a group G. These are called improper or trivial normal subgroups of group G and all other normal subgroups of G, if exist, are called proper normal subgroups of G.

2. H is a normal subgroup of G, iff $xH\,x^{-1} \subseteq H\ \forall\, x \in G$, where $xH\,x^{-1} = \{xh\,x^{-1} \mid h \in H, x \in G\}$.

3. H is a normal subgroup of G iff $x^{-1}\,Hx \subseteq H\ \forall\, x \in G$.

$$(\because\ x \in G \Rightarrow x^{-1} \in G,\ \forall\, h \in H,\ x^{-1}\,h\,(x^{-1})^{-1} \in H \Rightarrow x^{-1}\,hx \in H)$$

4. Any non-abelian group, whose every subgroup is normal, is called a Hamilton group.

Theorem 1. A subgroup H of a group G is normal iff $xH\,x^{-1} = H\ \forall\, x \in G$.

Proof : Suppose $xH\,x^{-1} = H.\ \forall\, x \in G$.

Let $x \in G, h \in H$, and $xh\,x^{-1} \in xH\,x^{-1}$

$\Rightarrow$ $xh\,x^{-1} \in H$.

$\therefore$ H is a normal subgroup of G.

Conversely suppose H is a normal subgroup of G.

Let $xh\,x^{-1} \in xH\,x^{-1}\ \forall\, x \in G, h \in H$.

$\Rightarrow$ $xh\,x^{-1} \in H$ since H is a normal subgroup of G.

$\Rightarrow$ $xH\,x^{-1} \subseteq H\ \forall\, x \in G$.

Also $x \in G \Rightarrow x^{-1} \in G$ and hence $\forall\, x \in G,\ x^{-1}\,H\,(x^{-1})^{-1} \in H$.

$\Rightarrow$ $x^{-1}\,Hx \subseteq H \Rightarrow x\,(x^{-1}\,Hx)\,x^{-1} \subseteq xH\,x^{-1} \Rightarrow H \subseteq xH\,x^{-1}\ \forall\, x \in G$

$\therefore$ $xH\,x^{-1} = H\ \forall\, x \in G$.

Theorem. Every subgroup of an abelian group is normal.

Proof : Let H be a subgroup of an abelian group (G, .).

Let $h \in H$, $x \in G$ and e be the identity in G.

$\therefore$ $h = eh = (xx^{-1})\,h = x\,(x^{-1}\,h) = x\,(hx^{-1})$

$$(\because\ \text{G is an abelian} \Rightarrow x^{-1}\,h = hx^{-1})$$

$\forall\, x \in G,\ h = xh\,x^{-1} \in H$ since $h \in H$.

Hence H is a normal subgroup of group (G, .)

Theorem. The intersection of any two normal subgroups of a group is a normal subgroup.

Proof : Let H, K be two normal subgroups of a group (G, .). Since H, K are subgroups of G, $H \cap K$ is also a subgroup of G.

Let $n \in H \cap K$ and $x \in G \Rightarrow n \in H$ and $n \in K$.

$\therefore$ $xn\,x^{-1} \in H$ and $xn\,x^{-1} \in K$ since H, K are normal subgroups in G.

$\Rightarrow \qquad xnx^{-1} \in H \cap K.$

Hence $H \cap K$ is a normal subgroup of group G.

Corallary : The arbitrary intersection of any number of normal subgroups of a group G is also a normal subgroup of G.

Simple Group :

Definition. *A group G is said to be simple if it has no proper normal subgroups.*

Example. $\{1, \omega, \omega^2\}$ is a simple group.

Theorem. Every group of prime order is simple.

Proof : First we shall prove every group of prime order has no proper subgroups. For this, let G be a finite group of order n where n is prime.

If possible, let H be a subgroup of order m.

Then $m \leq n$.

By Lagrange's theorem, m is a divisor of n (*i.e.,*) $m \mid n$.

Since n is prime, m is either 1 or n.

$\therefore$ $H = \{e\}$ or $H = G$, which are improper subgroups of G. Hence every group of prime order has no proper subgroups. Hence, every group of prime order has no proper normal subgroups.

$\therefore$ Every group of prime order is simple.

Note. Every group of prime order does not have any proper subgroup.

Theorem. No abelian group of composite order is simple.

Proof : We know that every abelian group of composite order possesses a proper subgroup. Also every subgroup of an abelian group is normal.

Therefore, every abelian group of composite order possesses a proper normal subgroup.

Hence no abelian group of composite order is simple.

4.7.1 Quotient Group

Definition. *Let H be a normal subgroup of a group (G,·). For $a \in G$, Ha is the right coset of H and aH is the left coset of H in G. Since H is normal, aH = Ha. Thus, $\frac{G}{H}$ is the set of all cosets H in G. Define an operation coset multiplication on $\frac{G}{H}$ such that Ha, Hb $\in \frac{G}{H} \Rightarrow$ Ha. Hb = Hab, and if $\frac{G}{H}$ is a group w.r.t. coset, multiplication, then this group is called the **quotient group** or **factor group** of G by H.*

Theorem. The set $\frac{G}{H}$ of all cosets of H in G w.r.t., coset multiplication is a group.

Proof : Let H be a normal subgroup of a group G.

Since H is normal in G, $Ha = aH \ \forall \ a \in G$.

$\therefore$ $\frac{G}{H}$ is the set of all cosets of H in G.

i.e., $$\frac{G}{H} = \{Ha \mid a \in G\}$$

For $a, b \in G$, we have $Ha, Hb \in \frac{G}{H}$.

We define coset multiplication on $\frac{G}{H}$ as $Ha \,.\, Hb = Hab$.

Closure Property : Let $Ha, Hb \in \frac{G}{H}$

Then $$Ha \,.\, Hb = H\,(aH)\,b = H\,(Ha)\,b = HHab = Hab.$$

Since $ab \in G$, $Hab \in \frac{G}{H}$. Thus, $Ha \,.\, Hb \in \frac{G}{H}$.

Associativity : Let $a, b, c \in G$, then $Ha, Hb, Hc \in \frac{G}{H}$.

We have
$$\begin{aligned}[(Ha\,(Hb)]\,(Hc) &= (Hab)\,(Hc) = H\,(ab)\,c \\ &= Ha\,(bc) \\ &= (Ha)\,(Hbc) = (Ha)\,(Hb \,.\, Hc)\end{aligned}$$

Existence of Identity : Let $Ha \in \frac{G}{H} \Rightarrow Ha \,.\, He = Hae = Ha$ and $He \,.\, Ha = Hea = Ha$.

$$\exists\, He \in \frac{G}{H} \ni Ha \,.\, He = Ha = He \,.\, Ha.$$

$\therefore$ He is called identity in $\frac{G}{H}$.

Existence of Inverse : Let $Ha \in \frac{G}{H}$.

Since $a \in G$, $a^{-1} \in G$, we have $Ha^{-1} \in \frac{G}{H}$.

$\therefore$ $Ha \,.\, Ha^{-1} = Ha\,a^{-1} = He$ and $Ha^{-1} \,.\, Ha = Ha^{-1}\,a = He$.

$\therefore$ Every element of $\frac{G}{H}$ is invertible and $(Ha)^{-1} = Ha^{-1}$.

Hence $\frac{G}{H}$ is a group w.r.t. coset multiplication.

Theorem. If H is a normal subgroup of a finite group G, then $0\left(\frac{G}{H}\right) = \frac{0(G)}{0(H)}$.

Proof : $0\left(\frac{G}{H}\right)$ = Number of distinct cosets of H in G.

$$= \frac{\text{Number of elements in G}}{\text{Number of elements in H}} = \frac{0(G)}{0(H)}.$$

Theorem. Every quotient group of an abelian group is abelian.

Proof : Let H be a subgroup of an abelian group G.

Let $\frac{G}{H}$ be the quotient group of G by H.

For any $a, b \in G$, $ab = ba$ since G is abelian.

For $Ha, Hb \in \frac{G}{H}$, $Ha \,.\, Hb = H\, ab = H\, ba = Hb \,.\, Ha$

$\therefore$ $\frac{G}{H}$ is abelian if G is abelian.

Example. *$H = \{1, -1\}$ is a normal subgroup of the group $G = \{1, -1, i, -i\}$ under multiplication and write $\frac{G}{H}$.*

Solution. Clearly, $H \subset G$, and H is a group under multiplication, then H is a subgroup of group (G, .).

$\therefore$ $1H = \{1, -1\} = H1, (-1)H = \{-1, 1\} = H(-1)$

and $iH = \{i, -i\}$ and $-iH = \{-i, i\} = H(-i)\, Hi$

$\therefore$ H is a normal subgroup of G.

Here $H1 = \{1, -1\} = H = H(-1)$ and

$Hi = \{i, -i\} = H(-i)$

Hence $\frac{G}{H} = \{H, Hi\}$.

Example. *If N is normal in the group G. Show that $\frac{G}{N}$ is abelian iff $xy\,x^{-1}\,y^{-1} \in N\ \forall\, x, y \in G$.*

Solution. Given N is normal in the group G let $x, y \in G$.

Let $Nx, Ny \in \frac{G}{H}$

$\therefore$ $xy\,x^{-1}\,y^{-1} \in N \Rightarrow (xy)\,(yx)^{-1} \in N$

$\Rightarrow$ $N\,xy = N\,yx$ $\quad (\because\ ab^{-1} \in H \Rightarrow Ha = Hb$, when H is subgroup)

$\Rightarrow \qquad \text{N}x \,.\, \text{N}y = \text{N}y \,.\, \text{N}x$

$\Rightarrow \quad \dfrac{\text{G}}{\text{H}}$ is abelian.

4.8 HOMOMORPHISMS, ISOMORPHISMS OF GROUPS

Definition. ***Homomorphism Into :*** *Let (G, .), (G′, .) be two groups and f : G → G′ be a function. If for a, b ∈ G, f (a . b) = f (a) . f (b), then f is said to be a homomorphism from G into G′.*

Definition. ***Homomorphism Onto :*** *Let (G, .), (G′, .) be two groups and f a mapping from G onto G′. If for a, b ∈ G, f (ab) = f (a). – f (b), then f is said to be a homomorphism from G onto G′.*

Homomorphism onto is also called as epimorphism.

Definition. ***Endomorphism :*** *A homomorphism of a group G into itself is called an* ***endomorphism.***

Definition. ***Monomorphism :*** *If the homomorphism into is one-one, then it is called* ***monomorphism.***

Definition. ***Isomorphism :*** *Let (G, 0), (G′, *) be any two groups and f : G → G′ is said to be isomorphism if*

(i) $f(a \circ b) = f(a) * f(b)\ \forall\, a, b \in G.$

(ii) f is one-one, and onto.

In this case, we say that G is isomorphic to G′ and we write (G, 0) ≅ (G′, *) (or) G ≅ G′.

Example. Let G be the additive group of integers and G′ be the multiplicative group whose elements are 2^m for $m \in Z$. A function f: G → G′ defined by $f(m) = 2^m$ for $m \in Z$. Then f is homomorphism G into G′.

Because for any $m, n \in Z$, $f(m+n) = 2^{m+n} = 2^m \,.\, 2^n = f(m)\, f(n)$.

Example. Let (Z, +) be a group. Define f: Z → Z by $f(a) = a + 2\ \forall\ a \in Z$

Then f is not a homomorphism, because for any $a, b \in Z$,

$$f(a+b) = a + b + 2,\ f(a) + f(b) = a + 2 + b + 2 \text{ and } f(a+b) \neq f(a) + f(b).$$

Example. *Let G be a multiplicative group with identity e. Define a mapping f : G → G such that f (a) = e for a ∈ G. Then f is an endomorphism on G.*

Solution. Given f: G → G ∋ $f(a) = e$ for $a \in G$

Let $\qquad a, b \in G \Rightarrow ab \in G$

$\Rightarrow \qquad f(ab) = e = ee$

$\qquad\qquad = f(a)\, f(b)$

∴ f is homomorphism from G to G.

⇒ f is an endomorphism in G.

Example. *Let G be the additive group of integers.*

and $G' = Z_m = \{\bar{0}, \bar{1}, \bar{2}, \bar{3}, \ldots\ldots, \overline{m-1}\}$ *be the group of residue classes modulo m w.r.t. addition of residue classes. Define a mapping f : G → G′ such that f (a) =* $\bar{a}$ *for a ∈ G. Then f is epimorphism from G to G′.*

Solution. Given f: G → G′ by $f(a) = \bar{a}$ for $a \in$ G.

Let $\bar{x} \in$ G′ ∃ at least one $x \in$ G ∋ $f(x) = \bar{x}$.

∴ f is onto.

And let $a, b \in$ G, $f(a + b) = \overline{a+b} = \bar{a} +_m \bar{b} = f(a) +_m f(b)$

∴ f is homomorphism from G to G′.

Hence f is epimorphism from G to G′.

4.8.1 Properties of Homomorphism

Theorem. Let (G, .), (G′, .) be two groups, and f be a homomorphism from G into G′. Then (*i*) $f(e) = e'$ where e is the identity in G, e' is the identity in G′.

(*ii*) $f(a^{-1}) = [f(a)]^{-1}$.

Proof : (*i*) $f(e) = f(ee) \Rightarrow f(ee) = f(e)$

$\Rightarrow \quad f(e)\, f(e) = e'\, f(e)$ (∵ f is a homomorphism and $e', f(e) \in$ G′)

$\Rightarrow \quad f(e) = e'$ (By right cancellation law in G′).

(*ii*) Let $a \in$ G $\Rightarrow a^{-1} \in$ G.

∴ $f(a)\, f(a^{-1}) = f(aa^{-1}) = f(e) = e'$ where $f(a), f(a'), e' \in$ G′.

∴ $f(a^{-1}) = (f(a))^{-1}$.

Theorem. If f is a homomorphism from a group (G, .) into (G′, .), then (f (G)), .) is a subgroup of G′. (or) the homomorphic image of a group is a group. [JNTU - Feb. 07 - Set 4]

Proof : By dcfinition, f(G) = $\{f(a) \mid a \in$ G }.

Let $a', b' \in f$(G), ∃ $a, b \in$ G ∋ $f(a) = a', f(b) = b'$.

∴ $a' (b')^{-1} = f(a)\, (f(b))^{-1} = f(a)\, f(b^{-1}) = f(ab^{-1})$ (∵ f is a homomorphism)

For $a, b \in$ G $\Rightarrow a, b^{-1} \in$ G $\Rightarrow ab^{-1} \in$ G.

For $ab^{-1} \in$ G, $f(ab^{-1}) = a^1 (b^1)^{-1}$

Since $ab^{-1} \in$ G, $f(ab^{-1}) \in f$(G)

$\Rightarrow \quad a^1 (b^1)^{-1} \in f$(G) for $a^1, b^1 \in f$(G).

Also f(G) ⊆ G′ ⇒ f(G) is a subgroup of (G′, .)

i.e., the homomorphic image of the group G is a subgroup.

i.e., the homomorphic image of a group is a group.

Theorem. Every homomorphic image of an abelian group is abelian.

[JNTU - Nov. 2006 - Set 3]

Proof : Let (G, .) be an abelian group and (G′, .) be a group.

Let f: G → G′ be a homomorphism and onto.

⇒ G′ = f(G) is the homomorphic image of G.

Let $f(a), f(b) \in$ G′

∴ $f(a)\, f(b) = f(ab) = f(ba)$ (by commutative in G)

$= f(b)\, f(a)$ $(\because\ f$ is homomorphism)

$\therefore$ G′ is an abelian group.

4.8.2 Kernel of a Homomorphism

Definition. *If f is a homomorphism of a group G into a group G′, then the set K of all those elements of G which are mapped by f onto the identity e′ n G′ is called the Kernel of the homomorphism f (i.e.,) Kernel of f = {x ∈ G | f (x) = e′} Kernel f is denoted by Ker f (or) K.*

Note 1. Ker $f \neq \phi$ because e is the identity in G.

Then $f(e) = e'$ (*i.e.,*) $e \in$ Ker f.

2. Ker $f \subseteq$ G.

Example 1. Let (Z, +) be the group of integers f: Z → Z defined by $f(x) = \hat{o}\ \forall\ x \in$ domain Z, and $\hat{o}$ is the zero in codomain Z. Then Ker f = Z.

Example 2. Let (Q_0, .) be the group of non-zero rational numbers f: $Q_0 \to Q_0$ defined by $f(x) = x$, then Ker f = {1}.

Example 3. Let (G, .) be the group such that f: G → G′ defined by $f(x) = x$, then Ker $f = \{e\}$

Theorem. If f is a homomorphism of a group G into a group G′, then Kernel of f is a normal subgroup of G.

Proof : Let e be the identity in G and e' be the identity in G′.

Given f: G → G′ is a homomorphism. Let K = Ker f.

$$\therefore \quad K = \{ x \in G \mid f(x) = e' \}.$$

Since $f(e) = e',\ e \in K \Rightarrow K \neq \phi$.

Let $a, b \in K \Rightarrow f(a) = e', f(b) = e'$.

$$\therefore \quad f(ab^{-1}) = f(a)\, f(b^{-1}) = f(a)\, (f(b))^{-1}$$

$$= e'\, (e')^{-1} = e'\, e' = e'.$$

$$\Rightarrow \quad ab^{-1} \in K \Rightarrow K \text{ is a subgroup of G.}$$

Let $x \in G, f(x\, a\, x^{-1}) = f(x)\, (f(a)\, f(x^{-1}) = f(x)\, e'\, (f(x))^{-1} = f(x)\, (f(x))^{-1} = e'$

$$\Rightarrow \quad x\, a\, x^{-1} \in K$$

$\Rightarrow$ K is normal in G $\Rightarrow$ K is a normal subgroup of G.

Theorem. (The necessary and sufficient condition for a homomorphism) : Let f be a homomorphism from a group (G, .) onto a group (G′, .) with kernel K is isomorphism iff Ker $f = \{e\}$.

Proof : Let f be a homomorphism of a group G onto a group G′. Let e, e' be the identities in G, G′ respectively. Let K be the kernel of f. Suppose that f is an isomorphim of G onto G′. Then f is one-one.

Let $a \in K \Rightarrow f(a) = e' \Rightarrow f(a) = f(e)$ $(\because\ f(e) = e')$

$$\Rightarrow \quad a = e.$$

$\therefore$ e is the only element of G which belongs to K.

Hence $\qquad K = \{e\}$.

Conversely, suppose $\qquad K = \{e\}$.

Let $\quad a, b \in G$, now $f(a) = f(b) \Rightarrow f(a)\,(f(b))^{-1} = f(b)\,(f(b))^{-1}$

$$\Rightarrow \quad f(a)\,f(b^{-1}) = f(b)\,f(b^{-1})$$

$$\Rightarrow \quad f(ab^{-1}) = f(bb^{-1}) = f(e) = e'.$$

$$\Rightarrow \quad ab^{-1} \in K$$

$$\Rightarrow \quad ab^{-1} = e$$

$$\Rightarrow \quad ab^{-1}\,b = eb$$

$$\Rightarrow \quad ae = b$$

$$\Rightarrow \quad a = b.$$

$\therefore\ f$ is one-one.

Hence f is an isomorphism of G onto G′.

Example. *If for a group (G, .), $f : G \to G$ is given by $f(a) = a^2\ \forall\ a \in G$ is homomorphism, prove that G is abelian.*

Solution. Let f: G → G, such that $f(a) = a^2$, $a \in G$ is a homomorphism.

Let $\qquad a, b \in G \Rightarrow ab \in G$.

$\therefore \qquad f(a) = a^2, f(b) = b^2, f(ab) = (ab)^2$

Since f is a homomorphism,

$$f(ab) = f(a)\,f(b)$$

$$\Rightarrow \quad (ab)^2 = a^2\,b^2 \Rightarrow (ab)\,(ab) = (aa)\,(bb)$$

$$\Rightarrow \quad a\,(ba)\,b = a\,(ab)\,b$$

$\Rightarrow \qquad ba = ab$ (by cancellation laws in G)

$\Rightarrow$ G is abelian.

Example. *Let G be a multiplicative group and $f : G \to G$, such that for $a \in G$, $f(a) = a^{-1}$. Prove that f is one-one, onto. Also prove that f is homomorphism iff G is commutative.*

Solution. Let f: G → G is a mapping such that $f(a) = a^{-1}$ for $a \in G$.

(*i*) Let $a, b \in G \Rightarrow a^{-1}, b^{-1} \in G$ and $f(a), f(b) \in G$.

Now $\qquad f(a) = f(b) \Rightarrow a^{-1} = b^{-1} \Rightarrow (a^{-1})^{-1} = (b^{-1})^{-1} \Rightarrow a = b$

$\therefore\ f$ is one-one.

(*ii*) Let $a \in G \Rightarrow a^{-1} \in G$ such that $f(a^{-1}) = (a^{-1})^{-1} = a$.

To each $a \in G$ (codomain) $\exists\ a^{-1} \in G$ (domain) $\ni f(a^{-1}) = a$

$\therefore\ f$ is onto.

(*iii*) Suppose f is a homomorphism.

For $\qquad a, b \in G, ab \in G$, now $f(ab) = f(a)\,f(b)$

$$\Rightarrow \quad (ab)^{-1} = a^{-1}\,b^{-1} \Rightarrow b^{-1}\,a^{-1} = a^{-1}\,b^{-1} \Rightarrow (b^{-1}\,a^{-1})^{-1} = (a^{-1}\,b^{-1})^{-1}$$

$$\Rightarrow \quad (a^{-1})^{-1}\,(b^{-1})^{-1} = (b^{-1})^{-1}\,(a^{-1})^{-1} \Rightarrow ab = ba.$$

$\Rightarrow$ G is abelian.

(*iv*) Suppose G is abelian.

For $a, b \in G, f(ab) = (ab)^{-1} = b^{-1}\, a^{-1} = a^{-1}\, b^{-1}$ ($\because$ G is abelian)

$= f(a)\, f(b)$

$\therefore$ f is a homomorphism.

Example. *Let (R, +) be a group of real numbers under addition and (R⁺, .) is a group of positive real numbers under multiplication. Let $f : R \to R^+$ be a mapping such that $f(x) = e^x$ for $x \in R$ show that f is an isomorphism.*

Solution. If x is a real number, $e^x > 0$ and hence $e^x \in R^+$.

Let $a, b \in R \Rightarrow e^a, e^b \in R^+$

Now $e^a = e^b \Rightarrow a = b \Rightarrow f$ is one-one.

Let $c \in R^+$ and log c is a real number, positive or negative or zero.

Also $\log c \in R$

$\therefore$ $f(\log c) = e^{\log c} = c$

Thus to each $c \in R^+ \exists \log c \in R \ni f(\log c) = c$

$\therefore$ f is onto.

Let $a, b \in R \Rightarrow a + b \in R$

Then $f(a+b) = e^{a+b} = e^a\, e^b = f(a)\, f(b)$

$\therefore$ f is a homomorphism which is one-one and onto.

$\therefore$ $f: R \to R^+$ is an isomorphism.

Note : Ker $f = \{0\}$ since 0 is the identity in R, 1 is the identity in R^+, $e^0 = 1$.

Example. *Show that the group ($G = \{0, 1, 2, 3\}, +_4$) and the group ($G' = \{1, -1, i, -i\}$, .) are isomorphic.*

Solution. We have to find an isomorphism f from G to G′ such that f is one-one onto.

Table 4.20.

$+_4$	0	1	2	3
0	0	1	2	3
1	1	2	3	0
2	2	3	0	1
3	3	0	1	2

Table 4.21.

·	1	−1	i	$-i$
1	1	−1	i	$-i$
−1	−1	1	$-i$	i
i	i	$-i$	−1	1
$-i$	$-i$	i	1	−1

Identity in G is 0 and identity in G′ is 1.

Let $f(0) = 1$ and $f(a^{-1}) = (f(a))^{-1}$ for $a \in$ G.

We note that an isomorphism is also a homomorphism.

Now define $f(1) = i, f(3) = -i, f(2) = -1$.

Since the image of an inverse must be equal to the inverse of the images

$$f(1) = f(3^{-1}) = (f(3))^{-1} = (-i)^{-1} = i,$$
$$f(2) = f(2^{-1}) = (f(2))^{-1} = (-1)^{-1} = -1.$$
$$f(3) = f(1^{-1}) = (f(1))^{-1} = i^{-1} = -i.$$

For $a, b \in$ G, $f(a +_4 b) = f(a) \cdot f(b)$ since $f(0 +_4 2) = f(2) = -1, f(0)\, f(2) = 1\,(-1)$
$= -1$, etc.

$\therefore$ f is a homomorphism. Also f is one-one and onto.

Thus, f: G → G′ is an isomorphism such that $f(0) = 1, f(1) = i, f(2) = -1, f(3) = -i$.

Hence G ≅ G′.

Theorem. (Fundamental theorem on homomorphism of groups)

Every homomorphic image of a group G is isomorphic to some quotient group of G.

(or) If ϕ is a homomorphism from a group G onto a group G′, then G / Ker ϕ is isomorphic with G′.

Proof : Let f be a homomorphism from G onto a group G′.

Since f is onto, $f(G) = G'$.

Let Ker f = K. Then K is a normal subgroup of G and $\frac{G}{K}$ is the quotient group of G by K.

Now we shall prove that $\frac{G}{K} \cong G'$.

For $a \in$ G, K$a \in \frac{G}{K}$ and $f(a) \in G'$.

Define a mapping $\phi : \frac{G}{K} \to G'$ such that $\phi\,(ka) = f(a)$ for $a \in$ G.

(*i*) For $a, b \in$ G, Ka = K$b \Rightarrow a^{-1} \in K \Rightarrow f(ab^{-1}) = e'$, where e' is the identity in G′.

$$\Rightarrow \quad f(a)\, f(b^{-1}) = e'$$
$$\Rightarrow \quad f(a)\, (f(b))^{-1} f(b) = e' f(b).$$
$$\Rightarrow \quad f(a)\, e' = f(b)$$
$$\Rightarrow \quad f(a) = f(b)$$
$$\Rightarrow \quad \phi\,(Ka) = \phi(Kb)$$

$\therefore$ ϕ is well defined.

(*ii*) For $a, b \in G$, $Ka, Kb \in \frac{G}{K}$. Now $\phi(Ka) = \phi(Kb) \Rightarrow f(a) = f(b)$

$$\Rightarrow \quad f(a)\,(f(b))^{-1} = f(b)\,(f(b))^{-1}$$

$$\Rightarrow \quad f(a)\,f(b^{-1}) = e'$$

$$\Rightarrow \quad f(ab^{-1}) = e' \Rightarrow ab^{-1} \in K$$

$$\Rightarrow \quad Ka = Kb$$

$\therefore$ ϕ is one-one.

(*iii*) Let $x \in G'$. Since $f : G \to G'$ is onto $\exists\ a \in G \ni f(a) = x$

$\therefore$ $Ka \in \frac{G}{K}$ and so $\phi(Ka) = f(a) = x$.

$\therefore$ ϕ is onto.

(*iv*) For $a, b \in G$, $Ka, Kb \in \frac{G}{K}$.

Now $\phi[(Ka)(Kb)] = \phi(Kab)$ ($\because$ coset multiplication is defined in $\frac{G}{K}$)

$$= f(ab) = f(a)\,f(b)$$

$$= \phi(Ka)\,\phi(Kb)$$

ϕ is homomorphism.

Hence ϕ is an isomorphism from $\frac{G}{K}$ onto G'.

$$\therefore \quad \frac{G}{K} \cong G'.$$

4.8.3 Automorphisms of a Group

Definition. *If $f : G \to G$ is an isomorphism from a group G to itself then f is called an* ***automorphism*** *of G.*

Example. *Let $f : Z \to Z$ such that $f(x) = -x$ for $x \in Z$ where Z is an additive group of integers is an automorphism.* [JNTU - February 07 - Set 3]

Solution.

(*i*) Let $x_1, x_2 \in Z$, $f(x_1) = f(x_2) \Rightarrow -x_1 = -x_2 \Rightarrow x_1 = x_2$

$\therefore$ f is one-one.

(*ii*) For $x \in Z$ (codomain), $\exists\ -x \in Z$ (domain) so that

$$f(-x) = -(-x) = x$$

$\therefore$ f is onto.

(*iii*) Let $x_1, x_2 \in Z \Rightarrow x_1 + x_2 \in Z$ and

$$f(x_1 + x_2) = -(x_1 + x_2) = (-x_1) + (-x_2) = f(x_1) + f(x_2)$$

$\therefore$ f is homomorphism from Z to Z.

Hence f is an automorphism of Z.

Example. *If G is an additive group of complex numbers, show that $f: G \to G$ such that $f(z) = pz$ (p is a non-zero complex number) is an automorphism of G.*

Solution. Let $z_1, z_2 \in$ G (domain)

$\therefore \quad f(z_1) = pz_1, f(z_2) = pz_2 \in G$ (codomain)

Now $\quad f(z_1) = f(z_2) \Rightarrow pz_1 = pz_2 \Rightarrow z_1 = z_2$

$\therefore$ f is one-one.

Let $z' \in$ G (codomain) $\exists \dfrac{z'}{p} \in$ G (domain) $(\because p \neq 0) \ni f\left(\dfrac{z'}{p}\right) = P\dfrac{z'}{p} = z'$.

$\therefore$ f is onto.

Also $\quad f(z_1 + z_2) = p(z_1 + z_2) = pz_1 + pz_2 = f(z_1) + f(z_2)$

$\therefore$ f is homomorphism from G to G.

Hence f is an automorphism of G.

Example. *Show that the mapping $f: G \to G$ such that $f(a) = a^{-1}\ \forall a \in G$, is an automorphism of a group G iff G is abelian.*

Solution. Let f: G → G such that $f(a) = a^{-1}$ for $a \in$ G.

(*i*) Let f be an automorphism of a group G.

Let $\quad x, y \in G \Rightarrow xy \in G$

$\therefore \quad f(xy) = (xy)^{-1} = y^{-1}x^{-1} = f(y)f(x) = f(yx)$ $(\because f$ is homomorphism)

$\Rightarrow \quad xy = yx$ $(\because f$ is one-one).

$\Rightarrow$ G is abelian.

(*ii*) Let G be abelian and let $x, y \in G \Rightarrow xy = yx$

$$f(x) = f(y) \Rightarrow x^{-1} = y^{-1} \Rightarrow (x^{-1})^{-1} = (y^{-1})^{-1} \Rightarrow x = y$$

$\therefore$ f is one-one.

For $\quad x \in G, x^{-1} \in G$ and $f(x^{-1}) = (x^{-1})^{-1} = x$.

For $\quad x \in$ G (co-domain), $\exists\, x^{-1} \in$ G (domain) $\ni f(x^{-1}) = x$

$\therefore$ f is onto.

For $\quad x, y \in G, f(xy) = (xy)^{-1} = y^{-1}x^{-1} = x^{-1}y^{-1}$ $(\because$ G is abelian)

$\quad = f(x)f(y)$.

$\therefore$ f is homomorphism.

Hence f is automorphism of G.

Example. *Let G be a group and H a subgroup of G. Let f be an automorphism of G and f (H) = f (h) | h ∈ H}. Prove that f (H) is a subgroup of G.*

Solution. Let H is a subgroup of the group G, f be an automorphism of G and $f(H) = \{f(h) \mid h \in H\} \Rightarrow f(H) \neq \phi$ since $f(e) \in f(H)$.

Clearly $f(\text{H}) \subseteq \text{G}$.

Let $a, b \in \text{H} \Rightarrow ab^{-1} \in \text{H}$ since H is a subgroup of G.

$\Rightarrow$ $f(ab^{-1}) \in f(\text{H})$.

$\Rightarrow$ $f(a)\, f(b^{-1}) \in f(\text{H})$

$\Rightarrow$ $f(a)\, (f(b))^{-1} \in f(\text{H})$ for $f(a), f(b) \in f(\text{H})$.

Since $\phi \neq f(\text{H}) \subseteq \text{G}$, $f(\text{H})$ is a subgroup of G.

EXERCISE 4.4

1. If $\text{H} = \{1, -1\}$ and $\text{G} = \{1, -1, i, -i\}$, then find all the right cosets of H in G.
2. Prove that $(\{0, 3, 6, 9, 12\}, +_{15})$ is a subgroup of $(\text{Z}_{15}, +_{15})$ where $\text{Z}_{15} = \{0, 1, 2, 3, 4, \ldots\ldots, 14\}$ and find all left cosets of the above subgroup in Z_{15}.
3. Let H be a subgroup of a group G and let $\text{T} = \{x \in \text{G} \mid x\text{H} = \text{H}x\}$. Show that T is a subgroup of G.
4. Use Lagrange's theorem prove that a finite group can not be expressed as the union of two of its proper subgroups.
5. If H is a subgroup of a group G such that $x^2 \in \text{H}$ for every $x \in \text{G}$, then prove that H is normal in G.
6. If G is a group and $\text{C}(\text{Z}) = \{x \in \text{G} \mid ax = xa \;\forall\; a \in \text{G}\}$ prove that $c(z)$ is normal in G.
7. Let (Z, +) be the group of integers and $f: \text{Z} \to \text{Z}$ by $f(a) = a + 2$. Is f homomorphism ?
8. Let G be the multiplicative group G of all 2×2 non-singular matrices whose elements are real numbers. Let G′ be the multiplicative group of non-zero real numbers. $f: \text{G} \to \text{G}'$ such that $f(\text{A}) = |\text{A}|$ for $\text{A} \in \text{G}$. Then prove that f is homomorphism from G onto G′.
9. Show that the mapping $f: \text{G} \to \text{G}'$ such that $f(x + iy) = x$, where G is a group of complex numbers under addition, G′ is a group of real numbers under addition, is a homomorphism onto and find ker f.
10. If G is a group and $\phi: \text{G} \to \text{G}$ is defined as $\phi(x) = x^{-1} \;\forall\; x \in \text{G}$. Then f is not a homomorphism.
11. Let (Z, +) be the group of integers. Prove that $f: \text{Z} \to \text{Z}$, where $f(x) = 2x \;\forall\; x \in \text{Z}$ is a homomorphism. Find ker f. Is it onto homomorphism ?
12. Let $\text{G}_1, \text{G}_2, \text{G}_3$ are three groups and $f: \text{G}_1 \to \text{G}_2$, $g: \text{G}_2 \to \text{G}_3$ are isomorphism, prove that $g \circ f: \text{G}_1 \to \text{G}_3$ is an isomorphism.

ANSWERS

1. H, Hi
2. H, 1 + H, 2 + H
7. Not homomorphism.
11. Ker $f = \{0\}$, f is not onto homomorphism.

4.9 PERMUTATION GROUPS

Definition. *A permutation is a one-one mapping of a non-empty set onto itself.*

Note. When a set S is a finite with n elements in it, we say that a permutation of n symbols.

Symbol for a permutation : When $S = \{a_1, a_2, \ldots\ldots, a_n\}$, let $f: S \to S$ be a permutation such that $f(a_1) = b_1, f(a_2) = b_2, \ldots\ldots, f(a_n) = b_n$, where $b_1, b_2, \ldots\ldots, b_n$ are nothing but the elements $a_1, a_2, \ldots\ldots, a_n$ of S in some order. So we write $f = \begin{pmatrix} a_1 & a_2 & \ldots\ldots & a_n \\ b_1 & b_2 & \ldots\ldots & b_n \end{pmatrix}$ where each element in the second row is the f-image of the corresponding element (element lying above) in the first row.

Definition. *Let S be a non-empty set. Two permutations f and g defined over a non empty set S are said to be equal if* $f(a) = g(a)\ \forall\ a \in S$.

Example. Let S = {1, 2, 3, 4, 5},

Let $f = \begin{pmatrix} 1 & 2 & 3 & 4 & 5 \\ 3 & 4 & 5 & 2 & 1 \end{pmatrix}$, $g = \begin{pmatrix} 5 & 1 & 3 & 2 & 4 \\ 1 & 3 & 5 & 4 & 2 \end{pmatrix}$ are two permutations.

We have $f(1) = 3 = g(1), f(2) = 4 = g(2), f(3) = 5 = g(3), f(4) = 2 = g(4),$
$f(5) = 1 = g(5).$

Therefore, $f(a) = g(a)\ \forall\ a \in S.$

Hence $f = g.$

Definition. *If* $S = \{a_1, a_2, \ldots\ldots, a_n\}$, *then a one-one mapping from S onto itself is called a permutation of degree n. The number n of elements in S is called the degree of the permutation. We have n ! permutations by using n elements of S. Then set of all these n ! permutations is called permutation group or symmetric group of degree n on n symbols.*

The permutation group on n symbols is denoted by S_n or P_n.

Note 1. The elements of S also can be denoted by 1, 2, 3, ..., n or by any other symbols.

Note 2. Order of S_n is n !.

Example. Let S = {1, 2, 3}, then we have 3 ! = 6. Permutations in S_3.

These are $\begin{pmatrix} 1 & 2 & 3 \\ 1 & 2 & 3 \end{pmatrix}, \begin{pmatrix} 1 & 2 & 3 \\ 1 & 3 & 2 \end{pmatrix}, \begin{pmatrix} 1 & 2 & 3 \\ 2 & 1 & 3 \end{pmatrix}, \begin{pmatrix} 1 & 2 & 3 \\ 2 & 3 & 1 \end{pmatrix}, \begin{pmatrix} 1 & 2 & 3 \\ 3 & 1 & 2 \end{pmatrix}, \begin{pmatrix} 1 & 2 & 3 \\ 3 & 2 & 1 \end{pmatrix}.$

Definition. *Let S be a finite non-empty set. An* ***identity permutation*** *on S denoted by I, is defined as* $I(a) = a\ \forall\ a \in S.$

Example 1. Let $S = \{a_1, a_2, \ldots\ldots, a_n\}$. Then $I = \begin{pmatrix} a_1 & a_2 & \ldots\ldots & a_n \\ a_1 & a_2 & \ldots\ldots & a_n \end{pmatrix}$ is the identity permutation on S.

Example 2. Let S = {1, 2, 3, 4, 5}, then $f = \begin{pmatrix} 1 & 2 & 3 & 4 & 5 \\ 1 & 2 & 3 & 4 & 5 \end{pmatrix}$ is the identity permutation on S.

Product of permutations or multiplication of permutations or composition of permutations in S_n.

Let $f = \begin{pmatrix} a_1 & a_2 & \ldots\ldots & a_n \\ b_1 & b_2 & \ldots\ldots & b_n \end{pmatrix}$ and $g = \begin{pmatrix} b_1 & b_2 & \ldots\ldots & b_n \\ c_1 & c_2 & \ldots\ldots & c_n \end{pmatrix}$

be two permutations (elements) of S_n. Here $b_1, b_2, \ldots\ldots, b_n$ or $c_1, c_2, \ldots\ldots, c_n$ are nothing but the elements $a_1, a_2, \ldots\ldots, a_n$ of S in some order.

Now $\quad f(a_1) = b_1,\ g(b_1) = c_1\ ;\ f(a_2) = b_2,\ g(b_2) = c_2$, etc.

By definition, we have $\quad c_1 = g(b_1) = g(f(a_1)) = (g \circ f)(a_1)$ or $(gf)(a_1)$

$$c_2 = g(b_2) = g(f(a_2)) = (g \circ f)(a_2) \text{ or } (gf)(a_2),$$

Similarly $\quad (g \circ f)(a_3) = c_3 = (gf)(a_3), \ldots\ldots, (g \circ f)(a_n) = c_n = (gf)(a_n).$

Then
$$g \circ f = gf = \begin{pmatrix} a_1 & a_2 & \ldots\ldots & a_n \\ c_1 & c_2 & \ldots\ldots & c_n \end{pmatrix},$$

Note 1. $g \circ f \in S_n$ for $g, f \in S_n$.

2. $f \circ g \neq g \circ f$.

Example. If $f = \begin{pmatrix} 1 & 2 & 3 & 4 & 5 \\ 5 & 3 & 2 & 4 & 1 \end{pmatrix}$ and $g = \begin{pmatrix} 1 & 2 & 3 & 4 & 5 \\ 4 & 3 & 1 & 2 & 5 \end{pmatrix}$ are two permutations on S, we compute $f \circ g$, $g \circ f$ as follows :

$$f \circ g = \begin{pmatrix} 1 & 2 & 3 & 4 & 5 \\ 4 & 2 & 5 & 3 & 1 \end{pmatrix},$$

since $\quad f \circ g(1) = f(g(1)) = f(4) = 4,\ (f \circ g)(2) = f(g(2) = f(3) = 2$, etc.

and
$$g \circ f = \begin{pmatrix} 1 & 2 & 3 & 4 & 5 \\ 5 & 1 & 3 & 2 & 4 \end{pmatrix},$$

since $\quad (g \circ f)(1) = g(f(1)) = g(5) = 5,\ (g \circ f)(2) = g(f(2)) = g(3) = 1$, etc.

Example. If $f = \begin{pmatrix} 1 & 2 & 3 & 4 & 5 \\ 5 & 3 & 2 & 4 & 1 \end{pmatrix}$ and $g = \begin{pmatrix} 1 & 2 & 3 & 4 & 5 \\ 1 & 2 & 3 & 4 & 5 \end{pmatrix}$

Then
$$f \circ g = \begin{pmatrix} 1 & 2 & 3 & 4 & 5 \\ 5 & 3 & 2 & 4 & 1 \end{pmatrix} \circ \begin{pmatrix} 1 & 2 & 3 & 4 & 5 \\ 1 & 2 & 3 & 4 & 5 \end{pmatrix} = f$$

and
$$g \circ f = \begin{pmatrix} 1 & 2 & 3 & 4 & 5 \\ 1 & 2 & 3 & 4 & 5 \end{pmatrix} \circ \begin{pmatrix} 1 & 2 & 3 & 4 & 5 \\ 5 & 3 & 2 & 4 & 1 \end{pmatrix} = f$$

$\therefore \quad f \circ g = g \circ f$ when $g = I$.

Inverse of a Permutation: It is also a permutation (bijection).

If $\quad f = \begin{pmatrix} a_1 & a_2 & \ldots\ldots & a_n \\ b_1 & b_2 & \ldots\ldots & b_n \end{pmatrix}$, then its inverse, denoted by f^{-1}, and

$$f^{-1} = \begin{pmatrix} b_1 & b_2 & \ldots\ldots & b_n \\ a_1 & a_2 & \ldots\ldots & a_n \end{pmatrix}$$

Note : $f \circ f^{-1} = I = f^{-1} \circ f$.

Example. If $f = \begin{pmatrix} 1 & 2 & 3 & 4 & 5 \\ 5 & 3 & 2 & 4 & 1 \end{pmatrix}$, then $f^{-1} = \begin{pmatrix} 5 & 3 & 2 & 4 & 1 \\ 1 & 2 & 3 & 4 & 5 \end{pmatrix} = \begin{pmatrix} 1 & 2 & 3 & 4 & 5 \\ 5 & 3 & 2 & 4 & 1 \end{pmatrix}$

Note. To write the inverse of a permutation, write the 2nd row as 1st row and 1st row as 2nd row.

Theorem. If S, of n symbols, then the S_n of all permutations on S in a group with respect to the operation product of two permutations.

Proof : Let $S = (a_1, a_2, \ldots\ldots, a_n)$.

Let $$f = \begin{pmatrix} a_1 & a_2 & \ldots\ldots & a_n \\ b_1 & b_2 & \ldots\ldots & b_n \end{pmatrix}, g = \begin{pmatrix} b_1 & b_2 & \ldots\ldots & b_n \\ c_1 & c_2 & \ldots\ldots & c_n \end{pmatrix} \in S_n$$

$$\Rightarrow \quad g\,o\,f = \begin{pmatrix} a_1 & a_2 & \ldots\ldots & a_n \\ c_1 & c_2 & \ldots\ldots & c_n \end{pmatrix} \in S_n.$$

Since $b_1, b_2, \ldots\ldots, b_n$ and $c_1, c_2, \ldots\ldots, c_n$ are nothing but the elements $a_1, a_2, \ldots\ldots, a_n$ of S in some order.

$\therefore$ S_n is closed w.r.t. '0'.

The product of permutations is associative in S_n since composition of mappings in associative.

Let $$f = \begin{pmatrix} a_1 & a_2 & \ldots\ldots & a_n \\ b_1 & b_2 & \ldots\ldots & b_n \end{pmatrix} \in S_n, I = \begin{pmatrix} a_1 & a_2 & \ldots\ldots & a_n \\ a_1 & a_2 & \ldots\ldots & a_n \end{pmatrix}$$

$$\rightarrow \quad f\,o\,I = \begin{pmatrix} a_1 & a_2 & \ldots\ldots & a_n \\ b_1 & b_2 & \ldots\ldots & b_n \end{pmatrix} = f$$

and $I\,o\,f = f$

$\therefore$ $f\,o\,I = f = I\,o\,f$

Then I is the identity permutation in S_n.

Let $$f = \begin{pmatrix} a_1 & a_2 & \ldots\ldots & a_n \\ b_1 & b_2 & \ldots\ldots & b_n \end{pmatrix} \in S_n.$$

$$\Rightarrow \quad f^{-1} = \begin{pmatrix} b_1 & b_2 & \ldots\ldots & b_n \\ a_1 & a_2 & \ldots\ldots & a_n \end{pmatrix} \in S_n.$$

Then $$f^{-1}\,o\,f = \begin{pmatrix} a_1 & a_2 & \ldots\ldots & a_n \\ a_1 & a_2 & \ldots\ldots & a_n \end{pmatrix} = I$$

Similarly $f\,o\,f^{-1} = I$

$\therefore$ $f\,o\,f^{-1} = I = f^{-1}\,o\,f$ and $f^{-1} \in S_n$.

Hence $(S_n, 0)$ is a group.

Note. The set S_n of all permutations on n symbols is a finite group of order $n\,!$ w.r.t. multiplication of permutations. For $n \leq 2$, the group is abelian and for $n \geq 3$, the group is non-abelian.

Definition. *Let $S = \{a_1, a_2, \ldots\ldots, a_n\}$ and a permutation*

$f=\begin{pmatrix} a_1 & a_2 & a_3 & \ldots\ldots & a_k & a_{k+1} & \ldots\ldots & a_n \\ a_2 & a_3 & a_4 & \ldots\ldots & a_1 & a_{k+1} & \ldots\ldots & a_n \end{pmatrix}$ is called a cyclic permutation of length k and degree n. It is represented by $(a_1, a_2, \ldots\ldots, a_k)$ or $(a_1\ a_2\ a_3 \ldots\ldots a_k)$, which is a cycle of length k or k-cycle.

Note. In this type of notation we ignore elements that are mapped onto themselves. Also we can write the cycle $(a_1\ \ a_2\ \ a_3\ \ \ldots\ldots\ \ a_k)$ as $(a_2\ \ a_3\ \ \ldots\ldots\ \ a_k\ \ a_1)$ or $(a_3\ \ a_4\ \ \ldots\ldots\ \ a_k\ \ a_1\ \ a_2)$ etc.

Example 1. If S = {1, 2, 3, 4, 5, 6}, the permutation f on S is $\begin{pmatrix} 1 & 2 & 3 & 4 & 5 & 6 \\ 3 & 1 & 4 & 6 & 5 & 2 \end{pmatrix}$. It can be written as (1 3 4 6 2), since $f(1) = 3, f(3) = 4, f(4) = 6, f(6) = 2, f(2) = 1$ and $f(5) = 5$. f is a cycle of length 5. f also can be written as (3 4 6 2 1) or (4 6 2 1 3), etc.

Example 2. Let S = {1, 2, 3, 4, 5, 6}, the permutation f on S is $\begin{pmatrix} 1 & 2 & 3 & 4 & 5 & 6 \\ 4 & 3 & 5 & 1 & 2 & 6 \end{pmatrix}$.
Here f is not a cyclic permutation since $f(1) = 4, f(4) = 1, f(2) = 3, f(3) = 5, f(5) = 2, f(6) = 6$.

Example 3. If f = (1 3 4 6) is a cyclic permutation on 6 symbols. Then f^{-1} = (6 4 3 1) = (1 6 4 3), ... etc.

Note. A cycle of length 1 is the identity permutation since $f(a_1) = a_1, f(a_2) = a_2, \ldots\ldots, f(a_n) = a_n$.

Definition. A cycle of length 2 is called a transposition.

Example. If S = {1, 2, 3, 4, 5} and a permutation f on S is $\begin{pmatrix} 1 & 2 & 3 & 4 & 5 \\ 1 & 3 & 2 & 4 & 5 \end{pmatrix}$, then f = (2, 3) is a cycle of length 2 and degree 5.

Here $\qquad f^{-1} = (3, 2) = (2, 3)$

i.e., $\qquad f^{-1} = f.$

Note. Inverse of a transposition is itself.

Definition. *Let* $S = \{a_1, a_2, \ldots\ldots, a_n\}$. *If* f, g *be two cycles on* S *such that they have no common elements, then* f, g *are said to be disjoint cycles.*

Example. Let S = {1, 2, 3, 4, 5, 6, 7} and

(*i*) If f = (1 3 7), g = (2 4 5) ; then f, g are disjoint cycles.

(*ii*) If f = (1 3 7), g = (2 3 4 5) then f, g are not disjoint cycles.

Note. The multiplication of disjoint cycles is commutative.

Theorem. Every permutation can be expressed as a product of disjoint cycles, which is unique (apart from the order of the factors).

Example. *Write down the following products as disjoinjt cycles. (1 3 2) (5 6 7) (2 6 1) (4 5) and find its inverse.*

Solution. $(1\ \ 3\ \ 2)(5\ \ 6\ \ 7)(2\ \ 6\ \ 1)(4\ \ 5)$

$$= \begin{pmatrix} 1 & 2 & 3 & 4 & 5 & 6 & 7 \\ 3 & 1 & 2 & 4 & 6 & 7 & 5 \end{pmatrix} \begin{pmatrix} 1 & 2 & 3 & 4 & 5 & 6 & 7 \\ 2 & 6 & 3 & 5 & 4 & 1 & 7 \end{pmatrix}$$

From the first two disjoint cycles — From the last two disjoint cycles

$$= \begin{pmatrix} 1 & 2 & 3 & 4 & 5 & 6 & 7 \\ 1 & 7 & 2 & 6 & 4 & 3 & 5 \end{pmatrix} = (1)\ (2\ \ 7\ \ 5\ \ 4\ \ 6\ \ 3)$$

Since 7 is the maximum number in any cycle, we take every cycle as a permutation of degree 7.

and inverse is $(1)\ (3\ \ 6\ \ 4\ \ 5\ \ 7\ \ 2)$.

Theorem. Every cycle can be expressed as a product of transpositions.

The truth of this theorem is only verified here.

Example. Let $f = (2\ \ 4\ \ 3)$ of degree 4.

Then $f = (2\ \ 3)(2\ \ 4)$

$$\left[\because \begin{pmatrix} 1 & 2 & 3 & 4 \\ 1 & 3 & 2 & 4 \end{pmatrix}\begin{pmatrix} 1 & 2 & 3 & 4 \\ 1 & 4 & 3 & 2 \end{pmatrix} = \begin{pmatrix} 1 & 2 & 3 & 4 \\ 1 & 4 & 2 & 3 \end{pmatrix} = (2\ \ 4\ \ 3)\right]$$

Also we have

$$f = (2\ \ 3)(1\ \ 2)(2\ \ 1)(2\ \ 4)$$
$$= (2\ \ 3)(1\ \ 2)(2\ \ 1)(1\ \ 3)(3\ \ 1)(2\ \ 4)$$
$$= (2\ \ 3)(1\ \ 2)(2\ \ 1)(1\ \ 3)(3\ \ 1)(1\ \ 4)(4\ \ 1)(2\ \ 4) \text{ etc.}$$

Thus every cycle can be expressed as a product of transpositions in many ways.

Theorem. Every permutation can be expressed as a product of transpositions in many ways.

It is a consequence of above two theorems

Definition. *A permutation is said to be an even (odd) permutation if it can be expressed as a product of an even (odd) number of transpositions.*

Note 1. If f is expressed as a product of n transpositions then either n is even or odd but not both.

Note 2. Every transposition is an odd permutation.

Result 1. Identity permutation I is always an even permutation, since I can be expressed as a product of two transpositions.

Result 2. A cycle of length n can be expressed as a product of $(n - 1)$ transpositions. If n is odd, then the cycle can be expressed as a product of even number of transpositions. If n is even, then the cycle can be expressed as a product of odd number of transpositions.

Result 3. The product of two odd permutations is an even permutation. Siunce f, g are two odd permutations. f can be expressed as a product of r (odd number) transpositions and g can be expressed as a product of s (odd number) transpositions. Then $g \circ f$ or gf can be expressed as rts (*i.e.,*) even number of transpositions. Hence $g \circ f$ or gf is even.

Result 4. The product of two even permutations is an even permutation.

Result 5. The product of an odd permutation and an even permutation is an odd permutation.

Result : The inverse of an odd permutation is an odd permutation.

Proof : Let f be an odd permutation and I be the identity permutation.

$\Rightarrow$ f^{-1} is also permutation and $f \circ f^{-1} = \mathrm{I} = f^{-1} \circ f$.

Since I is an even and f is an odd permutation, we must have f^{-1} to be odd permutation.

Result : The inverse of an even permutation is an even permutation.

Example. *Show that the following :*

(i) $f = \begin{pmatrix} 1 & 2 & 3 & 4 & 5 & 6 & 7 & 8 & 9 \\ 6 & 1 & 4 & 3 & 2 & 5 & 7 & 9 & 8 \end{pmatrix}$ *is odd permutation*

(ii) $g = \begin{pmatrix} 1 & 2 & 3 & 4 & 5 & 6 & 7 \\ 3 & 1 & 4 & 7 & 6 & 2 & 5 \end{pmatrix}$ *is even permutation.*

Solution.

$$(i)\ f = \begin{pmatrix} 1 & 2 & 3 & 4 & 5 & 6 & 7 & 8 & 9 \\ 6 & 1 & 4 & 3 & 2 & 5 & 7 & 9 & 8 \end{pmatrix} = (1\ \ 6\ \ 5\ \ 2)(3\ \ 4)(7)(8\ \ 9)$$

$$= (1\ \ 2)(1\ \ 5)(1\ \ 6)(3\ \ 4)(8\ \ 9)(7)$$

$\therefore$ which is the product 5 transpositions.

Then f is odd permutation.

$$(ii)\ g = \begin{pmatrix} 1 & 2 & 3 & 4 & 5 & 6 & 7 \\ 3 & 1 & 4 & 7 & 6 & 2 & 5 \end{pmatrix} = (1\ \ 3\ \ 4\ \ 7\ \ 5\ \ 6\ \ 2)$$

$\Rightarrow$ Cycle g length is 7.

$\Rightarrow$ g is the product of 6 transpositions.

$\Rightarrow$ g is even permutation.

Example. *Prove that* $(1\ \ 2\ \ 3\ \ \ldots\ldots\ \ n)^{-1} = (n\ \ (n-1)\ \ \ldots\ldots\ \ 3\ \ 2\ \ 1)$.

Solution.

$$(1\ \ 2\ \ 3\ \ \ldots\ldots\ \ n)^{-1} = \begin{pmatrix} 1 & 2 & 3 & \ldots\ldots & n-1 & n \\ 2 & 3 & 4 & \ldots\ldots & n & 1 \end{pmatrix}^{-1} = \begin{pmatrix} 2 & 3 & 4 & \ldots\ldots & n & 1 \\ 1 & 2 & 3 & \ldots\ldots & n-1 & n \end{pmatrix}$$

$$= \begin{pmatrix} 1 & 2 & 3 & 4 & \ldots\ldots & n \\ n & 1 & 2 & 3 & \ldots\ldots & n-1 \end{pmatrix}$$

$$= (1\ \ n\ \ (n-1)\ \ \ldots\ldots\ \ 4\ \ 3\ \ 2)$$

$$= (n\ \ (n-1)\ \ \ldots\ldots\ \ 4\ \ 3\ \ 2\ \ 1).$$

Theorem. ***Cayleys's theorem :***

Every finite group is isomorphic to a permutation group.

Definition. *A permutation group is said to be a regular permutation group if it is isomorphic to a finite group.*

Example. *Find the regular permutation group isomorphic to the multiplicative group* $\{1, \omega, \omega^2\}$.

Solution. Here, we use Cayley's theorem.

Let $G = \{1, \omega, \omega^2\}$ and $f_a : G \to G$ be a mapping defined by $f_a(x) = ax$ for each $a \in G$ and x is any element in G.

Then the regular permutation group $= \left\{F_1, F_\omega, F^{\omega^2}\right\}$.

where
$$f_1 = \begin{pmatrix} 1 & \omega & \omega^2 \\ 1.1 & 1.\omega & 1.\omega^2 \end{pmatrix} = \begin{pmatrix} 1 & \omega & \omega^2 \\ 1 & \omega & \omega^2 \end{pmatrix}$$

$$f_\omega = \begin{pmatrix} 1 & \omega & \omega^2 \\ \omega.1 & \omega.\omega & \omega.\omega^2 \end{pmatrix} = \begin{pmatrix} 1 & \omega & \omega^2 \\ \omega & \omega^2 & \omega^3 \end{pmatrix} = \begin{pmatrix} 1 & \omega & \omega^2 \\ \omega & \omega^2 & 1 \end{pmatrix}$$

$$f_{\omega^2} = \begin{pmatrix} 1 & \omega & \omega^2 \\ \omega^2.1 & \omega^2.\omega & \omega^2.\omega^2 \end{pmatrix} = \begin{pmatrix} 1 & \omega & \omega^2 \\ \omega^2 & 1 & \omega \end{pmatrix}$$

4.10 CYCLIC GROUPS

Definition. *A group (G, .) is said to be cyclic if every element of G can be expressed as some integral power of a where* $a \in G$.

Here a is called generator of G and cyclic group G is denoted by $G = \langle a \rangle$.

Definition. *A group (G, +) is said to be cyclic if every element of G can be expressed as some integral multiple of a, where* $a \in G$.

Here a is a called generator of G and cyclic group G is denoted by $G = \langle a \rangle$.

Example. $G = \{1, -1, i, -i\}$ is a cyclic group w.r.t. '.'.

Here $i, -i$ are two generators because $i^4 = 1, i^2 = -1, i^3 = -i, i = i$ and $1 = (-i)^4, -1 = (-i)^2, i = (-i)^3, -i = (-i)$.

Example. (Z, +) is a cyclic group, $Z = \langle 1 \rangle$ because $0 = 0\,(1), 1 = 1\,(1), 2 = 2\,(1), ..., -1 = -1\,(1), -2 = -2\,(1), ...$

and $Z = \langle -1 \rangle$ because $0 = 0\,(-1), 1 = -1\,(-1), -1 = 1\,(-1), 2 = -2\,(-1), -2 = 2\,(-1), ...$

Theorem. Every cyclic group is an abelian group. Is converse true ? Justify your answer.

Proof : Let $G = \langle a \rangle$ be a cyclic group.

We have $G = \{a^n \mid n \in Z\}$.

Let
$$a^r, a^s \in G \Rightarrow a^r . a^s = a^{r+s} = a^{s+r} = a^s . a^r$$

$\therefore$ G is an abelian group.

Converse is need not be true *i.e.*, every abelian group is need not be cyclic. For example, in Klein's group of order 4, $e^{-1} = e$, $a^{-1} = a$, $b^{-1} = b$, $c^{-1} = c$,

we have $e = e^2 = e^3 = e^4$, $a = a$, $a^2 = e$, $a^3 = a$, $a^4 = e$, $b = b$, $b^2 = e$, $b^3 = b$, $b^4 = e$, $c = c$, $c^2 = e$, $c^3 = c$, $c^4 = e$.

None of the elements of G generates G even though G is abelian *i.e.*, G is an abelian but not cyclic.

Theorem. If a is a generator of a cyclic group G, then a^{-1} is also a generator of G.

Proof : Let $G = \langle a \rangle$ be *a* cyclic group generated by *a* let $a^r \in G$ where $r \in Z$.

We have $$a^r = (a^{-1})^{-r}, \text{ where } -r \in Z.$$

$\therefore$ Each element of G is generated by a^{-1}.

Thus a^{-1} is also a generator of G.

Theorem. Every subgroup of a cyclic group is cyclic.

Proof : Let $G = \langle a \rangle$ be a cyclic group.

Let H be a subgroup of G.

Since H is a subgroup of G, we take that every element of H is an element of G. Thus it can be expressed as a^n for some $n \in Z$. Let *d* be the smallest of the positive integers such that $a^n \in H$. We will now prove that $H = \langle a^d \rangle$.

Let $$a^m \in H \text{ where } m \in Z.$$

By division algorithm we can find integers *q* and *r*, such that $m = dq + r$, $0 \le r < d$.

$$\therefore \quad a^m = a^{dq + r} = a^{dq} \,.\, a^r = (a^d)^q \,.\, a^r.$$

$$\text{But} \quad a^d \in H \Rightarrow (a^d)^q \in H \Rightarrow a^{dq} \in H \Rightarrow a^{-dq} \in H$$

$$\text{Now,} \quad a^m, a^{-dq} \in H \Rightarrow a^m \,.\, a^{-dq} \in H \Rightarrow a^{m - dq} \in H$$

$$\Rightarrow \quad a^r \in H.$$

But $0 < r < d$ and $a^r \in H$ is a contradiction to our assumption that *d* is the smallest integer such that $a^d \in H$.

$$\therefore \quad r = 0.$$

$$\therefore \quad m = dq$$

$$\Rightarrow \quad a^m = a^{dq} = (a^d)^q$$

$\therefore$ Every element of H can be written as $(a^d)^q$, $q \in Z$

$$\therefore \quad H = \langle a^d \rangle.$$

Hence a subgroup H of a cyclic group (G, .) is cyclic and a^d is a generator of H.

Theorem. Every isomorphic image of a cyclic group is again cyclic.

Proof : Let G be a cyclic group $\ni G = \langle a \rangle$.

$$\Rightarrow \quad G = \{a^n \mid n \in Z\}$$

Let $f: G \to G'$ is an isomorphism, G, G′ are groups

$$\Rightarrow \quad f(G) = G' \text{ is an isomorphic image.}$$

Now $\quad a^n \in G \Rightarrow f(a^n) \in G'$.

$\therefore \quad f(a^n) = f(a \,.\, a \,......\, a)$ (n times) where n is a positive integer.

$= f(a) \,.\, f(a) \,...\, f(a)$ (n times) $\quad (\because\ f$ is an isomorphism)

$= (f(a))^n$.

Hence every element $f(a^n) \in G'$ can be expressed as $(f(a))^n$.

$\therefore\ f(a)$ is a generator of G' implying that G' is cyclic.

Theorem. The quotient group of a cyclic group is cyclic.

Theorem. Every group of prime order is cyclic.

Theorem. If G is an infinite cyclic group, then G has exactly two generators.

Theorem. A cyclic group of order n has $\phi(n)$ generators.

Example. *Find the number of generators of a group of order 10.*

Solution. Since $(1, 10) = 1$, $(3, 10) = 1$, $(7, 10) = 1$, $(9, 10) = 1$, Group has 4 generators.

CHAPTER

5

Elementary Combinatorics

5.1 BASIC OF COUNTING

Two basic counting principles

There are two fundamental principles of all counting problems.

1. Sum Rule or Disjunctive Rule.
2. Product Rule or Sequential Rule.

Sum (Disjunctive) Rule:

If an event can occur in *m* ways and another event can occur in *n* ways, and if these two events cannot occur simultaneously, then one of the two events can occur in $m + n$ ways.

More generally, if $E_1, E_2, ..., E_n$ are mutually exclusive events and E_1 can happen n_1 ways, E_2 can happen n_2 ways, ..., E_n can happen n_n ways. Then one of the *n* events (E_1 or E_2 or ... or E_n) can occur in $n_1 + n_2 + ... + n_n$ ways.

The **sum rule** can also be formulated in terms of choices. If an object can be selected from a collection in n_1 ways and an object can be selected from a separate collection in n_2 ways, then the selection of one object from either one collection or the other can be made in $n_1 + n_2$ ways.

Example 1. If there are 5 boys and 4 girls in a class, then there are $5 + 4 = 9$ ways of selecting one student (either a boy or a girl) as class representative.

Example 2. A student can choose a computer project from one of three lists contain 23, 18, 10 possible projects. Then the number of possible projects are there to choose from are $23 + 18 + 10 = 51$.

Product (Sequential) Rule:

If an event can occur in m ways and a second event can occur in *n* ways, and if the number of ways the second event occurs does not depend upon how the first event occurs, then the two events can occur simultaneously in *mn* ways.

More generally, if events $E_1, E_2, ..., E_n$ can happen in $n_1, n_2, ..., n_n$ ways, then the sequence of events E_1 first, followed by E_2 ..., followed by E_n can happen in $n_1 \cdot n_2 \ldots n_n$ ways.

The product rule can also be stated in terms of choices :

If a first object can be chosen in n_1 ways, a second in n_2 ways, ..., and the nth object can be chosen in n_n ways, then a choice of a first, second, ... and the nth object can be made in $n_1 \cdot n_2$, ... n_n ways.

Example 1. If 2 distinguishable dice are rolled then the first die can fall (event E_1) in 6 ways and the second can fall (event E_2) in 6 ways. Hence there are $6 \cdot 6 = 36$ outcomes when 2 dice are rolled.

Example 2. If the chairs of an auditorium are to be labeled with a letter and a positive integer not exceeding 100, then the largest number of chairs that can be labeled differently in $26 \times 100 = 2600$.

Example. *How many ways can we get a sum of 4 or of 8 when two distinguishable dice are rolled ? and How many ways can we get an even sum ?*

Solution. Consider two distinguishable dice with different colours say Red, White.

(*i*) Then we see that the outcomes (1, 3), (2, 2) and (3, 1) are the only ones whose sum is 4. Thus there are 3 ways to obtain the sum 4.

Similarly, we obtain the sum 8 from the outcomes (2, 6), (3, 5), (4, 4), (5, 3) and (6, 2).

Thus, there are $3 + 5 = 8$ outcomes whose sum is 4 or 8.

(*ii*) The number of ways to obtain an even sum is the same as the number of ways to obtain either the sum 2, 4, 6, 8, 10 (or) 12 there is 1 way to obtain the sum 2, 3 ways to obtain the sum 4, 5 ways to obtain 6, 5 ways to obtain 8, 3 ways to obtain 10 and 1 way to obtain 12.

Hence the total number of ways to obtain an even sum are $1 + 3 + 5 + 5 + 3 + 1 = 18$.

Example. *From a well shefled pack of playing cards, find the following :*

(i) How many ways can we draw a heart or a spade ?

(ii) How many ways can we draw a heart or an ace ?

(iii) How many ways can we draw an ace or a king ?

(iv) How many ways can we draw a card numbered 2 through 10 ?

(v) How many ways can we draw a card or a king ?

Solution.

(*i*) Since there are 13 hearts and 13 spades, we may draw a heart or a spade in $13 + 13 = 26$ ways.

(*ii*) Since there are only 3 aces that are not hearts, we may draw a heart or an ace in $13 + 3 = 16$ ways.

(*iii*) We may draw an ace or a king in $4 + 4 = 8$ ways.

(*iv*) Since there are 9 cards numbered 2 through 10 in each of 4 suits, clubs, diamonds, hearts or spades, we may choose a numbered card in 36 ways.

Here note that we are counting aces as distinct from numbered cards.

(*v*) We may choose a numbered card or a king in $36 + 4 = 40$ ways.

Example. *A shoe store has 30 styles of shoes. If each style is available in 12 different lengths, 4 different widths, and 6 different colors, how many kinds of shoes must be kept in stock ?*

Solution. By product rule, 12 . 4 . 6 kinds of shoes in stock for each style.

Hence 30 . 12 . 4 . 6 = 8640 kinds of shoes must be kept in stock.

Example. *A new born child can be given 1, 2 or 3 names. In how many ways can a child be named if we can choose from 300 names (and no name can be repeated) ?*

Solution. One name can be given to a child in 300 ways.

Two names can be given to a child in 300×299 ways.

Three names can be given to a child in $300 \times 299 \times 288$ ways.

Hence the total number of ways = $300 + 300 \times 299 + 300 \times 299 \times 298$ = 2, 68, 20, 600.

Example. *How many different bit strings are there of length 7 ?*

Solution. Each of the 7 bits can be chosen in two ways, since each bit is either zero or one.

By the product, there are $2^7 = 128$ different bit strings of length 7.

Example. *How many different license plates are available if each plate contains a sequence of three letters followed by three digits (and no sequences of letters are prohibited, even if they are obscene) ?*

Solution. There are 26 choices for each of the three letters and 10 choices for each of the three digits. Hence, by the product rule there are a total of 26 . 26 . 26 . 10 . 10 .10 = 17576000 possible license plates.

Example. *In how many different ways 4 rings of a lock can be combined when each ring has 10 digits from 0 to 9 ?*

If the lock opens with only one combination of 4 digits, how many unsuccessful events are possible ?

Solution.

(*i*) The ways in which 4 digits can be combined are $10 \times 10 \times 10 \times 10 = 10^4$.

(*ii*) Since the lock opens with only one combination of 4 digits, the unsuccessful events (attempts) will be $10^4 - 1 = 9999$.

Example. *A six person committee composed of Shiva, Krishna, Raja, Rama, Jaya, Indra is to select a chair person, secretary and treasurer.*

(i) In how many ways can this be done ?

(ii) In how many ways can this be done if either Shiva or Krishna must be chair person ?

(iii) In how many ways can this be done Indra must hold one of the offices ?

Solution.

(*i*) The chair person can be selected in 6 ways, once the chair person has been selected, the secretary can be selected in 5 ways. After selection of the chair person and secretary, the treasurer can be selected in 4 ways. Therefore, the total number of ways is 6 . 5 . 4 = 120, by product rule.

(*ii*) If Shiva is chair person, then there are 5 . 4 = 20 ways to select the remaining officers. Similarly, if Krishna is chair person, there are 20 ways to select the remaining officers. Since these cases are disjoint, by the sum rule, there are 20 + 20 = 40 possibilities.

(*iii*) If Indra is chair person, there are 20 ways to select the remaining officers. Similarly, if Indra is secretary, there are 20 possibilities, and if Indra is treasurer, there are so possibilities. Since these 3 cases are pairwise disjoint, there are 20 + 20 + 20 = 60 possibilities, by the sum rule.

EXERCISE 5.1.

1. How many ways can we get a sum of 8 when two indistinguishable dice are rolled ? An even sum ?
2. A foot ball stadium has 5 gates on the eastern boundary and 4 gates on the western boundary.
   (*i*) In how many ways can a person enter through an east gate and leave by a western gate ?
   (*ii*) In how many different ways in all can a person enter and get out through different gates ?
3. If each of 20 questions in a objective type examination has 2 answers (1 correct and 1 wrong). In how many ways of answering all questions ?
4. There are five different roads from city A to city B, three different roads from city B to city C, and three different roads that go directly from A to C.
   (*a*) How many different ways are there to go from A to C via B ?
   (*b*) How many different ways are there from A to C altogether ?
   (*c*) How many different ways are there from A to C and then back to A ?
   (*d*) How many different trips are there from A to C and back again to A that visit B both going and coming ?
   (*e*) How many different trips are there that go from A to C via B and return directly from C to A ?
   (*f*) How many different trips are there from A to C and back to A that visit B atleast once ?
5. How many different license plates are there that involve 1, 2 or 3 letters followed by 4 digits ?
6. How many 2 digit or 3 digit numbers can be formed using the digits 1, 3, 4, 5, 6, 8 and 9 if no repetition is allowed.
7. How many 3-digit numbers can be formed using the digits 1, 3, 4, 5, 6, 8 and 9 when (*i*) digits may be repeated ; (*ii*) digits may not be repeated.
8. In a railway compartment 6 seats are vacant on a bench. In how many ways can 3 passengers can sit on them ?
9. There are 10 telegrams and 2 messenger boys.
   (*a*) In how many different ways can the telegrams can be distributed to the messenger boys if (*i*) the telegrams are distinguishable (*ii*) the telegrams are indistinguishable.
   (*b*) In how many different ways can the telegrams be distributed to the messenger boys and then delivered to 10 different people if the telegrams are distinguishable ?
10. Suppose that the license plates of a certain state require 3 English letters followed by 4 digits.
   (*i*) How many different plates can be manufactured if repetition of letters and digits are allowed ?
   (*ii*) How many plates are possible if only the letters can be repeated ?
   (*iii*) How many are possible if only the digits can be repeated ?
   (*iv*) How many are possible if no repetitions are allowed at all ?

ANSWERS

1. 3, 12
2. (*i*) $5 \times 4 = 20$ (*ii*) $9 \times 8 = 72$
3. 2^{20}
4. (*a*) $5 \,.\, 3 = 15$ (*b*) $15 + 3 = 18$
   (*c*) 18^2 (*d*) 15^2

(*e*) $15 . 3 = 45$ (*f*) $15 . 3 + 3 . 15 + 15^2 = 315$

5. $26 . 10^4 + 26^2 . 10^4 + 26^3 . 10^4 = (26 + 26^2 + 26^3)\ 10^4$

6. $7 . 6 + 7 . 6 . 5 = 252$

7. (*i*) 7^3 (*ii*) $7 . 6 . 5$

8. 120

9. (*a*) (*i*) 2^{10} (*ii*) 11

(*b*) 2^{10} (10 !).

10. (*i*) $26^3 . 10^4$ (*ii*) $26^3 . 10 . 9 . 8 . 7$

(*iii*) $26 . 25 . 24 . 10^4$ (*iv*) $26 . 25 . 24 . 10 . 9 . 8 . 7$.

5.2 COMBINATIONS AND PERMUTATIONS

Definition. *A combination of n objects taken r at a time called an* ***unordered selection*** *of r* ($r \le n$) *of the n objects (an r-combination of n objects)*

Definition. *A permutation of n objects taken r at a time called an* ***ordered selection*** *or* ***arrangement*** *of r of the n objects (an r-permutation of n objects).*

Note : The order of the things is not considered in combinations, and the order of the things is considered in permutations.

Example 1. Suppose there are 4 things *a, b, c, d.* The selections that can be made by taking two at a time without repetitions are *ab, ac, ad, bc, bd, cd.*

i.e., The 2-combinations of these 4 objects without repetitions are *ab, ac, ad, bc, bd, cd* and the arrangement that can be made by taking two at a time without repetitions from the given 4 things *a, b, c, d* are *ab, ac, ad, bc, bd, cd; ba, ca, da, cb, db, dc* which are also the 2-permutations of 4 objects *a, b, c, d* without repetitions.

Note that we are simply defining the terms r-combinations and *r*-permutations here and have not mentioned anything about the properties of the no objects. *i.e.,* these definitions say nothing about whether or not a given element may appear more than once in the list of *n* objects.

Notation : In this chapter, we use $\{p . a, q . b, r . c\}$ to indicate either

1. That we have $p + q + r$ objects including *p a's, q b's, r c's*
2. That we have 3 objects *a, b, c* with condition that a can be selected at most *p* times, *b* can be selected at most *q* times, *c* can be selected at most *r* times.

The numbers *p, q, r* are called repetition numbers.

Example 2. The 3-combinations of $\{3 . a, 2 . b, 5 . c\}$ are *aaa, aab, aac, abb, acc, bbc, abc, ccc, ccb.* Here the numbers 3, 2, 5 are called repetition numbers.

In order to include the case where there is no limit on the number of times an object can be repeated in a selection, we use the symbol ∞ as a repetition number to mean that an object can occur an infinite number of times.

Example 3. The 3-combinations of $\{\infty . a, 2 . b, \infty . c\}$ are the same as in Example 2 even though *a* and *c* can be repeated an infinite number of times. This is because, in 3-combinations, 3 is the limit on the number of objects to be chosen.

If we are considering selections where each object has ∞ as its repetition number then we designate such selections as selections with **unlimited repetitions**. In particular, a selection of *r* objects in this case will be called ***r*-combinations with unlimited repetitions** and any ordered arrangement of these *r* objects will be an ***r-permutation with unlimited repetitions***.

Example. The 3-combinations of *a*, *b*, *c*, *d* with unlimited repetitions are the 3-combinations of $\{\infty \,.\, a, \infty \,.\, b, \infty \,.\, c, \infty \,.\, d\}$ there are 20 such 3-combinations, namely

aaa, *aab*, *aac*, *aad*,
bbb, *bba*, *bbc*, *bbd*,
ccc, *cca*, *ccb*, *ccd*,
ddd, *dda*, *ddb*, *ddc*,
abc, *abd*, *acd*, *bcd*.

And there are $4^3 = 64$ of 3-*p*ermutations with unlimited repetitions since the first position can be filled 4 ways (with *a*, *b*, *c*, *d*), the second position can be filled 4 ways and likewise for the third position.

Example. *List all 3-combinations, and 3-permutations of* $\{\infty.\, a, \infty \,.\, b, \infty \,.\, c, \infty \,.\, d\}$ *in Tabular form.*

Solution.

Table 5.1

	3-Combinations with Unlimited Repetitions	*3-Permutations with Unlimited Repetitions*
	aaa	*aaa*
	aab	*aab, aba, baa*
	aac	*aac, aca, caa*
	aad	*aad, ada, daa*
	bbb	*bbb*
	bba	*bba, bab, abb*
	bbc	*bbc, bcb, cbb*
	bbd	*bbd, bdb, dbb*
	ccc	*ccc*
	cca	*cca, cac, acc*
	ccb	*ccb, cbc, bcc*
	ccd	*ccd, cdc, dcc*
	ddd	*ddd*
	dda	*dda, dad, add*
	ddb	*ddb, dbd, bdd*
	ddc	*ddc, dcd, cdd*
	abc	*abc, acb, bac, bca, cab, cba*
	abd	*abd, adb, bad, bda, dab, dba*
	acd	*acd, adc, dac, dca, cad, cda*
	bcd	*bcd, bdc, cbd, cdb, dbc, dcb*
Total number	20	64

5.3 ENUMERATION OF COMBINATIONS AND PERMUTATIONS (WITHOUT REPETITION)

Let P (*n*, *r*) denote the number of *r*-permutations without repetitions.

Theorem. (Enumerating r-permutations without repetitions)

$$P(n, r) = n(n-1)(n-2) \dots (n-r+1) = \frac{n!}{n-r!}$$

Proof : Since there are n distinct objects, the first position of an r-permutation may be filled in n ways. The second position can be filled in $n - 1$ ways since no repetitions are allowed and there are $n - 1$ objects left to choose from. The third can be filled in $n - 2$ ways and so on until the rth position is filled in $n - (r - 1) = n - r + 1$ ways (see the following figure).

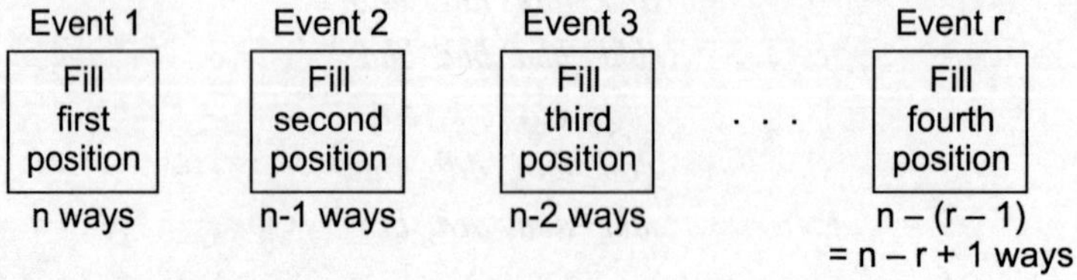

By applying the product rule, we conclude that

$$P(n, r) = n(n-1)(n-2) \dots (n-r+1)$$

$$= \frac{n!}{n-r!} \text{ (by the definition of factorials)}$$

Note. There are $n!$ permutations of n distinct objects.

Example. There are $3! = 6$ permutations of 3 distinct objects a, b, c and there are $4! = 24$ permutations of 4 objects a, b, c, d.

Example. *Compute P (8, 5).*

Solution. $P(8, 5) = \frac{8!}{8-5!} = \frac{8!}{3!} = 8 \,.\, 7 \,.\, 6 \,.\, 5 \,.\, 4 =$

Example. The number of 3-permutations of {a, b, c, d, e} is P (5, 3)

$$= \frac{5!}{5-3!} = \frac{5!}{2!} = 5 \,.\, 4 \,.\, 3 = 60.$$

The number of 5-letter words using the letters a, b, c, d and e is $P(5, 5) = \frac{5!}{0!} = 5! = 120.$

Example. *In how many ways are there to distribute 10 different books among 15 people if no person is to receive more than one book ?*

Solution. $P(15, 10) = \frac{15!}{5!}.$

Example. *Compute n and r if P (n, r) = 3024.*

Solution. Since P (n, r) is a product of consecutive integers.

We write $\quad P(n, r) = 3024 = 9 \times 8 \times 7 \times 6 = P(9, 4)$

$\Rightarrow \quad n = 9, r = 4.$

Example. *Find n if P (n – 1, 3) : P (n + 1, 3) = 5 : 12.*

Solution. $P(n-1, 3) : P(n+1, 3) = 5 : 12$

$\Rightarrow \quad 12\, P(n-1, 3) = 5\, P(n+1, 3)$

$$\Rightarrow \quad 12\,(n-1)\,(n-2)\,(n-3) = 5\,(n+1)\,n\,(n-1)$$

$$\Rightarrow \quad 12\,(n-2)\,(n-3) = 5\,(n+1)\,n$$

$$\Rightarrow \quad 7n^2 - 65n + 72 = 0$$

$$\Rightarrow \quad (7n-9)\,(n-8) = 0$$

$$\Rightarrow \quad n = 8 \quad \text{or} \quad \frac{9}{7}.$$

Since n is a positive integer, $n = \frac{9}{7}$ is rejected

$$\therefore \quad n = 8.$$

Example. *Find the number of different ways in which 4 boys and 6 girls may be arranged in a row so that no two boys shall be together.*

Solution. Since no two boys shall be together, the boys can occupy the place at the beginning or at the end or in between every two girls. With the given restriction the arrangement is as shown below where B is stand for boys and G's stand for girls.

B G B G B G B G B G B G B.

There are 6 girls and 6 places and hence they can be arranged in P (6, 6) (*i.e.,*) 6 ! ways. There are 7 places for 4 boys and hence they can be arranged in P (7, 4)ways. Since the two operations are independent of each other, the total number of arrangements is 6 ! . P (7, 4) = 720 × 840 = 604800.

Example. *In how many ways can the letters of the English alphabet be arranged so that there are exactly 5 letters between the letters a and b ?*

Solution. There are P (24, 5) ways to arrange the 5 letters between a and b, 2 ways to place *a* and *b*, and then 20 ! ways to arrange any 7-letter word treated as one unit along with the remaining 19 letters.

The total number of ways = P (24, 5) . (20 !) (2).

Example. *How many numbers of 4 digits can be formed from 0, 1, 2, 3, 5, 6 if no digit is to occur more than once in each number ? How many of these are divisible by 2 ?*

Solution.

(*i*) The number of four digit numbers formed with the six digits 0, 1, 2, 3, 5, 6 is P (6, 4).

Among these, the numbers that begin with 0 are P (5, 3).

∴ The number of 4 digit numbers formed = P (6, 4) – P (5, 3).

(*ii*) The number of 4 digit numbers when 0 occupies the one's place = P (5, 3)

The number of 4 digit numbers when 2 occupies the one's place = P (5, 3) – P (4, 2)

The number of 4 digit numbers when 6 occupies the one's place = P (5, 3) – P (4, 2) since the numbers that begin with 0 are P (4, 2).

Hence the total number of numbers divisible by 2.

$$= P(5, 3) + P(5, 3) - P(4, 2) + P(5, 3) - P(4, 2)$$
$$= 3 . P(5, 3) - 2 . P(4, 2).$$

Example. *Find the number of positive integers that can be formed from the digits 1, 2, 3, 4 and 5, if no digit is repeated in any integer.* [JNTU-Feb. 07 - Set 4]

Solution. ${}^5P_5 = 5! = 120$.

Example. *How many arrangements of the elements of {I, N, T, R, O, D, U, C, E} can be formed ? In how many of them (i) the vowels come together and (ii) the vowels do not come together.*

Solution. The number of arrangements that can be formed with 9 elements taken all at a time is P (9, 9) = 9 !.

(*i*) Since the vowels I, O, U, E are to be together, treat them as one entity. Now there are 6 elements and these can be arranged in 6 ! ways. Also the vowels can be arranged among themselves in 4 ! ways.
Since the two operations are independent, the total number of arrangements is 6 ! . 4 !.

(*ii*) The number of arrangements in which all the vowels do not come together = The total number of arrangements 9 ! the number of arrangements in which the vowels come together = 9 ! – 6 ! . 4 ! =

Example. *How many numbers between 1,000 and 10,000 can be formed with the digits 1, 3, 5, 7, 9 each digit being only once in each number ?*

Solution. Each number should consist of 4 digits and the required number is the same as the number of permutations of 5 different things taken 4 at a time = P (5, 4) = 120.

Example. *The letters of the word VICTORY are rearranged in all possible ways and the words thus obtained are arranged as in a dictionary, what is the rank of the given word ?*

Solution. The letters of the word VICTORY arranged in alphabetical order are CIORTVY. Since all the words are arranged as in a dictionary, the words which continue with C, I, O, R, T procede the word VICTORY.

The numbers of words that begin with	C = 6 !
The number of words that begin with	I = 6 !
The number of words that begin with	O = 6 !
The number of words that begin with	R = 6 !
The number of words that begin with	T = 6 !
The number of words that begin with	VC = 5 !
The number of words that begin with	VICO = 3 !
The number of words that begin with	VICR = 3 !

The next word is VICTORY = 1

Hence the rank of VICTORY is 5 . 6 ! + 5 ! + 2 . 3 ! + 1 = 3733.

Example. *In how many ways can the digits 0, 1, 2, 3, 4, 5, 6, 7, 8 and 9 be arranged so that (i) 0 and 1 are adjacent and in the order of 01 (ii) 0, 1 are adjacent.* [JNTU - Nov 06 - Set 3]

Solution. Let us take 0 and 1 as one entry and remaining 8 are eight entries. Total 9 entries.

(*i*) Then required number of ways are $^9P_9 = 9\ 1$

(*ii*) 0, 1 can be arranged amongst themselves in 2 ! ways then the total number of ways are 2 ! 9 !.

Example. *Using the digits 1, 3, 4, 5, 6, 8 and 9, how many 3-digit numbers can be formed.*

[J.N.T.U. - Nov. 06 - Set 2]

Solution. Total digits are 7.

Required 3-digit numbers are P (7, 3).

Example. *How many 3-digit numbers can be formed if no digit is repeated.*[JNTU-Nov 06 - Set 2]

Solution. The digits are 0, 1, 2, 3, 4,5, 6, 7, 8, 9.

Required 3-digit numbers are P (10, 3).

Example. *How many 3-digit numbers can be formed if 3 and 4 are adjacent to each other ?*

[JNTU-Nov 06 - Set 2]

Solution. Total number of digits are 10.

Let us consider 3 and 4 as one digit then total digits in this case are 9 and 3, 4 can be arranged amongst themselves in 2 ! ways. Hence the number of 3-digit numbers can be formed if 3 and 4 are adjacent to each other is 2 ! P (9, 3).

Example. *How many 3-digit numbers can be formed if 3 and 4 are not adjacent to each other.*

[JNTU - Nov 06 - Set 2]

Solution. Required number of numbers is P (10, 3) – 2 . P (9, 3) because 3 digit numbers are P (10, 3) and 3 digit numbcrs so that 3, 4 are adjacent to each other is 2 . P (9, 3), by previous example.

Example. *Find the sum of all the numbers formed by taking all the digits from {2, 3, 4, 5}, without repetitions.*

Solution. There are 4 ! = 24 numbers that can be formed with 4 digits.

Of these 3 ! numbers contain 2 in the units place.

3 ! numbers contain 2 in the tens place

3 ! numbers contain 2 in the hundreds place

and 3 ! numbers contain 2 in the thousands place.

The value contributed by 2 to the sum

$$= 3\,!\,(2) + 3\,!\,(20) + 3\,!\,(200) + 3\,!\,(2000)$$

$$= 3\,!\,(2222) = 2\,.\,3\,!\,(1111)$$

Similarly, the value contributed by 3 to the sum = 3 . 3 ! (1111)

the value contributed by 4 to the sum = 4 . 3 ! (1111)

and the value contributed by 5 to the sum = 5 . 3 ! (1111)

The required sum = 3 ! (1111) (2 + 3 + 4 + 5)

= 6 (1111) (14) = 93,324.

Note : The sum of all the numbers formed by taking all the digits of {*a*, *b*, *c*, *d*, *e*} is given by (5 – 1) ! (11, 111) $(a + b + c + d + e)$.

5.4 CIRCULAR PERMUTATIONS

We have considered so far linear permutations (*i.e.,*) permutations of things arranged in a row. There is another type of arrangement of things round a circle or in a circular order called circular permutation. In a circular permutation there is no first or last place of an object. Hence the principles of linear permutations are not applicable in circular permutations. In circular permutations the relative positions of the things alone need be taken into consideration and not actual positions.

Let us consider an arrangement of four different things A, B, C, D round a circle. Let A, B, C, D be one circular arrangement. This arrangement ABCD, when arranged in circular order, gives rise to the arrangement ABCD, DABC, CDAB, BCDA as shown in the following figure.

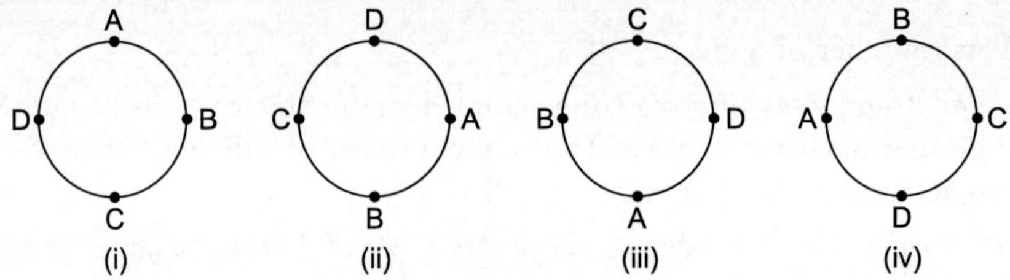

Figure 5.1.

The arrangements will be different when considered as arrangements in a row, but are alike as arrangements round a circle since the relative positions of things involved are exactly the same. Thus one circular permutation of four different things give rise to four different linear permutations.

Theorem. The number of circular permutations of n different things taken all at a time round a circle is $n - 1$!

Proof : Let n different things are arranged in n different positions round a circle. Let K denote the number of circular permutations. Each circular permutation (of n things) gives n different linear permutations give n ! linear permutations.

$$\therefore \qquad K \cdot n = n! \Rightarrow K = \frac{n!}{n} = n - 1\,!$$

Hence the number of circular permutations of n different things taken all at a time is $n - 1$!

Consider two circular permutations ABCD and ADCB as shown in the following figure.

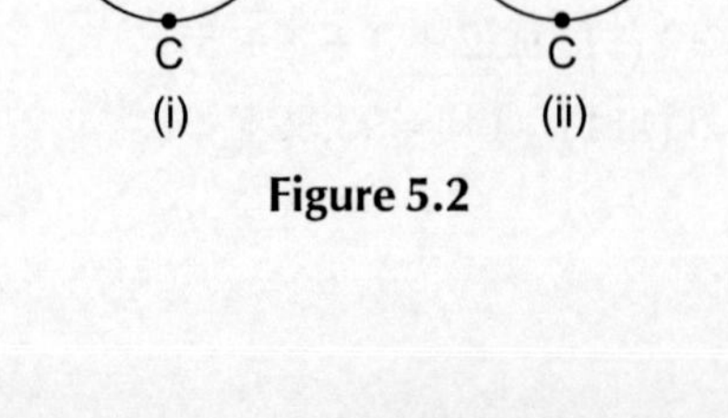

Figure 5.2

In either of these permutations each thing has same neighbours disregarding left or right. Thus if the direction (clockwise or anticlockwise) in which the things are positioned is ignored, we get the same circular permutation. Hence, by ignoring the direction of the arrangement of things, four different things give rise to $\frac{4-1!}{2} = \frac{3!}{2}$ circular permutations.

If any arrangement of n different things round a circle in the clockwise direction is considered to be not different from the similar arrangement in the anticlockwise direction, then the $n - 1$! arrangements are not all different, but each arrangement is counted twice. Therefore, in this case the number of circular permutations of n different things $= \frac{n-1!}{2}$.

Note. In the arrangement of flowers in a closed garland or beads in a-necklace, no distinction need be made between clockwise and anti-clockwise directions.

Example. *How many ways can 5 children arrange themselves in a ring.*

Solution. $5 - 1 ! = 4 ! = 24$ ways.

Example. *How many ways can 5 programmers and 3 software engineers sit around a table so that no two software engineers are together ?*

Solution. Let the programmers be seated first, then they can be seated in $5 - 1 ! = 4 !$ ways. Now 3 software engineers have to sit in between the programmers so that no two software engineers sit together. This means there are 3 ! ways for software engineers to fill up the 3 place between 5 programmers. Thus, the total number of ways in which both programmers and software engineers can sit are $4 ! . 3 ! = 24 \times 6 = 144$.

Example. *In how many ways can 7 differently coloured beads be strung on a necklace ?*

Solution. Since the arrangement is circular, the direction of the arrangements need not be considered, the number of ways required $= \frac{7-1!}{2} = \frac{6!}{2} = \frac{720}{2} = 360$.

Example. *How many ways are there to seat 10 boys and 10 girls around a circular table, if boys and girls seat alternatively.* [JNTU Nov - 06 - Set 3]

Solution. The number of ways to seat 10 boys first are 9 !, then there are 10 places to seat girls. The number of ways to seat 10 girls in 10 places are 10 ! therefore there are 9 ! 10 ! ways to seat 10 boys, 10 girls around a circular table, if boys and girls seat alternatively.

Now let us discuss a formula for enumerating r-combinations of n objects without repetition. Any r-permutation of n objects without repetition can be obtained by first choosing the r elements (this can be done in C (n, r) ways) and then arranging the r elements in all possible orders (this can be done in r ! ways). Thus, it follows that P $(n, r) = r$! C (n, r).

Theorem. (Enumerating r – combinations without repetitions).

$$C\,(n, r) = \frac{P\,(n, r)}{r\,!} = \frac{n\,!}{n - r\,!\,r\,!}.$$

Note. The number of n-combinations without repetitions is $C\,(n, n) = \frac{n\,!}{n - n\,!\,n\,!} = 1$.

Note. $C\,(n, r) = C\,(n, n - r)$

Note. If $C\,(n, r) = C\,(n, s)$ then either $r = s$ or $r + s = n$.

Note. $C\,(n, r)$ is greatest if $r = \frac{n}{2}$ when n is even

$$r = \frac{n-1}{2} \quad \text{or} \quad \frac{n+1}{2} \text{ when } n \text{ is odd.}$$

Example. *If C (n, r) = 126, find n.*

Solution. Since C (*n*, *r*) is a positive integer, we write

$$C(n, r) = 126 = 63 \times 2 = 9 \times 7 \times 2 = \frac{9 \times 8 \times 7}{4} = \frac{9 \times 8 \times 7 \times 6}{4 \times 6}$$

$$= \frac{9 \times 8 \times 7 \times 6}{4 \times 3 \times 2 \times 1} = C(9, 4)$$

$$\therefore \quad n = 9.$$

Example. *If C (n, 6) = C (n, 10) find C (n, 8).*

Solution. Since C (*n*, 6) = C (*n*, 10), 6 + 10 = *n* ⇒ *n* = 16.

$$\therefore \quad C(n, 8) = C(16, 8) = \frac{16!}{8!\,8!} = 12870.$$

Example. *Find n if C (n, 3) : C (n – 1, 4) = 8 : 5*

Solution. Given $\dfrac{C(n, 3)}{C(n-1, 4)} = \dfrac{8}{5}$

$$\Rightarrow \quad \frac{\dfrac{n!}{n-3!\,3!}}{\dfrac{n-1!}{n-5!\,4!}} = \frac{8}{5} \Rightarrow \frac{n!}{n-3!\,3!} \cdot \frac{n-5!\,4!}{n-1!} = \frac{8}{5}$$

$$\Rightarrow \quad \frac{4n}{(n-3)(n-4)} = \frac{8}{5} \Rightarrow 5n = 2(n-3)(n-4)$$

$$\Rightarrow \quad 2n^2 - 19n + 24 = 0 \Rightarrow (2n-3)(n-8) = 0 \Rightarrow n = \frac{3}{2}, 8$$

$\therefore \quad n = 8$. Since n is positive integer.

Example. *How many ways can a hand of 5 cards be selected from a deck of 52 cards ?*

Solution. Each hand is essentially a 5-combination of 52 cards.

Then the number such hands $= C(52, 5) = \dfrac{52!}{47!\,5!}$

$$= \frac{52.51.50.49.48}{5.4.3.2.1} = 25,98,960.$$

Example. *How many ways can a committee of 5 be chosen from 9 people?*

Solution. $C(9, 5) = \dfrac{9!}{4!\,5!} = \dfrac{9.8.7.6}{4.3.2.1} = 126.$

Example. *How many committees of 6 or more can be chosen from 9 people ?*

Solution. C (9, 6) + C (9, 7) + C (9, 8) + C (9, 9).

$$= \frac{9!}{6!\,3!} + \frac{9!}{7!\,2!} + \frac{9!}{8!\,1!} + \frac{9!}{9!\,0!} = 130.$$

Example. *How many ways can a committee of 4 teachers and 5 students be chosen from 9 teachers and 15 students.*

Solution. The teachers can be selected in C (9, 4) ways while the students can be chosen in C (15, 5) ways then the committee can be formed in C (9, 4) . C (15, 5) ways.

Example. *How many diagonals are there in a polygon of n sides ?*

Solution. A polygon of n sides has n vertices. Any two vertices determine either a side or a diagonal. Thus, the sum of number of sides, the number of diagonals is C (n, 2), then the number of diagonals is C $(n, 2) - n = \dfrac{n(n-1)}{2} - n = \dfrac{n(n-3)}{2}$.

Example. *The English alphabet contains 21 consonants and 5 vowels. Consider 7-letter words with 3 different vowels and 4 different consonants.*

(i) How many such words can be formed ?
(ii) How many such words contain the letter a ?
(iii) How many such words contain the letters a and b ?
(iv) How many such words begin with a and end with b ?
(v) How many such words contain the letters b and c ?
(vi) How many such words begin with b and end with c ?

Solution.

(*i*) The number of ways of chosen 3 vowels is C (5, 3). The number of ways of chosen 5 consonants is C (21, 4). The number of ways of arrangements with 7 letters is 7 !. Hence the number of 7-letter words with 3 different vowels and 4 different consonants is C (5, 3). C (21, 4) . 7 !

(*ii*) C (4, 2) . C (21, 4) . 7 ! (Choose the vowels, choose the consonants, and then arrange the 7 letters).

(*iii*) C (4, 2) . C (20, 3) . 7 !

(*iv*) C (4, 2) . C (20, 3) . 5 !

(*v*) C (5, 3) . C (19, 2) . 7 !

(*vi*) C (5, 3), C (19, 2) . 5 ! (Places of b, c are fixed, arrange remaining 5 letters in 5 ! way).

Example. *A cricket 11 is to be selected out of 14 players of whom 5 are bowlers. Find the number of ways in which this can be done so as to include at least 3 bowlers.*

Solution. The only possibilities regarding the selection are :

Table 5.2

Bowlers	*Others*	*Number of ways of Selections*
3	8	C (5, 3) . C (9, 8)
4	7	C (5, 4) . C (9, 7)
5	6	C (5, 5) . C (9, 6)

Thus, the total number of ways of forming cricket 11 = C (5, 3) . C (9, 8) + C (5, 4) . C (9, 7) + C (5, 5) . C (9, 6).

Example. *Suppose there are 15 red balls and 5 white balls. Assume that the balls are distinguishable and that a sample of 5 balls is to be selected.*

(i) How many samples of 5 balls are there ?
(ii) How many samples contain all red balls.
(iii) How many samples contain 3 red balls and 2 white balls ?
(iv) How many samples contain atleast 4 red balls ?

Solution.

(*i*) Total balls are 20.
The number of samples of 5 balls are C (20, 5).

(*ii*) Number of red balls are 15.
The number of samples contain all red balls are C (15, 5).

(*iii*) The number of samples contain 3 red balls are C (15, 3) and the number of samples contain 2 white balls are C (5, 2).
Total number of samples are C (15, 3) . C (5, 2).

(*iv*) Total number of samples are C (15, 4) C (5, 1) + C (15, 5) C (5, 0).

EXERCISE 5.2.

1. Find P (9, 4) and C (7, 4).
2. How many ways are there to distribute 12 different books among 15 people if no person is to receive more than 1 book.
3. Suppose there are six digits 1, 2, 4, 5, 6 and 7.
   (*i*) How many six digit numbers can be formed if no digit is repeated.
   (*ii*) How many six digit numbers can be formed if each number is divisible by 2.
   (*iii*) How many six digit numbers can be formed if digit 5 is in ten's place.
   (*iv*) How many six digit numbers can be formed if each number begins with 1 and ends with 5.
   (*v*) How many six digit numbers can be formed if 1, 5 are in extreme places.
   (*vi*) How many six digit numbers can be formed if each number greater than 4 lakhs.
4. (*i*) Find the number of ways in which 5 boys and 5 girls can be seated in a row if the boys and girls are to have alternative seats.
   (*ii*) Find the number of ways to seat them alternatively if boy A and girl B are to sit in adjacent seats.
   (*iii*) Find the number of ways to seat them alternatively if boy A and girl B must not sit in adjacent seats.
5. In how many ways can the digits 0, 1, 2, 3, 4, 5, 6, 7, 8 and 9 be arranged so that 0 and 1, 2 are adjacent.
6. (*i*) How many ways are there to seat 5 boys and 5girls around a circular table ?
   (*ii*) If boys and girls alternative, how many ways are there ?
7. Find the number of ways in which 5 different English books, 6 French books, 3 German books, and 7 Russian books can be arranged on a shelf so that all books of the same language are together.
8. How many 5-card poker handshave (*i*) 4 aces ?
   (*ii*) 4 of a kind ?
   (*iii*) exactly 2 pairs
   (*iv*) a full house (that is, 3 of a kind and 2 of another kind) ?
   (*v*) a pair of aces ?
   (*vi*) a flush (a set of 5 consecutive cards in one suit) ?
   (*vii*) a spade flush ?

9. In how many ways can a team of 5 be chosen from 10 players so as to

(*i*) include both the strongest and weakest player ?

(*ii*) include the strongest but exclude the weakest player ?

(*iii*) exclude both the strongest and the weakest player ?

10. There are 30 females and 35 males in the Junior Class while there are 25 females and 20 males in the senior class. In how many ways can a committee of 10 be chosen so that there are exactly 5 females and 3 Juniors on the committee ?

ANSWERS

1. 3024, 35
2. P(15, 12)
3. (*i*) 6 ! (*ii*) 3 (5 !) (*iii*) 5 ! (*iv*) 4 ! (*v*) 2 (4 !) (*vi*) 5 !
4. (*i*) 5 ! 6 ! (*ii*) 2 ! (4 ! 5 !) (*iii*) 5 (4 ! 4 !)
5. 3 ! 8 !
6. (*i*) 9 ! (*ii*) 5 ! 4 !
7. 4 ! (5 ! 6 ! 3 ! 7 !)
8. (*i*) 48. C (4, 4) (*ii*) 48. C(13, 1). C (4, 4) (*iii*) 44. C (13, 2). C (4, 2). C (4, 2) (*iv*) 13.12. C (4, 3). C (4, 2) (*v*) 4^3. C(4, 2). C (12, 3) (*vi*) C (4, 1). C (13, 5) (*vii*) C (13, 5)
9. (*i*) C (8, 3) (*ii*) C (8, 4) (*iii*) C (8, 5).
10. C (30, 0) C (35, 3) C (25, 5) C(20, 2) + C (30, 1) C (35, 2) C (25, 4) C (20, 3) + C (30, 2) C (35, 1) C (25, 3) C (20, 4) + C (30, 3) C (35, 0) C (25, 2) C (20, 5).

5.5 ENUMERATING COMBINATIONS AND PERMUTATIONS WITH REPETITIONS

Now let us learn to counting permutations and combinations with unlimited repetitions. Let U (n, r) denote the number of r-permutations of n objects with unlimited repetitions and V (n, r) denote the number of r-combinations of n objects with unlimited repetitions. That is, if a_1, a_2, ..., a_n are the n objects, we are counting r-combinations and r-permutations of $\{\infty . a, \infty . a_2, ..., \infty . a_n\}$

Theorem. (Enumerating r-permutations with unlimited repetitions).

$$U(n, r) = n^r.$$

Proof : Suppose there are r blank places. The first blank place can be filled in n ways. When the firstblank place is thus filled up, the second also can be filled up in n ways as repetition is allowed. The first two blank places can be filled up in $n \times n = n^2$ ways. The third blank place can also be filled up in n ways. Hence the first three blank places can be filled up in $n \times n \times n = n^3$ ways.

Proceeding in this way, the r blank places can be filled up in n^r ways.

Corollary : The number of permutations of n different things, taken not more than r at a time, when each thing may occur any number of times.

$$= n + n^2 + n^3 + ... + n^r = \frac{n\left(n^r - 1\right)}{n - 1}.$$

Example. *There are 25 true or false questions on an examination. How many different ways can a student do the examination if he or she can also choose to leave the answer blank ?*

Solution. 3^{25}.

Example. *The results of 40 foot ball games (win, lose or, tie) are to be predicted. How many different forecasts can contain exactly 25 correct results ?*

Solution. Choose 25 correct results in C (40, 25) ways. Each of the remaining 15 games has 2 wrong forecasts. Thus, there are C (40, 25) . 2^{15} forecasts with exactly 25 correct predictions.

Example. *A telegraph can transmit two different signals, a dot and a dash. What length of these symbols is needed to encode the 26 letters of the English alphabet and the ten digits 0, 1, 2, ..., 9 ?*

Solution. Since there are two choices for each character, the number of different sequences of length K is 2^K.

The number of non-trivial sequences of length n or less is $2 + 2^2 + 2^3 + ... + 2^n = 2^{n+1} - 2$. If $n = 4$, this total is 30, which is enough to encode the letters of the English alphabet, but not enough to also encode the digits. To encode the digits we need to allow sequences of length up to 5 for then there are possibly $2^{5+1} - 2 = 62$ total sequences, because in Morse Code all letters are transmitted by sequences of four or fewer characters while all digits are transmitted by sequences of length 5.

Theorem. (Enumerating *r*-combinations with unlimited repetitions).

V (n, r) = The number of *r*-combinations of *n* distinct objects with unlimited repetitions.

= The number of non-negative integral solutinos to $x_1 + x_2 + ... + x_n = r$

= The number of ways of distributing *r* similar balls into *n* numbered boxes.

= The number of binary numbers with $n - 1$ one's and *r* zeros.

$= C(n - 1 + r, r) = C(n - 1 + r, n - 1)$

$$= \frac{(n+r-1)!}{r!\,n-1!}.$$

Remark : Of Course, the number of r-combinations of $\{\infty . a_1, \infty . a_2, ..., \infty . a_n\}$ is the same as the number of *r*-combinations of $\{r . a_1, r . a_2, ..., r . a_n\}$.

The following examples will clarify the conclusions of above theorem.

Example. The number of 4-combinations of $\{\infty . a_1, \infty . a_2, \infty . a_3, \infty . a_4, \infty . a_5\}$ is C (5 – 1 + 4, 4) = C (8, 4) = 70.

Example. The number of 3-combinations of 5 objects with unlimited repetitions is C (5 – 1 + 3, 3) = C (7, 3) = 35.

Example. The number of non-negative integral solutions to $x_1 + x_2 + x_3 + x_4 + x_5 = 50$ is C (5 – 1 + 50, 50) = C (54, 50)

$$= \frac{54!}{4!\,50!} = 3,16,251.$$

Example. The number of binary numbers with ten 1's and five 0's is C (10 + 5, 5) = C (15, 5) = 3003.

Example. The number of ways of placing 10 similar balls in 6 numbered boxes is C (6 – 1 + 10, 10) = C (15, 10) = 3003.

Example. *(i) How many different outcomes are possible by tossing 10 similar coins ?*

(ii) How many different outcomes are possible from tossing 12 similar dice ?

(iii) How many ways can 20 similar books be placed on 5 different shelves ?

(iv) Out of a large supply of pennies, nickels, dimes and quarters, in how many ways can 10 coins be selected ?

Solution.

(*i*) This is the same as placing 10 similar balls into two boxes labeled "heads" and "tails". Total outcomes are C (2 – 1 + 10, 10) = C (11, 10) = 11.

(*ii*) This is the same as placing 10 similar balls into 6 numbered boxes. Therefore, there are C (17, 12) possibilities.

(*iii*) C (5 – 1 + 20, 20) = C (24, 20) ways.

(*iv*) C (4 – 1 + 10, 10) = C (13, 10), since this is equivalent to placing 10 similar balls in 4 numbered boxes labeled "pennies", "nickels", "dimes", and "quarters".

Example. *Enumerate the number of ways of placing 20 indistinguishable balls into 5 boxes where each box is non-empty.*

Solution. First, place one ball in each of the 5 boxes. Then we must count the number of ways of distributing the 15 remaining balls into 5 boxes with unlimited repetitions. By theorem, we can do this in C (5 – 1 + 15, 15) = C (19, 15) ways.

Alternatively, we can also model this problem as a solution of an equation problem. If x_i represents the number of balls in the ith box. Then we are asked to enumerate the number of integral solutions to $x_1 + x_2 + x_3 + x_4 + x_5 = 20$ where each $x_i > 0$. After distributing 1 ball into each of the 5 boxes, we then are to enumerate the number of integral solutions of $y_1 + y_2 + y_3 + y_4 + y_5 = 15$ where each $y_i \geq 0$. Hence the number of solutions is C (5 – 1 + 15, 15) = C (19, 15).

Example. *How many integral solutions are there to $x_1 + x_2 + x_3 + x_4 + x_5 = 20$, where $x_1 \geq 3$, $x_2 \geq 2$, $x_3 \geq 4$, $x_4 \geq 6$ and $x_5 \geq 0$?*

Solution. First, distribute 3 balls in box 1, 2 balls in box 2, 4 balls in box 3, 6 balls in box 4, and 0 balls in box 5. Then remaining balls are 5, we have to distributed these 5 balls into 5 boxes with unlimited repetition. Next we want the number of solutions of $y_1 + y_2 + y_3 + y_4 + y_5 = 5$ where each $y_i \geq 0$.

There are C (5 – 1 + 5, 5) = C (9, 5) such solutions.

Theorem. The number of integral solutions of $x_1 + x_2 + \ldots\ldots + x_n = r$, where each $x_i > 0$.

= The number of ways of distributing r similar balls into n numbered boxes with at least one ball in each box.

= C $(n - 1 + (r - n), r - n)$ = C $(r - 1, r - n)$ = C $(r - 1, n - 1)$.

Theorem. Suppose that $r_1, r_2, \ldots\ldots, r_n$ are integers.

Then the number of integral solutions of $x_1 + x_2 + \ldots\ldots + x_n = r$ where $x_1 \geq r_1$, $x_2 \geq r_2$, ..., $x_n \geq r_n$.

= The number of ways of distributing r similar balls into n numbered boxes where there are at least r_1 balls in the first box, at least r_2 balls in the second box, ..., and atleast r_n balls in the nth box.

$= C(n - 1 + r - r_1 - r_2 - ......, - r_n, r - r_1 - r_2 - ......, - r_n)$

$= C(n - 1 + r - r_1 - r_2 - ......, - r_n, n - 1)$ since $C(n, r) = C(n, n - r)$.

Example. *Find the number of integer solutions of $x_1 + x_2 + x_3 + x_4 + x_5 = 30$, where $x_1 \geq 2$, $x_2 \geq 3$, $x_3 \geq 4$, $x_4 \geq 2$, $x_5 \geq 0$.*

Solution. By above theorem, the number of integral solutions of $x_1 + x_2 + ... + x_n = r$, where $x_1 \geq r_1, ..., x_n \geq r_n$ is $C(n - 1 + r - r_1 - r_2 - ...... - r_n, n - 1)$

$\therefore$ The number of integer solutions of $x_1 + x_2 + x_3 + x_4 + x_5 = 30$,

where $x_1 \geq 2$, $x_2 \geq 3$, $x_3 \geq 4$, $x_4 \geq 2$, $x_5 \geq 0$ is

$$C(5 - 1 + 30 - 2 - 3 - 4 - 2 - 0, 5 - 1) = C(23, 4) = \frac{23!}{19!\,4!} = 8855.$$

Example. *Find the number of distinct triples (x_1, x_2, x_3) of non-negative integers satisfying the inequality $x_1 + x_2 + x_3 < 6$.*

Solution. First we count the number of solutions for $x_1 + x_2 + x_3 = n$, where $n = 0, 1, 2, 3, 4, 5$.

$\therefore$ The number of solutions for $x_1 + x_2 + x_3 = 0$

is $\quad C(3 - 1 + 0, 2) = C(2, 2)$

and the number of integral solutions for $x_1 + x_2 + x_3 = n$ for $n = 1, 2, 3, 4, 5$ are $C(3 - 1 + 1, 2)$, $C(3 - 1 + 2, 2)$, $C(3 - 1 + 3, 2)$, $C(3 - 1 + 4, 2)$, $C(3 - 1 + 5, 2)$ respectively. Hence the number of non-negative integers for $x_1 + x_2 + x_3 < 6$ (*i.e.,*) ≤ 5 is $C(2, 2) + C(3, 2) + C(4, 2) + C(5, 2) + C(6, 2) + C(7, 2)$.

Example. *Find the number of integral solutions to $x_1 + x_2 + x_3 + x_4 = 50$ where $x_1 \geq -4$, $x_2 \geq 7$, $x_3 \geq -14$, $x_4 \geq 10$.*

Solution. Let us take $x_1 = y_1 - 4$, $x_2 = y_2 + 7$, $x_3 = y_3 - 14$, $x_4 = y_4 + 10$, then y_1, y_2, y_3, y_4 are non-negative integers and given equation reduces to $y_1 - 4 + y_2 + 7 + y_3 - 14 + y_4 + 10 = 50$

$\Rightarrow \quad y_1 + y_2 + y_3 + y_4 = 51$, where $y_1 \geq 0$, $y_2 \geq 0$, $y_3 \geq 0$, $y_4 \geq 0$.

Hence the number of integral solutions is

$$C(4 - 1 + 51, 4 - 1) = C(54, 3).$$

5.6 ENUMERATING PERMUTATIONS WITH CONSTRAINED REPETITIONS

Suppose that we are given a particular selection of r objects where there are some repetitions. We desire a formula for the number of permutations on this given selection of r objects. Let us see the following example.

Example. *How many 10-permutations are there of $\{3 . a, 4 . b, 2 . c, 1 . d\}$.*

Solution. Let x be the number of such permutations we see that there are $3!\,x$ permutations if we replace the *a's* by a_1, a_2, and a_3. Likewise, there are $(4!)(3!\,x)$ permutations if we also replace the *b's* by b_1, b_2, b_3 and corresponding to the number of ways of arranging b_1, b_2, b_3, b_4. Continuing we see that there are $(2!)(4!)(3!)$ x permutations if we also replace the *c's* by c_1, c_2. But then we know that there are $10!$ permutations of $\{a_1, a_2, a_3, b_1, b_2, b_3, b_4, c_1, c_2, d\}$.

Hence $\quad (2!)(4!)(3!)\,x = 10!$

$$\Rightarrow \qquad x = \frac{10!}{2!\,4!\,3!} = 12600.$$

Now let us introduce the following notation.

Suppose that $q_1, q_2, \ldots\ldots, q_r$ are non-negative integers such that $n = q_1 + q_2 + \ldots\ldots + q_r$ and suppose that $a_1, a_2, \ldots\ldots, a_r$ are r distinct objects. Let P $(n\,;\, q_1, q_2, \ldots\ldots, q_r)$ denote the number of n-permutations of the n-combination $\{q_1 \cdot a_1, q_2 \cdot a_2, \ldots\ldots, q_r \cdot a_r\}$.

With above example and this notation, we pove the following theorem.

Theorem. (Enumerating n-permutations with constrained repetitions). P $(n\,;\, q_1, q_2, \ldots\ldots, q_r) = \dfrac{n!}{q_1!\,q_2!\ldots\ldots q_r!}$

Proof : Let $x = \text{P}\,(n\,;\, q_1, q_2, \ldots\ldots, q_r)$.

If the q_1 a_1's were all different there would be $(q_1\,!)\ x$ permutatinos since each odd permutation would give rise to $q_1\,!$ new permutations corresponding to the number of ways of arranging the q_1 distinct objects in a row. If the q_2 a_2's were all replaced by distinct objects, then by similar reasoning there would be $(q_2\,!)\ (q_1\,!)\ x$ permutations. If we repeat this procedure until all the objects are distinct we will have $(q_r\,!)\ \ldots\ldots\ (q_2\,!)\ (q_1\,!)\ x$ permutations.

We know that there are $n\,!$ permutations of n distinct objects.

Hence $\qquad n\,! = q_1\,!\ q_2\,!\ \ldots\ldots,\ q_r\,!\ x$

$$\Rightarrow \qquad x = \frac{n!}{q_1!\,q_2!\ldots\ldots q_r!}$$

$$\text{P}\,(n\,;\, q_1, q_2, \ldots\ldots, q_r) = \frac{n!}{q_1!\,q_2!\ldots\ldots q_r!}$$

Note. P $(n\,;\, q_1, q_2, \ldots\ldots, q_r) = \text{C}\,(n, q_1)\ \text{C}\,(n - q_1, q_2)\ \text{C}\,(n - q_1 - q_2, q_3)\ \ldots\ldots\ \text{C}\,(n - q_1 - q_2 - \ldots\ldots, - q_{r-1}, q_r)$

Example. The number of arrangements of letters in the word ENGINEERING is

P $(11\,;\, 3, 3, 2, 2, 1) = \dfrac{11!}{3!\,3!\,2!\,2!\,1!}$ since this equals the number of permutations of {3 . E, 3 . N, 2 . G, 2 . I, 1 . R}.

Example. *How many different arrangements of letters of MISSISSIPPI ?*

[JNTU - Nov 06 - Set 4]

Solution. The word contains 11 letters. Of which S's are 4, I's are 4 and the rest all different.

$$\therefore \qquad \text{P}\,(11\,;\, 4, 4, 1, 1, 1) = \frac{11!}{4!\,4!\,1!\,1!\,1!}.$$

Example. *In how many ways can 28 different books be given to 6 students so that 2 of the students will have 4 books each and the other 4 will have 5 books each ?*

Solution. Choose the 2 students to receive 4 books each in C (6, 2) ways. Then to each such choice the 28 books can be distributed in P $(28\,;\, 4, 4, 5, 5, 5, 5) = \dfrac{28!}{4!\,4!\,5!\,5!\,5!\,5!}$ ways.

Thus, there are C (6, 2) $\left(\dfrac{28!}{(4!)^2\,(5!)^4}\right)$ total distributions.

Example. *Find the number of arrangements of letters in the word TALLAHASSEE.*

Solution. The word contain 11 letters of which 3 are A's, 2 are E's, 2 are L's, 2 are S's, 1 is H, and 1 is T. The *n* number of arrangements of these letters is $\dfrac{11!}{3!2!2!2!1!1!}$.

Note : The number of arrangements of these letters that begin with T and end with S is $\dfrac{9!}{3!1!2!2!1!}$.

Partition and Cross Partitions

Let S be a non-empty set. The collection $\{A_1, A_2, \ldots\ldots, A_n\}$ of subsets of S is a partition of S iff

(*i*) $S = A_1 \cup A_2 \cup \ldots\ldots \cup A_r$

(*ii*) For any A_i, $A_j \in S$, either $A_i = A_j$ or $A_i \cap A_j = \phi$, $A_1, A_2, \ldots\ldots, A_r$ are called cells of the partition and each A_i a non empty subset of S.

If $\{A_1, A_2, \ldots\ldots, A_r\}$ and $\{B_1, B_2, \ldots\ldots, B_r\}$ are both partitions of a non-empty set S, then $\{A_i \cap B_j\}$ forms a partition of S $\{A_i \cap B_j\}$ is called a cross partition of S.

Example. Let $S = \{a, b, c, d, e, f, g\}$ and $A_1 = \{a, d, e\}$, $A_2 = \{b, c\}$, $A_3 = \{f\}$, $A_4 = \{g\}$ then $\{A_1, A_2, A_3, A_4\}$ forms a partition of S. We observe that

(*i*) $A_1 \cup A_2 \cup A_3 \cup A_4 = S$ and (*ii*) $A_i \cap A_j = \phi$ for all A_i, $A_j \in S$.

Ordered Partition:

Definition. *A partition a non empty set is called an* **ordered partition** *of S if there is a specified order on the subsets of s.*

Definition. *An ordered r-tuple of sets $\{A_1, A_2, \ldots\ldots, A_r\}$ is called a r-part ordered partition if the sets $A_1, A_2, \ldots\ldots, A_r$ forms a partition of S.*

Example. Let $S = \{a, b, c, d, e\}$ and $A_1 = \{a, b,\}$, $A_2 = \{c\}$, $A_3 = \{d\}$, $A_4 = \{e\}$, then $\{A_1, A_2, A_3, A_4\}$ is an ordered partition of S. It is a 4-part ordered partition.

Theorem : (Enumerating ordered partitions of a set). The number of ordered partitions of a set S of type $(q_1, q_2, \ldots\ldots, q_r)$ where $|S| = n$ is

$$P(n\,;\, q_1, q_2, \ldots\ldots, q_r) = \frac{n!}{q_1!\,q_2!\ldots\ldots q_r!}.$$

Proof : By choosing the q_1 elements to occupy the first subset in $C(n, q_1)$ ways, the q_2 elements for the second subset in $C(n - q_1, q_2)$ ways, etc., then the number of ordered partitions of type $(q_1, q_2, \ldots\ldots, q_r)$ is

$C(n, q_1)\, C(n - q_1, q_2) \ldots\ldots C(n - q_1 - q_2 - \ldots\ldots - q_{r-1}, q_r)$ which is equal to $P(n\,;\, q_1, q_2, \ldots\ldots, q_r)$ and it also equal to $\dfrac{n!}{q_1!\,q_2!\ldots\ldots q_r!}$.

Example. *In the game of bridge four players (usually called North, East, South, and West) seated in a specified order are each dealt a "hand" of 13 cards.*

(i) How many ways can 52 cards be dealt to the four players ?

(ii) In how many ways will one player be dealt all four kings ?

(iii) In how many deals will North be dealt 7 hearts and South the other 6 hearts ?

Solution. (*i*) $\dfrac{52\,!}{(13\,!)^4}$

(*ii*) Choose the player to receive the Kings in 4 ways, then partition the remaining cards. There are $4\left(\dfrac{48\,!}{9\,!\,13\,!\,13\,!\,13\,!}\right)$ or 4C (48, 9) . C (39, 13). C (26, 13) . C (13, 13) ways.

(*iii*) Choose the 7 hearts for North (and automatically give the other 6 hearts to South) in C (13, 7) ways, then partition the remaining cards. There are C (13, 7) . $\dfrac{39\,!}{6\,!\,7\,!\,13\,!\,13\,!}$ hands.

For finding the number of unordered partitions, we will give a formula only in the case that all subsets have the same number of elements.

Theorem. (Enumerating unordered partitions of equal cell size). Let S be *a* set with n elements where $n = q\,.\,t$. Then the number of unordered partitions of S of type $(q, q, \ldots\ldots, q)$ is $\dfrac{n\,!}{t\,!\,(q\,!)^t}$ where t equals to the number of subsets.

Proof : We know that the number of ordered partitions of S of type $(q_1, q_2, \ldots\ldots, q_t)$ is

$$P\,(n\,;\,q_1, q_2, \ldots\ldots, q_t) = \frac{n\,!}{q_1\,!\,q_2\,!\ldots\ldots q_t\,!}$$

Hence the number of unordered partitions of S of type

$(q, q, \ldots\ldots, q)$ is $P\,(n\,;\,q, q, \ldots\ldots, q)$ (t times) $= \dfrac{n\,!}{t\,!\,q\,!\,q\,!\ldots\ldots q\,!}$ (t times) $= \dfrac{n\,!}{t\,!\,(q\,!)^t}$.

Example. *In the game of bridge, four players seated and are each dealt a "hand" of 13 cards. How many ways can the 52 cards be dealt to the four players by disregarding dealing order.*

Solution. There are $\dfrac{52\,!}{4\,!\,(13\,!)^4}$ ways. ($\because$ By considering order, there are $\dfrac{52\,!}{(13\,!)^4}$ ways)

Example. *In how many ways can 12 of the 14 people be distributed inito 3 teams of 4 each ?*

Solution. First, count the number of ways to choose the 12 people, then count the number of unordered partitions of type (4, 4, 4). There are C (14, 12) $\dfrac{12\,!}{3\,!\,(4\,!)^3}$ ways.

EXERCISE 5.3.

1. In how many ways can 5 similar books be placed in 3 different selves ?
2. How many outcomes are obtained from rolling n indistinguishable dice ?
3. In how many ways can 5 glasses be filled with 10 different kinds of kool-aid if no mixing is allowed and glasses are (*i*) indistinguishable ? (*ii*) distinguishable ?
4. In how many ways can we distribute 12 identical marbles among 6 distinct containers ?
5. How many integral solutions are there of $x_1 + x_2 + x_3 + x_4 + x_5 = 30$, where for each (*i*) $x_i \geq 0$; (*ii*) $x_i \geq 1$; and (*iii*) $x_i > i$.
6. Find the number of non-negative integral solutions to $x_1 + x_2 + x_3 + x_4 + x_5 = 10$.
7. Find the number of non-negative integer solutions of the equality $x_1 + x_2 + \ldots\ldots + x_6 \leq 10$.
8. In how many ways can we distribute 7 apples and 6 oranges among 4 children so that each child gets at least 1 apple ?
9. Find (*i*) P (8 ; 4, 3, 2, 1)
   (*ii*) P (15 ; 4, 0, 2, 3, 6).
10. In how many different 8-digit numbers can be formed by arranging the digits 1, 1, 1, 1, 2, 3, 3, 3 ?
11. Suppose the J.N.T University has a residence hall that has 5 single rooms, 5 double rooms, and 3 rooms for 3 students. In how many ways can 24 students be assigned to the 13 rooms ?
12. Suppose that a set S has *n* distinct elements. How many *n*-part ordered partitions ($A_1, A_2, \ldots\ldots, A_n$) are there in which each set A_i has exactly 1 element ?
13. Find the number of arrangements of the letters of
    (*i*) TENNESSEE
    (*ii*) TRIANNUAL.
14. A chess player places black and white chess pieces (2 knights, 2 bishops, 2 rooks, 1 queen and 1 king of each color) in the first two rows of a chess board. In how many ways can this be done ?
15. How many anagrams (arrangements of the letters) are there of 7 . *a*, 5 .*c*, 1 . *d*, 5 . *e*, 1 . *g*, 1 . *h*, 7 . *i*, 3 . *m*, 9 . *n*, 4 . 0, 5 . *t* ?

ANSWERS

1. C (3 – 1 + 5, 5) = C (7, 5)
2. C (6 – 1 + *n*, *n*) = C (*n* + 5, *n*)
3. (*i*) C (10 – 1 + 5, 5) = C (14, 5)
   (*ii*) 10^5.
4. C (6 – 1 + 12, 12) = C (17, 12).
5. (*i*) C (34, 30) (*ii*) C (29, 25)
   (*iii*) C (14, 10)
6. C (10 + 5 – 1, 5 – 1) = C (14, 4)
7. C (6 – 1 + 0, 5) + C (6 – 1 + 1, 5) + ... + C (6 – 1 + 10, 5) = C (5, 5) + C (6, 5) + ... + C (15, 5)
8. C (6, 3) . C (9, 6) = 1680.
9. (*i*) 140 (*ii*) $\dfrac{15!}{4!0!2!3!6!}$

10. $\dfrac{8!}{4!1!3!}$

11. $P(24\,;\,1, 1, 1, 1, 1, 2, 2, 2, 2, 2, 3, 3, 3) = \dfrac{24!}{(2!)^5\,(3!)^3}$

12. $P(n\,;\,1, 1, 1, \ldots\ldots, 1) = n!$

13. (*i*) $\dfrac{9!}{1!4!2!2!}$ (*ii*) $\dfrac{9!}{1!1!1!2!2!1!1!}$

14. $P(16\,;\,2, 2, 2, 2, 2, 2, 1, 1, 1, 1) = \dfrac{16!}{(2!)^6}$

15. $P(48\,;\,7, 5, 1, 5, 1, 1, 7, 3, 9, 4, 5) = \dfrac{48!}{(7!)^2\,(5!)^3\,3!9!4!}$

5.7 BINOMIAL COEFFICIENTS

In this section we will present some basic identities involving binomial coefficients. In formulas arising from the analysis of algorithms in Computer Science, the binomial coefficients occur over and over again, so that a facility for manipulating them is a necessity. The study of identities is itself a major field of study in combinatorial mathematics.

Combinatorial Reasoning :

The symbol $C(n, r)$ has two meanings : These are the combinatorial, and the factorial. In other words, $C(n, r)$ represents the number of ways of choosing r objects from n distinct objects (the combinatorial meaning) and, as well, $C(n, r)$ equals to $\dfrac{n!}{r!(n-r)!}$ (the factorial or algebraic meaning). Therefore, as a general rule, all theorems and identities about factorials and binomial coefficients can be viewed as two kinds of statements for which two kinds of proofs can be given : a combinatorial proof and an algebraic proof. Hence a combinatorial proof will be based on decomposing a set into subsets in a certain prescribed manner and then counting the number of ways of selecting these subsets, while an algebraic proof will be patterned mainly on the manipulation of factorials.

5.8 COMBINATORIAL IDENTITIES

1. Symmetry Property : $C(n, r) = C(n, n-r)$.

Solution. $C(n, r)$ denotes the number of selections of n objects when r-things are selected, and the remaining $(n-r)$ objects are not selected. Therefore, selecting of r-things is the same as discarding $n-r$ objects. For each selection of $r-$ things *i.e.,* for each $r-$ combination there is a discarding of $n-r$ remaining objects. The number of ways in which we discard $n-r$ objects is $C(n, n-r)$.

Since the set of $r-$ combinations and the set of discarding of $n-r$ objects all in one-one corresponding we get, $C(n, r) = C(n, n-r)$.

2. Pascals Identity : $C(n, r) = C(n-1, r-1) + C(n-1, r)$.

Solution. Method 1 :

$$C(n-1, r-1) + C(n-1, r) = \frac{n-1!}{r-1!(n-1-r+1)!} + \frac{n-1!}{(n-1-r)!r!}$$

$$= \frac{n-1!r + n-1!(n-r)}{n-r!r!}$$

$$= \frac{n-1!(r+n-r)}{r!n-r!} = \frac{n-1!.n}{r!n-r!} = \frac{n!}{r!n-r!}$$

$$= C(n, r)$$

Method 2 : $C(n, r)$ is the number of r-combinations of n objects. Let n be any one of the n objects. Fix the object x the $C(n, r)$ combinations can be grouped into :

(*i*) Those $r-$ combinations that contain the object x.

(*ii*) Those $r-$ combinations that do not contain the object x.

Since the objects x in all $r-$ combinations of (*i*), we are to choose $(r-1)$ objects in each combinations of (*i*) from the remaining $(n-1)$ objects, this can be done in $C(n-1, r-1)$ ways.

∴ The number of combinations that contain x, is $C(n-1, r-1)$.

To count the number of $r-$ combinations of (*ii*). *i.e.,* the number of $r-$ combinations which do not contain x, we omit the objects x from n objects and choose r objects from the remaining $(n-1)$ objects. Therefore, the number $r-$ combinations that do not contain x is $C(n-1, r)$. Since (*i*), (*ii*) are disjoint, by sum rule ; $C(n, r) = C(n-1, r-1) + C(n-1, r)$.

3. Pascals row sum identity :

$C(n, 0) + C(n, 1) + C(n, 2) + ... + C(n, n) = 2^n$.

Solution. This is equivalent to finding the number of subsets of A with n-elements. The number of subsets of A can be considered follows :

The number of subsets having no elements (*i.e.,*) subsets with 0 element is $C(n, 0)$.

The number of subsets having 1 element is $C(n, 1)$, and so on.

The number of subsets with n-elements is $C(n, n)$. Adding above all, we get, the number of elements in the power set of A (*i.e.,*) $|A| = C(n, 0) + C(n, 1) + ... + C(n, n)$, but the number of elements in the power set of A is 2^n.

∴ $C(n, 0) + C(n, 1) + ... + C(n, n) = 2^n$.

Note : The identity (2), gives us an alternative method for determining the numerical values of $C(n, r)$. For example, if we know $C(5, 0)$, $C(5, 1)$, $C(5, 2)$, $C(5, 3)$, $C(5, 4)$, $C(5, 5)$, we can determine $C(6, 1)$, $c(6, 2)$, $C(6, 3)$, $C(6, 4)$, $C(6, 5)$ simply by addition.

Using identity (2) and the fact that $C(n, 0) = 1 = C(n, n)$ for all non-negative integers n, we can build successive rows in the table of binomial coefficients, called Pascal's triangle. (Figure 5.3)

The Pascal's identity says that the numbers on the rth row are found by adding the two nearest binomial coefficients in the row above it. For instance, $C(6, 4)$ is the sum of the two circled binomial coefficients above it.

i.e., $C(6, 4) = C(5, 3) + C(5, 4)$.

Let's list Pascal's triangle again (Figure 5.3) with numerical values of the binomial coefficients entered. The number on the rth row along the rth diagonal is $C(n, r)$.

We can construct the rows from $n = 0$ to $n = 8$ in Figure 5.4

Row number

$n = 0$ C (0, 0)

$n = 1$ C (1, 0) C (1, 1)

$n = 2$ C (2, 0) C (2, 1) C (2, 2)

$n = 3$ C (3, 0) C (3, 1) C (3, 2) C (3, 3)

$n = 4$ C (4, 0) C (4, 1) C (4, 2) C (4, 3) C (4, 4)

$n = 5$ C (5, 0) C (5, 1) C (5, 2) [C (5, 3)] [C (5, 4)] C (5, 5)

$n = 6$ C (6, 0) C (6, 1) C (6, 2) C (6, 3) [C (6, 4)] C (6, 5) C (6, 6)

Figure 5.3.

$n = 0$ 1

$n = 1$ 1 1

$n = 2$ 1 2 1

$n = 3$ 1 3 3 1

$n = 4$ 1 4 6 4 1

$n - 5$ 1 5 10 10 5 1

$n = 6$ 1 6 15 20 15 6 1

$n = 7$ 1 7 21 35 35 21 7 1

$n = 8$ 1 8 28 56 70 56 28 8 1

Figure 5.4.

4. Show that $P(n, r) = nP(n - 1, r - 1)$

Solution. Method 1. $P(n, r) = \dfrac{n!}{n - r!}$

$$\Rightarrow \text{ and } \quad P(n-1, r-1) = \frac{n-1!}{n-1-r+1!} = \frac{n-1!}{n-r!}$$

$$\Rightarrow \quad n\,P(n-1, r-1) = \frac{n(n-1!)}{n-r!} = \frac{n!}{n-r!} = P(n, r).$$

Method 2. The identity holds because in arranging n objects we can fill the first position n ways and then arrange the remaining $n - 1$ objects in $r - 1$ positions.

5. Show that $P(n, r) = P(n, r - 1)(n - r + 1)$.

Solution. Method 1 : $(n - r + 1)\,P(n, r - 1) = (n - r + 1)\,\dfrac{n!}{n - r + 1!}$

$$= \frac{(n-r+1).n!}{(n-r+1)(n-r)!} = \frac{n!}{n-r!} = P(n, r)$$

Method 2. We observe that any r-permutations of n objects can also be attained by first arranging $r - 1$ of the objects in some order and then filling the rth position. The first $r - 1$ positions can be filled in P $(n, r - 1)$ ways while the rth position can be filled in $n - (r - 1) = n - r + 1$ ways.

Hence $\quad$ $P(n, r) = (n - r + 1)\ P(n, r - 1)$.

6. Newton's Identity : $C(n, r)\ C(r, k) = C(n, k)\ C(n - k, r - k)$ for integers $n \geq r \geq k \geq 0$.

Solution. Method 1 :

$$C(n, r)\ C(r, k) = \frac{n!}{n-r!r!} \cdot \frac{r!}{r-k!k!} = \frac{n!}{n-r!} \cdot \frac{1}{r-k!k!}$$

$$= \frac{n!}{n-k!k!} \cdot \frac{n-k!}{r-k!(n-r)!}$$

$$= C(n, k) \cdot \frac{n-k!}{r-k!(n-k-(r-k))!}$$

$$= C(n, k) \cdot C(n - k, r - k).$$

Method 2 : The Left hand side counts the number of ways of selecting two sets : First a set A of r objects and then from A, a set B of k objects. For example, we may be counting the number of ways to select a committee of r people and then to select a subset of k leaders from this committee. On other hand, the right-hand side counts the number of ways we could select the group of k leaders from the n people first, and then select the remaining r-k people for the committee from the remaining n-k people. Hence the result is true.

5.9 BINOMIAL THEOREM

Any sum of two unlike symbols, such as $x + y$, is called a binomial (or) a binomial expression.

The binomial theorem is a formula for the powers of a binomial.

By actual multiplication and simplification it is clear that

$$(x + y)^4 = (x + y)^2\ (x + y)^2 = (x^2 + 2xy + y^2)\ (x^2 + 2xy + y^2)$$
$$= x^4 + 4x^3y + 6x^2y^2 + 4xy^3 + y^4.$$

The expansion on the right side can be written with the help of the knowledge of combinations as

$$(x + y)^4 = C(4, 0)\ x^4 + C(4, 1)\ x^3y + C(4, 2)\ x^2y^2 + C(4, 3)\ xy^3 + C(4, 4)\ y^4.$$

Note 1. The expansion of this expression is the product of four factors each equal to $x + y$. The expansion is a homogeneous expression of 4th degree, since each term is a product of x and y whose sum of the powers is equal to 4.

Note 2. The coefficients of successive terms, beginning from the first, are the number of combinations formed by selecting r (= 0, 1, 2, 3, 4) at a time from n(= 4) dissimilar things.

Theorem. (The Binomial Theorem) :

If n is a positive integer, then $(x+y)^n = \sum_{r=0}^{n} C(n, r)\, x^{n-r} y^r$

(*i.e.,*) $(x+y)^n = C(n, 0)\, x^n + C(n, 1)\, x^{n-1} y + C(n, 2)\, x^{n-2} y^2 + \ldots\ldots + C(n, r)\, x^{n-r} y^r + \ldots\ldots + C(n, n)\, y^n.$

Proof : We apply induction

If $n = 1$, $(x+y)^n = (x+y)^1 = x + y = C(n, 0)\, x + C(n, n)\, y$

The theorem is true for $n = 1$.

Let us assume that the theorem is true for $n = m$

Then $(x+y)^m = C(m, 0)\, x^m + C(m, 1)\, x^{m-1} y + C(m, 2)\, x^{m-2} y^2 + \ldots\ldots + C(m, r)\, x^{m-r} y^r + \ldots\ldots + C(m, m)\, y^m.$

Next we shall show that the theorem is true for $n = m + 1$.

$$\begin{aligned}(x+y)^{m+1} &= (x+y)\,[C(m, 0)\, x^m + C(m, 1)\, x^{m-1} y + C(m, 2)\, x^{m-2} y^2 + \ldots\ldots + C(m, r)\, x^{m-r} y^r + \ldots\ldots + C(m, m)\, y^m].\\ &= C(m, 0)\, x^{m+1} + (1 + C(m, 1))\, x^m y + (C(m, 1) + C(m, 2))\, x^{m-1} y^2 + \ldots\ldots + (C(m, r-1) + C(m, r))\, x^{m-r+1} y^r + \ldots\ldots + C(m, m)\, y^{m+1}\\ &= C(m+1, 0)\, x^{m+1} + (C(m, 0) + C(m, 1))\, x^m y + (C(m, 1) + C(m, 2))\, x^{m-1} y^2 + \ldots\ldots + (C(m, r-1) + C(m, r))\, x^{m-r+1} y^r + \ldots\ldots + C(m+1, m+1)\, y^{m+1}.\end{aligned}$$

Since $C(m, 0) = 1 = C(m+1, 0)$, $C(m, m) = 1 = C(m+1, m+1)$

$\therefore$ $(x+y)^{m+1} = C(m+1, 0)\, x^{m+1} + C(m+1, 1)\, x^m y + C(m+1, 2)\, x^{m-1} y^2 + \ldots\ldots + C(m+1, r)\, x^{m+1-r} y^r + C(m+1, m+1)\, y^{m+1}.$

Hence the theorem is true for $n = m + 1$.

By finite mathematical induction, the theorem is true for all positive integers n.

Note 1. The expansion $(x+y)^n$ contains $n + 1$ terms.

2. The sum of the powers of x and y in each term is equal to n.

3. the $r + 1^{\text{th}}$ term of the expansion $(x+y)^n$ is $C(n, r)\, x^{n-r} y^r$, which is called the general term and it is denoted by T_{r+1}.

4. The integers $C(n, 0)$, $C(n, 1)$, $C(n, 2)$, ..., $C(n, n)$ are called the binomial coefficients of the expansion $(x+y)^n$.

Simply they are denoted by C_0, C_1, C_2, ..., C_n.

Note 5. The expansion $(x+y)^n$ and $(y+x)^n$ are equal.

6. If we take $y = 1$ in the expansion $(y+x)^n$, we have

$$(1+x)^n = C(n, 0) + C(n, 1)\, x + C(n, 2)\, x^2 + \ldots\ldots + C(n, r)\, x^r + \ldots\ldots + C(n, n)\, x^n.$$

7. The expansion $(y-x)^n$ is $C(n, 0)\, y^n - C(n, 1)\, y^{n-1} x + C(n, 2)\, y^{n-2} x^2 - \ldots\ldots + (-1)^r C(n, r)\, y^{n-r} x^r + \ldots\ldots + (-1)^n x^n$.

8. If we take $y = 1$ in the expansion of $(y-x)^n$, we get $(1-x)^n = C(n, 0) - x \,.\, C(n, 1) + x^2\, C(n, 2) - \ldots\ldots + (-1)^n x^n$.

Example. *Expand* $(x + y)^7$

Solution. $(x+y)^7 = C(7, 0)\, x^7 + C(7, 1)\, x^6y + C(7, 2)\, x^5y^2 + C(7, 3)\, x^4y^3 + C(7, 4)\, x^3y^4 + C(7, 5)\, x^2y^5 + C(7, 6)\, xy^6 + C(7, 7)\, y^7 = x^7 + 7x^6y + 21x^5y^2 + 35x^4y^3 + 35x^3y^4 + 21x^2y^5 + 7xy^6 + y^7$

Example. *Expand* $(2a + 5b)^6$ *by using binomial theorem.*

Solution. $(2a+5b)^6 = C(6, 0)\,(2a)^6 + C(6, 1)\,(2a)^5(5b) + C(6, 2)\,(2a)^4(5b)^2 + C(6, 3)\,(2a)^3(5b)^3 + C(6, 4)\,(2a)^2(5b)^4 + C(6, 5)\,(2a)(5b)^5 + C(6, 6)\,(5b)^6.$

$= 2^6 a^6 + 6 . 2^5 . 5 a^5b + 15 . 2^4 . 5^2 a^4b^2 + 20 . 2^3 . 5^3 a^3b^3 + 15 . 2^2 . 5^4 a^2b^4 + 6 . 2 . 5^5 ab^5 + 5^6 . b^6.$

Example. *Find the middle term (s) of (i)* $\left(2x - \frac{1}{3x}\right)^{10}$ *(ii)* $\left(x - \frac{3}{y}\right)^9$.

Solution. (*i*) Expansion of $\left(2x - \frac{1}{3x}\right)^{10}$ contains 11 terms.

Middle term is 6th term.

The $r + 1^{\text{th}}$ term in the expansion is $T_{r+1} = C(n, r)\,(2x)^{n-r}\left(-\frac{1}{3x}\right)^r$

Put $\quad r = 5, \quad n = 10$

$$\Rightarrow \quad T_{5+1} = C(10, 5)\,(2x)^5\left(-\frac{1}{3x}\right)^5.$$

$$T_6 = \frac{-896}{27}.$$

(*ii*) Expansion of $\left(x - \frac{3}{y}\right)^9$ contains 10 terms.

Then middle terms are 5th, 6th terms.

$$T_{r+1} = C(n, r)\,(x)^{n-r}\left(-\frac{3}{y}\right)^r$$

Put $\quad r = 4, n = 9$ in T_{r+1}, we get

$$T_{4+1} = C(9, 4)\,(x)^5\left(-\frac{3}{y}\right)^4.$$

$$T_5 = 10206\,\frac{x^5}{y^4}$$

and put $\quad r = 5, n = 9$ in T_{r+1}, we get

$$T_{5+1} = C(9, 5)\,(x)^4 \left(-\frac{3}{y}\right)^5$$

$$T_6 = -30618\,\frac{x^4}{y^5}$$

Example. *Find the coefficient of $x^9 y^3$ in the expansion of $(2x - 3y)^{12}$.*

Solution. By the binomial theorem, $(2x - 3y)^{12} = \sum_{r=0}^{12} \binom{12}{r} (2x)^r (-3y)^{12-r}$

$$= \sum_{r=0}^{12} \binom{12}{r} 2^r (-3)^{12-r} x^r y^{12-r}$$

In this expansion, the coefficient of $x^9 y^3$ (by taking $r = 9$) is $\binom{12}{9} 2^9 (-3)^3 = \frac{+12!}{9!3!}(-1)\,2^9 . 3^3$

$= -1946.$

Example. *Find the term independent of x in the expansion of $\left(x^2 + \frac{1}{x}\right)^{12}$.*

Solution. We write $\left(x^2 + \frac{1}{x}\right)^{12} = (x^2 + x^{-1})^{12}$.

The general term in the expansion of $(x^2 + x^{-1})^{12}$ is

$$t_{r+1} = \binom{12}{r} \left(x^2\right)^{12-r} \left(x^{-1}\right)^r$$

$$= \binom{12}{r} x^{24-2r}\, x^{-r}$$

$$= \binom{12}{r} x^{24-3r}$$

Take $24 - 3r = 0 \Rightarrow 3r = 24 \Rightarrow r = 8.$

Hence the coefficient of x^0 is $\binom{12}{8}$.

The term independent of x is $\binom{12}{8} = 495.$

Example. *Find the two successive terms in the expansion of $(1 + x)^{24}$ whose coefficients are in the ratio 4 : 1.*

Solution. Let T_r and T_{r+1} be two successive terms.

$$T_{r+1} = C(24, r)\, x^r,\ T_r = c(24, r-1)\, x^{r-1}$$

The coefficients of the two successive terms are respectively C (24, r – 1) and C (24, r),

then $$\frac{C(24, r-1)}{C(24, r)} = \frac{4}{1}$$

$$\Rightarrow \quad \frac{(24)!}{(r-1)!\,(24-r+1)!} \times \frac{r!\,(24-r)!}{24!} = 4$$

$$\Rightarrow \quad \frac{r}{24-r+1} = 4$$

$$\Rightarrow \quad r = 100 - 4r$$

$$\Rightarrow \quad 5r = 100$$

$$\Rightarrow \quad r = 20.$$

5.10 PROPERTIES OF THE BINOMIAL COEFFICIENTS (IDENTITIES)

1. Prove that C (n, 0) + C (n, 1) + C (n, 2) + ... + C (n, n) = 2^n

(or) $C_0 + C_1 + C_2 + \ldots\ldots + C_n = 2^n$.

Proof : We have $(1 + x)^n = C(n, 0) + C(n, 1)\, x + C(n, 2)\, x^2 + \ldots\ldots + C(n, n)\, x^n$.

Put $x = 1$, we get

$$2^n = C(n, 0) + C(n, 1) + C(n, 2) + \ldots\ldots + C(n, n)$$

(*i.e.,*) $C_0 + C_1 + C_2 + \ldots\ldots + C_n = 2^n$.

2. Prove that C (n, 0) + C (n, 2) + C (n, 4) + ...

$$= C(n, 1) + C(n, 3) + C(n, 5) + \ldots\ldots = 2^{n-1}$$

(or) $C_0 + C_2 + C_4 + \ldots\ldots = C_1 + C_3 + C_5 + \ldots\ldots = 2^{n-1}$.

Proof : We have $(1 + x)^n = C(n, 0) + C(n, 1)\, x + C(n, 2)\, x^2 + \ldots\ldots + C(n, n)\, x^n$.

Put $x = -1$, we get

$$0 = C(n, 0) - C(n, 1) + C(n, 2) - C(n, 3) + \ldots\ldots$$

$$\Rightarrow \quad C(n, 0) + C(n, 2) + (C(n, 4) + \ldots\ldots = C(n, 1) + C(n, 3) + \ldots\ldots$$

$$= \frac{1}{2}\left[C(n, 0) + C(n, 1) + C(n, 2) + C(n, 3) + \ldots\ldots\right]$$

$$= \frac{1}{2}(2^n) \text{ (by Identity 1)}$$

$$= 2^{n-1}.$$

Therefore, C (n, 0) + C (n, 2) + C (n, 4) + ... = C (n, 1) + C (n, 3) + C (n, 5) + ...

$$= 2^{n-1}$$

i.e., $C_0 + C_2 + C_4 + ... = C_1 + C_3 + C_5 + ... = 2^{n-1}$.

3. Prove that :

(*i*) $C(n, 1) + 2C(n, 2) + 3C(n, 3) + ... + n\,C(n, n) = n \,.\, 2^{n-1}$ (*i.e.,*) $C_1 + 2 \,.\, C_2 + 3 \,.\, C_3 + ...$ $+ n \,.\, C_n = n \,.\, 2^{n-1}$.

(*ii*) $C(n, 1) - 2 \,.\, C(n, 2) + 3 \,.\, C(n, 3) - ...... + (-1)^{n-1} \,.\, n\,c(n, n) = 0$ for $n \geq 2$

(*i.e.,*) $C_1 - 2 \,.\, C_2 + 3 \,.\, C_3 - ...... + (-1)^{n-1}\, n \,.\, C_n = 0$ for $n \geq 2$

(*i.e.,*) $$\sum (-1)^{n-1}\, n \,.\, C_n = 0 \text{ for } n \geq 2.$$

Proof : We have $C(n, 0) + C(n, 1)\,x + C(n, 2)\,x^2 + ... + C(n, n)\,x^n$
$= (1 + x)^n$.

Differentiating both sides w.r.t. 'x', we get

$$C(n, 1) + C(n, 2)\,2x + ... + C(n, n)\,nx^{n-1} = n(1 + x)^{n-1} \quad ...(1)$$

Put $x = 1$ in (1), we get

$$C(n, 1) + 2 \,.\, C(n, 2) + ... + n\,C(n, n) = n\,2^{n-1}$$

$$\Rightarrow \quad C_1 + 2 \,.\, C_2 + ... + n\,C_n = n\,2^{n-1}$$

and put $x = -1$ in (1), we get

$$C(n, 1) - 2 \,.\, C(n, 2) + 3 \,.\, C(n, 3) - ...... + C(n, n)\,n(-1)^{n-1} = 0$$

$$\Rightarrow \quad C_1 - 2 \,.\, C_2 + 3 \,.\, C_3 - ...... + (-1)^{n-1}\, n \,.\, C_n = 0$$

$$\Rightarrow \quad \sum (-1)^{n-1}\, n \,.\, C_n = 0 \text{ for } n \geq 2.$$

5. Prove that $a \,.\, C_0 + (a + d) \,.\, C_1 + (a + 2d) \,.\, C_2 + ... + (a + nd) \,.\, C_n = (2a + nd)\, 2^{n-1}$.

Proof : Let $S = a \,.\, C_0 + (a + d) \,.\, C_0 + (a + 2d) \,.\, C_2 + ... + (a + nd) \,.\, C_n \quad ...(1)$

It is also can be written as

$$S = (a + nd) \,.\, C_n + \left(a + \overline{n-1}d\right) \,.\, C_{n-1} + \left(a + \overline{n-2}d\right) \,.\, C_{n-2} + ... + a \,.\, C_0$$

$$\Rightarrow \quad S = (a + nd) \,.\, C_0 + \left(a + \overline{n-1}d\right) \,.\, C_1 + \left(a + \overline{n-2}d\right) \,.\, C_2 + ... + a \,.\, C_n \quad ...(2)$$

Since $C_{n-r} = C_r$

$(-1) + (2) \Rightarrow 2S = (2a + nd) \,.\, C_0 + (2a + nd) \,.\, C_1 + (2a + nd) \,.\, C_2 + ... + (2a + nd) \,.\, C_n$.

$$\Rightarrow \quad 2S = (2a + nd) \,.\, [C_0 + C_1 + C_2 + ... + C_n]$$

$$\Rightarrow \quad 2S = (2a + nd) \,.\, 2^n$$

$$\Rightarrow \quad S = (2a + nd) \,.\, 2^{n-1}$$

$$\Rightarrow \quad a \,.\, C_0 + (a + d) \,.\, C_1 + (a + 2d) \,.\, C_2 + ... + (a + nd) \,.\, C_n = (2a + nd) \,.\, 2^{n-1}.$$

Example. *Show that*

$C_0 + 3 \,.\, C_1 + 5 \,.\, C_2 + ... + (2n + 1) \,.\, C_n = (n + 1) \,.\, 2^n$.

Solution. Let $S = C_0 + 3 \,.\, C_1 + 5 \,.\, C_2 + ... + (2n + 1) \,.\, C_n \quad ...(1)$

It is also can be written as

$$S = (2n + 1) \,.\, C_n + (2n - 1) \,.\, C_{n-1} + (2n - 3) \,.\, C_{n-2} + ... + C_0.$$

$\Rightarrow \qquad S = (2n+1) . C_0 + (2n-1) . C_1 + (2n-3) . C_2 + \ldots\ldots + C_n \qquad \ldots(2)$

Since $\qquad C_{n-r} = C_r.$

$$(1) + (2) \Rightarrow 2S = (2n+2) . C_0 + (2n+2) . C_1 + (2n+2) . C_2 + \ldots\ldots + (2n+2) . C_n$$
$$= (2n+2) [C_0 + C_1 + C_2 + \ldots\ldots + C_n]$$

$\Rightarrow \qquad 2S = (2n+2)\, 2^n = 2\,(n+1)\, 2^n$

$\Rightarrow \qquad S = (n+1) . 2^n$

$\Rightarrow \qquad C_0 + 3 . C_1 + 5 . C_2 + \ldots\ldots + (2n+1) . C_n = (n+1) . 2^n$

(or) we have $\qquad a . C_0 + (a+d) . C_1 + (a+2d) . C_2 + \ldots\ldots + (a+nd) . C_n = (2a+nd) . 2^{n-1}.$

Put $\qquad a = 1, d = 2$, we get

1. $C_0 + 3 . C_1 + 5 . C_2 + \ldots\ldots + (2n+1) . C_n = (2 + n . 2)\, 2^{n-1} = (n+1) . 2^n.$

Prove that $C_0 + \frac{C_1}{2} x + \frac{C_2}{3} x^2 + \ldots\ldots + \frac{C_n}{n+1} x^n = \frac{(1+x)^{n+1} - 1}{(n+1)\,x}$

and hence deduce that

(*i*) $C_0 + \frac{C_1}{2} + \frac{C_2}{3} + \ldots\ldots + \frac{C_n}{n+1} = \frac{2^{n+1} - 1}{n+1}$

(*ii*) $C_0 - \frac{C_1}{2} + \frac{C_2}{3} - \ldots\ldots (-1)^n \frac{C_n}{n+1} = \frac{1}{n+1}.$

(*iii*) $2 . C_0 + 2^2 . \frac{C_1}{2} + 2^3 . \frac{C_2}{3} + \ldots\ldots + 2^{n+1} . \frac{C_n}{n+1} = \frac{3^{n+1} - 1}{n+1}.$

Proof : Consider, $C_0 + \frac{C_1}{2} x + \frac{C_2}{3} x^2 + \ldots\ldots + \frac{C_n}{n+1} x^n$

$$= 1 + \frac{n}{1!} . \frac{x}{2} + \frac{n\,(n-1)}{2!} \frac{x^2}{3} + \ldots\ldots + \frac{1}{n+1} x^n$$

Since $\; C_r = {}^nC_r,\; C_0 = C_n = 1.$

$$= 1 + \frac{n}{2!} x + \frac{n\,(n-1)}{3!} x^2 + \ldots\ldots + \frac{1}{n+1} x^n$$

$$= \frac{1}{(n+1)\,x} \left[\frac{(n+1)\,x}{1!} + \frac{(n+1)\,n}{2!} x^2 + \frac{(n+1)\,n\,(n-1)}{3!} x^3 + \ldots\ldots + x^{n+1} \right]$$

$$= \frac{1}{(n+1)x} \left[C(n+1,1)x + C(n+1,2)x^2 + C(n+1,3)x^3 + \ldots\ldots + C(n+1, n+1)x^{n+1} \right]$$

$$= \frac{1}{(n+1)\,x} \left[\left(1 + C\,(n+1, 1)\,x + C\,(n+1, 2)\,x^2 + \ldots\ldots + C\,(n+1, n+1)\,x^{n+1}\right) - 1 \right]$$

$$= \frac{1}{(n+1)x}\left[(1+x)^{n+1} - 1\right] \qquad ...(1) \ (\because \text{ by taking first 1 as C } (n+1, 0))$$

Hence the result.

(*i*) Put $x = 1$ in (1), we get

$$C_0 + \frac{C_1}{2} + \frac{C_2}{3} + ... + \frac{C_n}{n+1} = \frac{2^{n+1} - 1}{n+1}.$$

(*ii*) Put $x = -1$ in (1), we get

$$C_0 - \frac{C_1}{2} + \frac{C_2}{3} - ......(-1)^n \frac{C_n}{n+1} = \frac{0-1}{(n+1)(-1)} = \frac{1}{n+1}$$

(*iii*) Put $x = 2$ in (1), we get

$$C_0 + \frac{C_1}{2}(2) + \frac{C_2}{3} . 2^2 + ... + \frac{C_n}{n+1} . 2^n = \frac{3^{n+1} - 1}{(n+1)^2}$$

$$\Rightarrow \quad 2 . C_0 + 2^2 . \frac{C_1}{2} + 2^3 . \frac{C_2}{3} + ... + 2^{n+1} . \frac{C_n}{n+1} = \frac{3^{n+1} - 1}{n+1}$$

Vandermonde's Identity:

C $(n, 0)$. C (m, r) + C $(n, 1)$. C $(m, r-1)$ + ... + C (n, r) C $(m, 0)$ = C $(n + m, r)$ for integers $n \geq r \geq 0$ and $m \geq r \geq 0$.

Proof : Consider $(1 + x)^{n+m} = (1 + x)^n (1 + x)^m$

$(1 + x)^{n+m} = C(n+m, 0) + C(n+m, 1)\, x + ... + C(n+m, r)\, x^r + ... + x^{n+m}$.

$\Rightarrow$ The coefficient of x^r in $(1 + x)^{n+m}$ is $c(n + m, r)$ ...(1)

and $\quad (1 + x)^n = C(n, 0) + C(n, 1)\, x + C(n, 2)\, x^2 + ... + C(n, r)\, x^r + ... + x^n$

$\quad (1 + x)^m = C(m, 0) + C(m, 1)\, x + ... + C(m, r-1)\, x^{r-1} + C(m, r)\, x^r + ... + x^m$.

The coefficient of x^r in the product of $(1 + x)^n$, $(1 + x)^m$ is

C $(n, 0)$ C (m, r) + C $(n, 1)$ C $(m, r-1)$ + ... + C (n, r) C $(m, 0)$ ...(2)

From (1), (2) results, we have

$$C(n+m, r) = C(n, 0)\, C(m, r) + C(n, 1)\, C(m, r-1) + ... + C(n, r)\, C(m, 0).$$

5.11 THE MULTINOMIAL THEOREM

The sum of r unlike things, $x_1 + x_2 + ... + x_r$ is called a **multinomial**. In particular $x_1 + x_2$ is a binomial, and $x_1 + x_2 + x_3$ is a trinomial.

The binomial theorem formula can be extended to give a formula for powers of trinomials $(x_1 + x_2 + x_3)^n$ or more generally for powers of multinomials $(x_1 + x_2 + ... + x_r)^n$.

Theorem. (Multinomial theorem) : Let n be a positive integer. Then for all $x_1, x_2, ..., x_r$ we have

$$(x_1 + x_2 + \ldots\ldots + x_r)^n = \sum P(n; q_1, q_2, \ldots\ldots, q_r)\, x_1^{q_1}\, x_2^{q_2} \ldots\ldots x_r^{q_r}$$

Where the summation extends over all sets of non-negative integers $q_1, q_2, \ldots\ldots, q_r$ where $q_1 + q_2 + \ldots\ldots + q_r = n$ and $P(n; q_1, q_2, \ldots\ldots, q_r) = \dfrac{n!}{q_1!\, q_2! \ldots\ldots q_r!}$

Proof : We note that in the expansion of $(x_1 + x_2 + \ldots\ldots + x_r)^n$, the coefficient of $x_1^{q_1}\, x_2^{q_2} \ldots\ldots x_r^{q_r}$ is the number of ways we can select x_1 from q_1 of the n factors, x_2 from q_2 of the remaining $n - q_1$ factors, x_3 from q_3 of $n - q_1 - q_2$ remaining factors, and so on. Therefore, the coefficient of $x_1^{q_1}\, x_2^{q_2} \ldots\ldots x_r^{q_r}$ is

$$C(n, q_1) \,.\, c(n - q_1, q_2) \,.\, C(n - q_1 - q_2, q_3) \ldots\ldots C(n - q_1 - q_2 - \ldots\ldots n_{r-1}, n_r)$$

$$= \frac{n!}{q_1!(n-q_1)!} \cdot \frac{(n-q_1)!}{q_2!(n-q_1-q_2)!} \cdot \frac{(n-q_1-q_2)!}{q_3!(n-q_1-q_2-q_3)!} \ldots\ldots \frac{(n-q_1-q_2-\ldots\ldots-q_{r-1})!}{q_r!(n-q_1-q_2-\ldots\ldots-q_{r-1}-q_r)!}$$

$$= \frac{n!}{q_1!\, q_2!\, q_3! \ldots\ldots q_r!}.$$

Note 1. Alternative statement of Multinomial theorem is the general term in the expansion of $(x_1 + x_2 + x_3 + \ldots\ldots + x_r)^n$ is $\dfrac{n!}{q_1!\, q_2! \ldots\ldots q_r!} \,.\, x_1^{q_1}\, x_2^{q_2} \ldots\ldots x_r^{q_r}$ where $q_1, q_2, \ldots\ldots, q_r$ are non-negative integers not exceeding n and

$q_1 + q_2 + q_3 + \ldots\ldots + q_r = n.$

The expression $\dfrac{n!}{q_1!\, q_2! \ldots\ldots q_r!}$ is called a multinomial coefficient.

Note 2. There are $C(n + r - 1, n)$ terms in the expansion of $(x_1 + x_2 + \ldots\ldots + x_r)^n$.

Example. *Evaluate P (18 ;4, 5, 6, 3)*

Solution. $P(18; 4, 5, 6, 3) = \dfrac{18!}{4!5!6!3!}.$

Example. *Find the coefficient of $x^5\, y^{10}\, z^5\, w^5$ in the expansion of (i) $(x + 7y + 3z + w)^{25}$; and (ii) $(x - 7y + 3z - w)^{25}$.*

Solution. (*i*) By Multinomial theorem, the general term in the expansion of $(x + 7y + 3z + w)^{25}$ is

$$P(25; q_1, q_2, q_3, q_4)\, x^{q_1}\, (7y)^{q_2}\, (3z)^{q_3}\, \omega^{q_4}.$$

Put $q_1 = 5,\ q_2 = 10,\ q_3 = 5,\ q_4 = 5$, we get

The general term is $P(25; 5, 10, 5, 5)\, x^5\, (7y)^{10}\, (3z)^5\, w^5$

$$= \frac{25!}{5!10!5!5!} 1^5 \,.\, 7^{10} \,.\, 3^5 \,.\, 1^5\, x^5\, y^{10}\, z^5\, \omega^5.$$

Hence coefficient of $x^5\ y^{10}\ z^5\ w^5$ is $\dfrac{25!\,7^{10}\,.\,3^5}{10!\,(5!)^3}$.

(*ii*) By Multinomial theorem, the general term in $(x - 7y + 3z - w)^{25}$ is P (25 ; 5, 10, 5, 5) x^5 $(-7y)^{10}\ (3z)^5\ (-w)^5$ (Same as in (*i*))

$$= \text{P}\ (25\ ;\ 5,\ 10,\ 5,\ 5)\ x^5\ (-7)^{10}\ y^{10}\ 3^5\ z^5\ (-1)^5\ w^5.$$

The coefficient of $x^5\ y^{10}\ z^5\ w^5$ in the expansion of

$(x - 7y + 3z - w)^{25}$ is P (25 ; 5, 10, 5, 5) 7^{10} (-1)

$$= (-1)\ .\ 7^{10}\ \frac{25!}{5!\,10!\,5!\,5!}$$

$$= (-1)\ 7^{10}\ \frac{25!}{10!\,(5!)^3}.$$

Example. *Determine the number of terms in the expansion of* $(x - 7y + 3z - w)^{25}$.

Solution. We have the number of terms in the expansion of $(x_1 + x_2 + \ldots\ldots + x_r)^n$ are C $(n + r - 1, n)$.

Here $\qquad r = 4,\ n = 25.$

$$\text{C}\ (n + r - 1,\ n) = \text{C}\ (25 + 4 - 1,\ 25) = \text{C}\ (28,\ 25).$$

Example. *Find the term which contains* x^{11} *and* y^4 *in the expansion of* $(2x^3 + 3xy^2 + z^2)^6$.

Solution. By the multinomial theorem, the general term in the expansion of $(2x^3 + 3xy^2 + z^2)^6$ is

$$\text{C}\ (6\ ;\ q_1,\ q_2,\ q_3)\ \left(2x^3\right)^{q_1}\left(3xy^2\right)^{q_2}\left(z^2\right)^{q_3}$$

$$= \text{C}\ (6\ ;\ q_1,\ q_2,\ q_3)\ 2^{q_1}\ 3^{q_2}\ x^{3q_1 + q_2}\ y^{2q_2}\ z^{2q_3} \qquad \ldots(1)$$

For getting required term, choose $3q_1 + q_2 = 11,\ 2q_2 = 4$

$\Rightarrow \qquad q_2 = 2,\ q_1 = 3$

Since $\qquad q_1 + q_2 + q_3 = 6,\ q_3 = 1.$

Hence the coefficient of $x^{11}\ y^4\ z$ is

$$\text{C}\ (6\ ;3,\ 2,\ 1)\ 2^3\ 3^2\ \text{(From (1))}$$

$$= \frac{6!}{3!\,2!\,1!}\,.\,8\,.\,9 = 4320.$$

Example. *Find the number of distinct terms in the expansion of* $(x_1 + x_2 + x_3 + x_4 + x_5)^{16}$.

Solution. Each term in the expansion is of the form (by multinomial theorem) is

P (16 ; $n_1,\ n_2,\ n_3,\ n_4,\ n_5$) $x_1^{n_1}\ x_2^{n_2}\ x_3^{n_3}\ x_4^{n_4}\ x_5^{n_5}$.

Where each n_i is a non-negative integer, and these n_i 's sum to 16.

Therefore, the number of distinct terms in the expansion is precisely equal to the number of non-negative integer solutions of the equation $n_1 + n_2 + n_3 + n_4 + n_5 = 16$.

This number is C (5 – 1 + 16, 16) = C (20, 16) = 4845.

EXERCISE 5.4.

1. Expand (*i*) $(x + y)^5$; (*ii*) $(2x - 3y)^7$.
2. Find 10th term of $\left(\frac{a}{2b} - \frac{2b}{a^2}\right)^{14}$
3. Find middle term of $\left(x + \frac{1}{x}\right)^{20}$
4. Find the coefficient of

   (*i*) x^8 in $\left(3x^2 + \frac{5}{x^3}\right)^9$

   (*ii*) x^{-6} in $\left(x^4 - \frac{1}{x^2}\right)^{15}$.
5. If the coefficient of x in the expansion of $\left(x^2 + \frac{k}{x}\right)^5$ is 270, find the value of k.
6. Find the term independent of x in the expansion of (*i*) $\left(x^2 - \frac{1}{x}\right)^9$; and (*ii*) $\left(\frac{4x^2}{3} - \frac{3}{2x}\right)^9$
7. If the coefficient of the rth term is to that of $r + 1^{th}$ term in the expansion of $(1 + x)^{20}$ is the ratio 1 : 2 find the value of r.
8. Find the coefficient of $x_1^2\ x_3\ x_4^3\ x_5^4$ in the expansion of $(x_1 + x_2 + x_3 + x_4 + x_5)^{10}$.
9. Find the coefficient of $x_1^3\ x_2^2\ x_3^2\ x_5^3$ in $(x_1 + x_2 + x_3 + x_4 + x_5)^{10}$.
10. Find the coefficient of $x^3y^3z^2$ in $(2x - 3y + 5z)^8$.
11. Prove that C $(n + 3, r)$ – 3C $(n + 2, r)$ + 3C $(n + 1, r)$ – C (n, r) = C $(n, r - 3)$.
12. Find relations by equating the coefficients of x^k in the following :

    (*i*) $(1 + x)^{n+1} = (1 + x)$ (C $(n, 0)$ + C $(n, 1)$ x + ... + C (n, n) x^n)

    (*ii*) $(1 + x)^{n+2} = (1 + x^2 + 2x)$ (C $(n, 0)$ + C $(n, 1)$ x + ... + C (n, n) x^n).
13. Determine the number of terms in the expansion of $(x_1 - 2x_2 + 3x_3 + x_4 - x_5)^{20}$.

ANSWERS

1. (*i*) $x^5 + 5x^4y + 10x^3y^2 + 10x^2\ y^3 + 5xy^4 + y^5$

   (*ii*) $(2x)^7 - 7\ (2x)^6\ (3y) + 21\ (2x)^5\ (3y)^2 - 35\ (2x)^4\ (3y)^3 + 35\ (2x)^3\ (3y)^4 - 21\ (2x)^2\ (3y)^5 + 7\ (2x)\ (3y)^6 - (3y)^7$
2. – C (14, 5) . $2^4 . \frac{b^4}{a^{13}}$.
4. (*i*) C (9, 2) . $3^7 . 5^2$ (*ii*) – C (15, 4)
5. 3
6. (*i*) C (9, 3) (*ii*) C (9, 3) 3^3
7. 7
8. 12600

9. $P(10\,;\,2, 1, 3, 4) = \dfrac{10!}{2!1!3!4!}$

10. $2^3\,(-3)^3 \,.\, 5^2 \,.\, 560.$

11. (*i*) $C(n+1, k) = C(n, k-1) + C(n, k)$

(*ii*) $C(n+2, k) = C(n, k) + 2C(n, k-1) + C(n, k-2)$

13. $C(24, 20)$.

5.12 THE PRINCIPLE OF INCLUSION-EXCLUSION

In previous section we discussed the sum rule by which we can count the number of elements in the union of disjoint sets. However, if the sets are not disjoint we must refine the statement of the sum rule to a rule commonly called the **principle of inclusion-exclusion** (it is also called the **Sieve method**).

Theorem. (The **principle of inclusion-exclusion for two sets**)

If A and B are subsets of the Universal Set U,

Then $|A \cup B| = |A| + |B| - |A \cap B|$.

Proof : We prove this result by Venn diagram.

Let A, B are subsets of the universe set U. Since in counting the elements of A and the elements of B, we have counted the elements of $A \cap B$ twice. But $A \cup B$ is the union of the three disjoint sets $A \cap \bar{B}$, $A \cap B$, $\bar{A} \cap B$.

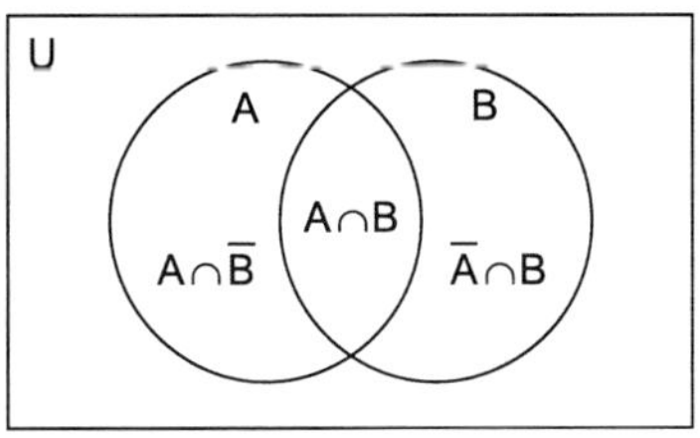

Figure 5.5.

By the sum rule, $|A \cup B| = |A \cap \bar{B}| + |A \cap B| + |\bar{A} \cap B|$ ...(1)

Since $(A \cap \bar{B}) \cup (A \cap B) = A$ and

$(\bar{A} \cap B) \cup (A \cap B) = B$, $|A| = |A \cap \bar{B}| + |A \cap B|$ and

$$|B| = |\bar{A} \cap B| + |A \cap B|$$

$$\Rightarrow \quad |A| + |B| = |A \cap \bar{B}| + |A \cap B| + |A \cap B| + |A \cap B|$$

$$= |A \cup B| + |A \cap B| \qquad \text{(From (1))}$$

$$\Rightarrow \quad |A \cup B| = |A| + |B| - |A \cap B|.$$

Note. If $A \cap B = \phi$, then this is the sum rule.

Example. *A sample of 80 people revealed that 25 like cinema and 60 like television programmes. Find the number of people who like both cinema and television programmes.*

Solution. Let A be the set of people who like cinema and B be the set of people who like television programmes.

$$\Rightarrow \qquad |A| = 25, |B| = 60, |A \cup B| = 80$$

Principle of inclusion-exclusion is

$$|A \cup B| = |A| + |B| - |A \cap B|.$$

$$80 = 25 + 60 - |A \cap B|$$

$$\Rightarrow \qquad |A \cap B| = 85 - 80 = 5.$$

Example. *A certain computer center employs 100 computer programmers of these 47 can program in FORTRAN, 35 in* ***pascal*** *and 23 can program in both languages. How many can program in neither of these 2 languages ?"*

Solution. Let A_1 be the set of programmers in FORTRAN language, and let A_2 be the set of programmes in Pascal language.

$$\Rightarrow \qquad |A_1| = 47, |A_2| = 35, |A_1 \cap A_2| = 23.$$

$\therefore$ by the principle of inclusion-exclusion,

$$|A_1 \cup A_2| = |A_1| + |A_2| - |A_1 \cap A_2|$$

$$= 47 + 35 - 23 = 59.$$

$\therefore$ The number of programmers, who can program in either FORTRAN (or) pascal = 59.

Hence the number of programmers, who can program in neither FORTRAN nor pascal = 100 − 59 = 41.

Example. *From a group of 10 professors how many ways can a committee of 5 members be formed so that atleast one of professor A and professor B will be included ?*

Solution. Here we apply the principle of inclusion-exclusion. Let A_1 and A_2 be the set of committees that include professor A and professor B, respectively.

$$\therefore \qquad |A_1| = C(9, 4) = 126 = |A_2| \text{ and } |A_1 \cap A_2| = C(8, 3) = 56.$$

$$\therefore \qquad |A_1 \cup A_2| = |A_1| + |A_2| - |A_1 \cap A_2| = 126 + 126 - 56 = 196.$$

Theorem. If A_1, A_2, A_3 are finite sets, then $|A_1 \cup A_2 \cup A_3| = |A_1| + |A_2| + |A_3| - |A_1 \cap A_2| - |A_1 \cap A_3| - |A_2 \cap A_3| + |A_1 \cap A_2 \cap A_3|$.

Note. $\left|\overline{A_1} \cap \overline{A_2} \cap \overline{A_3}\right| = \left|\overline{A_1 \cup A_2 \cup A_3}\right| = |U| - |A_1 \cup A_2 \cup A_3|$ by De-Morgan's Law.

Theorem. (Generalized statement of the principle of inclusion-exclusion) : If A_i are finite subsets of a universal set U, then $|A_1 \cup A_2 \cup \ldots\ldots \cup A_n| = \sum_{i=1}^{n} |A_i| - \sum_{i,j} \left|A_i \cap A_j\right| + \sum_{i,j,k} \left|A_i \cap A_j \cap A_k\right| + \ldots\ldots + (-1)^{n-1} |A_1 \cap A_2 \cap \ldots\ldots, \cap A_n|$.

Where the second summation is taken over all 2-combinations $\{i, j\}$ of the integers $\{1, 2, \ldots\ldots, n\}$, the third summation is taken over all 3-combinations $\{i, j, k\}$ of $\{1, 2, 3, \ldots\ldots, n\}$ and so on.

Cor : If A_1, A_2, A_3, A_4 are finite sets, then $|A_1 \cup A_2 \cup A_3 \cup A_4| = |A_1| + |A_2| + |A_3| + |A_4| - |A_1 \cap A_2| - |A_1 \cap A_3| - |A_1 \cap A_4| - |A_2 \cap A_3| - |A_2 \cap A_4| - |A_3 \cap A_4| + |A_1 \cap A_2 \cap A_3| + |A_1 \cap A_2 \cap A_4| + |A_1 \cap A_3 \cap A_4| + |A_2 \cap A_3 \cap A_4) - |A_1 \cap A_2 \cap A_3 \cap A_4|$.

Note 1. In general there are C $(n, 1)$ + C $(n, 2)$ + C $(n, 3)$ + ... + C $(n, n) = 2^{n-1}$ terms on the right side of generalized statement of the principle of inclusion-exclusion for n-subsets of U.

Note 2. $|\overline{A_1} \cap \overline{A_2} \cap \ldots\ldots \cap \overline{A_n}| = |\overline{A_1 \cup A_2 \cup \ldots\ldots \cup A_n}| = |U| - |A_1 \cup A_2 \cup \ldots\ldots \cup A_n|$.

Example. *In a language survey of students it is formed that 80 students know English, 60 know French, 50 know German, 30 know English and French, 20 know French and German, 15 know English and German and 10 students know all the three languages. How many students know*

(i) at least one language

(ii) English only

(iii) French and one but not both out of English and German

(iv) at least two languages ?

Solution. Let E, F, G denote, the sets of students knowing English, French and German respectively.

Method I : (By using Venn diagram).

The various regions in Venn diagram represent the various subsets obtained by taking intersections of E, F, G and their complements. For example, the shaded region in Figure 5.6 represents $G \cap E' \cap F'$ where′ stands for complementation. Now in each region we go on putting its cardinality. We start with $E \cap F \cap G$ which is given to contain 10 elements.

Since $E \cap F$ is the disjoint union of $E \cap F \cap G$ and $E \cap F \cap G'$

and $|E \cap F| = 30$, it follows that $|E \cap F \cap G'| = 20$.

Since $F \cap G$ is the disjoint union of $F \cap G \cap E$ and $F \cap G \cap E'$

and $|F \cap G| = 20$, it follows that $|F \cap G \cap E'| = 10$.

Since $|E \cap F \cap G| = 10$

and since $E \cap G$ is the disjoint union of $E \cap G \cap F$ and $E \cap G \cap F'$ and $|E \cap G| = 15$, it follows that $|E \cap G \cap F'| = 5$

Since $|E \cap F \cap G| = 10$.

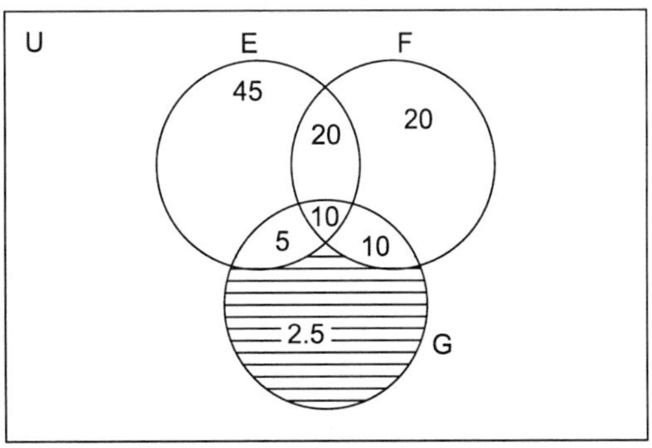

Figure 5.6.

Hence by Venn diagram,

(*i*) The number of students know at least one language = 45 + 20 + 25 + 20+ 10 + 5 + 10 = 135.

(*ii*) The number of students know only English = 45.

(*iii*) The number of students know French and one but not both out of English and German.

$$= 20 + (20 - 10) + (10 - 10) = 30.$$

(*iv*) The number of students know at least two languages = 20 + 10 + 5 + 10 = 45.

Method II : Here we apply the principle of inclusion-exclusion.

We have $|E| = 80, |F| = 60, |G| = 50,$

$$|E \cap F| = 30, |F \cap G| = 20, |G \cap E| = 15, |E \cap F \cap G| = 10$$

(*i*) The number of students know at least one language is

$$|E \cup F \cup G| = |E| + |F| + |G| - |E \cap F| - |F \cap G| - |G \cap E| + (E \cap F \cap G)$$
$$= 80 + 60 + 50 - 30 - 20 - 15 + 10 = 135.$$

(*ii*) Let $S_1 = E \cap F, S_2 = E \cap G \Rightarrow S_1 \cap S_2 = E \cap F \cap G$

$$\Rightarrow \quad |S_1 \cup S_2| = |S_1| + |S_2| - |S_1 \cap S_2|$$
$$= 30 + 15 - 10 = 35$$
$$\Rightarrow \quad |E - (S_1 \cup S_2)| = 80 - 35 = 45.$$

The number of students know only English is 45.

(*iii*) Let $S_1 = F \cap E, S_2 = F \cap G \Rightarrow S_1 \cap S_2 = E \cap F \cap G$

$$\Rightarrow \quad |S_1 \cup S_2| = |S_1| + |S_2| - |S_1 \cap S_2| = 30 + 20 - 10 = 40.$$
$$\Rightarrow \quad |(S_1 \cup S_2) - (S_1 \cap S_2) = |S_1 \cup S_2| - |S_1 \cap S_2| = 40 - 10 = 30.$$

∴ The number of students know French and one but not both English and German = 30.

(*iv*) Let $S_1 = E \cap F, S_2 = F \cap G, S_3 = E \cap G \Rightarrow S_1 \cap S_2 \cap S_3 = E \cap F \cap G$ and $S_1 \cap S_2 = S_2 \cap S_3 = S_3 \cap S_1 = E \cap F \cap G.$

Then
$$|S_1 \cup S_2 \cup S_3| = |S_1| + |S_2| + |S_3| - |S_1 \cap S_2| - |S_2 \cap S_3| - |S_3 \cap S_1| + |S_1 \cap S_2 \cap S_3|$$
$$= 30 + 20 + 15 - 10 - 10 - 10 + 10$$
$$= 45.$$

Hence the number of students know at least two languages is 45.

Example. *Find the number of integers between 1 and 250 which are divisible by any of the integers 2, 3, 5 or 7 and hence find the number of integers between 1, 250 which are not divisible by 2, 3, 5 or 7.*

Solution. Let U be the set of integers x such that $1 \le x \le 250$.

Let A_1, A_2, A_3, A_4 be the set of elements of U divisible by 2, 3, 5 and 7 respectively. Thus $A_1 \cap A_2$ denotes those positive integers ≤ 250 that divisible by $2 \times 3 = 6$, $A_1 \cap A_3$ those divisible by $2 \times 5 = 10$, $A_2 \cap A_3$ those divisible by $3 \times 5 = 15$ and $A_1 \cap A_4 = \{$all integers $x \,/\, 1 \le x \le 250$ and x is divisible by 14$\}$.

We have
$$|A_1| = \left\lfloor \frac{250}{2} \right\rfloor = 125, \; |A_2| = \left\lfloor \frac{250}{3} \right\rfloor = 83, \; |A_3| = \left\lfloor \frac{250}{5} \right\rfloor = 50,$$
$$|A_4| = \left\lfloor \frac{250}{7} \right\rfloor = 35, \; |A_1 \cap A_2| = \left\lfloor \frac{250}{6} \right\rfloor = 41,$$

$$|A_1 \cap A_3| = \left\lfloor \frac{250}{10} \right\rfloor = 25,\ |A_1 \cap A_4| = \left\lfloor \frac{250}{14} \right\rfloor = 17,$$

$$|A_2 \cap A_3| = \left\lfloor \frac{250}{15} \right\rfloor = 25,\ |A_2 \cap A_4| = \left\lfloor \frac{250}{21} \right\rfloor = 11,$$

$$|A_3 \cap A_4| = \left\lfloor \frac{250}{35} \right\rfloor = 7,\ |A_1 \cap A_2 \cap A_3| = \left\lfloor \frac{250}{2\times 3\times 5} \right\rfloor = 8,$$

$$|A_1 \cap A_2 \cap A_4| = \left\lfloor \frac{250}{2\times 3\times 7} \right\rfloor = 5,\ |A_1 \cap A_3 \cap A_4| = \left\lfloor \frac{250}{2\times 5\times 7} \right\rfloor = 3,$$

$$|A_2 \cap A_3 \cap A_4| = \left\lfloor \frac{250}{3\times 5\times 7} \right\rfloor = 2 \text{ and } |A_1 \cap A_2 \cap A_3 \cap A_4| = \left\lfloor \frac{250}{2\times 3\times 5\times 7} \right\rfloor = 1$$

(where $\lfloor x \rfloor$ means the greatest integer $\leq x$).

Then the number of integers between 1 and 250, which are divisible by any of the integers 2, 3, 5, 7 are $|A_1 \cup A_2 \cup A_3 \cup A_4) = |A_1| + |A_2| + |A_3| + |A_4| - |A_1 \cap A_2| - |A_1 \cap A_3| - |A_1 \cap A_4| - |A_2 \cap A_3| - |A_2 \cap A_4| - |A_3 \cap A_4 + \cap A_3| + |A_1 \cap A_3 \cap A_4| + |A_1 \cap A_2 \cap A_4| + |A_2 \cap A_3 \cap A_4| - |A_1 \cap A_2 \cap A_3 \cap A_4|$

$$= 125 + 83 + 50 + 35 - 41 - 25 - 17 - 25 - 11 - 7 + 8 + 5 + 3 + 2 - 1 = 193.$$

and the number of integers between 1 and 250, which are not divisible by 2, 3, 5, or 7 are $250 - 193 = 57$.

Example. *In a survey of 800 voters, the following information was found : 300 were college educated, 260 were from high-income families, 325 were registered Demo-crats, 184 were college educated and from high-income families, 155 were college educated and registered Democrats, 165 were from high-income families and were registered Democrats, 94 were college educated, from high-income families, and were registered Democrats.*

Let E = {*Voters who were college educated*}

I = {*Voters who were from high-income families*}

D = {*Voters who were registered Democrats*}.

Draw a Venn diagram and list the number of elements in the 8 different regions of the diagram.

Solution. Given $|U| = 800$, where U = {Voters in a survey},

$$|E| = 300,\ |I| = 260,\ |D| = 325,\ |E \cap I| = 184,\ |E \cap D| = 155,$$
$$|I \cap D| = 165,\ |E \cap I \cap D| = 94.$$

Now in each region of Venn diagram, we go on putting its cardinality we start with $E \cap I \cap D$ which is given to contain 94 elements.

(*i*) Since $E \cap I$ is the disjoint union of $E \cap I \cap D$ and $E \cap I \cap D'$ and $|E \cap I| = 184$, it follows that $|E \cap I \cap D'| = 184 - 94 = 90$.

(*ii*) Since $I \cap D$ is the disjoint union of $E \cap I \cap D$ and $E' \cap I \cap D$, and $|I \cap D| = 165$, it follows that $|E' \cap I \cap D| = |I \cap D| - |E \cap I \cap D| = 165 - 94 = 71$

(*iii*) and Since $E \cap D$ is the disjoint union of $E \cap I \cap D$ and $E \cap I' \cap D$, and $|E \cap D| = 155$, it follows that $|E \cap I' \cap D| = |E \cap D| - |E \cap I \cap D| = 155 - 94 = 61$.

(*iv*) The number of voters who were only college educated

$$= 300 - (90 + 61 + 94)$$
$$= 300 - 245 = 55.$$

(*v*) The number of voters who were only from high-income families

$$= 260 - (90 + 71 + 94)$$
$$= 260 - 255 = 5.$$

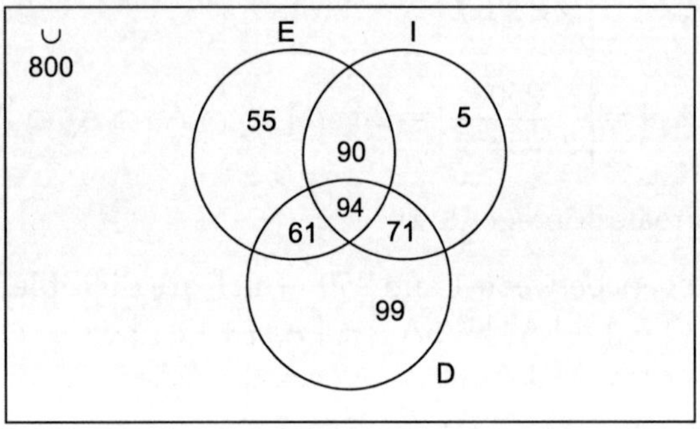

Figure 5.7.

(*vi*) The number of voters who were only from registered Democrates $= 325 - (61 + 71 + 94)$
$= 325 - 226 = 99.$

∴ The number of elements in the 8 different regions of the Venn diagram are shown above Figure 5.7.

Example. *In how many ways can the letters {5 . a, 4 . b, 3 . c} be arranged so that all the letters of the same kind are not in a single block ?*

Solution. Let U be the set of $\frac{12!}{5!4!3!}$ permutations of these letters. Let A_1 be the arrangements of the letters where the 5a's are in a single block, A_2 be the arrangements where the 4*b*'s are in a single block, and A_3 be the arrangements where the 3c's are in one block. Then,

$$|A_1| = \frac{8!}{4!3!}, |A_2| = \frac{9!}{5!3!}, |A_3| = \frac{10!}{5!4!},$$

$$|A_1 \cap A_2| = \frac{5!}{3!}, |A_2 \cap A_3| = \frac{7!}{5!}, |A_1 \cap A_3| = \frac{6!}{4!}$$

and $\quad |A_1 \cap A_2 \cap A_3| = 3!$

∴ $\quad |\overline{A_1} \cap \overline{A_2} \cap \overline{A_3}| = |U| - (|A_1| + |A_2| + |A_3|) + (|A_1 \cap A_2| + |A_2 \cap A_3| + |A_1 \cap A_3|) - |A_1 \cap A_2 \cap A_3|$

$$= \frac{12!}{5!4!3!} - \left(\frac{8!}{4!3!} + \frac{9!}{5!3!} + \frac{10!}{5!4!}\right) + \left(\frac{5!}{3!} + \frac{7!}{5!} + \frac{6!}{4!}\right) - 3!$$

Example. *Count the number of integral solutions to $x_1 + x_2 + x_3 = 20$ where $2 \le x_1 \le 5$, $4 \le x_2 \le 7$, $-2 \le x_3 \le 9$.*

Solution. Let U be the set of solutions (x_1, x_2, x_3) where $2 \le x_1, 4 \le x_2, -2 \le x_3$

We know that $|U| = C(20 - 2 - 4 + 2 + 3 - 1, 3 - 1)$

$= C(18, 2)$.

Let $A_1 = \{(x_1, x_2, x_3) \in U / x_1 \ge 6\}$,

$A_2 = \{(x_1, x_2, x_3) \in U \mid x_2 \ge 8\}$, $A_3 = \{(x_1, x_2, x_3) \in U / x_3 \ge 10\}$.

We wish to count the number of elements in

$$\overline{A_1} \cap \overline{A_2} \cap \overline{A_3} = \left(\overline{A_1 \cup A_2 \cup A_3}\right).$$

By the principle of inclusion-exclusion, $\left|\overline{A_1 \cup A_2 \cup A_3}\right| = |U| - |A_1 \cup A_2 \cup A_3| = |U| - [|A_1| + |A_2| + |A_3| - |A_1 \cap A_2| - |A_1 \cap A_3| - |A_2 \cap A_3| + |A_1 \cap A_2 \cap A_3|]$

Now A_1 is the set of solutions of given equation

where $6 \le x_1, 4 \le x_2 - 2 \le x_3$.

Thus, $|A_1| = C(20 - 6 - 4 + 2 + 3 - 1, 3 - 1) = C(14, 2)$.

Similarly, A_2 is the set of solutions of given equation $x_1 + x_2 + x_3 = 20$, where $2 \le x_1, 8 \le x_2, -2 \le x_3$.

Therefore, $|A_2| = C(20 - 2 - 8 + 2 + 3 - 1, 3 - 1) = C(14, 2)$

and A_3 is the set of solutions of $x_1 + x_2 + x_3 = 20$ where $2 \le x_1, 4 \le x_2, 10 \le x_3$.

Therefore, $|A_3| = C(20 - 2 - 4 - 10 + 3 - 1, 3 - 1) = C(6, 2)$.

Now $A_1 \cap A_2$ is the set of solutions where $6 \le x_1, 8 \le x_2 - 2 \le x_3$.

Thus $|A_1 \cap A_2| = C(20 - 6 - 8 + 2 + 3 - 1, 3 - 1) = C(10, 2)$.

Likewise $|A_1 \cap A_3| = C(20 - 6 - 4 - 10 + 3 - 1, 3 - 1) = C(2, 2)$

and $|A_2 \cap A_3| = C(20 - 2 - 8 - 10 + 3 - 1, 3 - 1) = C(2, 2)$.

$|A_1 \cap A_2 \cap A_3| = 0$ since 20 does not exceed $6 + 8 + 10 = 24$.

Therefore, $\left|\overline{A_1 \cup A_2 \cup A_3}\right| = C(18, 2) - C(14, 2) - C(14, 2) - C(6, 2) + C(10, 2) + C(2, 2) + C(2, 2) - 0$

$\therefore$ $\left|\overline{A_1 \cup A_2 \cup A_3}\right| = C(18, 2) - 2C(14, 2) - C(6, 2) + C(10, 2) + 2C(2, 2)$.

EXERCISE 5.5.

1. If there are 1000 faculty members. Suppose that 200 faculty members can speak French and 50 can speak Russian, while only 20 can speak both French and Russian. How many faculty members can speak either French or Russian ? and how many faculty members speak neither French nor Russian.
2. If there are 200 faculty members that speak French, 50 that speak Russian, 100 that speak Spanish, 20 that speak French and Russian, 60 that speak French and Spanish, 35 that speak Russian and Spanish, while only 10 speak French, Russian and Spanish, how many speak either French or Russian or Spanish ?

3. Find the number of integers between 1 and 1000 inclusive that are divisible by none of 5, 6 and 8. Hint: the intersection of the set of integers divisible by 6 with the set of integers divisible by 8 is the set of integers divisible by 24.
4. In a survey of students at Florida State University the following information was obtained : 260 were taking a statistics course, 208 were taking a mathematics course, 160 were taking a Computer Programming Course, 76 were taking Statistics and Mathematics, 48 were taking Statistics and Computer Programming, 62 were taking Mathematics Course and Computer Programming, 30 were taking all 3 kinds of courses and 150 were taking none of the 3 courses.
   (*a*) How many students were surveyed ?
   (*b*) How many students were taking a statistics and a mathematics course but not a Computer Programming Course ?
   (*c*) How many were taking a Statistics and a Computer Course but not a mathematics course ?
   (*d*) How many were taking a Computer Programming and a Mathematics Course but not a Statistics Course ?
   (*e*) How many were taking a Statistics Course but not taking a Course in Mathematics or in Computer Programming ?
   (*f*) How many were taking a Mathematics Course but not taking a Statistics Course or a Computer Programming Course ?
   (*g*) How many were taking a Computer Programming Course but not taking a Course in Mathematics or in Statistics ?

   And draw a Venn diagram and list the number of elements in all different regions of the diagram.
5. Suppose that, a certain Computer Center employs 100 Computer Programmers. Of these 47 can program in FORTRAN, 35 in pascal, 20 in COBOL, and 23 can program in FORTRAN and pascal, 12 in COBOL and FORTRAN, 11 in pascal and COBOL and 5 in FORTRAN, pascal and COBOL. How many can program in none of these 3 languages ?
6. How many integral solutions are there of $x_1 + x_2 + x_3 + x_4 = 20$ if $1 \leq x_1 \leq 6$, $1 \leq x_2 \leq 7$, $1 \leq x_3 \leq 8$ and $1 \leq x_4 \leq 9$?
7. Find the number of integer solutions of the equation $x_1 + x_2 + x_3 = 20$ such that $2 \leq x_1 \leq 5$, $4 \leq x_2 \leq 7$, $-2 \leq x_3 \leq 9$.
8. In how many ways can the 26 letters of the English alphabet be permuted so that none of the patterns CAR, PUN, DOG or BYTE occurs ?

ANSWERS

1. 230, 770	**2.** 245	**3.** 600
4. (*a*) 622	(*b*) 46	(*c*) 18
(*d*) 32	(*e*) 166	(*f*) 100
(*g*) 80		
5. 39	**6.** 217	**7.** 3

8. $26! - (3 \times 24! + 23!) + (3 \times 22! + 3 \times 21!) - (20! + 3 \times 19!) + 17!$.

CHAPTER

6

Recurrence Relations

6.1 GENERATING FUNCTIONS OF SEQUENCES

The objects of interest in this unit are sequences of real numbers $(a_0, a_1, a_2, ..., a_n, ...)$ that is, functions whose domain is the set of non-negative integers and whose range is the set of real numbers.

In other words, a function $a : Z^+ \to R$ is called **sequence** or **sequence of real** numbers.

We shall use expressions like $A = \{a_n\}_{n=0}^{\infty}$ to denote such sequences.

Example 1. $A = \{3^n\}_{n=0}^{\infty}$ is the sequence and

$$A = (1, 3, 9, 27, 81, ..., 2^n, ...)$$

Example 2. $B = \{B_n\}_{n=0}^{\infty}$ where $B_n = n + 1$ and $B = (1, 2, 3, 4, ..., n + 1, ...)$

Next we introduce the concept of generating function, a generalization of the concept of polynomial, to allow for that eventuality and it is one of the most powerful tool for solving recurrence relations.

Definition. *Let $A = \{a_n\}_{n=0}^{\infty}$ be the sequence of real numbers, then $a_0 + a_1 x + a_2 x^2 + ... + a_r x^r + ... = \sum_{n=0}^{\infty} a_n x^n$ is called power series and we write it as $A(x)$.*

$$\therefore \qquad A(x) = \sum_{n=0}^{\infty} a_n x^n = a_0 + a_1 x + a_2 x^2 + ... + a_n x^n + ...$$

where x is an indeterminate.

The expression A (x) is called a *formal power series* a_i is the coefficient of x^i, the term $a_i x^i$ is the term of degree i and a_0 is called the constant term. We use the word "formal" to distinguish between the abstract symbol $A(x) = \sum_{n=0}^{\infty} a_n x^n$ and the concept of power series.

Definition. *Let* $A = \{a_n\}_{n=0}^{\infty} = (a_0, a_1, a_2, \ldots\ldots, a_n, \ldots\ldots)$ *be the sequence. Then its generating function is defined to be* $A(x) = \sum_{n=0}^{\infty} a_n x^n = a_0 + a_1 x + a_2 x^2 + \ldots\ldots$

Note 1. We can easily obtain generating function from sequence and its converse. For example, the generating function of $\{a_n\}_{n=0}^{\infty} = \{\alpha^n\}_{n=0}^{\infty}$ is

$$\alpha^0 + \alpha x + \alpha^2 x + \alpha^3 x^3 + \ldots\ldots \ (1)$$

The infinite series in (1) can be written in closed form as $\frac{1}{1-\alpha x}$ which is a rather compact way to represent the sequence $\{a_n\} = (1, \alpha, \alpha^2, \ldots\ldots)$

Note 2. If $a_k \neq 0$ and $a_i = 0$ for $i \geq k + 1$ in A (x), then

$$A(x) = a_0 + a_1 x + a_2 x^2 + \ldots\ldots + a_k x^k$$

is a polynomial of degree k and it is also called generating function of finite sequence $(a_0, a_1, \ldots\ldots, a_k)$.

Example 1. The generating function of the sequence $\{3^n\}_{n=0}^{\infty}$ is

$$A(x) = 1 + 3x + 3^2 x^2 + 3^3 x^3 + \ldots\ldots + 3^n x^n + \ldots\ldots$$

Example 2. *Find the generating function for the sequence (1, 1, 1, 1, 1, 1).*

Solution. Generating function for the sequence (1, 1, 1, 1, 1, 1) is

$$A(x) = 1 + 1x + 1x^2 + 1x^3 + 1x^4 + 1x^5$$
$$= 1 + x + x^2 + x^3 + x^4 + x^5.$$

Example 3. *Find the generating functions for the following sequences.*

(i) $\{n\}$ *or (1, 2, 3, 4, ...)*

(ii) $\{(-1)^{n-1} n\}$ *or (1, – 2, 3, – 4, ...)*

Solution. (*i*) Generating function for sequence $\{n\}$ is

$$A(x) = \sum_{n=0}^{\infty} a_n x^n, \text{ where } a_n = n.$$

$$= a_0 + a_1 x + a_2 x^2 + \ldots\ldots$$
$$= 1 + 2x + 3x^2 + \ldots\ldots$$
$$= (1 - x)^{-2}.$$

(*ii*) Generating function for the sequence 1, – 2, 3, – 4, ... is

$$A(x) = 1 - 2x + 3x^2 - 4x^3 + \ldots\ldots$$
$$= (1 + x)^{-2}$$

Example 4. *Find a generating function for the sequence*

$$A = \{a_r\}_{r=0}^{\infty}, \text{ where } a_r = \begin{cases} 1 & \text{if} \quad 0 \le r \le 2 \\ 3 & \text{if} \quad 3 \le r \le 5 \\ 0 & \text{if} \quad r \ge 6 \end{cases}$$

Solution. $A(x) = \sum_{r=0}^{\infty} a_r x^r$

$$= a_0 + a_1 x + a_2 x^2 + a_3 x^3 + a_4 x^4 + a_5 x^5 + a_6 x^6 + \ldots\ldots$$
$$= 1 + 1x + 1x^2 + 3x^3 + 3x^4 + 3x^5 + 0x^6 + \ldots\ldots$$
$$= 1 + x + x^2 + 3x^3 + 3x^4 + 3x^5.$$

Example 5. *Find the generating function for the sequence (1, 1, 1, 1, 1, 1) in closed form.*

Solution. The generating function of (1, 1, 1, 1, 1, 1) is

$$A(x) = 1 + 1x + 1x^2 + 1x^3 + 1x^4 + 1x^5 = 1 + x + x^2 + x^3 + x^4 + x^5$$
$$= \frac{x^6 - 1}{x - 1}.$$

Example 6. *Let n be a positive integer and $a_k = C(n, k)$ for $k = 0, 1, 2, \ldots\ldots, n$. Find the generating function for the sequence $(a_0, a_1, \ldots\ldots, a_n)$*

Solution. The generating function for the given sequence is

$$A(x) = C(n, 0) + C(n, 1)\, x + C(n, 2)\, x^2 + \ldots\ldots + C(n, n)\, x^n$$
$$= (1 + x)^n, \text{ by binomial theorem.}$$

Example 7. *Find the closed form expression for the generating function of the sequences*

(i) $a_n = 9 \,.\, 5^n,\ n \ge 0$

(ii) $a_n = 4,\ n \ge 0.$

Solution.

(*i*) The generating function for the sequence $\{a_n\} = \{9 \,.\, 5^n\}$ is

$$A(x) = 9 \,.\, 5^0 + 9 \,.\, 5^1\, x + 9 \,.\, 5^2\, x^2 + 9 \,.\, 5^3\, x^3 + \ldots\ldots$$
$$= 9\,(1 + 5x + (5x)^2 + (5x)^3 + \ldots\ldots)$$
$$= 9\,(1 - 5x)^{-1} = \frac{9}{1 - 5x}.$$

(*ii*) The generating function for the sequence $\{a_n\} = \{4\}$ is

$$A(x) = 4 + 4x + 4x^2 + 4x^3 + \ldots\ldots$$
$$= 4\,(1 + x + x^2 + x^3 + \ldots\ldots) = 4\,(1 - x)^{-1} = \frac{4}{1 - x}.$$

Example 8. *Find the generating function of the sequence* $\{a_n\} = \left\{\frac{1}{n!}\right\}$.

Solution. The generating function of the sequence $\left\{\frac{1}{n!}\right\}$ is

$$A(x) = \frac{1}{0!} + \frac{1}{1!}x + \frac{1}{2!}x^2 + \frac{1}{3!}x^3 + \ldots\ldots$$

$$= 1 + \frac{x}{1!} + \frac{x^2}{2!} + \frac{x^3}{3!} + \ldots\ldots$$

$$= e^x.$$

6.1.1 Closed Form Expressions for Generating Functions

The most basic tool used to express generating functions in closed form is the closed form expression for the geometric series, which is an expression of the form $a + ar + ar^2 + \ldots\ldots$

Note that, it can be finite or infinite.

Example 1. Find the generating function for $a_n = 3^n$, $n \geq 0$ in closed form.

Solution. The generating function for this sequence is

$$A(x) = 1 + 3x + 3^2 x^2 + 3^3 x^3 + \ldots\ldots$$

$$\Rightarrow \quad A(x) - 1 = 3x + (3x)^2 + (3x)^3 + \ldots\ldots$$

$$\Rightarrow \quad A(x) - 1 = (3x)\,[1 + (3x) + (3x)^2 + \ldots\ldots]$$

$$\Rightarrow \quad \frac{A(x) - 1}{3x} = 1 + 3x + (3x)^2 + \ldots\ldots = A(x)$$

$$\Rightarrow \quad \frac{A(x) - 1}{3x} = A(x) \Rightarrow A(x) - 1 = 3x\,A(x)$$

$$\Rightarrow \quad A(x)(1 - 3x) = 1 \Rightarrow A(x) = \frac{1}{1 - 3x}$$

(or)

$$A(x) = 1 + 3x + (3x)^2 + (3x)^3 + \ldots\ldots$$

$$= (1 - 3x)^{-1} = \frac{1}{1 - 3x}.$$

Example 2. *Find the generating function for the sequence* $(1, \alpha, \alpha^2, \alpha^3, \ldots\ldots)$ *where* α *is fixed constant in closed form.*

Solution. The generating function for this sequence is

$$A(x) = 1 + \alpha x + \alpha^2 x^2 + \alpha^3 x^3 + \ldots\ldots$$

$$= (1 - \alpha x)^{-1} = \frac{1}{1 - \alpha x}.$$

Example 3. *Find the generating function (in closed form) of the fibonacci sequence* $\{f_n\}$ *defined by* $f_n = f_{n-1} + f_{n-2}\,;\, f_0 = 0\,;\, f_1 = 1$.

Solution. The generating function is

$$A(x) = f_0 + f_1x + f_2x^2 + f_3x^3 + \ldots\ldots = \sum_{n=0}^{\infty} f_n x^n \qquad \ldots(1)$$

Consider $\qquad f_n = f_{n-1} + f_{n-2},\ n \geq 0$

Multiply both sides by x^n, and sum over all $n \geq 2$, we get

$$\sum_{n=2}^{\infty} f_n x^n = \sum_{n=2}^{\infty} f_{n-1} x^n + \sum_{n=2}^{\infty} f_{n-2} x^n \qquad \ldots(2)$$

Consider the first sum

$$\sum_{n=2}^{\infty} f_n x^n = f_2x^2 + f_3x^3 + \ldots\ldots$$

$$= A(x) - f_0 - f_1x \qquad \text{(From (1))}$$

Similarly, $$\sum_{n=2}^{\infty} f_{n-1} x^n = f_1x^2 + f_2x^3 + \ldots\ldots$$

$$= x(f_1x + f_2x^2 + \ldots\ldots)$$
$$= x(A(x) - f_0) \qquad \text{(From (1))}$$

and $$\sum_{n=2}^{\infty} f_{n-2} x^n = f_0x^2 + f_1x^3 + f_2x^4 + \ldots\ldots$$

$$= x^2(f_0 + f_1x + f_2x^2 + \ldots\ldots)$$
$$= x^2 A(x)$$

Substituting these three expressions in (2), we get

$$A(x) - f_0 - f_1 x = x(A(x) - f_0) + x^2 A(x)$$

Since $\qquad f_0 = 0$ and $f_1 = 1$, we get

$$A(x) - x = x A(x) + x^2 A(x)$$

$\Rightarrow \qquad A(x)(1 - x - x^2) = x$

$\Rightarrow \qquad A(x) = \dfrac{x}{1 - x - x^2}$

Therefore, the required generating function is $A(x) = \dfrac{x}{1 - x - x^2}$

Example 4. *Find the sequences generated by the following functions :*

(i) $(2 + x)^3$ $\qquad$ *(ii)* $2x^2(1 - x)^{-1}$ $\qquad$ *(iii)* $3x^3 + e^{2x}$.

Solution. (*i*) We have $(2 + x)^3 = 8\left(1 + \dfrac{x}{2}\right)^3$

$$= 8\left(1 + C(3,1)\frac{x}{2} + C(3,2)\frac{x^2}{4} + C(3,3)\frac{x^3}{8}\right)$$

$$= 8\left(1 + \frac{3}{2}x + \frac{3}{4}x^2 + \frac{1}{8}x^3\right)$$

$$= 8 + 12x + 6x^2 + 1x^3.$$

This shows that the sequence generated by $(2 + x)^3$ is

8, 12, 12, 1, 0, 0, 0, ...

(*ii*) We have $2x^2 (1 - x)^{-1} = 2x^2 (1 + x + x^2 + x^3 + ...)$

$$= 0 + 0x + 2x^2 + 2x^3 + 2x^4 + 2x^5 + ...$$

∴ The required sequence is 0, 0, 2, 2, 2, 2, ...

(*iii*) We have $3x^3 + e^{2x} = 3x^3 + 1 + \frac{2x}{1!} + \frac{(2x)^2}{2!} + \frac{(2x)^3}{3!} + \frac{(2x)^4}{4!} + ...$

$$= 1 + 2x + 2x^2 + \left(3 + \frac{8}{6}\right)x^3 + \frac{16}{24}x^4 + ...$$

$$= 1 + 2x + 2x^2 + \frac{13}{3}x^3 + \frac{2}{3}x^4 + ...$$

∴ The sequence generated by $3x^3 + e^{2x}$ is $1, 2, 2, \frac{13}{3}, \frac{2}{3}, ...$

Example 5. *Find the generating functions for the following sequences :*

(*i*) *0, 1, 2, 3, ...*
(*ii*) *$1^2, 2^2, 3^2$, ...*
(*iii*) *$0^2, 1^2, 2^2, 3^2$, ...*
(*iv*) *$1^3, 2^3, 3^3$, ...*
(*v*) *$0^3, 1^3, 2^3, 3^3$, ...*

Solution. (*i*) Generating function is $0 + 1x + 2 \cdot x^2 + 3x^3 + ...$

$$= x(1 + 2x + 3x^2 + ...) = x(1 - x)^{-2}$$

$$= \frac{x}{(1-x)^2}.$$

(*ii*) We have $x + 2x^2 + 3x^3 + ... = \frac{x}{(1-x)^2}$ by (*i*) differentiating, we get

$$1 + 2(2x) + 3(3x^2) + ... = \frac{d}{dx}\frac{x}{(1-x)^2} = \frac{1+x}{(1-x)^3}$$

$\Rightarrow$ $1^2 + 2^2 x + 3^2 x^2 + ... = \frac{1+x}{(1-x)^3}$ which is the generating function of $1^2, 2^2, 3^2$, ...

(*iii*) We have $0^2 + 1^2 x + 2^2 x^2 + 3^2 x^3 + ... = x[1^2 + 2^2 x + 3^2 x^2 + ...] = \dfrac{x(1+x)}{(1-x)^3}$ (by above result (*ii*))

$\therefore$ Generating function of $0^2, 1^2, 2^2, 3^2, ...$ is $\dfrac{x(1+x)}{(1-x)^3}$

(*iv*) We have $1^2 x + 2^2 x^2 + 3^2 x^3 + ... = \dfrac{x(1+x)}{(1-x)^3}$ (by (*iii*))

Differentiating w.r.t. on both sides, we get

$$1^2 + 2^2\, 2x + 3^2\, 3x^2 + ... = \frac{d}{dx}\frac{x(1+x)}{(1-x)^3}$$

$\Rightarrow$ $1^3 + 2^3 x + 3^3 x^2 + ... = \dfrac{x^2 + 4x + 1}{(1-x)^4}$; which is required generating function.

(*v*) We have $1^3 + 2^3 x + 3^3 x^2 + ... = \dfrac{x^2 + 4x + 1}{(1-x)^4}$

$\Rightarrow$ $0^3 + 1^3 x + 2^3 x^2 + 3^3 x^3 + ... = \dfrac{x\left(x^2 + 4x + 1\right)}{(1-x)^4}$

Then generating function for $0^3, 1^3, 2^3, 3^3, ...$ is $\dfrac{x\left(x^2 + 4x + 1\right)}{(1-x)^4}$.

Example 6. *Find the generating function for the sequence* $\{a_n\}_{n=0}^{\infty}$ *if* $a_n = n + n^2$.

(or) Find the generating function for the sequence 0, 2, 6, 12, 20, 30, ...

Solution. We have the generating function for the sequence $\{n\}$ is

$$A(x) = \frac{x}{(1-x)^2} \qquad \text{(See example 3)}$$

and the generating function for the sequence $\{n^2\}$ is

$$B(x) = \frac{x(1+x)}{(1-x)^3} \qquad \text{(See example 5 (iii))}$$

Then, the generating function for the given sequence $\{n + n^2\}$ is

$$A(x) + B(x) = \frac{x}{(1-x)^2} + \frac{x(1+x)}{(1-x)^3} = \frac{2x}{(1-x)^3}.$$

Example 7. If n is a positive integer, prove that $C(-n, r) = (-1)^r C(n+r-1, r) = (-1)^r C(n+r-1, n-1)$ and hence find generating function for $C(n+r-1, r)\, a^r$

Proof : We have $C(-n, r) = \dfrac{(-n)(-n-1)(-n-2)\ldots\ldots(-n-r+1)}{r!}$, by definition

$$= (-1)^r \frac{n(n+1)(n+2)\ldots\ldots(n+r-1)}{r!} = (-1)^r \frac{\overline{n+r-1}!}{\overline{n-1}!\,r!}$$

$$= (-1)^r C(n+r-1, r) = (-1)^r C(n+r-1, n-1)$$

$$\Rightarrow \qquad C(n+r-1, r) = (-1)^r C(-n, r)$$

Therefore,

$$\sum_{r=0}^{\infty} C(n+r-1, r)\, a^r x^r = \sum_{r=0}^{\infty} (-1)^r C(-n, r)\, a^r x^r$$

$$= \sum_{r=0}^{\infty} C(-n, r)(-ax)^r$$

$$= (1-ax)^{-n}$$

$\therefore$ Generating function for $C(n-r+1, r)\, a^r$ is $(1-ax)^{-n}$.

Note. The generating function for $C(n-r+1, r)(-a)^r$ is $(1+ax)^{-n}$.

Table 6.1 Table of Generating Functions

S. No.	*Sequence*	*Generating Function*
1.	$C(k, n)$	$(1+x)^k$
2.	1	$\dfrac{1}{1-x}$
3.	a^n	$\dfrac{1}{1-ax}$
4.	$(-1)^n$	$\dfrac{1}{1+x}$
5.	$(-1)^n a^n = (-a)^n$	$\dfrac{1}{1+ax}$
6.	$C(k-1+n, n)$ (k is a fixed positive integer)	$\dfrac{1}{(1-x)^k}$
7.	$C(k-1+n, n)\, a^n$	$\dfrac{1}{(1-ax)^k}$
8.	$C(k-1+n, n)(-a)^n$	$\dfrac{1}{(1+ax)^k}$
9.	$n+1$	$\dfrac{1}{(1-x)^2}$

10.	n	$\frac{x}{(1-x)^2}$
11.	$(n + 2)\ (n + 1)$	$\frac{2}{(1-x)^3}$
12.	$(n + 1)\ n$	$\frac{2x}{(1-x)^3}$
13.	n^2	$\frac{x\,(1+x)}{(1-x)^3}$
14.	$n\ a^n$	$\frac{ax}{(1-ax)^2}$
15.	$n^2\ a^n$	$\frac{ax\,(1+ax)}{(1-ax)^3}$.

Definition. *Let* $A\ (x) = \sum_{r=0}^{\infty} a_r x^r$, $B\ (x) = \sum_{s=0}^{\infty} b_s x^s$ *be two formal power series. Then,*

1. **Equality :** $A\ (x) = B\ (x)$ iff $a_n = b_n$ for each $n \geq 0$

2. **Multiplication by a scalar number C :** $C\ A(x) = \sum_{r=0}^{\infty} (Ca_r)\, x^r$

3. **Sum :** $A(x) + B(x) = \sum_{n=0}^{\infty} (a_n + b_n)\, x^n$.

4. **Product :** $A(x)\ B(x) = \sum_{n=0}^{\infty} P_n\, x^n$ where $P_n = \sum_{i+j=n} a_i\, b_j$

$$= a_0\ b_n + a_1\ b_{n-1} + a_2\ b_{n-2} + \ldots\ldots + a_{n-1}\ b_1 + a_n\ b_0$$

Then
$$A(x)\ B(x) = a_0\ b_0\ x^0 + (a_0\ b_1 + a_1\ b_0)\ x^1 + (a_0\ b_2 + a_1\ b_1 + a_2\ b_0)\ x^2 + \ldots\ldots + (a_0\ b_n + a_1\ b_{n-1} + a_2\ b_{n-2} + \ldots\ldots + a_n\ b_0)\ x^n + \ldots\ldots$$

6.2 CALCULATING COEFFICIENT OF GENERATING FUNCTION

Rule of Counting :

Suppose we wish to determine the number of integer solutions of the equation $e_1 + e_2 + \ldots\ldots + e_n = r$ where $0 \leq r \leq n$ under the constraints that

e_1 can take the integer values $a_{11}, a_{12}, a_{13}, \ldots\ldots$

e_2 can take the integer values $a_{21}, a_{22}, a_{23}, \ldots\ldots$

....

e_n can take the integer values $a_{n1}, a_{n2}, a_{n3}, \ldots\ldots$

To solve this problem, first we define the functions

$f_1(x), f_2(x), \ldots\ldots, f_n(x)$ as follows :

$$f_1(x) = x^{a_{11}} + x^{a_{12}} + x^{a_{13}} + \ldots\ldots$$

$$f_2(x) = x^{a_{21}} + x^{a_{22}} + x^{a_{23}} + \ldots\ldots$$

....

$$f_n(x) = x^{a_{n1}} + x^{a_{n2}} + x^{a_{n3}} + \ldots\ldots$$

and consider $f(x)$ as $f(x) = f_1(x)\, f_2(x) \ldots\ldots f_n(x)$.

and determine the coefficient of x^r in this function.

This coefficient happens to be equal to the number of solutions that we desired to find. The function $f(x)$ is called the generating function for the problem.

Example 1. The coefficient of x^9 in $(1 + x^3 + x^8)^{10}$ is C (10, 3) $= \dfrac{10!}{3!7!}$ because the only solutions of $e_1 + e_2 + \ldots\ldots + e_{10} = 9$ where each $e_i = 0, 3, 8$ are those solutions where 3 values are equal to three, and the remaining values are 0 and the coefficient of x^{25} is $\dfrac{10!}{3!2!5!}$ because each solution to the equation will involve three 3's, two 8's, and five 0's.

Example 2. *Find the generating function that determines the number of non-negative integer solutions of the equation* $e_1 + e_2 + e_3 + e_4 + e_5 = 20$ *under the constraints* $0 \le e_1 \le 3$, $0 \le e_2 \le 4$, $2 \le e_3 \le 6$, $2 \le e_4 \le 5$, e_5 *is odd with* $e_5 \le 9$.

Solution.

Take

$$f_1(x) = x^0 + x^1 + x^2 + x^3$$

$$f_2(x) = x^0 + x^1 + x^2 + x^3 + x^4$$

$$f_3(x) = x^2 + x^3 + x^4 + x^5 + x^6$$

$$f_4(x) = x^2 + x^3 + x^4 + x^5$$

$$f_5(x) = x^1 + x^3 + x^5 + x^7 + x^9.$$

Hence the generating function for the number of non-negative integer solutions of the given equation under the given constraints is

$$f(x) = f_1(x)\, f_2(x)\, f_3(x)\, f_4(x)\, f_5(x).$$

$$= \frac{1-x^4}{1-x}.\frac{1-x^5}{1-x}.x^2.\frac{1-x^5}{1-x}.x^2.\frac{1-x^4}{1-x}.x.\left(1+x^2+x^4+x^6+x^8\right)$$

$$= x^5\left(1-x^4\right)^2\left(1-x^5\right)^2\left(1-x\right)^{-4}.\frac{1-\left(x^2\right)^5}{1-x^2}$$

$$= x^5\left(1-x^4\right)^2\left(1-x^5\right)^2\left(1-x\right)^{-5}\left(1+x\right)^{-1}\left(1-x^{10}\right)$$

Example 3. *Find the coefficient of* x^{32} *in* $(1 + x^5 + x^9)^{10}$.

Solution. We want the solutions of $e_1 + e_2 + \ldots\ldots + e_{10} = 32$

where each $e_i = 0, 5, 9$. This can be done only with three 9's, one 5, and six 0's. Thus, the coefficient of x^{32} is $\dfrac{10!}{3!1!6!}$.

Example 4. *Find a generating function for a_r = the number of ways the sum r can be obtained when*

(i) 2 distinguishable dice are tossed

(ii) 2 distinguishable dice are tossed and the first shows an even number and the second shows an odd number.

Solution.

(*i*) We want the number of integral solutions to $e_1 + e_2 = r$ where $1 \le e_i \le 6$. Then a_r is the coefficient of x^r in the generating function $(x + x^2 + x^3 + x^4 + x^5 + x^6)^2$.

(*ii*) We want the number of integral solutions to $e_1 + e_2 = r$ where $1 \le e_1 \le 6$ and e_1 is even while $1 \le e_2 \le 6$ and e_2 is odd. Then a_r is the coefficient of x^r in the generating function $(x^2 + x^4 + x^6)\,(x + x^3 + x^5)$.

Example 5. *Build a generating function for a_r = the number of integral solutions to the equation $e_1 + e_2 + e_3 = r$ if $0 < e_i$ for each i.*

Solution. Given that, for each i, $0 < e_i$ *i.e.,* $1 \le e_i$ generating function is $(x + x^2 + x^3 + \ldots\ldots)^3$.

Example 6. *Find the coefficient of x^{27} in $(x^4 + x^5 + x^6 + \ldots\ldots)^5$*

Solution. We have $(x^4 + x^5 + x^6 + \ldots\ldots)^5 = (x^4)^5\,(1 + x + x^2 + \ldots\ldots)^5$

$$= x^{20}\,[(1 - x)^{-1}]^5$$
$$= x^{20}\,(1 - x)^{-5}$$
$$= x^{20} \sum_{r=0}^{\infty} C\,(5 + r - 1, r)\, x^r$$
$$= x^{20} \sum_{r=0}^{\infty} C\,(4 + r, r)\, x^r$$
$$= \sum_{r=0}^{\infty} C\,(4 + r, r)\, x^{20 + r}$$

Take $20 + r = 27$, then $r = 7$

$\therefore$ The coefficient of x^{27} is $C\,(4 + 7, 7) = C\,(11, 7) = \dfrac{11!}{7!\,4!} = 330.$

Example 7. *Find the coefficient of x^{20} in $(x + x^2 + 2x^3 + x^4)\,(x^2 + x^3 + x^4 + \ldots\ldots)^5$*

Solution. We have $(x + x^2 + 2x^3 + x^4)\,(x^2 + x^3 + x^4 + \ldots\ldots)^5$.

$$= x\,(1 + x + 2x^2 + x^3)\,(x^2)^5\,(1 + x + x^2 + \ldots\ldots)^5$$
$$= x^{11}\,(1 + x + 2x^2 + x^3)\,[(1 - x)^{-1}]^5$$
$$= x^{11}\,(1 + x + 2x^2 + x^3)\,(1 - x)^{-5}$$
$$= x^{11}\,(1 + x + 2x^2 + x^3) \sum_{r=0}^{\infty} C\,(4 + r, r)\, x^r$$

$\therefore$ Coefficient of x^{20} in $(x + x^2 + 2x^3 + x^4)\,(x^2 + x^3 + x^4 + \ldots..)^5$ is

$C\,(4 + 9, 9) + C\,(4 + 8, 8) + 2\,C\,(4 + 7, 7) + C\,(4 + 6, 6)$

$$= C\,(13, 9) + C\,(12, 8) + 2\,C\,(11, 7) + C\,(10, 6).$$

Example 8. *Find the coefficient of* x^{12} *in* $(1 + 2x + 3x^2)\,(x - x^2 + x^3 - x^4 + ...)^3$.

Solution. We have $(1 + 2x + 3x^2)\,(x - x^2 + x^3 - x^4 + ...)^3$

$$= (1 + 2x + 3x^2)\, x^3\, (1 - x + x^2 - x^3 + ...)^3$$
$$= x^3\, (1 + 2x + 3x^2)\, [(1 + x)^{-1}]^3$$
$$= x^3\, (1 + 2x + 3x^2)\, (1 + x)^{-3}$$
$$= x^3\, (1 + 2x + 3x^2) \sum_{r=0}^{\infty} C\,(3 + r - 1, r)\, x^r\, (-1)^r$$

∴ The coefficient of x^{12} is $-\,C\,(2 + 9, 9) + 2\,C\,(2 + 8, 8) - 3\,C\,(2 + 7, 7)$

$$= -\,C\,(11, 9) + 2\,C\,(10, 8) - 3\,C\,(9, 7)$$

Example 9. *Find the coefficient of* x^{10} *in (i)* $(1 + x + x^2 + ...)^2$ *; (ii)* $\dfrac{1}{(1-x)^3}$ *; and (iii)* $\dfrac{1}{(1+x)^3}$

Solution. (*i*) $(1 + x + x^2 + ...)^2 = [(1 - x)^{-1}]^2 = (1 - x)^{-2}$

$$= \sum_{r=0}^{\infty} C\,(r + 1, r)\, x^r$$
$$= \sum_{r=0}^{\infty} (r + 1)\, x^r .$$

∴ The coefficient of x^{10} is 11.

(*ii*) $$\frac{1}{(1-x)^3} = (1 - x)^{-3} = \sum_{r=0}^{\infty} C\,(r + 2, r)\, x^r$$
$$= \sum_{r=0}^{\infty} \frac{(r+2)\,(r+1)}{2}\, x^r$$

∴ The coefficient of x^{10} is $\dfrac{(12)\,(11)}{2} = 66.$

(*iii*) $$\frac{1}{(1+x)^3} = (1 + x)^{-3} = \sum_{n=0}^{\infty} C\,(r + 2, r)\,(-1)^r\, x^r$$
$$= \sum_{r=0}^{\infty} \frac{(r+2)\,(r+1)}{2}\,(-1)^r\, x^r$$

∴ The coefficient of x^{10} is $(-\,1)^{10}\,\dfrac{(12)\,(11)}{2} = 66.$

Example 10. *Find the coefficient of* x^{12} *in (i)* $\dfrac{x^2}{(1-x)^{10}}$ *; (ii)* $\dfrac{x^2 - 3x}{(1-x)^4}$.

Solution. (*i*) $\dfrac{x^2}{(1-x)^{10}} = x^2\,(1 - x)^{-10}$

$$= x^2 \sum_{r=0}^{\infty} C\,(r + 9, r)\, x^r$$

∴ The coefficient of x^{12} is C (19, 10) (Here $r = 10$).

(*ii*) $\dfrac{x^2 - 3x}{(1-x)^4} = (x^2 - 3x)(1 - x)^{-4} = (x^2 - 3x)\sum_{r=0}^{\infty} C(3 + r, r)\, x^r$

The coefficient of x^{12} is C (13, 10) – 3 C (14, 11).

Example 11. *Using generating function find the number of*

(i) non-negative ; and

(ii) positive integral solutions of the equation $e_1 + e_2 + e_3 + e_4 = 25$.

Solution. (*i*) In the case of non-negative integer solutions, e_1'*s* can take the values 0, 1, 2, 3, ..., accordingly we can write $f_i(x) = x^0 + x^1 + x^2 + x^3 + x^4 + \ldots\ldots$

for $i = 1, 2, 3, 4$.

The generating function for the number of non-negative integer solution is $f(x) = f_1(x) f_2(x) f_3(x) f_4(x)$

$$\Rightarrow \quad f(x) = (1 + x + x^2 + \ldots\ldots)(1 + x + x^2 + \ldots\ldots)(1 + x + x^2 + \ldots\ldots) \times (1 + x + x^2 + \ldots\ldots)$$

$$= (1 + x + x^2 + x^3 + \ldots\ldots)^4.$$

$$= [(1 - x)^{-1}]^4 = (1 - x)^{-4} = \sum C(3 + r, r)\, x^r.$$

Taking $x^r = x^{25}$, we get $r = 25$

∴ The coefficient of x^{25} is C (3 + 25, 25) = C (28, 25).

$$= \frac{28!}{25!\,3!} = 3276.$$

Thus, the given equation has 3276 non-negative integer solutions.

(*ii*) In the case of positive integer solutions, e_i'*s* can take values 1, 2, 3, 4, ... accordingly. We can write $f_i(x) = x^1 + x^2 + x^3 + x^4 + \ldots\ldots$ for $i = 1, 2, 3, 4$.

Therefore, the generating function is $f(x) = f_1(x) f_2(x) f_3(x) f_4(x)$.

$$\therefore \quad f(x) = (x + x^2 + x^3 + x^4 + \ldots\ldots)^4$$

$$= x^4 (1 + x + x^2 + x^3 + \ldots\ldots)^4$$

$$= x^4 [(1 - x)^{-1}]^4$$

$$= x^4 (1 - x)^{-4} = x^4 \sum_{r=0}^{\infty} C(3 + r, r)\, x^r$$

∴ The coefficient of x^{25} C (3 + 21, 21)

$$= \frac{24!}{21!\,3!} = 2024.$$

Thus, the given equation has 2024 positive integer solutions.

Example 12. *In how many ways can we distribute 24 pens to 4 children so that each child gets at least 3 pens but not more than eight.*

Solution. The given problem is equivalent to the problem of finding the number of integer solutions of the equation $x_1 + x_2 + x_3 + x_4 = 24$ with $3 \leq x_i \leq 8$ for each x_i. Keeping the constraints on x_i in mind, let us consider the functions $f_i(x) = x^3 + x^4 + x^5 + x^6 + x^7 + x^8$

for $i = 1, 2, 3, 4$.

Therefore, the generating function for the problem is

$$f(x) = f_1(x)\, f_2(x)\, f_3(x)\, f_4(x)$$
$$= (x^3 + x^4 + x^5 + x^6 + x^7 + x^8)^4$$

The required number is the coefficient of x^{24} in this function.

Now, we find that $f(x) = x^{12}(1 + x + x^2 + x^3 + x^4 + x^5)^4$

$$= x^{12}\left(\frac{1-x^6}{1-x}\right)^4 = x^{12}\left(1-x^6\right)^4 (1-x)^{-4}$$

$$= x^{12}\left[\sum_{r=0}^{4} C(4, r)\left(-x^6\right)^r . \sum_{s=0}^{\infty} C(4+s-1, s)\, x^s\right]$$

$$= x^{12}\left[\sum_{r=0}^{\infty}(-1)^r C(4, r)\, x^{6r} . \sum_{s=0}^{\infty} C(3+s, s)\, x^s\right]$$

From this, we find that the coefficient of x^{24} in $f(x)$ is C (4, 0) C (15, 12) – C (4, 1) C (9, 6) + C (4, 2) C (3, 0).

= 455 – 366 + 6 = 125, which is the required number.

Example 13. *Using the generating function, prove that the number of ways of choosing with repetitions, r of n objects is C (n + r – 1, r).*

Solution. The problem is equivalent to that of finding the number of ways of selecting r similar balls from n numbered boxes. If x_i is the number of balls selected from the box i, then x_i satisfy the equation

$$x_1 + x_2 + x_3 + \ldots\ldots + x_n = r \qquad \ldots(1)$$

and the constraints $x_i \geq 0$ for $i = 1, 2, 3, \ldots\ldots, n$ ...(2)

Thus, we have to find the number of integer solutions of the equation (1) under the constraints (2).

We take $\quad f_i(x) = x^0 + x^1 + x^2 + \ldots\ldots$

$$= (1 - x)^{-1} \text{ for } i = 1, 2, \ldots\ldots, n$$

so that $\quad f(x) = f_1(x)\, f_2(x) \ldots\ldots f_n(x)$

$\Rightarrow \quad f(x) = [(1 - x)^{-1}]^n = (1 - x)^{-n}$

$$= \sum_{r=0}^{\infty} C(n+r-1, r)\, x^r \qquad \ldots(3)$$

This is the generating function for the given problem, and the number of integer solutions of the equation (1) under the constraints (2) is the coefficient of x^r in this function. This coefficient, is evidently,

$$C(n + r - 1, r)$$

This proves the required result.

Example 14. *Find a generating function for a_r = the number of ways distributing r similar balls into n numbered boxes where each box is non-empty.*

Solution. First we model this problem as an integral solution of an equation problem : namely, we are to count the number of integral solutions to $e_1 + e_2 + \ldots\ldots + e_n = r$, where each $e_i \geq 1$.

Then the generating function is $(x + x^2 + x^3 + \ldots\ldots)^n$

$$= \left(\sum_{r=1}^{\infty} x^r\right)^n.$$

EXERCISE 6.1.

1. Find the sequence generated by
   (*a*) $x^2 (1 - x)^{-1}$ (*b*) $(1 - x)^{-1} - x^2$.
2. Find a generating function for the following sequences :
   (*a*) 1, 1, 0, 1, 1, ... (*b*) 1, 0, 1, 0, 1, 0, ... (*c*) $0^4, 1^4, 2^4, 3^4, \ldots\ldots$
3. Find a generating function for the sequence $\{a_r\}$, where $a_r = 0$ for $0 \leq r \leq n - 1$
   $= C(r - 1, n - 1)$ for $r \geq n$.
4. Find a generating function for the numeric function $a_r = \begin{cases} 2^r & \text{if} \quad r \text{ is even} \\ -2^r & \text{if} \quad r \text{ is odd} \end{cases}$.
5. Find the coefficient of x^5 in $(1 - 2x)^{-7}$.
6. Find the coefficient of x^{12} in
   (*a*) $(1 - 4x)^{15}$ (*b*) $(1 + x)^{-20}$
7. Find the coefficient of x^{50} in $(x^{10} + x^{11} + \ldots\ldots + x^{25})(x + x^2 + \ldots\ldots + x^{15})(x^{20} + x^{21} + \ldots\ldots + x^{45})$.
8. Find the coefficient of x^{25} in $(x^2 + x^3 + x^4 + x^5 + x^6)^7$.
9. Find the coefficient of x^8 in $\dfrac{1}{(x-3)(x-2)^2}$.
10. Determine the coefficient of x^{18} in the product $(x + x^2 + x^3 + x^4 + x^5)(x^2 + x^3 + x^4 + \ldots\ldots)^5$.
11. Find the coefficient of x^n in $(x^2 + x^3 + x^4 + \ldots\ldots)^4$
12. Determine the coefficient of x^{10} in $(1 + x + x^2)(x + 2x^2 + 3x^3 + 4x^4 + \ldots\ldots)^4$
13. Find a generating function for a_r = the number of non-negative integral solutions to $e_1 + e_2 + \ldots\ldots + e_n = r$, where $0 \leq e_i \leq 1$.
14. In how many ways can 12 oranges be distributed among three children X, Y, Z, so that X gets at least four, Y and Z get at least two, but Z gets no more than five ?
15. Find the coefficient of x^{23} in $(1 + x^5 + x^9)^{10}$.
16. Find a generating function for a_r = the number of non-negative intregral solutions of $e_1 + e_2 + e_3 + e_4 + e_5 = r$ where $0 \leq e_1 \leq 3$, $0 \leq e_2 \leq 3$, $2 \leq e_3 \leq 6$, $2 \leq e_4 \leq 6$, e_5 is odd and $1 \leq e_5 \leq 9$.

17. Build a generating function for a_r = the number of integral solutions to the equation $e_1 + e_2 + e_3 = r$ if (a) $0 \le e_i \le 3$ for each i (*b*) $2 \le e_i \le 5$ for each i (*c*) $0 \le e_1 \le 6$ and e_i is even, $2 < e_2 \le 7$ and e_2 is odd ; $5 \le e_3 \le 7$.
18. Find a generating function to count the number of integral solutions to $e_1 + e_2 + e_3 = 10$ if for each i, $e_i \ge 0$.

ANSWERS

1. (*a*) 0, 0, 1, 1, 1, ... (*b*) 1, 1, 0, 1, 1, 1, ...
2. (*a*) $(1 - x)^{-1} - x^2$ (*b*) $(1 + x^2)^{-1}$ (*c*) $\dfrac{x^4 + 11x^3 + 11x^2 + x}{(1-x)^5}$
3. $\left(\dfrac{x}{1-x}\right)^n$
4. $(1 + 2x)^{-1}$
5. $25 \cdot \dfrac{11!}{5!6!}$
6. (*a*) 4^{12} C (15, 12) (*b*) C (31, 12)
7. C (21, 19) – C (6, 4) – C (5, 3)
8. C (17, 11) – 7 C (12, 6) + C (7, 2) C (7, 1)
9. $-\dfrac{1}{12}\sum_{0}^{\infty}\left(\dfrac{1}{3}\right)^r\left(\dfrac{1}{2}\right)^{8-r} C(9-r, 8-r)$
10. C (11, 7) + C (10, 6) + C (9, 5) + C (8, 4) + C (7, 3)
11. C (n – 5, 3)
12. C (16, 9) + C (15, 8) + C (14, 7)
13. $(1 + x)^n$
14. 14
15. $\dfrac{10!}{3!1!6!}$
16. $(1 + x + x^2 + x^3)^2 (x^2 + x^3 + x^4 + x^5 + x^6)^2 (x + x^3 + x^5 + x^7 + x^9)$
17. (*a*) $(1 + x + x^2 + x^3)^3$ (*b*) $(x^2 + x^3 + x^4 + x^5)^3$
    (*c*) $(1 + x + x^4 + x^6) (x^3 + x^5 + x^7) (x^5 + x^6 + x^7)$
18. $(1 + x + x^2 + ... + x^{10})^3$.

6.3 RECURRENCE RELATIONS

The number of bacteria in a colony doubles every hour. If a colony begins with five bacteria, how many will be present in n hours ? To solve this problem, let a_n be the number of bacteria at the end of n hours. Since the number of bacteria doubles every hour, the relationship $a_n = 2a_{n-1}$ holds whenever n is a positive integer. This relationship, together with the initial condition $a_0 = 5$, uniquely determines a_n for all non-negative integers n, we can find a formula for a_n from this information.

Recursive definitions can be used to solve counting problems.

Definition. *A* ***recurrence relation*** *for the sequence* $\{a_n\}$ *is an equation that expresses* a_n *in terms of one or more of the previous terms of the sequence, namely,* $a_0, a_1, a_2, \ldots\ldots, a_{n-1}$, *for all integers* $n \geq 1$.

Note. Recurrence relation is also called **difference equation**.

Example 1. If a_n denotes the sum of the first n positive integers, then $s_n = n + s_{n-1}$ which is a recurrence relation.

Example 2. If s_n denotes the nth term of a geometric progression with common ratio r, then $s_n = r\, s_{n-1}$ which is a recurrence relation.

Example 3. $a_n - 3a_{n-1} + 2a_{n-2} = 0$

Example 4. $a_n - 5a_{n-1} + 6a_{n-2} = n^2 + 1$

Example 5. $a_n^2 + a_{n-1}^2 = -1$

Example 6. $a_n - (n-1)\, a_{n-1} + (n-1)\, a_{n-2} = 0$

Definition. *Suppose n and k are non-negative integers. A recurrence relation of the form* $C_0(n)\, a_n + C_1(n)\, a_{n-1} + \ldots\ldots + C_k(n)\, a_{n-k} = f(n)$ ... *(1) for* $n \geq k$, *where* $C_0(n), C_1(n), \ldots\ldots, C_k(n)$ *and* $f(n)$ *are functions of n is said to be a* ***linear*** *recurrence relation.*

If $C_0(n)$ and $C_k(n)$ are not identically zero, then k is called degree of above linear recurrence relation. If $C_0(n), C_1(n), \ldots\ldots, C_k(n)$ are constants, then the recurrence relation (1) is known as a linear recurrence relation with constant coefficients.

If $f(n) = 0$, then (1) is said to be homogeneous recurrence relation.

If $f(n) \neq 0$, then (1) is said to be inhomogeneous recurrence relation.

Note 1. All the examples above are linear recurrence relations except (5) because it contains square term.

Note 2. The relations in (1), (2), (3), (4) are linear with constant coefficients.

Note 3. Relations (1), (2) have degree 1 ; (3), (4) have degree 2.

Note 4. Relations (2), (3), (6) are homogeneous.

Note 5. Relations (1), (4) are inhomogeneous.

6.3.1 The Fibonacci Recurrence Relation

Definition. *The recurrence relation* $F_n = F_{n-1} + F_{n-2}$, $n \geq 2$ *with initial conditions* $F_0 = F_1 = 1$ *is known as Fibonacci recurrence relation (or) Fibonacci relation.*

The numbers F_n generated by the Fibonacci relation with initial conditions $F_0 = F_1 = 1$ are called the Fibonacci numbers. The sequence of Fibonacci numbers $\{F_n\}_{n=0}^{\infty}$ is the Fibonacci sequence **properties of Fibonacci numbers**.

1. $F_0 + F_1 + F_2 + \ldots\ldots + F_n = F_{n+2} - 1$

Proof : We have $F_n = F_{n-1} + F_{n-2}$; $n \geq 2$, $F_0 = F_1 = 1$

Replace n by $n + 2$, we get $F_{n+2} = F_{n+1} + F_n$

$\Rightarrow \qquad F_n = F_{n+2} - F_{n+1}$

Put $\qquad n = 0, 1, 2, \ldots.., n$; we get

$$F_0 = F_2 - F_1$$
$$F_1 = F_3 - F_2$$
$$F_2 = F_4 - F_3$$
$$\ldots\ldots$$
$$\ldots\ldots$$
$$F_n = F_{n+2} - F_{n+1}$$

Adding, $\quad F_0 + F_1 + F_2 + \ldots\ldots + F_n = F_{n+2} - F_1$

$\Rightarrow \quad F_0 + F_1 + F_2 + \ldots\ldots + F_n = F_{n+2} - 1$

2. $F_0 + F_2 + F_4 + \ldots.. + F_{2n} = F_{2n+1}$

3. $F_0^2 + F_1^2 + F_2^2 + \ldots\ldots + F_n^2 = F_n F_{n+1}$

The above (2) and (3) properities can be proved by mathematical induction. Also, there is an interesting relation between Pascal's triangle and the Fibonacci numbers, illustrated below :

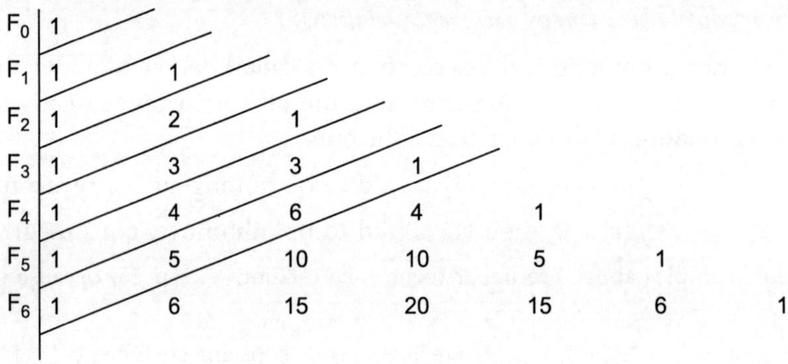

Clearly, the sum of the elements lying on the diagonal running upward from the left are Fibonacci numbers.

With this introductory exposure about recurrence relations, our main aim is to solve the recurrence relations.

Example. *Prove that* $F_n^2 = F_{n-1} F_{n+1} + (-1)^n, n \geq 2$

Solution. Using induction on n.

When $n = 2$, $\quad F_2^2 = 4 = 1 \,.\, 3 + 1 = F_1 F_3 + (-1)^2$

Now assume that the result is true for $n = k$

Therefore $\quad F_k^2 = F_{k-1} F_{k+1} + (-1)^k \quad \ldots(1)$

Consider

$$\begin{aligned} F_k F_{k+2} + (-1)^{k+1} &= F_k (F_{k+1} + F_k) + (-1)^{k+1} \\ &= F_k F_{k+1} + F_k^2 + (-1)^{k+1} \\ &= F_k F_{k+1} + F_{k-1} F_{k+1} + (-1)^k + (-1)^{k+1} \text{ (by (1))} \\ &= F_{k+1} (F_k + F_{k-1}) \end{aligned}$$

$= F_{k+1}\, F_{k+1}$ (By definition of recurrence relation)

$= F_{k+1}^2$

The result is true for $n = k + 1$

$\Rightarrow$ Hence the result by induction.

Example. Prove that $F_1 + F_3 + ... + F_{2n-1} = F_{2n} - 1,\ n \geq 1$.

Proof : Here, we use induction on n.

When $n = 1$, $F_1 = 1 = 2 - 1 = F_2 - 1$

Now assume that the result is true for $n = k$.

We have

$$F_1 + F_3 + ... + F_{2k-1} = F_{2k} - 1 \qquad ...(1)$$

Consider

$$F_1 + F_3 + ... + F_{2k-1} + F_{2k+1} = F_{2k} - 1 + F_{2k+1} \text{ by (1)}$$

$$= F_{2k+2} - 1 \quad \text{(by definition of Fibonacci Recurrence Relation)}$$

Therefore, the result is true for $n = k + 1$ by finite mathematical induction, the result is true for all $n \geq 1$.

6.4 SOLUTIONS OF RECURRENCE RELATIONS

Definition. *A sequence* $\{a_n\}_{n=0}^{\infty}$ *is said to be a* ***solution of a recurrence*** *relation if each value* a_n *(i.e.,)* $a_0, a_1, a_2, ..., a_n, ...$ *satisfies the recurrence relation.*

Example 1. $\{a_n\}_{n=0}^{\infty}$ where $a_n = 2^n$ is the solution of the recurrence relation $a_n = 2a_{n-1},\ n \geq 1$ and the sequence $\{c\,2^n\}_{n=0}^{\infty}$ where c is constant is also solution of recurrence relation $a_n = 2a_{n-1}$, $n \geq 1$.

Example 2. For the recurrence relation $a_n - 7\,a_{n-1} + 10a_{n-2} = 0,\ n \geq 2,\ a_n = c_1\,2^n + c_2\,5^n$ is the solution, where c_1, c_2 are arbitrary constants, because $a_n - 7a_{n-1} + 10a_{n-2} = (c_1\,2^n + c_2\,5^n) - 7\,(c_1\,2^{n-1} + c_2\,5^{n-1}) + 10\,(c_1\,2^{n-2} + c_2\,5^{n-2}) = (c_1\,2^n - 7c_1\,2^{n-1} + 10c_1\,2^{n-2}) + (c_1\,5^n - 7c_2\,5^{n-1} + 10c_2\,5^{n-2}) = c_1\,2^{n-2}\,(2^2 - 7\,(2) + 10) + c_2\,5^{n-2})\,(5^2 - 7\,(5) + 10)$

$$= c_1\,2^{n-2}\,(0) + c_2\,5^{n-2}\,(0) = 0.$$

Note 1. In each of the above examples there are infinitely many different solutions, one for each specific value of the constants. Suppose if we want the solution of the recurrence relation $a_n - 7a_{n-1} + 10a_{n-2} = 0,\ n \geq 2$ for which $a_0 = 10$ and $a_1 = 41$. These so called boundary conditions are requirements that must be satisfied in addition to that of satisfying the recurrence relation.

Note 2. We can compute a_n in terms of $a_{n-1}, ..., a_1, a_0$; then we can compute a_{n+1} in terms of a_n, $a_{n-1}, ..., a_1, a_0$ and so on, provided the value of the sequence at one or more points is given so that the computation can be initiated. Therefore, we need some values of the sequence. Usually the values for $a_0, a_1, a_2, ..., a_{k-1}$ are given and then it would be appropriate to call these initial conditions.

Note 3. Like in algebra, recurrence relations may have no solution. For example, $a_n^2 + a_{n-1}^2 = -1$ has no solution, since there are no real-valued functions f such that $f_n^2 + f_{n-1}^2 = -1$, because the squares of real numbers are always non-negative.

Example. $a_n = 3 \cdot 2^n + 7 \cdot 5^n$ is the solution of recurrence relation $a_n - 7a_{n-1} + 10a_{n-2} = 0$; $n \geq 2$ with initial conditions $a_0 = 10$, $a_1 = 41$.

Example 1. Find a recurrence relation for a_n the number of different ways to distribute either *a* \$1 bill, a \$2 bill, a \$5 bill, or a \$10 dollar bill on successive days until a total of n dollars has been distributed.

Solution : If, on the first day, we distribute a \$1 bill, then we are to distribute $n - 1$ dollars on the succeeding days and there are a_{n-1} ways to do that. If, on the oher hand, the first days distribution was a \$2 bill, there remains the problem of distributing $n - 2$ dollars ; this can be done a_{n-2} ways and so on. Thus, we see that $a_n = a_{n-1} + a_{n-2} + a_{n-5} + a_{n-10}$.

Example 2. *Find a recurrence relation for the number of ways to arrange flags on a flag-pole n-feet tall using 4 types of flags : red flags 2 feat high, or white, blue, and yellow flags each, foot high.*

Solution. Let a_n = the number of ways to arrange the flags on the flag-pole n-feet tall. If the first flag is 1 foot high, then there are a_{n-1} ways to arrange the flags on the other $n - 1$ feet of the pole. Since there are 3 colors for the first flag, in this case, there are $3a_{n-1}$ ways to arrange the pole. In case the first flag is red, and hence, 2 feet high, there are a_{n-2} ways to arrange the flags on the other $n - 2$ feet of the pole. Thus, $a_n = 3a_{n-1} + a_{n-2}$.

Example 3. *Suppose that a school principal decides to give a prize away each day. Suppose further that the principal has 3 different kinds of prizes worth \$1 each and 5 different kinds of prizes worth \$4 each. Find a recurrence relation for a_n = the number of different ways to distribute prizes worth n dollars.*

Solution. If, on the first day, *a* \$1 prize is given, the prize could have been chosen from one of the 3 different kinds of \$1 prizes, and then there are $n - 1$ dollars worth of prizes to be given away later. Thus, there are $3a_{n-1}$ ways to do this. If, on the other hand, the prize given on the first day was a \$4 prize then there are 5 different ways to choose the prize and a_{n-4} ways to distribute the remaining prizes.

Thus, $$a_n = 3a_{n-1} + 5a_{n-4}.$$

6.5 SOLVING RECURRENCE RELATIONS BY SUBSTITUTION AND GENERATING FUNCTIONS

We shall consider three methods of solving recurrence relations in this and the next two sections :

1. Substitution (Iteration)
2. Generating functions
3. Characteristic roots and

6.5.1 Method I (Substitution Method)

In this method, the recurrence relation for a_n is used repeated to solve for a general expression for a_n in terms of *n*.

Example 1. *Solve the recurrence relation $a_n = a_{n-1} + f(n)$ for $n \geq 1$ by substitution method.*

Solution. Here $a_1 = a_0 + f(1)$,

$$a_2 = a_1 + f(2) = a_0 + f(1) + f(2)$$
$$a_3 = a_2 + f(3) = a_0 + f(1) + f(2) + f(3)$$
$$\vdots$$
$$a_n = a_{n-1} + f(n) = a_0 + (f(1) + f(2) + \ldots\ldots + f(n))$$
$$= a_0 + \sum_{k=1}^{n} f(k).$$

Example 2. *Solve the recurrence relation $a_n = a_{n-1} + n^2$, where $a_0 = 7$ by substitution method.*

Solution. Initial condition is $a_0 = 7$.

We have
$$a_1 = a_0 + 1^2 = 7 + 1^2$$
$$a_2 = a_1 + 2^2 = 7 + 1^2 + 2^2$$
$$a_3 = a_2 + 3^2 = 7 + 1^2 + 2^2 + 3^2$$
$$\ldots\ldots\ldots\ldots\ldots\ldots\ldots\ldots$$
$$a_n = 7 + (1^2 + 2^2 + 3^2 + \ldots\ldots + n^2) = 7 + \frac{n(n+1)(2n+1)}{6}$$

which is required solution.

Example 3. *Solve the following recurrence relations by substitution.*

(i) $a_n = a_{n-1} + \dfrac{1}{n(n+1)}$, *where* $a_0 = 1$

(ii) $a_n = a_{n-1} + 3^n$, *where* $a_0 = 1$ [JNTU-MCA-Feb. 07]

Solution. (*i*) Initial condition is $a_0 = 1$.

We have
$$a_1 = a_0 + \frac{1}{1.2} = 1 + \frac{1}{1.2}$$
$$a_2 = a_1 + \frac{1}{2.3} = 1 + \frac{1}{1.2} + \frac{1}{2.3}$$
$$a_3 = a_2 + \frac{1}{3.4} = 1 + \frac{1}{1.2} + \frac{1}{2.3} + \frac{1}{3.4}$$
$$\ldots\ldots\ldots\ldots\ldots\ldots\ldots\ldots$$
$$a_n = 1 + \frac{1}{1.2} + \frac{1}{2.3} + \frac{1}{3.4} + \ldots\ldots + \frac{1}{n(n+1)}.$$
$$= 1 + \left(1 - \frac{1}{2}\right) + \left(\frac{1}{2} - \frac{1}{3}\right) + \left(\frac{1}{3} - \frac{1}{4}\right) + \ldots\ldots + \left(\frac{1}{n} - \frac{1}{n+1}\right)$$
$$a_n = 1 + 1 - \frac{1}{n+1} = 2 - \frac{1}{n+1} \text{ is the solution.}$$

(*ii*) Initial condition is $a_0 = 1$.

We have,
$$a_1 = a_0 + 3^1 = a_0 + 3 = 1 + 3$$
$$a_2 = a_1 + 3^2 = 1 + 3 + 3^2$$

$$a_3 = a_2 + 3^3 = 1 + 3 + 3^2 + 3^3$$

....

$$a_n = 1 + 3 + 3^2 + 3^3 + \ldots\ldots + 3^n.$$

$$a_n = \frac{3^{n+1} - 1}{3 - 1} = \frac{3^{n+1} - 1}{2} \text{ is the solution.}$$

Example 4. *Solve the recurrence relation $a_r = 2a_{r-1} + 1$ with $a_1 = 7$ for $r > 1$ by substitution method.*

Solution. The initial condition is $a_1 = 7$

We have
$$a_2 = 2a_1 + 1 = 2\,(7) + 1$$
$$a_3 = 2a_2 + 1 = 2\,(2\,.\,7 + 1) + 1 = 2^2\,.\,7 + 2 + 1$$
$$a_4 = 2a_3 + 1 = 2\,(2^2\,.\,7 + 2 + 1) + 1$$
$$= 2^3\,.\,7 + 2^2 + 2 + 1$$

....

$$a_r = 2^{r-1}\,.\,7 + 2^{r-2} + 2^{r-3} + \ldots\ldots + 2 + 1$$

$$= 7\,.\,2^{r-1} + \left(\frac{2^{r-1} - 1}{2 - 1}\right) \qquad \left(\because 1 + 2 + 2^2 + \ldots\ldots + 2^{r-2} = \frac{2^{r-1} - 1}{2 - 1}\right)$$

$\Rightarrow$
$$a_r = 7\,.\,2^{r-1} + 2^{r-1} - 1 \text{ for } r \geq 1$$

$\therefore$ The solution is $a_r = 7\,.\,2^{r-1} + 2^{r-1} - 1.$

Example 5. *Solve $a_n = a_{n-1} + n\,.\,3^n$, given $a_0 = 1$.*

Solution. We have,

$$a_n = a_{n-1} + n\,.\,3^n,\ a_0 = 1$$

$\therefore$
$$a_1 = a_0 + 1\,.\,3^1 = 1 + 1\,.\,3$$
$$a_2 = a_1 + 2\,.\,3^2 = 1 + 1\,.\,3 + 2\,.\,3^2$$
$$a_3 = a_2 + 3\,.\,3^3 = 1 + 1\,.\,3 + 2\,.\,3^2 + 3\,.\,3^3$$

....

$$a_n = 1 + 1\,.\,3 + 2\,.\,3^2 + 3\,.\,3^3 + \ldots\ldots\ n\,.\,3^n$$

$$a_n = 1 + \sum_{r=1}^{n} r\,.\,3^r, \text{ which is the required solution.}$$

Example 6. *If a_n is a solution of $a_{n+1} = ka_n$ for $n \geq 0$ and $a_3 = \frac{153}{49}$ and $a_5 = \frac{1377}{2401}$, find the value of k.*

Solution. The general solution of $a_{n+1} = ka_n$ is

$$a_n = k^n\, a_0,\ n \geq 1 \text{ (verify)}$$

Here
$$a_3 = k^3\,.\,a_0 \Rightarrow k^3\, a_0 = \frac{153}{49}$$

and $$a_5 = k^5 a_0 \Rightarrow k^5 a_0 = \frac{1377}{2401}$$

$\therefore$ $$\frac{k^5 a_0}{k^3 a_0} = \frac{1377}{2401} \times \frac{49}{153} = \frac{9}{49}$$

$$k^2 = \frac{9}{49} \quad \Rightarrow \quad k = \pm \frac{3}{7}.$$

Example 7. *Solve the recurrence relation* $a_n = a_{n-1} + \frac{n(n+1)}{2}$, $n \geq 1$.

[JNTU – November, 2006 – Set 4]

Solution. Given that $a_n = a_{n-1} + \frac{n(n+1)}{2}$

We have
$$\begin{aligned} a_1 &= a_0 + 1 \\ a_2 &= a_1 + 3 \\ &= a_0 + 1 + 3 \\ a_3 &= a_2 + 6 \\ &= a_0 + 1 + 3 + 6 \\ &\ldots\ldots\ldots\ldots\ldots\ldots \\ a_n &= a_{n-1} + \frac{n(n+1)}{2} \end{aligned}$$

$\therefore$ $$a_n = a_0 + 1 + 3 + 6 + \ldots\ldots + \frac{n(n+1)}{2}$$ is the required solution.

6.5.2 Method of Generating Functions

Recurrence relations can also be solved by using generating functions. Some equivalent expressions used are given below.

$$\text{If } A(x) = \sum_{n=0}^{\infty} a_n x^n \text{, then}$$

1. $\displaystyle\sum_{n=k}^{\infty} a_n x^n = A(x) - a_0 - a_1 x - \ldots\ldots - a_{k-1} x^{k-1}$

2. $\displaystyle\sum_{n=k}^{\infty} a_{n-1} x^n = x [A(x) - a_0 - a_1 x - \ldots\ldots - a_{k-2} x^{k-2}]$

3. $\displaystyle\sum_{n=k}^{\infty} a_{n-2} x^n = x^2 [A(x) - a_0 - a_1 x - \ldots\ldots - a_{k-3} x^{k-3}]$

4. $\sum_{n=k}^{\infty} a_{n-3}x^n = x^3 [A(x) - a_0 - a_1 x - \ldots\ldots - a_{k-4} x^{k-4}]$,

....

5. $\sum_{n=k}^{\infty} a_{n-k}x^n = x^k [A(x)]$

Where A (x) is called a generating function for a given recurrence relation.

Now we are ready to describe how to solve linear recurrence relation with constant coefficients by the use of generating functions. The process is best illustrated by examples.

Example 1. *Solve the recurrence relation : $a_n - 7a_{n-1} + 10a_{n-2} = 0$ for $n \geq 2$.*

[JNTU - MCA - Feb-07]

Solution. Here we use numbers for steps of the procedure.

1. Let $A(x) = \sum_{n=0}^{\infty} a_n x^n$.

2. Next multiply each term in the recurrence relation by x^n and sum from 2 to ∞ :

$$\sum_{n=2}^{\infty} a_n x^n - 7\sum_{n=2}^{\infty} a_{n-1}x^n + 10\sum_{n=2}^{\infty} a_{n-2}x^n = 0$$

3. Replace each infinite sum by an expression from the equivalent expressions :
$[A(x) - a_0 - a_1 x] - 7x[A(x) - a_0] + 10x^2 [A(x)] = 0$
4. Then Simplify :
$A(x)(1 - 7x + 10x^2) = a_0 + a_1 x - 7a_0 x$

$$\Rightarrow \qquad A(x) = \frac{a_0 + a_1 x - 7a_0 x}{1 - 7x + 10x^2} = \frac{a_0 + (a_1 - 7a_0)x}{(1-2x)(1-5x)}.$$

5. Decompose A (x) as a sum of partial fractions :

$$A(x) = \frac{c_1}{1-2x} + \frac{c_2}{1-5x},$$

where c_1 and c_2 are constants, as yet undetermined.
6. Express A (x) as a sum of familiar series :

$$A(x) = \frac{c_1}{1-2x} + \frac{c_2}{1-5x} = c_1 \sum_{n=0}^{\infty} 2^n x^n + c_2 \sum_{n=0}^{\infty} 5^n x^n.$$

7. Express a_n as the coefficient of x^n in A (x) and in the sum of the other series :

$$a_n = c_1 2^n + c_2 5^n.$$

Example 2. *Find a general expression for a_n using generating functions to $a_n - 7a_{n-1} + 12a_{n-2} = 0$; $n \geq 2$.*

Solution. Let $A(x) = \sum_{n=0}^{\infty} a_n x^n . = c_1 \sum_{n=0}^{\infty} (3x)^n + c_2 \sum_{n=0}^{\infty} (4x)^n$

Consider the given recurrence relation

$$a_n - 7a_{n-1} + 12a_{n-2} = 0 \; ; \; n \geq 2 \quad ...(1)$$

Multiply each term in (1) by x^n and sum from 2 to ∞, we get

$$\sum_{n=2}^{\infty} a_n x^n - 7 \sum_{n=2}^{\infty} a_{n-1} x^n + 12 \sum_{n=2}^{\infty} a_{n-2} x^n = 0$$

$$\Rightarrow \quad (A(x) - a_0 - a_1 x) - 7(A(x) - a_0)x + 12(A(x))x^2 = 0$$

$$\Rightarrow \quad A(x)(1 - 7x + 12x^2) = a_0 + a_1 x - 7a_0 x$$

$$\Rightarrow \quad A(x) = \frac{a_0 + (a_1 - 7a_0)x}{1 - 7x + 12x^2} = \frac{a_0 + (a_1 - 7a_0)x}{(1-3x)(1-4x)} = \frac{c_1}{1-3x} + \frac{c_2}{1-4x}$$

$$= c_1 \sum_{n=0}^{\infty} (3n)^n + c_2 \sum_{n=0}^{\infty} (4x)^n$$

$$\therefore \quad a_n = c_1 3^n + c_2 4^n$$

Example 3. *Solve the recurrence relation $a_n - 7a_{n-1} + 10a_{n-2} = 0$ for $n \geq 2$ with initial conditions $a_0 = 10$ and $a_1 = 41$ by using generating functioins.;* [JNTU - Nov 06 - Set 3]

Solution. From previous example 1 general solution of the recurrence relation $a_n - 7a_{n-1} + 10a_{n-2} = 0$, $n \geq 2$ without using initial conditions is

$$a_n = c_1 2^n + c_2 5^n \quad ...(1)$$

Put $n = 0$ in (1), we get

$$a_0 = c_1 2^0 + c_2 5^0 \Rightarrow 10 = c_1 + c_2 \quad ...(2)$$

Put $n = 1$ in (1), we get

$$a_1 = c_1 (2) + c_2 (5) \Rightarrow 41 = 2c_1 + 5c_2 \quad ...(3)$$

By solving (2), (3), we get

$$c_1 = 3, \; c_2 = 7.$$

$\therefore$ The unique solution of the recurrence relation is

$$a_n = (3)\, 2^n + (7)\, 5^n.$$

Example 4. *Using generating functions, solve the recurrence relation $a_n - 4a_{n-1} + 3a_{n-2} = 0$, $n \geq 2$ with initial conditions $a_0 = 2$ and $a_1 = 4$.*

Solution. Let $A(x) = \sum_{n=0}^{\infty} a_n x^n$

Consider the given recurrence relation

$$a_n - 4a_{n-1} + 3a_{n-2} = 0 \quad ...(1)$$

Multiply each term in (1) by x^n and sum from 2 to ∞, we get

$$\sum_{n=2}^{\infty} a_n x^n - 4\sum_{n=2}^{\infty} a_{n-1} x^n + 3\sum_{n=2}^{\infty} a_{n-2} x^n = 0$$

$\Rightarrow \quad (A(x) - a_0 - a_1 x) - 4(A(x) - a_0)x + 3x^2(A(x)) = 0$

$\Rightarrow \quad A(x)(1 - 4x + 3x^2) = a_0 + a_1 x - 4a_0 x$

$$\Rightarrow \quad A(x) = \frac{a_0 + a_1 x - 4a_0 x}{1 - 4x + 3x^2}$$

Put $\quad a_0 = 2,\ a_1 = 4$, we get

$$A(x) = \frac{2 + 4x - 8x}{1 - 4x + 3x^2} = \frac{2 - 4x}{(1-x)(1-3x)}$$

Using partial fractions,

$$A(x) = \frac{1}{1-x} + \frac{1}{1-3x}$$

$$= (1-x)^{-1} + (1-3x)^{-1}$$

$$= \sum_{n=0}^{\infty} x^n + \sum_{n=0}^{\infty} (3x)^n$$

$$= \sum_{n=0}^{\infty} \left(1^n + 3^n\right) x^n$$

$$\therefore \quad \sum_{n=0}^{\infty} a_n x^n = \sum_{n=0}^{\infty} \left(1^n + 3^n\right) x^n$$

Therefore, $\quad a_n = 1^n + 3^n$

$= 1 + 3^n$, which is required solution.

Example 5. *Solve the recurrence relation $a_n + 3a_{n-1} - 10a_{n-2} = 0$; $n \geq 2$, given $a_0 = 1$, $a_1 = 4$, using generating functions.* [JNTU - Nov 06 - Set 1]

Solution. Let $A(x) = \sum_{n=0}^{\infty} a_n x^n$

Consider the given recurrence relation

$$a_n + 3a_{n-1} - 10a_{n-2} = 0 \qquad \text{...(1)}$$

Multiplying each term in (1) by x^n and sum from 2 to ∞, we get

$$\sum_{n=2}^{\infty} a_n x^n + 3\sum_{n=2}^{\infty} a_{n-1} x^n - 10\sum_{n=2}^{\infty} a_{n-2} x^n = 0$$

$\Rightarrow \quad (A(x) - a_0 - a_1 x) + 3(A(x) - a_0)x - 10x^2(A(x)) = 0$

$\Rightarrow \quad A(x)(1 + 3x - 10x^2) = a_0 + 3a_0 x + a_1 x$

$$\Rightarrow \qquad A(x) = \frac{a_0 + 3a_0 x + a_1 x}{1 + 3x - 10x^2}$$

Put $a_0 = 1,\ a_1 = 4$, we get

$$A(x) = \frac{1 + 7x}{1 + 3x - 10x^2} = \frac{1 + 7x}{(1 - 2x)(1 + 5x)}$$

Using partial fractions,

$$A(x) = \frac{\frac{9}{7}}{1 - 2x} + \frac{\frac{-2}{7}}{1 + 5x}$$

$$\Rightarrow \qquad A(x) = \frac{9}{7}(1 - 2x)^{-1} - \frac{2}{7}(1 + 5x)^{-1}$$

$$= \frac{9}{7}\sum_{n=0}^{\infty}(2x)^n - \frac{2}{7}\sum_{n=0}^{\infty}(-1)^n (5x)^n$$

$$= \sum_{n=0}^{\infty}\frac{9}{7}(2x)^n - \frac{2}{7}(-1)^n (5x)^n$$

$$\therefore \qquad \sum_{n=0}^{\infty} a_n x^n = \sum_{n=0}^{\infty}\left(\frac{9}{7}2^n - \frac{2}{7}(-1)^n 5^n\right)x^n$$

Therefore, $a_n = \frac{9}{7} . 2^n - \frac{2}{7}(-1)^n 5^n$ which is required solution.

Example 6. *Solve $a_n - 6a_{n-1} + 12a_{n-2} - 8a_{n-3} = 0$ by generating functions.*

Solution. Given recurrence relation is

$$a_n - 6a_{n-1} + 12a_{n-2} - 8a_{n-3} = 0 \qquad ...(1)$$

Let $A(x) = \sum_{n=0}^{\infty} a_n x^n$

Multiply each term in (1) by x^n and sum from 3 to ∞, we get

$$\sum_{n=3}^{\infty} a_n x^n - 6\sum_{n=3}^{\infty} a_{n-1} x^n + 12\sum_{n=3}^{\infty} a_{n-2} x^n - 8\sum_{n=3}^{\infty} a_{n-3} x^n = 0$$

Replace the infinite sums by equivalent expressions, we get

$$[A(x) - a_0 - a_1 x - a_2 x^2] - 6x[A(x) - a_0 - a_1 x] + 12x^2[A(x) - a_0] - 8x^3 A(x) = 0.$$

$$\Rightarrow \quad A(x)[1 - 6x + 12x^2 - 8x^3] = a_0 + a_1 x - 6a_0 x + a_2 x^2 - 6a_1 x^2 + 12a_0 x^2$$

$$\Rightarrow \qquad A(x) = \frac{a_0 + a_1 x - 6a_0 x + a_2 x^2 - 6a_1 x^2 + 12a_0 x^2}{1 - 6x + 12x^2 - 8x^3}$$

$$= \frac{a_0 + (a_1 - 6a_0)\,x + (a_2 - 6a_1 + 12a_0)\,x^2}{(1-2x)^3}$$

By using partial fractions,

$$A\,(x) = \frac{c_1}{1-2x} + \frac{c_2}{(1-2x)^2} + \frac{c_3}{(1-2x)^3}, \text{ where } c_1,\ c_2,\ c_3 \text{ are constants.}$$

$$= c_1\,(1-2x)^{-1} + c_2\,(1-2x)^{-2} + c_3\,(1-2x)^{-3}$$

$$= c_1 \sum_{n=0}^{\infty} (2x)^n + c_2 \sum_{n=0}^{\infty} c\,(n+1, n)\,(2x)^n + \sum_{n=0}^{\infty} c_3\, c\,(n+2, n)\,(2x)^n$$

$$= \sum_{n=0}^{\infty} c_1\, 2^n x^n + \sum_{n=0}^{\infty} c_2\,(n+1)\,2^n x^n + \sum_{n=0}^{\infty} c_3 \frac{(n+2)\,(n+1)}{2}\,2^n x^n$$

$$\therefore \qquad \sum_{n=0}^{\infty} a_n x^n = \sum_{n=0}^{\infty}\left[c_1\, 2^n + c_2\,(n+1)\,2^n + c_3 \frac{(n+2)\,(n+1)}{2}\,2^n\right] x^n$$

Hence
$$a_n = c_1\, 2^n + c_2\,(n+1)\,2^n + c_3 \frac{(n+2)\,(n+1)}{2}\,2^n,$$

is the required solution.

Example 7. *Solve* $a_n - 8a_{n-1} + 21a_{n-2} - 18a_{n-3} = 0$ *for* $n \geq 3$.

Solution. Given recurrence relation is

$$a_n - 8a_{n-1} + 21a_{n-2} - 18a_{n-3} = 0 \qquad \text{...(1)}$$

Let
$$A\,(x) = \sum_{n=0}^{\infty} a_n x^n$$

Multiply each term in (1) by x^n and sum from 3 to ∞, we get

$$\sum_{n=3}^{\infty} a_n x^n - 8 \sum_{n=3}^{\infty} a_{n-1} x^n + 21 \sum_{n=3}^{\infty} a_{n-2} x^n - 18 \sum_{n=3}^{\infty} a_{n-3} x^n = 0$$

Replace the infinite sums by equivalent expressions, we get

$$[A\,(x) - a_0 - a_1\, x - a_2\, x^2] - 8x\,[A\,(x) - a_0 - a_1\, x] + 21x^2\,[A\,(x) - a_0] - 18x^3\,[A\,(x)] = 0$$

$$\Rightarrow \quad A\,(x)\,[1 - 8x + 21x^2 - 18x^3] = a_0 - 8a_0\, x + 21a_0\, x^2 + a_1\, x - 8a_1\, x^2 + a_2\, x^2$$

$$\Rightarrow \qquad A\,(x) = \frac{a_0 + (-8a_0 + a_1)\,x + (21a_0 - 8a_1 + a_2)\,x^2}{1 - 8x + 21x^2 - 18x^3}$$

$$= \frac{a_0 + (-8a_0 + a_1)\,x + (21a_0 - 8a_1 + a_2)\,x^2}{(1-2x)\,(1-3x)^2}$$

We see that there are constants c_1, c_2, c_3 such that

$$
\begin{aligned}
A(x) &= \frac{c_1}{1-2x} + \frac{c_2}{1-3x} + \frac{c_3}{(1-3x)^2} \\
&= c_1 (1-2x)^{-1} + c_2 (1-3x)^{-1} + c_3 (1-3x)^{-2} \\
&= c_1 \sum_{n=0}^{\infty} 2^n x^n + c_2 \sum_{n=0}^{\infty} 3^n x^n + c_3 \sum_{n=0}^{\infty} c(n+1, n)\, 3^n x^n \\
&= \sum_{n=0}^{\infty} c_1 2^n x^n + \sum_{n=0}^{\infty} c_2 3^n x^n + \sum_{n=0}^{\infty} (n+1) . c_3 3^n x^n
\end{aligned}
$$

$$\Rightarrow \qquad \sum_{n=0}^{\infty} a_n x^n = \sum_{n=0}^{\infty} \left[c_1 2^n + c_2 3^n + (n+1)\, c_3 3^n\right] x^n .$$

Hence $\qquad a_n = c_1 2^n + c_2 3^n + (n+1)\, c_3 3^n$ is the required solution.

Example 8. *Solve the recurrence relation $a_n - 9a_{n-1} + 26a_{n-2} - 24a_{n-3} = 0$ for $n \geq 3$ with initial conditions $a_0 = 0$, $a_1 = 1$ and $a_2 = 10$.*

Solution. Given recurrence relation is

$$a_n - 9a_{n-1} + 26a_{n-2} - 24a_{n-3} = 0 \qquad \text{...(1)}$$

Let $\qquad A(x) = \sum_{n=0}^{\infty} a_n x^n .$

Multiply each term in (1) by x^n and sum from 3 to ∞, we get

$$\sum_{n=3}^{\infty} a_n x^n - 9 \sum_{n=3}^{\infty} a_{n-1} x^n + 26 \sum_{n=3}^{\infty} a_{n-2} x^n - 24 \sum_{n=3}^{\infty} a_{n-3} x^n = 0$$

Replace the infinite sums by equivalent expressions, we get

$$[A(x) - a_0 - a_1 x - a_2 x^2] - 9x [A(x) - a_0 - a_1 x] + 26x^2 (A(x) - a_0) - 24x^3 A(x) = 0.$$

Put $\qquad a_0 = 0,\ a_1 = 1,\ a_2 = 10,$

$$[A(x) - x - 10x^2] - 9x [A(x) - x] + 26x^2 [A(x)] - 24x^3 A(x) = 0$$

$$\Rightarrow \quad A(x) [1 - 9x + 26x^2 - 24x^3] = x + 10x^2 - 9x^2 = x + x^2$$

$$\Rightarrow \qquad A(x) = \frac{x + x^2}{1 - 9x + 26x^2 - 24x^3} = \frac{x + x^2}{(1-2x)(1-3x)(1-4x)}$$

Using partial fractions, we get

$$A(x) = \frac{\frac{3}{2}}{1-2x} + \frac{-4}{1-3x} + \frac{\frac{5}{2}}{1-4x}$$

$$= \frac{3}{2}(1-2x)^{-1} - 4(1-3x)^{-1} + \frac{5}{2}(1-4x)^{-1}$$

$$= \frac{3}{2}\sum_{n=0}^{\infty} 2^n x^n - 4\sum_{n=0}^{\infty} 3^n x^n + \sum_{n=0}^{\infty} 4^n x^n$$

$$\therefore \qquad \sum_{n=0}^{\infty} a_n x^n = \sum_{n=0}^{\infty}\left[\frac{3}{2}(2)^n - 4(3)^n + \frac{5}{2}(4)^n\right] x^n.$$

Hence $\quad a_n = \frac{3}{2}(2)^n - 4(3)^n + \frac{5}{2}(4)^n$ is the required solution.

Example 9. *Solve the recurrence relation $a_n - 9a_{n-1} + 20a_{n-2} = 0$ for $n \geq 2$, given that $a_0 = -3$, $a_1 = -10$.*

Solution. Given recurrence relation is

$$a_n - 9a_{n-1} + 20a_{n-2} = 0 \qquad \text{...(1)}$$

Let $\quad A(x) = \sum_{n=0}^{\infty} a_n x^n$

Multiply each term in (1) by x^n and sum from 2 to ∞, we get

$$\sum_{n=2}^{\infty} a_n x^n - 9\sum_{n=2}^{\infty} a_{n-1} x^n + 20\sum_{n=2}^{\infty} a_{n-2} x^n = 0.$$

Replace the infinite sums by equivalent expressions, we get

$$[A(x) - a_0 - a_1 x] - 9x[A(x) - a_0] + 20x^2[A(x)] = 0$$

$$\Rightarrow \quad A(x)[1 - 9x + 20x^2] = a_0 + a_1 x - 9x a_0$$

Put $\quad a_0 = -3,\ a_1 = -10$, we get

$$A(x)[1 - 9x + 20x^2] = -3 + (-10)x - 9x(-3)$$

$$= 17x - 3$$

$$\Rightarrow \quad A(x) = \frac{17x - 3}{1 - 9x + 20x^2} = \frac{17x - 3}{(1-4x)(1-5x)}$$

$$= \frac{-5}{1-4x} + \frac{2}{1-5x}.$$

$$= -5(1-4x)^{-1} + 2(1-5x)^{-1}$$

$$= -5\sum_{n=0}^{\infty}(4x)^n + 2\sum_{n=0}^{\infty}(5x)^n$$

$$= -\sum_{n=0}^{\infty} 5 \cdot 4^n x^n + \sum_{n=0}^{\infty} 2 \cdot 5^n x^n$$

$$\therefore \qquad \sum_{n=0}^{\infty} a_n x^n = \sum_{n=0}^{\infty} \left(-5.4^n + 2.5^n\right) x^n.$$

Hence $\qquad a_n = -5.4^n + 2.5^n$

$= 2.5^n - 5.4^n$, is the required solution.

Example 10. *Solve $a_n - 3a_{n-1} + 2, n \geq 1$ with $a_0 = 1$ by generating functions.*

[JNTU – Feb 07 – Set 3]

Solution. Given recurrence relation is $a_n = 3a_{n-1} + 2$ ...(1)

Let $\qquad A(x) = \sum_{n=0}^{\infty} a_n x^n$ be the generating function.

Multiply each term in (1) by x^n and sum from 1 to ∞, we get

$$\sum_{n=1}^{\infty} a_n x^n = \sum_{n=1}^{\infty} 3a_{n-1} x^n + 2\sum_{n=1}^{\infty} x^n$$

Replace the infinite sums by equivalent expressions, we get

$$[A(x) - a_0] = 3x[A(x)] + 2x\sum_{n=1}^{\infty} x^{n-1}$$

$$\Rightarrow \qquad A(x)(1-3x) = a_0 + 2x(1-x)^{-1}$$

Put $\qquad a_0 = 1$, we get

$$A(x)(1-3x) = 1 + \frac{2x}{1-x} = \frac{1-x+2x}{1-x} = \frac{1+x}{1-x}$$

$$\Rightarrow \qquad A(x) = \frac{1+x}{(1-x)(1-3x)} = \frac{-1}{1-x} + \frac{2}{1-3x}$$

$$= -(1-x)^{-1} + 2(1-3x)^{-1}$$

$$= -\sum_{n=0}^{\infty} x^n + 2\sum_{n=0}^{\infty} (3x)^n$$

$$\therefore \qquad \sum_{n=0}^{\infty} a_n x^n = \sum_{n=0}^{\infty} \left[(-1) + 2.3^n\right] x^n$$

Hence $\qquad a_n = -1 + 2(3)^n$

$= 2(3)^n - 1$, is the required solution.

Example 11. *Solve the recurrence relation $a_n - 6a_{n-1} = 0$ for $n \geq 1$ and $a_0 = 1$ by using generating functions.*

Solution. Given recurrence relation is $a_n - 6a_{n-1} = 0$ ...(1)

Let $\quad A(x) = \sum_{n=0}^{\infty} a_n x^n$ be the generating function.

Multiply each term in (1) by x^n and sum from 1 to ∞, we get

$$\sum_{n=1}^{\infty} a_n x^n - 6 \sum_{n=1}^{\infty} a_{n-1} x^n = 0.$$

Replace the infinite sums by equivalent expressions, we get

$$[A(x) - a_0] - 6x\,[A(x)] = 0$$

$$A(x)(1 - 6x) = a_0$$

$$\Rightarrow \quad A(x) = \frac{a_0}{1 - 6x}$$

Substituting a_0 value, we get

$$A(x) = \frac{1}{1-6x} = (1 - 6x)^{-1} = \sum_{n=0}^{\infty} (6x)^n = \sum_{n=0}^{\infty} 6^n x^n$$

$$\Rightarrow \quad \sum_{n=0}^{\infty} a_n x^n = \sum_{n=0}^{\infty} 6^n x^n.$$

Hence $\quad a_n = 6^n$, which is required solution.

6.5.3 Method of Characteristic Roots

In this section, we explain the method of solving homogeneous linear recurrence relation of degree k by the method of characteristic roots. For this, we require the definition of the characteristic equation of a homogeneous linear recurrence relation.

Definition. *Let* $a_n + c_1 a_{n-1} + c_2 a_{n-2} + \ldots\ldots + c_k a_{n-k} = 0,\ n \geq k,\ c_k \neq 0$ *... (1) be a linear homogeneous recurrence relation of* ***degree k****. Then the equation,* $t^k + c_1 t^{k-1} + c_2 t^{k-2} + \ldots\ldots + c_k = 0$ *... (2) is called the* ***characteristic equation*** *of the given recurrence relation (1).* ***Degree of (2) is k****, therefore it has k roots.*

Let $\alpha_1, \alpha_2, \ldots\ldots, \alpha_k$ be the roots of the equation (2) and the roots $\alpha_1, \alpha_2, \ldots\ldots, \alpha_k$ are called **characteristic roots.**

Example. The characteristic equation of the recurrence relation

$$a_n - 3a_{n-1} + 2a_{n-2} = 0 \text{ is } t^2 - 3t + 2 = 0$$

$$\Rightarrow \quad (t-1)(t-2) = 0 \Rightarrow t = 1, 2 \text{ are characteristic roots.}$$

Here we have two types of characteristic roots

1. If the characteristic equation of a linear homogeneous recurrence relation of degree k has k distinct roots say $\alpha_1, \alpha_2, \ldots\ldots, \alpha_k$, then $a_n = c_1\alpha_1^n + c_2\alpha_2^n + \ldots\ldots + c_k\alpha_k^n$.

   Where $c_1, c_2, \ldots\ldots, c_k$ are constants, is the general solution of the given recurrence relation.
2. If the characteristic equation of a linear homogeneous recurrence relation of degree k has a root 'α' repeated k times then

$$a_n = (D_1 + D_2\, n + D_3\, n^2 + \ldots\ldots + D_k\, n^{k-1})\, \alpha^n$$

where D_1, D_2, ..., D_k are constants, is the general solution of the given recurrence relation.

Example 1. Solve $a_n - 3a_{n-1} + 2a_{n-2} = 0$ for $n \geq 2$.

Solution. The characteristic equation is $t^2 - 3t + 2 = 0$

$\Rightarrow \quad (t-1)(t-2) = 0$

$\Rightarrow \quad t = 1, 2.$

$\therefore$ The general solution is $a_n = c_1\, 1^n + c_2\, 2^n$

$= c_1 + c_2\, 2^n.$

Example 2. *Solve* $a_n - 3a_{n-1} - 4a_{n-2} = 0$ *for* $n \geq 2$.

Solution. The characteristic equation is $t^2 - 3t - 4 = 0$

$\Rightarrow \quad (t-4)(t+1) = 0 \Rightarrow t = 4, -1.$

The general solution is $a_n = c_1 (-1)^n + c_2\, 4^n$.

Example 3. *Solve* $a_n - 7a_{n-1} + 12a_{n-2} = 0$ *for* $n \geq 2$

Solution. The characteristic equation is $t^2 - 7t + 12 = 0$

$\Rightarrow \quad (t-3)(t-4) = 0 \Rightarrow t = 3, r.$

The general solution is $a_n = c_1\, 3^n + c_2\, 4^n$.

Example 4. Solve $a_n - 6a_{n-1} + 9a_{n-2} = 0$.

Solution. The characteristic equation is $t^2 - 6t + 9 = 0$

$\Rightarrow \quad (t-3)^2 = 0 \Rightarrow t = 3, 3.$

$\therefore$ The general solution is $a_n = (D_1 + D_2\, n)\, 3^n$.

Example 5. *Write the general form of the solution to*

(i) $a_n - 3a_{n-1} + 3a_{n-2} - a_{n-3} = 0$

(ii) $a_n - 9a_{n-1} + 27a_{n-2} - 27a_{n-3} = 0.$

Solution. (*i*) Characteristic equation is $t^3 - 3t^2 + 3t - 1 = 0$

$\Rightarrow \quad (t-1)^3 = 0 \Rightarrow t = 1, 1, 1.$

$\therefore$ The general solution is $a_n = (D_1 + D_2\, n + D_3\, n^2)\, 1^n$

$= D_1 + D_2\, n + D_3\, n^2.$

(*ii*) The characteristic equation is

$$f(t) = t^3 - 9t^2 + 27t - 27 = 0$$

Here $\quad f(3) = 108 - 108 = 0$

$\Rightarrow \quad t - 3$ is a factor of $f(t)$

3	1	−9	27	−27
	0	3	−18	27
	1	−6	9	0

$\therefore \quad f(t) = (t-3)(1t^2 - 6t + 9) = 0$

$= (t-3)(t-3)^2 = 0 \Rightarrow (t-3)^3 = 0$

$\Rightarrow \quad t = 3, 3, 3.$

General solution is $\quad a_n = (D_1 + D_2 n + D_3 n^2) 3^n.$

Example 6. *Solve the recurrence relation $a_n - 7a_{n-1} + 16a_{n-2} - 12a_{n-3} = 0$ for $n \geq 3$ with the initial conditions $a_0 = 1$, $a_1 = 4$ and $a_2 = 8$.*

Solution. The characteristic equation is $t^3 - 7t^2 + 16t - 12 = 0$

Take $\quad f(t) = t^3 - 7t^2 + 16t - 12$

Here $\quad f(1) = -2 \neq 0$

$f(-1) = -36 \neq 0$

$f(2) = 0$

$\therefore \quad t - 2$ is one factor of $f(t)$

2	1	−7	16	−12
	0	2	−10	12
	1	−5	6	0

$$f(t) = (t-2)(t^2 - 5t + 6) = (t-2)(t-2)(t-3) = (t-2)^2(t-3)$$

If $\quad f(t) = 0 \Rightarrow (t-2)^2 (t-3) = 0 \Rightarrow t = 2, 2, 3.$

The general solution is $a_n = c_1 2^n + c_2 n 2^n + c_3 3^n$ ...(1)

Put $\quad n = 0$ in (1), we get

$a_0 = c_1 2^0 + c_2 (0) 2^0 + c_3 3^0 \Rightarrow a_0 = c_1 + c_3$ ...(2)

Put $\quad n = 1$ in (1), we get

$a_1 = c_1 2 + c_2 (1) 2 + c_3 3 \Rightarrow 4 = 2c_1 + 2c_2 + 3c_3$ ...(3)

Put $\quad n = 2$ in (1), we get

$a_2 = c_1 2^2 + c_2 2 . 2^2 + c_3 3^2 \Rightarrow 8 = 4c_1 + 8c_2 + 9c_3$ ...(4)

By solving (2), (3), (4) equations, we get

$c_1 = 5, c_2 = 3, \quad \text{and} \quad c_3 = -4.$

Hence the unique solution of the recurrence relation is

$$a_n = (5) 2^n + 3 (n 2^n) - 4 (3^n).$$

Example 7. *Solve $a_n - 7a_{n-1} + 10a_{n-2} = 0$, $n \geq 2$ with $a_0 = 10$, $a_1 = 41$.*

Solution. Characteristic equation is $t^2 - 7t + 10 = 0$

$\Rightarrow \quad (t-2)(t-5) = 0 \Rightarrow t = 2, 5.$

$\therefore \quad a_n = c_1 2^n + c_2 5^n$ ...(1)

Put $\quad n = 0$ in (1) $\Rightarrow a_0 = c_1 2^0 + c_2 5^0 = c_1 + c_2$

$\Rightarrow \quad 10 = c_1 + c_2$ ...(2)

Put $\qquad n = 1$ in (1), we get

$$a_1 = c_1\, 2^1 + c_2\, 5^1$$

$\Rightarrow \qquad 41 = 2c_1 + 5c_2 \qquad \ldots(3)$

By solving (2), (3) we get

$$c_1 = 3,\ c_2 = 7$$

$\therefore$ Required solution is $a_n = (3)\ 2^n + (7)\ 5^n$.

Example 8. *Solve* $a_n + a_{n-1} - 5a_{n-2} + 3a_{n-3} = 0$

Solution. Characteristic equation is

$$f(t) = t^3 + t^2 - 5t + 3 = 0.$$

Since $\qquad f(1) = 0,\ t - 1$ is factor of $f(t)$.

$$\begin{array}{c|cccc} 1 & 1 & 1 & -5 & 3 \\ & 0 & 1 & 2 & -3 \\ \hline & 1 & 2 & -3 & \underline{|0} \end{array}$$

$\therefore \qquad f(t) = (t-1)\,(1t^2 + 2t - 3)$

$$= (t-1)\,(t+3)\,(t-1) = (t-1)^2\,(t+3)$$

$\because \qquad f(t) = 0$

$\therefore \qquad (t-1)^2\,(t+3) = 0$

$\Rightarrow \qquad t = 1,\ 1,\ -3$

$\therefore$ General solution is $a_n = C_1 1^n + C_2 n\, 1^n + C_3\,(-3)^n$

$$= C_1 + C_2 n + C_3\,(\ \ 3)^n.$$

Example 9. *Find the characteristic polynomial, characteristic equation for the homogeneous recurrence relations whose general solution has the form*

(i) $a_n = B_1 + n\,B_2$

(ii) $a_n = B_1 + n\,B_2 + n^2\,B_3$

(iii) $a_n = B_1\,2^n + B_2\,3^n$

(iv) $a_n = B_1\,2^n + B_2\,n\,2^n$

(v) $a_n = B_1\,2^n + B_2\,n\,2^n + B_3\,n^2\,2^n$

(vi) $a_n = C_1\,2^n + C_2\,n\,2^n + C_3\,3^n + C_4\,n\,3^n + C_5\,6^n$.

Solution. (*i*) Given general solution is $a_n = B_1 + n\,B_2$

$$= B_1\,1^n + n\,B_2\,1^n$$

$\therefore$ Characteristic polynomial is $(t-1)^2$

Characteristic equation is $(t-1)^2 = 0$.

(*ii*) Given general solution is $a_n = B_1 + n\,B_2 + n^2\,B_n$

$$= B_1\,1^n + n\,B_2\,1^n + n^2\,B_n\,1^n$$

$\therefore$ Characteristic polynomial is $(t-1)^3$

Characteristic equation is $(t-1)^3 = 0$.

(*iii*) Given general solution is $a_n = B_1\,2^n + B_2\,3^n$

Characteristic polynomial is $(t-2)\,(t-3) = t^2 - 5t + 6$

Characteristic equation is $t^2 - 5t + 6 = 0$.

(*iv*) Given general solution is $a_n = B_1 2^n + B_2 n 2^n$
Characteristic polynomial, equation are $(t-2)^2$, $(t-2)^2 = 0$.

(*v*) Given general solution is $a_n = B_1 2^n + B_2 n 2^n + B_3 n^2 2^n$
Characteristic polynomial, equation are $(t-2)^3$, $(t-2)^3 = 0$

(*vi*) Given general solution is $a_n = C_1 2^n + C_2 n 2^n + C_3 3^n + C_4 n 3^n + C_5 6^n$.
Characteristic polynomial, equation are $(t-2)^2 (t-3)^2 (t-6)$, $(t-2)^2 (t-3)^2 (t-6) = 0$.

Example 10. *Solve the Fibonacci relation $a_n = a_{n-1} + a_{n-2}$ with $a_0 = 0$, $a_1 = 1$ as initial conditions.*

Solution. The characteristic equation is $t^2 = t + 1$

$$\Rightarrow \quad t^2 - t - 1 = 0$$

$$\Rightarrow \quad t = \frac{1 \pm \sqrt{5}}{2}$$

The complete solution is $a_n = C_1 \left(\frac{1+\sqrt{5}}{2}\right)^n + C_2 \left(\frac{1-\sqrt{5}}{2}\right)^n$ ...(1)

Put $n = 0$ in (1), we get

$$a_0 = C_1 + C_2$$

$$\Rightarrow \quad 0 = C_1 + C_2 \qquad \text{...(2)}$$

Put $n = 1$ in (1), we get

$$a_1 = C_1 \left(\frac{1+\sqrt{5}}{2}\right) + C_2 \left(\frac{1-\sqrt{5}}{2}\right)$$

$$1 = C_1 \left(\frac{1+\sqrt{5}}{2}\right) + C_2 \left(\frac{1-\sqrt{5}}{2}\right) \qquad \text{...(3)}$$

By solving (2), (3), we get

$$C_1 = \frac{1}{\sqrt{5}}, \; C_2 = -\frac{1}{\sqrt{5}}$$

The complete solution of the Fibonacci relation is given by

$$a_n = \frac{1}{\sqrt{5}} \left(\frac{1+\sqrt{5}}{2}\right)^n - \frac{1}{\sqrt{5}} \left(\frac{1-\sqrt{5}}{2}\right)^n.$$

Example 11. *Solve $a_n = 2(a_{n-1} - a_{n-2})$ for $n \geq 2$ with $a_0 = 1$, and $a_1 = 2$.*

Solution. The characteristic equation is $t^2 = 2(t-1)$

$$\Rightarrow \quad t^2 - 2t + 2 = 0$$

$$\Rightarrow \qquad t = \frac{2 \pm \sqrt{(-2)^2 - 4(1)(2)}}{2(1)} = \frac{2 \pm \sqrt{-4}}{2}$$

$$= \frac{2 \pm 2i}{2} = 1 \pm i.$$

The general solution is $a_n = r^n (C_1 \cos n\theta + C_2 \sin n\theta)$ ...(1)

Where C_1, C_2 are arbitrary constants.

$$r = |1 \pm i| = \sqrt{2}, \theta = \tan^{-1}\frac{1}{1} = \frac{\pi}{4}$$

$$\therefore \qquad a_n = \left(\sqrt{2}\right)^n \left[C_1 \cos\frac{n\pi}{4} + C_2 \sin\frac{n\pi}{4}\right] \qquad ...(2)$$

The initial conditions are $a_0 = 1$, and $a_1 = 2$.

Put $n = 0$ in (2), we get

$$a_0 = \left(\sqrt{2}\right)^0 [C_1 \cos 0 + C_2 \sin 0]$$

$$\Rightarrow \qquad 1 = C_1$$

and put $n = 1$ in (2), we get

$$a_1 = \left(\sqrt{2}\right)^1 \left[C_1 \cos\frac{\pi}{4} + C_2 \sin\frac{\pi}{4}\right] = \sqrt{2}\left[1 . \frac{1}{\sqrt{2}} + C_2 . \frac{1}{\sqrt{2}}\right]$$

$$\Rightarrow \qquad 2 = 1 + C_2 \Rightarrow C_2 = 1$$

The general solution is

$$a_n = \left(\sqrt{2}\right)^n \left[\cos\frac{n\pi}{4} + \sin\frac{n\pi}{4}\right], \; n \geq 2.$$

6.6 SOLUTION OF INHOMOGENEOUS LINEAR RECURRENCE RELATIONS

In this section, we now to learn how to solve the inhomogeneous relation (IHR) : $a_0 + C_1 a_{n-1} + \ldots\ldots + C_k a_{n-k} = f(n)$ for $n \geq k$, where $C_k \neq 0$ and $f(n)$ is some specified function of n. We call it as forcing function of IHR.

For solving IHR, we use 2 methods :

These are (1) Characteristic roots method. 2. Generating functions method.

6.6.1 First Method : (Characteristic Roots Method)

The solution of a linear inhomogeneous recurrence relation with constant coefficients is the sum of the two parts. These are homogeneous solution and particular solution.

A solution, which satisfies the recurrence relation when the right hand side of the equation (relation) is set to 0, is called homogeneous solution.

A solution, which satisfies the recurrence relation with $f(n)$ on the right hand side is called particular solution. The homogeneous solution is denoted by $a_n^{(h)}$ and the particular solution is denoted by $a_n^{(p)}$. We follow the same procedure as in solving homogeneous recurrence relations for determining the homogeneous solution.

There is no general procedure for determining the particular solution. It depends on the nature of $f(n)$. To determine the particular solution, we use the following rules :

Rule 1. If $f(n)$ is of the form of a polynomial of degree m in n (*i.e.,*) $b_0 + b_1 n + b_2 n^2 + \ldots\ldots + b_{m-1} n^{m-1} + b_m n^m$

Then particular solution will be of the form

$Q_0 + Q_1 n + Q_2 n^2 + \ldots\ldots + Q_{m-1} n^{m-1} + Q_m n^m$. Provided one is not a characteristic root of the recurrence relation.

Rule 2. If $f(n)$ is of the form $(b_0 + b_1 n + b_2 n^2 + \ldots\ldots + b_{m-1} n^{m-1} + b_m n^m) a^n$, then particular solution is of the form $(Q_0 + Q_1 n + Q_2 n^2 + \ldots\ldots Q_{m-1} n^{m-1} + a_m n^m) a^n$, when a is not a characteristic root of the recurrence relation.

Rule 3. If a is the characteristic root of the multiplicity $r - 1$, when $f(n)$ is of the form $(b_0 + b_1 n + b_2 n^2 + \ldots\ldots + b_{m-1} n^{m-1} + b_m n^m) a^n$, then particular solution is of the form.

$$n^{r-1} (Q_0 + Q_1 n + Q_2 n^2 + \ldots\ldots + Q_{m-1} n^{m-1} + Q_m n^m) a^n.$$

Note 1. General solution = Homogeneous solution + particular solution.

i.e., $$a_n = a_n^{(h)} + a_n^{(p)}.$$

Note 2. If no initial conditions are given, then you have finished. If m initial conditions are given, we can get m linear equations in m unknowns and solve the system, to get a complete solution.

Example 1. *Solve the recurrence relation* $a_n - 9a_{n-1} + 20a_{n-2} = 1$.

Solution. Given recurrence relation is $a_n - 9a_{n-1} + 20a_{n-2} = 1$ ...(1)

Consider homogeneous relation $a_n - 9a_{n-1} + 20a_{n-2} = 0$.

The characteristic equation is $t^2 - 9t + 20 = 0$

$$t = 4, 5$$

$\Rightarrow$ $a_n^{(h)} = C_1 4^n + C_2 5^n$. Here one is not a characteristic root.

Since the right side $f(n) = 1$ is a constant, by rule 1, the particular solution will also be a constant, say Q. Substituting Q in (1), we get

$$Q - 9Q + 20Q = 1 \Rightarrow 12Q = 1 \Rightarrow Q = \frac{1}{12}$$

$\therefore$ $$a_n^{(p)} = \frac{1}{12}.$$

Hence the general solution $a_n = a_n^{(h)} + a_n^{(p)}$.

$$\Rightarrow \qquad a_n = C_1\, 4^n + C_2\, 5^n + \frac{1}{12}.$$

Example 2. *Solve $a_n - a_{n-1} - 6a_{n-2} = -30$ with $a_0 = 20$, $a_1 = -5$.*

Solution. Consider homogeneous relation $a_n - a_{n-1} - 6a_{n-2} = 0$.

The characteristic equation is $t^2 - t - 6 = 0$

$$\Rightarrow \qquad (t-3)(t+2) = 0 \Rightarrow t = -2, 3.$$

$\Rightarrow a_n^{(h)} = C_1(-2)^n + C_2\, 3^n$. Here 1 is not a ch. root.

Since the right side $f(n) = -30$ is a constant, by rule 1, the particular solution will also be a constant, say Q. Put $a_n = Q$ (In this case $a_{n-1} = Q = a_{n-2}$) in $a_n - a_{n-1} - 6a_{n-2} = -30$, we get

$$Q - Q - 6Q = -30 \Rightarrow -6Q = -30 \Rightarrow Q = 5$$

$$\Rightarrow \qquad a_n^{(p)} = 5.$$

Therefore, general solution $a_n = a_n^{(h)} + a_n^{(p)}$

$$a_n = C_1(-2)^n + C_2\, 3^n + 5 \qquad ...(1)$$

Put $n = 0$ in (1), we get

$$a_0 = C_1(-2)^0 + C_2\, 3^0 + 5$$

$$\Rightarrow \qquad 20 = C_1 + C_2 + 5 \Rightarrow C_1 + C_2 = 15 \qquad ...(2)$$

and put $n = 1$ in (1), we get

$$a_1 = C_1(-2)^1 + C_2\, 3^1 + 5$$

$$\Rightarrow \qquad -5 = -2C_1 + 3C_2 + 5 \Rightarrow -2C_1 + 3C_2 = -10 \qquad ...(3)$$

By solving (2), (3), we get

$$C_1 = 11,\ C_2 = 4.$$

Hence complete solution is $a_n = 11(-2)^n + 4(3)^n + 5$.

Example 3. *Solve $a_r + 5a_{r-1} = 9$ with initial condition $a_0 = 6$.*

Solution. The homogeneous recurrence relation is $a_r + 5a_{r-1} = 0$

∴ The characteristic equation is $t + 5 = 0$

$$\Rightarrow \qquad t = -5$$

$\Rightarrow \qquad a_r^{(h)} = C_1(-5)^r$. Here 1 is not a ch. root.

Since the right hand side of the given relation is a constant, the particular solution will also be a constant, say Q. Substituting in the given recurrence relation, we get

$$Q + 5Q = 9 \quad \Rightarrow \quad 6Q = 9 \Rightarrow Q = \frac{3}{2}$$

$$\Rightarrow \qquad a_r^{(p)} = \frac{3}{2}.$$

$\therefore$ The general solution $a_r = a_r^{(h)} + a_r^{(p)}$

$$a_r = C_1(-5)^r + \frac{3}{2} \qquad ...(1)$$

Put $r = 0$ in (1), we get

$$a_0 = C_1(-5)^0 + \frac{3}{2}$$

$$\Rightarrow \quad 6 = C_1 + \frac{3}{2} \Rightarrow C_1 = \frac{9}{2}.$$

Hence the complete solution is $a_r = \frac{9}{2}(-5)^r + \frac{3}{2}$.

Example 4. *Solve* $a_n - 5a_{n-1} + 6a_{n-2} = 1$. [JNTU - Feb 07 - Set 1 and Set 2]

Solution. Given recurrence relation is $a_n - 5a_{n-1} + 6a_{n-2} = 1$ ...(1)

Homogeneous relation is $a_n - 5a_{n-1} + 6a_{n-2} = 0$

The characteristic equation is $t^2 - 5t + 6 = 0$

$$\Rightarrow \quad (t-2)(t-3) = 0 \quad \Rightarrow \quad t = 2, 3.$$

$\therefore$ $a_n^{(h)} = C_1 2^n + C_2 3^n$. Here 1 is not a ch. root.

By Rule (1), the particular solution is Q.

Substituting this solution in (1) we get

$$Q - 5Q + 6Q = 1 \quad \Rightarrow \quad 2Q = 1 \Rightarrow Q = \frac{1}{2}$$

$$\therefore \quad a_n^{(p)} = \frac{1}{2}.$$

The general solution is $a_n = a_n^{(h)} + a_n^{(p)}$

$$= C_1 2^n + C_2 3^n + \frac{1}{2}.$$

Example 5. *Solve* $a_n - 7a_{n-1} + 10a_{n-2} = 4^n$. [JNTU - Feb 07 - Set 2]

Solution. Given recurrence relation is $a_n - 7a_{n-1} + 10a_{n-2} = 4^n$ ...(1)

Homogeneous relation is $a_n - 7a_{n-1} + 10a_{n-2} = 0$.

The characteristic equation is $t^2 - 7t + 10 = 0$

$$\Rightarrow \quad (t-2)(t-5) = 0 \quad \Rightarrow \quad t = 2, 5.$$

$$\therefore \quad a_n^{(h)} = C_1 2^n + C_2 5^n.$$

Since 4 is not characteristic root, by Rule 2, the particular solution is $Q\,4^n$.

Substituting this solution in (1), we get

$$Q\,4^n - 7Q\,4^{n-1} + 10Q\,4^{n-2} = 4^n$$

$$\Rightarrow \quad Q\,4^{n-2}[4^2 - 7 \times 4 + 10] = 4^n$$

$\Rightarrow \quad -2Q\,4^{n-2} = 4^n \quad \Rightarrow \quad Q = -8$

$\therefore \qquad a_n^{(p)} = -\,8.4^n.$

Hence the general solution is $a_n = a_n^{(h)} + a_n^{(p)}$

$= C_1\, 2^n + C_2\, 5^n + (-8)\, 4^n.$

Example 6. *Find the particular solution of the recurrence relation $a_n + 5a_{n-1} + 6a_{n-2} = 42\,(4)^n$.*

Solution. Given relation is $a_n + 5a_{n-1} + 6a_{n-2} = 42\,(4)^n$ ...(1)

Here the characteristic equation for homogeneous relation is $t^2 + 5t + 6 = 0$

$\Rightarrow \qquad (t+3)\,(t+2) = 0$

$\Rightarrow \qquad t = -\,3,\ -\,2$ are characteristic roots.

Since 4 is not a characteristic root, by Rule 2, the particular solution is $Q\,.\,4^n$.

Substituting this solution in (1), we get

$Q\,.\,4^n + 5Q\,.\,4^{n-1} + 6Q\,.\,4^{n-2} = 42\,.\,4^n$

$\Rightarrow \qquad Q\,.\,4^{n-2}\,(16 + 20 + 6) = 42\,.\,4^n$

$\Rightarrow \qquad 42Q = 42\,.\,4^2 \Rightarrow Q = 16$

$\Rightarrow \qquad a_n^{(p)} = 16\,.\,4^n = 4^2\,.\,4^n = 4^{n+2}$

Example 7. *Solve the recurrence relation $a_n + a_{n-1} = 3n\,2^n$.*

Solution. Given relation is $a_n + a_{n-1} = 3n\,.\,2^n$ ...(1)

Homogeneous relation is $a_n + a_{n-1} = 0$

The characteristic equation $t + 1 = 0 \Rightarrow t = -\,1$

$\therefore \qquad a_n^{(h)} = C_1\,(-1)^n.$

Since 2 is not a characteristic root, by Rule 2, the particular solution is $(Q_0 + Q_1\,n)\,2^n$.

Substituting this solution in (1), we obtain

$(Q_0 + Q_1\,n)\,2^n + (Q_0 + Q_1\,(n-1))\,2^{n-1} = 3n\,2^n$

which simplifies to $\left(\frac{3}{2}Q_0 - \frac{1}{2}Q_1\right)2^n + \frac{3}{2}Q_1\,n\,2^n = 3n\,2^n$

Comparing two sides, we get $\frac{3}{2}\,Q_1 = 3,\ \frac{3}{2}\,Q_0 - \frac{1}{2}\,Q_1 = 0$

$\Rightarrow \qquad Q_0 = \frac{2}{3},\ Q_1 = 2$

$\Rightarrow \qquad a_n^{(p)} = \left(\frac{2}{3} + 2n\right)2^n.$

Hence complete solution is $a_n = a_n^{(h)} + a_n^{(p)} = C_1\,(-1)^n + \left(\frac{2}{3} + 2n\right)2^n.$

Example 8. Find the general solution of $a_n - 3a_{n-1} - 4a_{n-2} = 4^n$.

Solution. Given recurrence relation is $a_n - 3a_{n-1} - 4a_{n-2} = 4^n$ ...(1)

Homogeneous relation is $a_n - 3a_{n-1} - 4a_{n-2} = 0$

The characteristic equation is $t^2 - 3t - 4 = 0$

$\Rightarrow \qquad (t+1)(t-4) = 0 \Rightarrow t = -1, 4.$

$\therefore \qquad a_n^{(h)} = C_1(-1)^n + C_2 4^n.$

Since 4 is a characteristic root, by Rule 3, the particular solution is $Q\, n\, 4^n$.

Substituting this solution in (1), we get

$$Q\,n\,4^n - 3Q(n-1)\,4^{n-1} - 4Q(n-2)\,4^{n-2} = 4^n$$

$$\Rightarrow \qquad Q\,n\,4^n - 3Q\,n\,4^{n-1} + 3Q\,4^{n-1} - 4Q\,n\,4^{n-2} + 8Q\,4^{n-2} = 4^n$$

$$\Rightarrow \qquad (16Q\,n - 12Q\,n + 12Q - 4Q\,n + 8Q)\,4^{n-2} = 16 \,.\, 4^{n-2}$$

$$\Rightarrow \qquad 20Q = 16 \Rightarrow Q = \frac{16}{20} = \frac{4}{5}$$

$$\therefore \qquad a_n^{(p)} = \frac{4}{5}\,n\,4^n.$$

The general solution is $a_n = a_n^{(h)} + a_n^{(p)}$

$$= C_1(-1)^n + C_2 4^n + \frac{4}{5}\,n\,4^n.$$

Example 9. *Solve the recurrence relation* $a_n - 7a_{n-1} + 10a_{n-2} = 6 + 8n$. *with* $a_0 = 1$, $a_1 = 2$.

Solution. Given recurrence relation is $a_n - 7a_{n-1} + 10a_{n-2} = 6 + 8n$ ...(1)

Homogeneous relation is $a_n - 7a_{n-1} + 10a_{n-2} = 0$

The characteristic equation is $t^2 - 7t + 10 = 0$

$\Rightarrow \qquad (t-2)(t-5) = 0 \Rightarrow t = 2, 5$

$\therefore \qquad a_n^{(h)} = C_1 2^n + C_2 5^n$. Here 1 is not a ch. root.

By Rule 1, the particular solution is of the form $Q_0 + Q_1 n$.

Substituting this solution in (1), we get

$$(Q_0 + Q_1 n) - 7(Q_0 + Q_1(n-1)) + 10(Q_0 + Q_1(n-2)) = 6 + 8n.$$

$$\Rightarrow \qquad (4Q_0 - 13Q_1) + 4Q_1 n = 6 + 8n.$$

By comparing like power coefficients on both sides, we get

$$4Q_0 - 13Q_1 = 6,$$

$$4Q_1 = 8$$

$$\Rightarrow \qquad Q_1 = 2,\ Q_0 = 8.$$

$$\therefore \qquad a_n^{(p)} = 8 + 2n.$$

Then general solution is $a_n = a_n^{(h)} + a_n^{(p)}$

$$a_n = C_1 2^n + C_2 5^n + 8 + 2n \qquad \text{...(2)}$$

Put $\qquad n = 0$ in (2), we get

$$a_0 = C_1 + C_2 + 8 \Rightarrow 1 = C_1 + C_2 + 8 \Rightarrow C_1 + C_2 = -7$$

and put $n = 1$ in (2), we get

$$a_1 = 2C_1 + 5C_2 + 10 \Rightarrow 2 = 2C_1 + 5C_2 + 10$$

$$\Rightarrow \quad 2C_1 + 5C_2 = -8$$

By solving these two equations, we get

$$C_1 = -9, C_2 = 2.$$

Hence complete solution is $a_n = -9\,(2)^n + 2\,(5)^n + 8 + 2n.$

Example 10. *Solve the recurrence relation* $a_n + 5a_{n-1} + 6a_{n-2} = 3n^2 - 2n + 1.$

[JNTU - Feb 07 - Set 1]

Solution. Given recurrence relation is $a_n + 5a_{n-1} + 6a_{n-2} = 3n^2 - 2n + 1$ ... (*i*)

The homogeneous relation is $a_n + 5a_{n-1} + 6a_{n-2} = 0$

The characteristic equation is $t^2 + 5t + 6 = 0$

$$\Rightarrow \quad (t + 2)(t + 3) = 0 \quad \Rightarrow \quad t = -2, -3.$$

$$\therefore \quad a_n^{(h)} = C_1(-2)^n + C_2(-3)^n.$$ Here 1 is not a ch. root.

Since the right hand side of (1) is $f(n)$ which is a polynomial by rule 1, the particular solution is of the form $Q_0 + Q_1 n + Q_2 n^2$.

Here $a_n = Q_0 + Q_1 n + Q_2 n^2$

$$\Rightarrow \quad a_{n-1} = Q_0 + Q_1(n-1) + Q_2(n-1)^2 \text{ and}$$

$$a_{n-2} = Q_0 + Q_1(n-2) + Q_2(n-2)^2$$

Substituting a_n, a_{n-1}, a_{n-2} values in (1), we get

$$(Q_0 + Q_1 n + Q_2 n^2) + 5(Q_0 + Q_1(n-1) + Q_2(n-1)^2) + 6(Q_0 + Q_1(n-2) + Q_2(n-2)^2)$$
$$= 3n^2 - 2n + 1$$

Which simplifies to

$$(12Q_0 - 17Q_1 + 29Q_2) + (12Q_1 - 34Q_2)\,n + 12Q_2\,n^2 = 3n^2 - 2n + 1$$

Comparing like power coefficients on both sides, we get

$$12Q_2 = 3,\ 12Q_1 - 34Q_2 = -2,\ 12Q_0 - 17Q_1 + 29Q_2 = 1$$

By solving, we get $Q_1 = \frac{13}{24}, Q_2 = \frac{1}{4}, Q_0 = \frac{71}{288}$

Therefore, $a_n^{(p)} = \frac{71}{288} + \frac{13}{24}n + \frac{1}{4}n^2$

Hence complete solution is $a_n = a_n^{(h)} + a_n^{(p)}$

$$\Rightarrow \quad a_n = C_1(-2)^n + C_2(-3)^n + \frac{71}{288} + \frac{13}{24}n + \frac{n^2}{4}.$$

Example 11. *Solve* $a_r - 10a_{r-1} + 25a_{r-2} = 2^r.$

Solution. Given recurrence relation is $a_r - 10a_{r-1} + 25a_{r-2} = 2^r$ ...(1)

Homogeneous relation is $a_r - 10a_{r-1} + 25a_{r-2} = 0$

The characteristic equation is $t^2 - 10t + 25 = 0$

$$\Rightarrow \qquad (t-5)^2 = 0 \;\Rightarrow\; t = 5, 5.$$

$$\therefore \qquad a_r^{(h)} = (C_1 + C_2\, r)\, 5^r.$$

Since 2 is not characteristic root, by Rule 2, the particular solution of the form $a_r^{(p)} = Q\, 2^r$.

Substituting this solution in (1), we get

$$Q\, 2^r - 10Q\, 2^{r-1} + 25Q\, 2^{r-2} = 2^r.$$

$$\Rightarrow \quad 2^r \left[Q - \frac{10Q}{2} + \frac{25Q}{4} \right] = 2^r$$

$$\Rightarrow \quad \frac{4Q - 20Q + 25Q}{4} = 1$$

$$\Rightarrow \qquad 9Q = 4 \;\Rightarrow\; Q = \frac{4}{9}$$

$$\therefore \qquad a_r^{(p)} = \frac{4}{9}\, 2^r.$$

Hence the general solution is $a_r = a_r^{(h)} + a_r^{(p)}$

$$= (C_1 + C_2 r)\, 5^r + \frac{4}{9}\, 2^r.$$

Example 12. *Solve* $a_n - 6a_{n-1} + 8a_{n-2} = n \,.\, 4^n$, *where* $a_0 = 8$, $a_1 = 22$.

Solution. Given recurrence relation is $a_n - 6a_{n-1} + 8a_{n-2} = n \,.\, 4^n$ ...(1)

Homogeneous relation is $a_n - 6a_{n-1} + 8a_{n-2} = 0$

The characteristic equation is $t^2 - 6t + 8 = 0$

$$\Rightarrow \qquad (t-2)(t-4) = 0 \;\Rightarrow\; t = 2, 4.$$

$$\therefore \qquad a_n^{(h)} = C_1\, 2^n + C_2\, 4^n.$$

Since 4 is the characteristic root with multiplicity 1 by Rule 3, the particular solution is of the form $n\,(Q_0 + Q_1\, n) \,.\, 4^n$.

Substituting this solution in (1), we get

$$n\,(Q_0 + Q_1\, n)\, 4^n - 6\,[(n-1)\,(Q_0 + Q_1\,(n-1)) \,.\, 4^{n-1}] + 8\,[(n-2)\,(Q_0 + Q_1\,(n-2))\, 4^{n-2}] = n \,.\, 4^n \qquad ...(2)$$

The equation (2) holds for all values of n and in particular when $n = 0$, $Q_0 + Q_1 = 0$ ...(3)

Putting $n = 1$ in (2), we get $Q_0 + 3Q_1 = 2$ ...(4)

By solving (3), (4), we get

$$Q_0 = -1,\; Q_1 = 1$$

$$\therefore \qquad a_n^{(p)} = n\,(-1 + n)\, 4^n = n\,(n-1)\, 4^n.$$

The general solution is $a_n = a_n^{(h)} + a_n^{(p)}$

$$a_n = C_1\, 2^n + C_2\, 4^n + n\,(n-1)\, 4^n \qquad ...(5)$$

The initial conditions $a_0 = 8$, $a_1 = 22$, gives $C_1 = 5$, $C_2 = 3$.

Hence complete solution is

$$a_n = 5 \,.\, 2^n + 3 \,.\, 4^n + n\,(n-1)\, 4^n.$$

Example 13. *Solve the recurrence relation*

$$a_{n+2} - 6a_{n+1} + 9a_n = 3 \,.\, 2^n + 7 \,.\, 3^n \text{ for } n \geq 0$$

given $\qquad a_0 = 1,\ a_1 = 4.$

Solution. Given recurrence relation is $a_{n+2} - 6a_{n+1} + 9a_n = 3 \,.\, 2^n + 7 \,.\, 3^n$ ...(1)

Homogeneous relation is $a_{n+2} - 6a_{n+1} + 9a_n = 0$

The characteristic equation is $t^2 - 6t + 9 = 0$

$$\Rightarrow \qquad (t-3)^2 = 0 \quad \Rightarrow \quad t = 3, 3.$$

$$\therefore \qquad a_n^{(h)} = (C_1 + C_2\, n)\, 3^n.$$

Since 3 is the characteristic root with multiplicity 2, by Rule 3, the particular solution is of the form

$$C \,.\, 2^n + D \,.\, n^2 \,.\, 3^n.$$

Substituting this solution in (1), we get

$$[C \,.\, 2^{n+2} + D \,.\, (n+2)^2 \,.\, 3^{n+2}] - 6\,[C \,.\, 2^{n+1} + D\,(n+1)^2 \,.\, 3^{n+1}] + 9\,[C \,.\, 2^n + D \,.\, n^2 \,.\, 3^n] = 3 \,.\, 2^n + 7 \,.\, 3^n.$$

Equating the corresponding terms on both sides, we get

$$C \,.\, 2^2 - 6C \,.\, 2 + 9C = 3 \text{ and } D\,(n+2)^2 \,.\, 3^2 - 6D\,(n+1)^2 \,.\, 3 + 9D\,n^2 = 7.$$

$$\Rightarrow \qquad C = 3,\ D = \frac{7}{9\,(n+2)^2 - 18\,(n+1)^2 + 9n^2} = \frac{7}{18}$$

$$\therefore \qquad a_n^{(p)} = 3 \,.\, 2^n + \frac{7}{18}\, n^2 \,.\, 3^n.$$

Therefore, the general solution is $a_n = a_n^{(h)} + a_n^{(p)}$.

$$\Rightarrow \qquad a_n = (C_1 + C_2\, n)\, 3^n + 3 \,.\, 2^n + \frac{7}{18}\, n^2 .\, 3^n \qquad ...(2)$$

Put $\qquad n = 0$ in (2), we get

$$a_0 = C_1 \,.\, 3^0 + 3 \,.\, 2^0 + \frac{7}{18} \,.\, 0 = C_1 + 3$$

$$\Rightarrow \qquad 1 = C_1 + 3 \Rightarrow C_1 = -2$$

Put $\qquad n = 1$ in (2), we get

$$a_1 = (C_1 + C_2)\, 3^1 + 6 + \frac{7}{6}$$

$$\Rightarrow \qquad 4 = (-2 + C_2)\, 3 + 6 + \frac{7}{6}$$

$$\Rightarrow \qquad C_2 = \frac{17}{18}.$$

Substituting C_1, C_2 values in (2), we get

$$a_n = \left(-2 + \frac{17}{18} n\right) 3^n + 3 . 2^n + \frac{7}{18} n^2 . 3^n.$$

6.6.2 Second Method : (Generating functions method)

We follow the same procedure as in solving homogeneous recurrence relations.

Procedure :

1. Let $A(x) = \sum_{n=0}^{\infty} a_n x^n$
2. Multiply each term in inhomogeneous relation by x^n, Sum from k to ∞.
3. And replace the infinite sums by expressions from the table of equivalent expressions for $A(x)$.

Example 1. *Find a solution to $a_n - a_{n-1} = 3(n-1)$ where $n \geq 1$, and $a_0 = 2$.*

Solution. Given recurrence relation is $a_n - a_{n-1} = 3(n-1)$ ...(1)

Let $$A(x) = \sum_{n=0}^{\infty} a_n x^n.$$

Multiplying each term in (1) by x^n and sum from 1 to ∞

$$\sum_{n=1}^{\infty} a_n x^n - \sum_{n=1}^{\infty} a_{n-1} x^n = 3 \sum_{n=1}^{\infty} (n-1) x^n$$

$$\Rightarrow \qquad (A(x) - a_0) - x A(x) = 3 \frac{x^2}{(1-x)^2}$$

$$\Rightarrow \qquad A(x)(1-x) = a_0 + \frac{3x^2}{(1-x)^2}$$

$$\Rightarrow \qquad A(x) = \frac{a_0}{1-x} + \frac{3x^2}{(1-x)^3} = \frac{2}{1-x} + \frac{3x^2}{(1-x)^3}$$

$$= 2(1-x)^{-1} + \frac{3}{2} \sum_{n=0}^{\infty} n(n-1) x^n \text{ by using table ....}$$

$$\therefore \qquad A(x) = 2 \sum_{n=0}^{\infty} x^n + \frac{3}{2} \sum_{n=0}^{\infty} n(n-1) x^n.$$

Hence $$a_n = 2 + \frac{3}{2} n(n-1).$$

Example 2. *Find a general expression for a solution to the recurrence relation $a_n - 5a_{n-1} + 6a_{n-2} = n(n-1)$ for $n \geq 2$, $a_0 = 1$, $a_1 = 5$.*

Solution. Given recurrence relation is $a_n - 5a_{n-1} + 6a_{n-2} = n(n-1)$ for $n \geq 2$ ...(1)

Let $$A(x) = \sum_{n=0}^{\infty} a_n x^n$$

Multiplying each term in (1) by x^n, and sum from 2 to ∞, we get

$$\sum_{n=2}^{\infty} a_n x^n - 5\sum_{n=2}^{\infty} a_{n-1} x^n + 6\sum_{n=2}^{\infty} a_{n-2} x^n = \sum_{n=2}^{\infty} n(n-1)x^n$$

$$\Rightarrow \quad (A(x) - a_0 - a_1 x) - 5x(A(x) - a_0) + 6x^2 A(x)$$

$$= \sum_{n=2}^{\infty} n(n-1)x^n.$$

Using the shifting properties of generating functions and table, 6.1, we see that

$$\sum_{n=2}^{\infty} n(n-1)x^n = \frac{2x^2}{(1-x)^3}.$$

Thus $$A(x)(1 - 5x + 6x^2) = a_0 + (a_1 - 5a_0)x + \frac{2x^2}{(1-x)^3}$$

$$\Rightarrow \quad A(x) = \frac{a_0 + (a_1 - 5a_0)x}{1 - 5x + 6x^2} + \frac{2x^2}{(1-x)^3(1-5x+6x^2)}$$

Put $a_0 = 1$, $a_1 = 5$, we get

$$A(x) = \frac{1}{1-5x+6x^2} + \frac{2x^2}{(1-x)^3(1-5x+6x^2)}$$

$$= \frac{1 - 3x + 5x^2 - x^3}{(1-x)^3(1-2x)(1-3x)}$$

By partial fractions,

$$A(x) = \frac{A}{1-x} + \frac{B}{(1-x)^2} + \frac{C}{(1-x)^3} + \frac{D}{1-2x} + \frac{E}{1-3x}$$

We can see that

$$A(x) = \frac{39}{12}\frac{1}{1-x} + \frac{3}{2}\cdot\frac{1}{(1-x)^2} + \frac{1}{(1-x)^3} - \frac{10}{1-2x} + \frac{21}{4}\cdot\frac{1}{1-3x}$$

$$= \frac{39}{12}(1-x)^{-1} + \frac{3}{2}(1-x)^{-2} + (1-x)^{-3} - 10(1-2x)^{-1} + \frac{21}{4}(1-3x)^{-1}$$

$$\Rightarrow \quad \sum_{n=0}^{\infty} a_n x^n = \frac{39}{12}\sum_{n=0}^{\infty} x^n + \frac{3}{2}\sum_{n=0}^{\infty}(n+1)x^n + \sum_{n=0}^{\infty}\frac{(n+2)(n+1)}{2}x^n - 10\sum_{n=0}^{\infty} 2^n x^n + \frac{21}{4}\sum_{n=0}^{\infty} 3^n x^n$$

Hence $$a_n = \frac{39}{12} + \frac{3}{2}(n+1) + \frac{(n+2)(n+1)}{2} - 10(2)^n + \frac{21}{4}(3)^n.$$

6.7 SOLVING NON-LINEAR RECURRENCE RELATIONS

Sometimes non-linear recurrence relations can be made into a linear recurrence by a substitution.

Example 1. *Solve the recurrence relation* $a_{n+2}^2 - 5a_{n+1}^2 + 4a_n^2 = 0$ *for* $n \geq 0$*, given* $a_0 = 4$, $a_1 = 13$.

Solution. Let $b_n = a_n^2$, then given relation becomes

$$b_{n+2} - 5b_{n+1} + 4b_n = 0,\ n \geq 0$$

$$\Rightarrow \quad b_n - 5b_{n-1} + 4b_{n-2} = 0,\ n \geq 2.$$

The characteristic equation for this relation is $t^2 - 5t + 4 = 0$

$$\Rightarrow \quad (t-4)(t-1) = 0 \Rightarrow t = 4, 1.$$

The general solution for b_n is $b_n = A \cdot 4^n + B \cdot 1^n$ ...(1)

where A, B are arbitrary constants.

Given that $a_0 = 4$, $a_1 = 13$

$$\Rightarrow \quad b_0 = a_0^2 = 16,$$

$$b_1 = a_1^2 = 169.$$

Using these in (1), we get A + B = 16, 4A + B = 169

Solving these, we get A = 51, and B = – 35.

Then putting these into (1), we get

$$b_n = 51 \cdot 4^n - 35 \Rightarrow a_n^2 = 51 \cdot 4^n - 35$$

$$\Rightarrow \quad a_n = \sqrt{51 \cdot 4^n - 35}, \text{ which is required solution for } a_n.$$

Example 2. *Solve the recurrence relation* $a_n^2 - 2a_{n-1}^2 = 1$ *; for* $n \geq 1$*, where* $a_0 = 2$.

Solution. Let $b_n = a_n^2$, then given relation becomes $a_n^2 - 2a_{n-1}^2 = 1$ to a linear recurrence relation $b_n - 2b_{n-1} = 1$, where $b_0 = a_0^2 = 2^2 = 4$.

The characteristic equation for this relation is

$$t - 2 = 0 \Rightarrow t = 2$$

$$\therefore \quad b_n^{(h)} = c\,2^n.$$

By Rule 1, particular solution of $b_n - 2b_{n-1} = 1$ is a constant, say Q.

Substituting Q for b_n in $b_n - 2b_{n-1} = 1$, we get

$$Q - 2Q = 1 \Rightarrow Q = -1$$

$$\therefore \quad b_n^{(p)} = -1$$

$$\therefore \quad b_n = b_n^{(h)} + b_n^{(p)}$$

$$b_n = C \cdot 2^n - 1$$

Put $n = 0$, we get

$$b_0 = C \cdot 2^0 - 1$$

$$\Rightarrow \quad 4 = C - 1 \Rightarrow C = 5$$

$$\therefore \quad b_n = 5 \cdot (2)^n - 1$$

$$\Rightarrow \quad a_n^2 = 5 \cdot 2^n - 1$$

$$\Rightarrow \quad a_n = \sqrt{5 \cdot 2^n - 1}$$

Example 3. *Solve the recurrence relation* $a_{n+2}^2 - 5a_{n+1}^2 + 6a_n^2 = 7n$ *for* $n \geq 0$, *given* $a_0 = 1 = a_1$.

Solution. Let $b_n = a_n^2$, then the given relation becomes $b_{n+2} - 5b_{n+1} + 6b_n = 7n$ and $b_0 = a_0^2 = 1^2 = 1$; $b_1 = a_1^2 = 1^2 = 1$.

$\therefore$ The recurrence relation is $b_{n+2} - 5b_{n+1} + 6b_n = 7n$ ...(1)

where $b_0 = 1 = b_1$.

For the homogeneous part of this relation, the characteristic equation is $t^2 - 5t + 6 = 0$

$$\Rightarrow \quad (t-2)(t-3) = 0 \Rightarrow t = 2, 3.$$

$$b_n^{(h)} = C_1 2^n + C_2 3^n$$

and the particular solution is of the form P + Q n, substituting this value for b_n in (1), we get $[P + Q(n+2)] - 5[P + Q(n+1)] + 6[P + Qn] = 7n$.

Equating the corresponding terms on both sides, we get

$$Q - 5Q + 6Q = 7, (2Q + P) - 5(Q + P) + 6P = 0$$

These give $Q = \dfrac{7}{2}, P = \dfrac{21}{4}$

$\therefore$ Particular solution, $b_n^{(p)} = \dfrac{7}{2}n + \dfrac{21}{4}$

$\therefore$ The general solution of (1) is $b_n = b_n^{(h)} + b_n^{(p)}$

$$= C_1 2^n + C_2 3^n + \frac{7}{2}n + \frac{21}{4} \quad ...(2)$$

Using $b_0 = 1$, $b_1 = 1$ in (2), we get

$$C_1 + C_2 + \frac{21}{4} = 1, 2C_1 + 3C_2 + \frac{7}{2} + \frac{21}{4} = 1$$

By solving we get, $C_2 = \dfrac{3}{4}, C_1 = -5$

$$\therefore \quad b_n = -5 \cdot 2^n + \frac{3}{4} \cdot 3^n + \frac{7}{2}n + \frac{21}{4}$$

$$\therefore \quad a_n = \sqrt{b_n} = \sqrt{\frac{3}{4} \cdot 3^n - 5 \cdot 2^n + \frac{7}{2}n + \frac{21}{4}}$$

which is the required solution.

EXERCISE 6.2.

1. Solve the following recurrence relations by substitution
   (i) $a_n = a_{n-1} + n$, where $a_0 = 2$.
   (ii) $a_n = a_{n-1} + n^3$, where $a_0 = 5$.
   (iii) $a_n = a_{n-1} + n(n-1)$, where $a_0 = 1$.
   (iv) $a_n = a_{n-1} + 3n^2 + 3n + 1$, where $a_0 = 1$.
   (v) $a_n = a_{n-1} + n\,3^n$, where $a_0 = 1$.
2. Write a general expression for $A(x) = \dfrac{P(x)}{Q(x)}$ specifying the coefficients for P (x) and Q (x), where A (x) generates the sequences $\{a_n\}$ and a_n satisfies the following recurrence relations.
   (i) $a_n + 5a_{n-1} + 3a_{n-2} = 0$ for $n \geq 2$, $a_0 = 1$, $a_1 = 2$.
   (ii) $a_n - 5a_{n-1} + 8a_{n-2} - 4a_{n-3} = 0$ for $n \geq 3$, $a_0 = 1$, $a_1 = 0$, $a_2 = 1$.
3. Solve the following recurrence relations using generating functions :
   (i) $a_n - 6a_{n-1} = 0$ for $n \geq 1$ and $a_0 = 1$.
   (ii) $a_n - 9a_{n-1} + 20a_{n-2} = 0$ for $n \geq 2$, and $a_0 = -3$, $a_1 = -10$.
   (iii) $a_n - 4a_{n-2} = 0$ for $n \geq 2$ and $a_0 = 0$, $a_1 = 1$.
   (iv) $a_n - a_{n-1} - 9a_{n-2} + 9a_{n-3} = 0$ for $n \geq 3$, and $a_0 = 0$, $a_1 = 1$, $a_2 = 2$.
   (v) $a_n - 10a_{n-1} + 33a_{n-2} - 36a_{n-3} = 0$ for $n \geq 3$ and $a_0 = 1$, $a_1 = 1$, $a_2 = -23$.
4. Find a general expression for a_n using generating functions
   (i) $a_n - 4a_{n-1} + 3a_{n-2} = 0$, $n \geq 2$.
   (ii) $a_n - 7a_{n-1} + 16a_{n-2} - 12a_{n-3} = 0$, $n \geq 3$.
5. Solve the following recurrence relations using the characteristic roots :
   (i) $a_r - 5a_{r-1} = 0$ for $r \geq 1$.
   (ii) $a_n - 3a_{n-1} - 4a_{n-2} = 0$ for $n \geq 2$.
   (iii) $a_n - a_{n-1} - 6a_{n-2} = 0$ for $n \geq 2$.
6. Solve the following recurrence relations using characteristic roots.
   (i) $a_n - 2a_{n-1} + a_{n-2} = 5n$ for $n \geq 2$.
   (ii) $a_n - 7a_{n-1} + 10a_{n-2} = 4^n$ for $n \geq 2$, $a_0 = 8$, $a_1 = 36$.
7. Find C_1, C_2, C_3 if the recurrence relation.
   $a_n + C_1\,a_{n-1} + C_2\,a_{n-2} + C_3\,a_{n-3} = 0$ for $n \geq 3$ has a general solution of the form.
   (i) $a_n = A_1\,3^n + A_2\,6^n$
   (ii) $a_n = A_1\,3^n + A_2\,n\,3^n + A_3\,2^n$.
8. Solve the following recurrence relations using characteristic roots.
   (i) $a_n - 4a_{n-1} - 12a_{n-2} = 0$
   (ii) $a_n - 4a_{n-1} + 4a_{n-2} = 0$
   (iii) $a_n + 5a_{n-1} + 6a_{n-2} = 0$
   (iv) $a_n - 7a_{n-1} + 12a_{n-2} = 4^n$
   (v) $a_n - 7a_{n-1} + 12a_{n-2} = 3^n$

9. Solve the following.

(*i*) $a_n^2 - 2\,a_{n-1}^2 = 0, n \geq 1, a_o = 2$

(*ii*) $\sqrt{a_n} - \sqrt{a_{n-1}^2} - 2\sqrt{a_{n-2}} = 0$ where $a_o = 1 = a_1$

(*iii*) $a_n = \sqrt{\dfrac{a_{n-2}}{a_{n-1}}}, a_o = 8, a_1 = \dfrac{1}{2\sqrt{2}}$

ANSWERS

1. (*i*) $a_n = \dfrac{n(n+1)}{2} + 2$ (*ii*) $a_n = \dfrac{n^2(n+1)^2}{4} + 5$

(*iii*) $a_n = 1 + \dfrac{n(n+1)(2n+1)}{6} - \dfrac{n(n+1)}{2}$ (*iv*) $a_n = (n+1)^3$

(*v*) $a_n = 1 + \sum n\,.\,3^n$

2. (*i*) $A(x) = \dfrac{1-5x-9x^2}{1-5x+8x^2-4x^3}$ (*ii*) $A(x) = \dfrac{1+7x}{1+5x+3x^2}$

3. (*i*) $a_n = 6n$ (*ii*) $a_n = 2\,.\,5^n - 5\,.\,4^n$

(*iii*) $a_n = 0$ if n is even $= 2^{n-1}$ if n is odd. (*iv*) $a_n = \dfrac{1}{12}\,[-3 + 4\,.\,3^n - (-3)^n]$

(*v*) $a_n = 15\,.\,3^n + 6\,(n+1)\,3^n - 20\,.\,4^n$

4. (*i*) $A(x) = \dfrac{a_0 + a_1 x - 4a_0 x}{1-4x+3x^2}$

(*ii*) $A(x) = \dfrac{a_0 + (a_1 - 7a_0)\,x + (a_2 - 7a_1 + 16a_0)\,x^2}{1-7x+16x^2-12x^3}$

5. (*i*) $a_r = C\,5^r$

(*ii*) $a_n = C_1\,4^n + C_2\,(-1)^n$

(*iii*) $a_n = C_1\,3^n + C_2\,(-2)^n$

6. (*i*) $a_n = C_1 + C_2\,n + \dfrac{5}{6}\,n^2\,(n+3)$

(*ii*) $a_n = (-8)\,4^n + 4\,.\,2^n + 12\,.\,5^n$

7. (*i*) $C_1 = -9, C_2 = 18, C_3 = 0$

(*ii*) $C_1 = -8, C_2 = 21, C_3 = -18$

8. (*i*) $a_n = c_1\,(-2)^n + c_2\,6^n$

(*ii*) $a_n = c_1\,2^n + c_2\,n\,2^n$

(*iii*) $a_n = c_1(-2)^n + c_2\,(-3)^n$

(*iv*) $a_n = c_1\,3^n + c_2\,4^n + n\,4^{n+1}$

(*v*) $a_n = c_1\,3^n + c_2\,4^n - n.3^{n+1}$

9. (*i*) $a_n = 2\sqrt{2^n}$

(*ii*) $a_n = \dfrac{\left[2^{n+1} + (-1)^n\right]^2}{3^2}$

(*iii*) $a_n = 2^{bn}$ where $b_n = 2(-1)^n + \left(\dfrac{1}{2}\right)^n$

CHAPTER

7

Graph Theory

7.1. INTRODUCTION

Graph theory was born in 1736, with Euler's paper on Konigsberg bridge problem. Euler, the father of graph theory, solved this long pending problem by means of a graph.

The term graph is used frequently to denote the diagram of the real valued function $y = f(x)$. But in graph theory we use the term graph as an object. Graphs are considered as discrete mathematical structures consisting of vertices and edges connecting the vertices.

7.2. REPRESENTATION OF GRAPH

Definition. *A **graph** G is a pair of sets (V, E) where V is a non-empty set and the elements of E are pairs of elements of V.*

If the elements of E are ordered pairs denoted as (u, v) then G is called a directed graph (digraph).

If the elements of E are unordered pairs denoted as $\{u, v\}$ then G is called a non-directed graph.

Example 1. G = (V, E)

where V = $\{a, b, c\}$ and E = $\{(a, b), (b, c), (a, c)\}$ is a directed graph.

Example 2. G = (V, E)

where V = $\{a, b, c\}$ and E = $\{\{a, b\}, \{b, c\}, \{a, c\}\}$ is an undirected graph.

1. **Graphical Representation**

   A graph may be represented by a diagram in which each vertex is denoted by a point (or a small circle) and each edge is represented by a straight line (or a curve) joining the points.

Example 1. Write the graphical representation of the digraph in example 1.1.

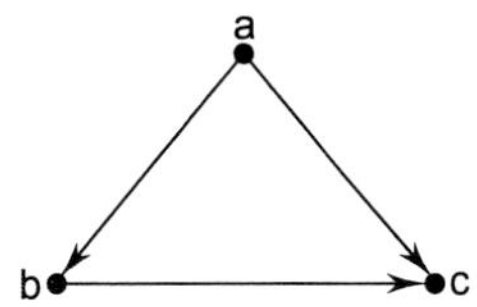

Example 2. Write the diagram of example 1.2.

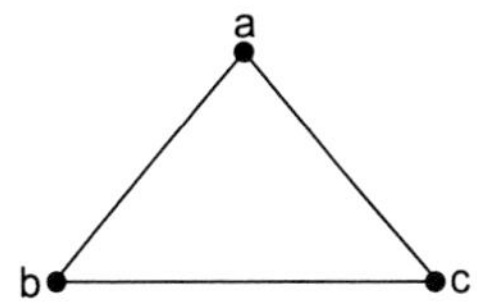

Example 3. Four main blood groups are A, B, AB and O. Persons with O blood group can donate it to any of the four types of blood group. Persons with A or B type blood group can only donate it to self or people with AB blood group. But persons with AB blood group can only donate it to the same group persons only. Draw a digraph to model the information.

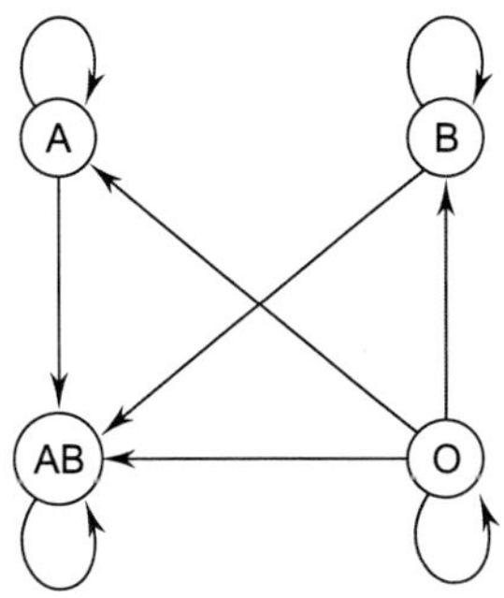

Example 4. Draw the graph of air routes between the 4 cities Delhi, Mumbai, Chennai and Hyderabad.

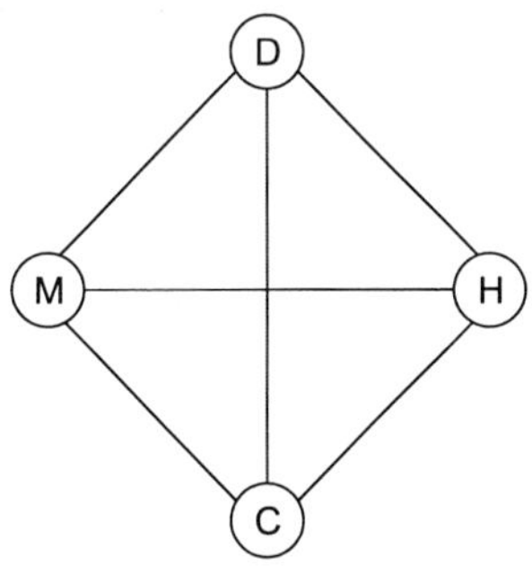

Example 5. Draw the graph of chemical compound C_2 H_6.

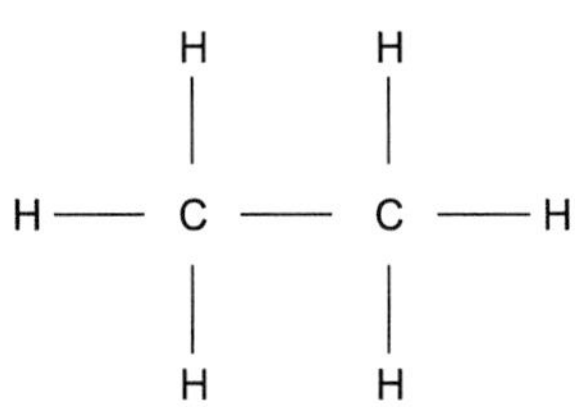

2. Adjacency Matrix Representation

Let G = (V, E) be a graph with '*n*' vertices ordered from v_1 to v_n.
An $n \times n$ matrix A = (a_{ij}), where

$$a_{ij} = \begin{cases} 1 & \text{if } v_i \text{ is adjacent to } v_j \\ 0 & \text{otherwise} \end{cases}$$

is called the adjacency matrix of G.

Note 1. The adjacency matrix of a non-directed graph is always symmetric since v_i is adjacent to v_j is as same as v_j is adjacent to v_i whereas the adjacency matrix of a directed graph may not be symmetric.

2. The advantage of matrix representation of graphs are : (*i*) it is a convenient way of representation when the number of vertices and edges are more (*ii*) Many known results of matrix algebra can be applied to study the concepts of graph theory.

Example 1. Find the adjacency matrix of the following graph :

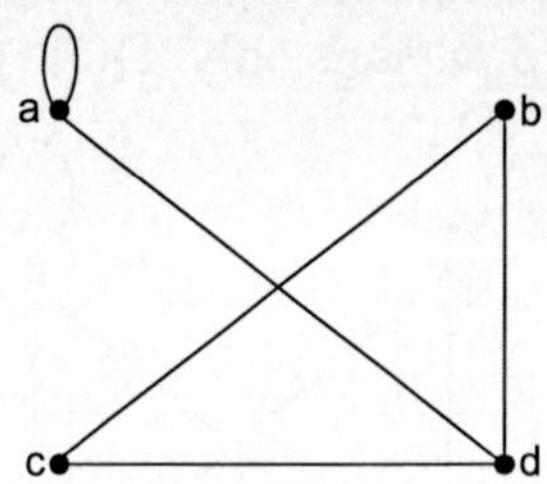

G

A_G =

	a	b	c	d
a	1	0	0	1
b	0	0	1	1
c	0	1	0	1
d	1	1	1	0

Example 2. Write the adjacency matrix of the following digraph.

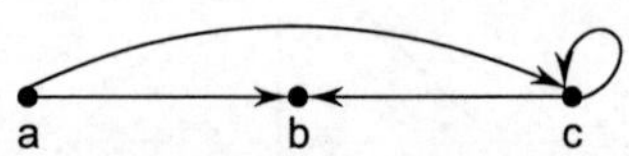

G

A_G =

	a	b	c
a	0	1	1
b	0	0	0
c	0	1	1

3. Incidence Matrix Representation

Let G = (V, E) be a non-directed graph, where V = $\{v_1, v_2, \ldots\ldots, v_n\}$ and E = $\{e_1, e_2, \ldots\ldots, e_m\}$. The $n \times m$ matrix I = (m_{ij}) where

$$m_{ij} = \begin{cases} 1 & \text{if } v_i \text{ is incident with } e_j \\ 0 & \text{otherwise} \end{cases}$$

is called the incidence matrix of G.

Example 1. Write the incidence matrix for the graph.

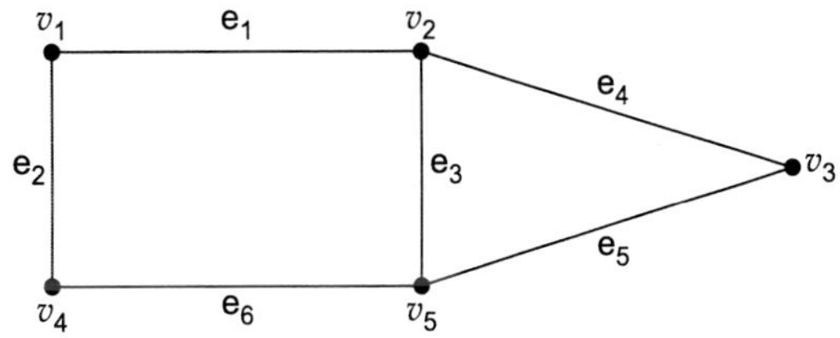

$I_G =$

	e_1	e_2	e_3	e_4	e_5	e_6
v_1	1	1	0	0	0	0
v_2	1	0	1	1	0	0
v_3	0	0	0	1	1	0
v_4	0	1	0	0	0	1
v_5	0	0	1	0	1	1

7.3. SPANNING TREES

To know the concept of spanning trees here we introduce some basic concepts informally through examples. However, the formal definitions of these would be learn by the reader in the next unit.

Any subpart of a graph G is called subgraph.

Example 1. Let the graph G, H_1 and H_2

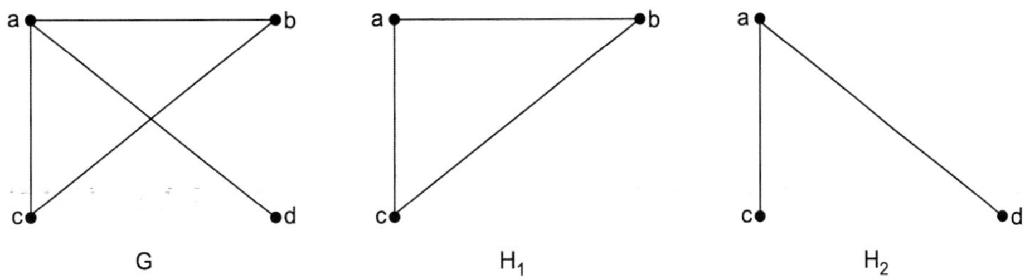

H_1, H_2 are subgraphs of G.

A graph is connected if there is a tour between every pair of its vertices.

Example 2. The graph G_1

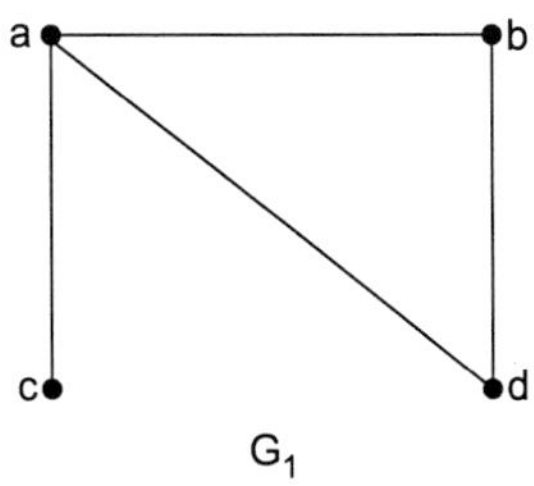

is connected

Example 3. The graph G_2

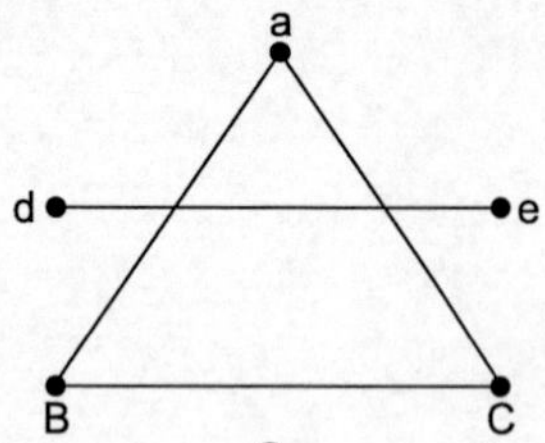

G_2

is not connected, since we can not find a tour from '*a*' to '*d*'.

However, the subgraphs H_1, H_2 of G_2 shown below are connected.

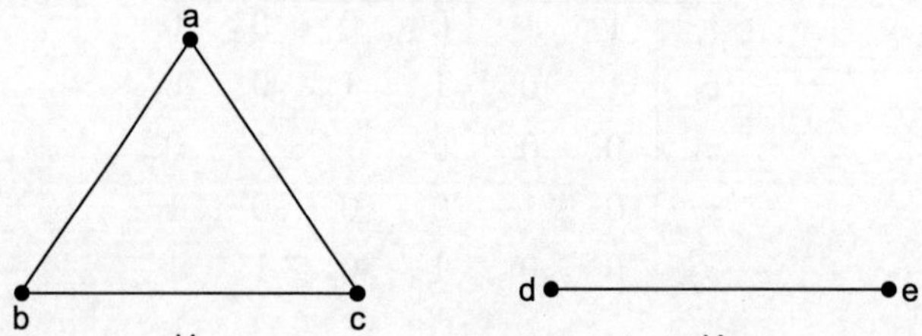

H_1 H_2

A circular tour through different vertices and edges is called a cycle.

Example 4. Consider the graph G,

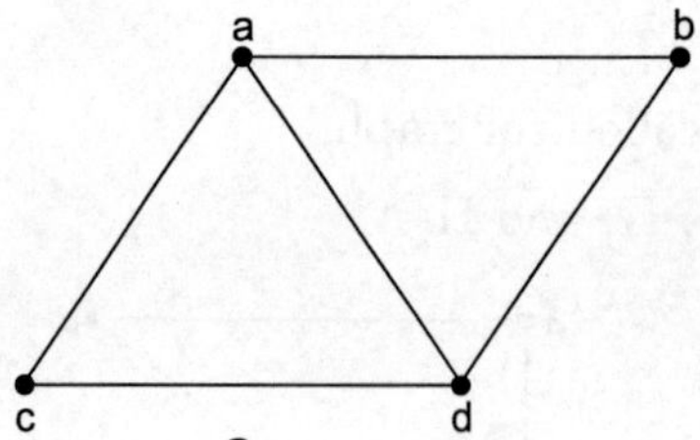

G

The following 3 are cycles in G.

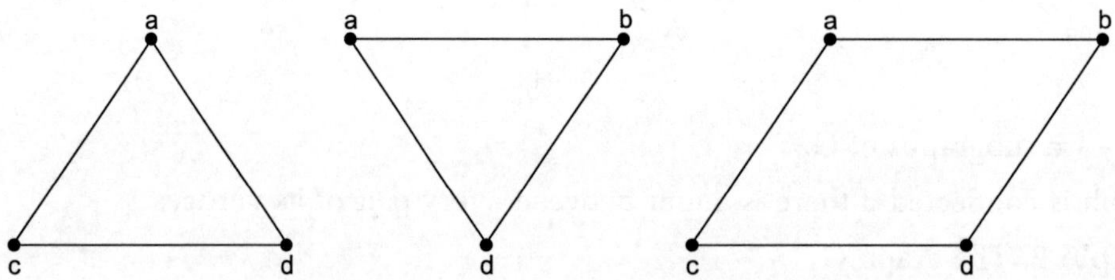

(i) (ii) (iii)

Example 5. The following graph has no cycles.

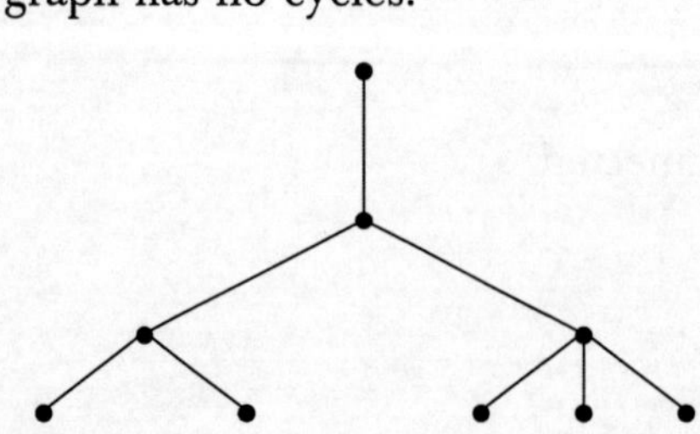

Definition : *A **tree** T is a connected graph without cycles.*

Example 6. Identify which of the following are trees :

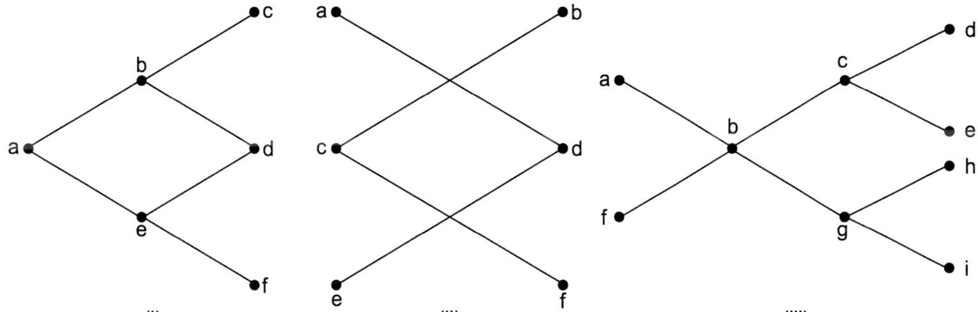

(i) (ii) (iii)

(*i*) is a connected graph and it has a cycle *a-b-d-e-d*. Hence it is not a tree.

(*ii*) has no cycles but it is not connected, since there is no tour from '*a*' to '*c*'. Hence it is not a tree.

(*iii*) is a tree, since it is a connected graph and has no cycles.

Definition : *A **rooted tree** is a tree in which a particular vertex is designated as the root.*

Example 7. Consider the following tree :

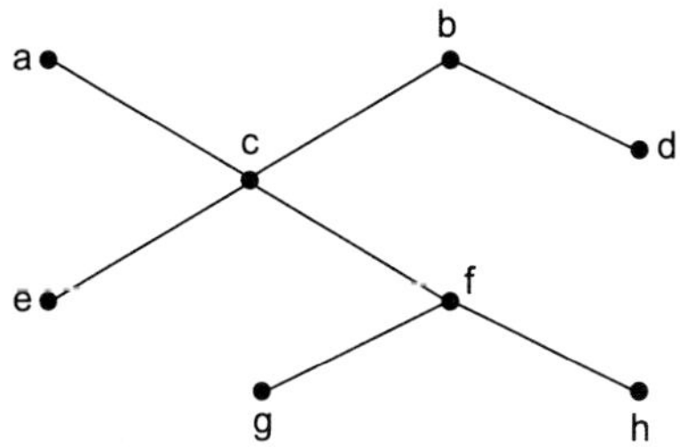

Let '*a*' be designated as root. Then the rooted tree can be written as

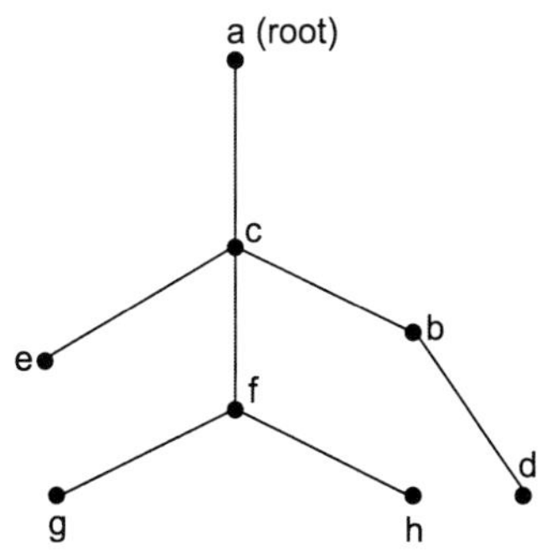

Parent	Children
a	c
c	e, f, b
f	g, h
b	d

Let '*b*' be designated as root. Then the rooted tree can be written as

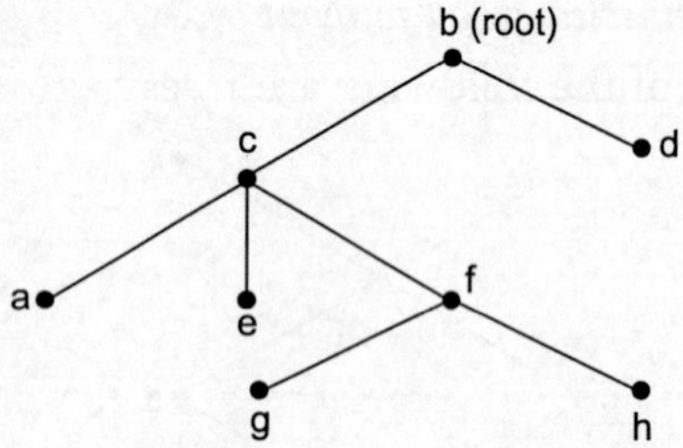

Parent	Children
b	c, d
c	a, e, f
f	g, h

Definition : *A rooted tree is a* ***directed tree*** *if there is a root from which there is a directed path to each vertex.*

Example 8. The following graph is a directed tree.

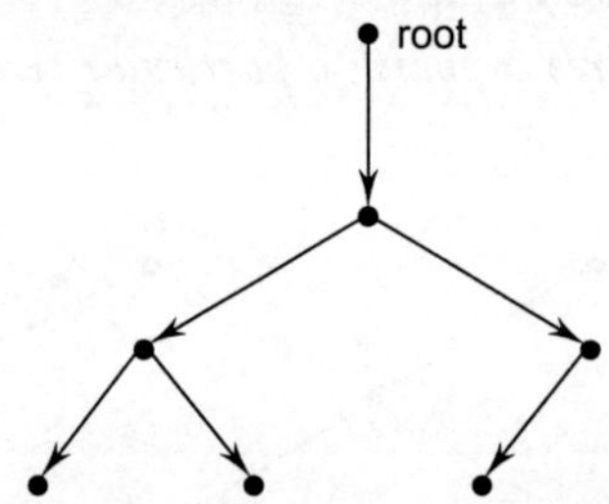

Theorem : A tree T with 'n' vertices has exactly $n - 1$ edges.

Proof : Let T be a tree with 'n' vertices. Let 'v' be designated as root. From root v (or parent) when we take edges we get one child for each edge. Thus, the number of edges is equal to the number of children. But each vertex in T except the root 'v' is a child to some parent. Therefore the number of edges in T is one less (root v) to the number of vertices in T. Thus, we have $n - 1$ edges in T.

Definition : *A subgraph H of a graph G is called a* ***spanning tree*** *of G if*

(*i*) H is a tree, and

(*ii*) H contains all the vertices of G.

A spanning tree that is a directed tree is called a directed spanning tree of G.

Example 9. A graph G and its spanning tree H is shown in the following figures :

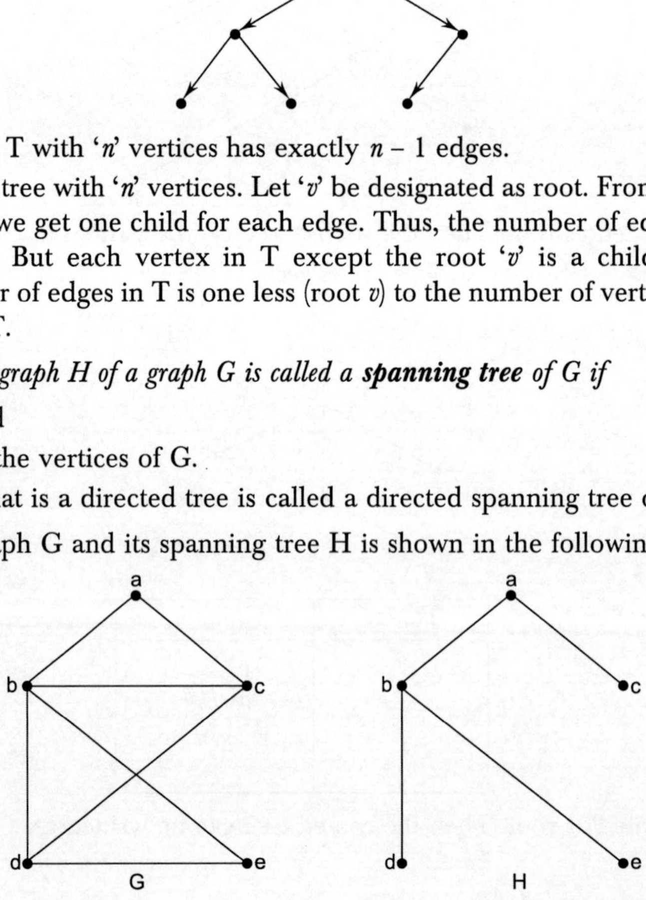

Example 10. Find all spanning trees of the following graph G.

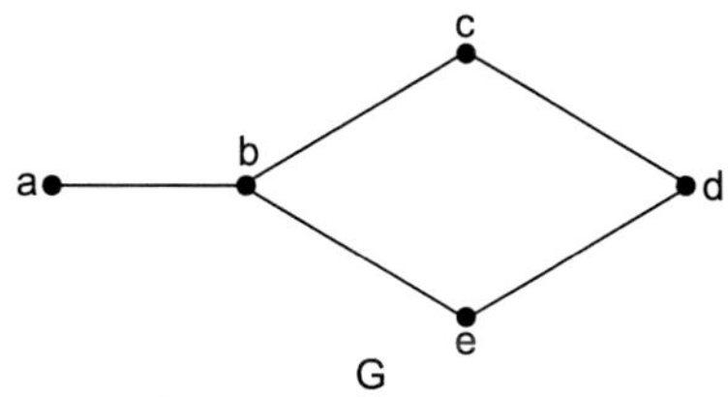

G

G has 5 vertices and 5 edges. It's spanning tree should contain 4 edges. We have to remove one edge which breaks the cycle *b–c–d–e–b.* Thus, the spanning trees are :

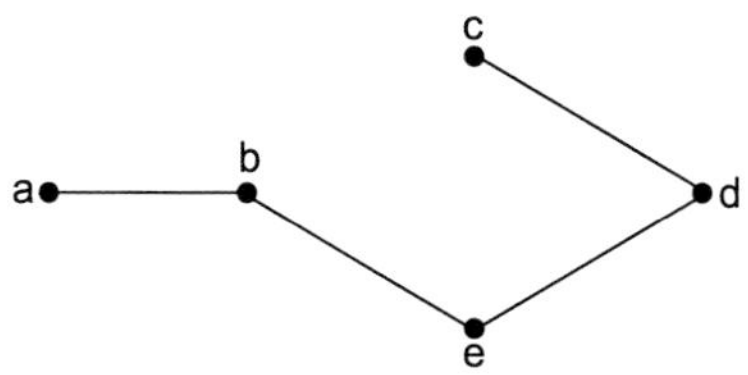

(i)

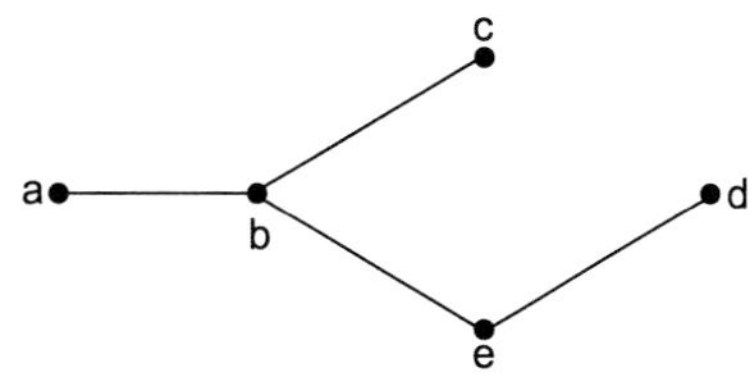

(ii)

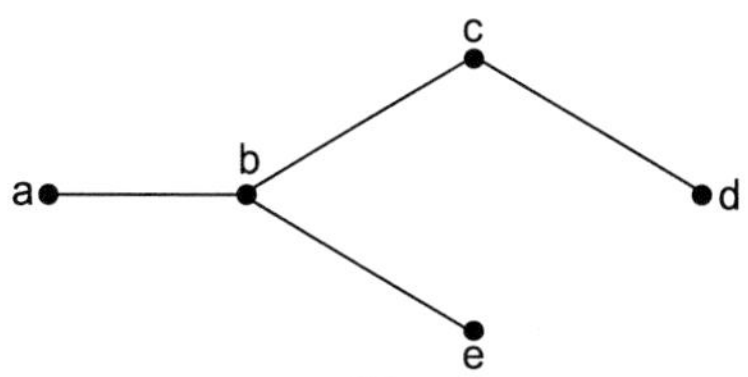

(iii)

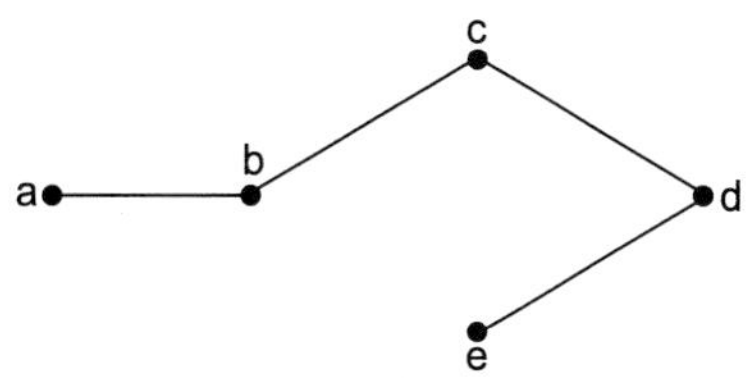

(iv)

Example 11. Draw all the spanning trees of the graph G.

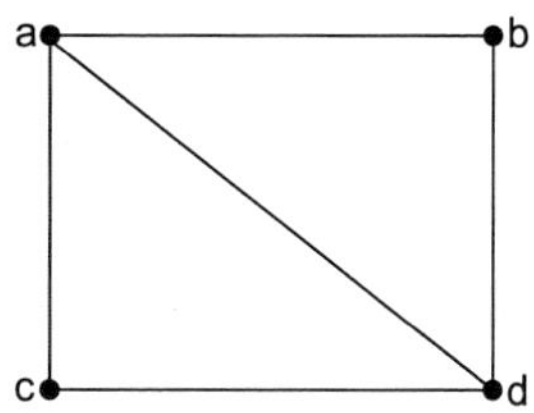

We have 4 vertices and 5 edges in G. A spanning tree of G should contain $4 - 1 = 3$ edges. We therefore need to remove 2 edges from G such that the resulting graph has no cycles. Following are the possible spanning trees of G.

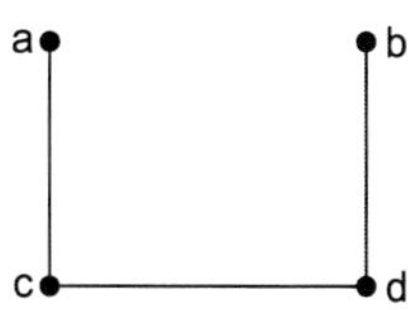

(i)

(ii)

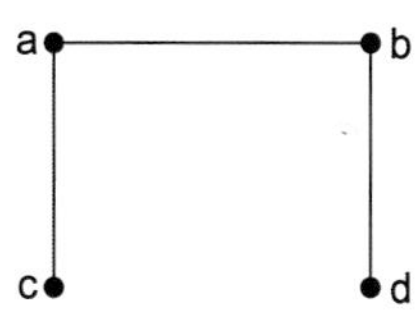

(iii)

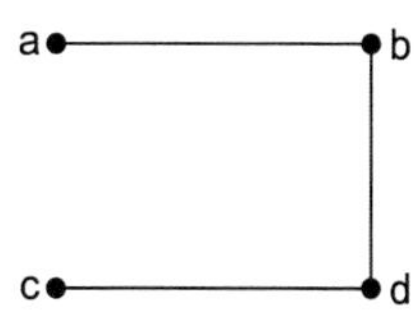

(iv)

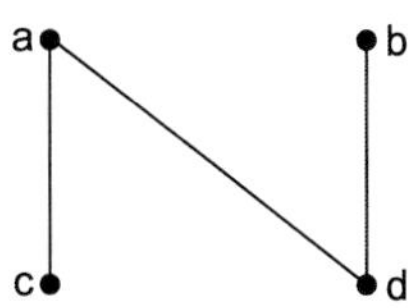

(v)

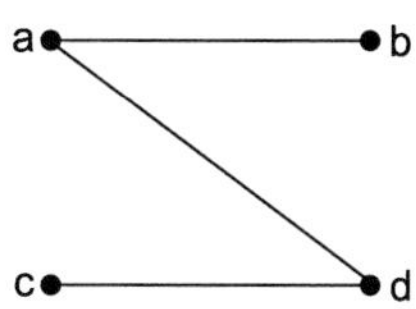

(vi)

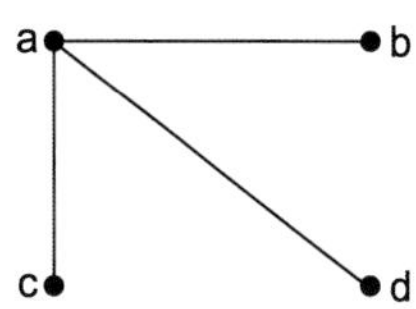

(vii)

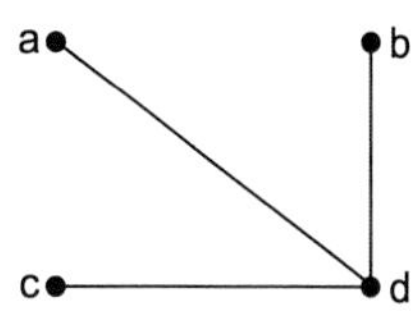

(viii)

Example 12. Find all directed spanning trees with root '*b*' of the following directed graph.
(JNTU, Feb 2007, Set 4)

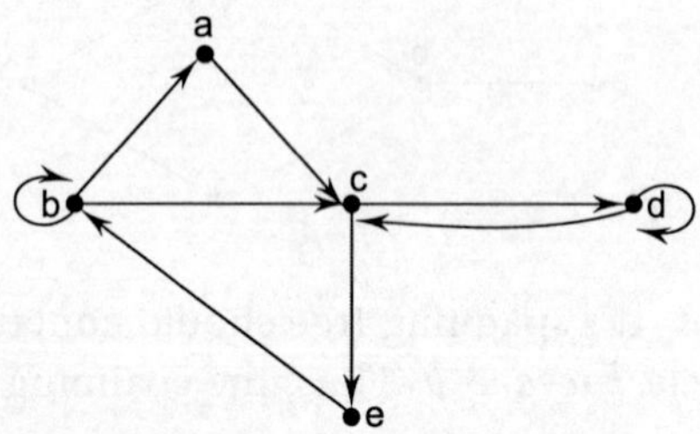

The graph has 5 vertices. Its spanning tree should contain 5 – 1 = 4 edges. Following are the directed spanning trees with root '*b*'.

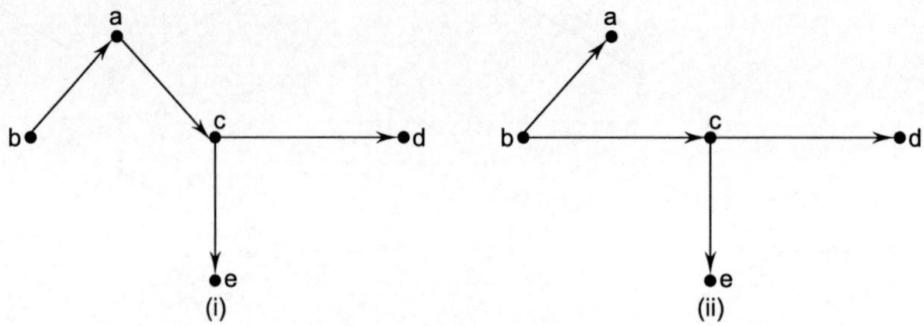

7.4 BREADTH-FIRST SEARCH AND DEPTH-FIRST SEARCH

To find spanning trees in graphs we remove edges which destroy the cycles in the graphs. But such an algorithm is not very efficient. We have two efficient algorithms for finding a spanning tree of a connected graph. They are Breadth-First Search (BFS) and Depth-First Search (DFS). In these algorithms spanning trees are constructed by adding edges successively.

Breadth-First Search Method (BFS)

We start at any vertex, taken as a root, and find edges incident on it. The vertices added at this stage become the vertices at level 1 in the spanning tree, arbitrarily order them. Next for each vertex at level 1, taken in order, we add edges incident in it so long as it does not produce a cycle. This leads to the vertices at level 2. We repeat the procedure till no further edge can be added to any of the vertex at the last level reached.

Example 1. Use BFS to find a spanning tree for the graph shown in the following figure with the vertex ordering abcdefg and *gfedcba.*

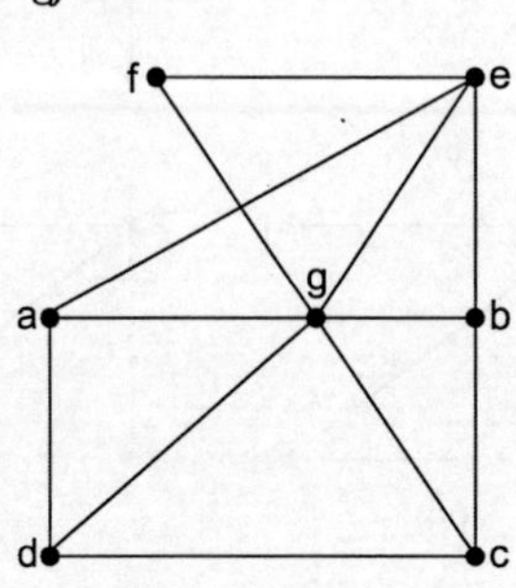

I. Let the vertex ordering be *abcdefg*.

(i) Start with the vertex 'a' as a root.

• a

(ii) Add all the edges incident on 'a'. They are $\{a, d\}$, $\{a, e\}$, $\{a, g\}$. The vertices d, e, g are at level 1 in the tree.

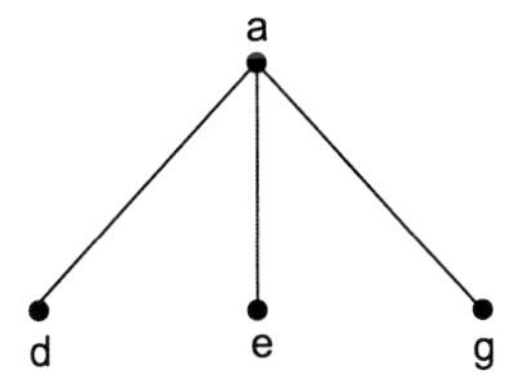

(iii) Add d to $\{d, c\}$ but not $\{d, g\}$ as it forms a cycle.

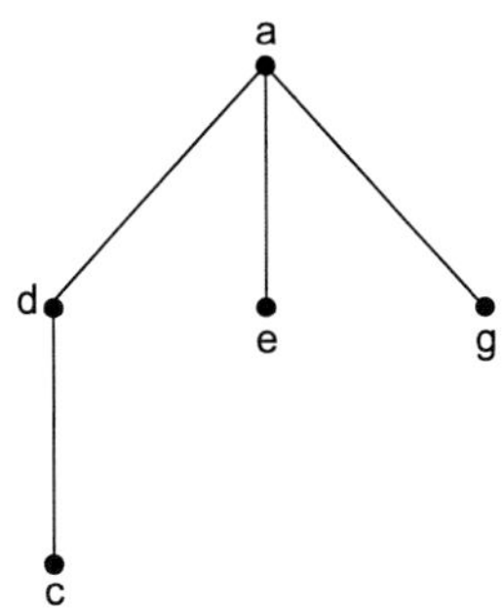

(iv) at e add $\{e, b\}$ and $\{e, f\}$ but not $\{e, g\}$ as it forms a cycle.

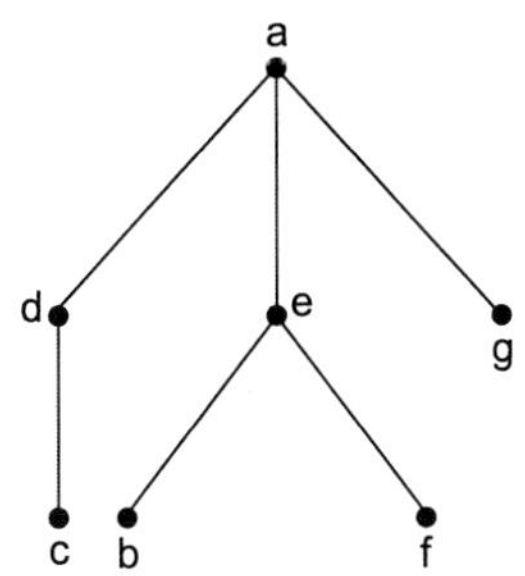

It contains all the vertices. Hence it is a spanning tree. We can also write this spanning tree as follows :

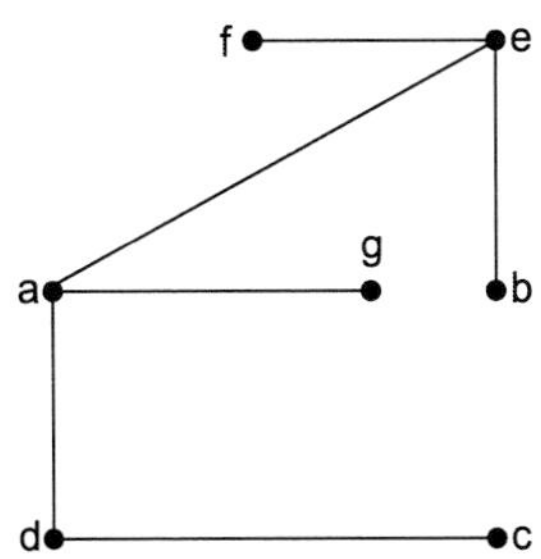

II. Let the vertex ordering be *gfedcba*

(i) We start with the vertex 'g' as root.

• g

(*ii*) Add all the edges incident on '*g*'. They are *f, e, d, c, b, a*.

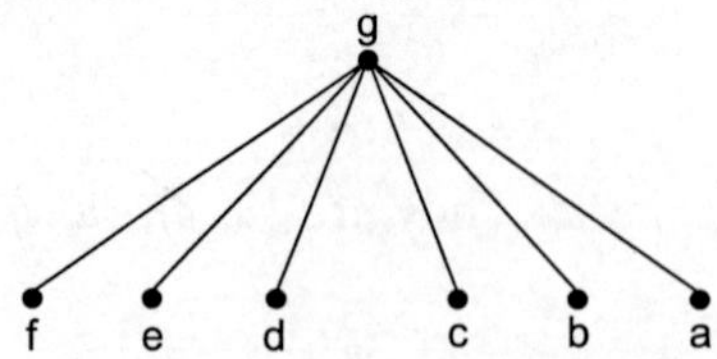

Since it contains all the vertices, it is a spanning tree. Also this can be written as follows:

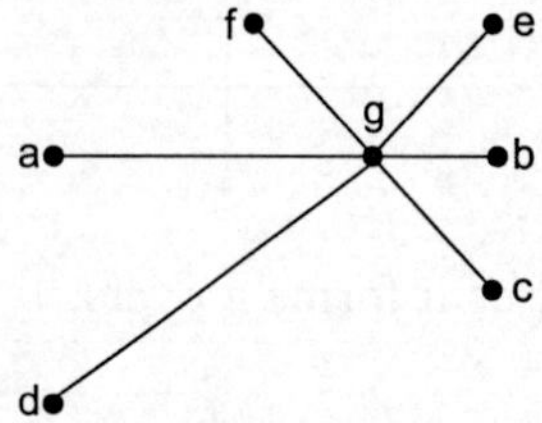

Breadth-First Search Algorithm

Input : A connected graph G with vertices labelled v_1, v_2, ..., v_n.

Output : A spanning tree T for G.

Method :

1. (Start) Let v_1 be the root of T. Form the set $V = \{v_1\}$.
2. (Add new edges) consider the vertices of V in order consistent with the original labelling. Then for each vertex '*x*' of V, add the edge $\{x, v_k\}$ to T where k is the minimum index such that adding the edge $\{x, v_k\}$ to T does not produce a cycle. If no edge can be added, then stop ; T is spanning tree for G. After all the vertices of V have been considered in order, go to step 3.
3. (Update V) Replace V by all the children v in T of the vertices x of V where the edges $\{x, v\}$ were added in Step 2. Go back and repeat step 2 for the new set V.

Depth-First Search Method (DFS) :

Arbitrarily choose a vertex as the root of the spanning tree T. Form a path starting at this vertex by successively adding edges where each new edge is incident with the last vertex in the path and the vertex not already in the path. Continue this process by adding edges to this path as long as possible without producing any cycles.

If the path goes through all the vertices in the graph, then the tree become a spanning tree. Otherwise more edges must be added, move back to next vertex to last vertex in the graph and if possible, form a new path starting at this vertex passing through vertices that are not already visited. If still all the vertices are not visited, move back to another vertex and try again. Repeat this process until no more edges can be added.

Example 2. Use DFS to find a spanning tree for the graph in example 3.1 with the vertex ordering abcdefg and gfedcba.

I. Let the vertex ordering be *abcdefg*.

(*i*) Start with the vertex '*a*' as a root.

(*ii*) We can not add *b* and *c* to *a*. Add *d* to *a*.

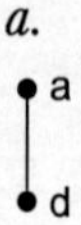

(*iii*) Now start with *d*. We can not add *b*. Add *c* to *d*.

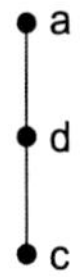

(*iv*) We start with *c*. We can add *b* to *c*.

(*v*) Start at *b*. We remain with vertices *e*, *f*, *g*. We can add *e* to *b*.

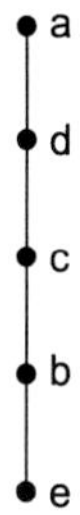

(*vi*) We can add *f* to *e* and then *g* to *f*.

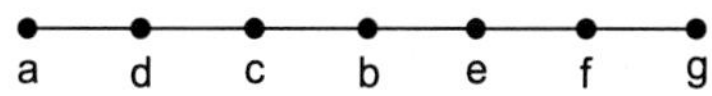

It contains all the vertices, hence it is the spanning tree as follows :

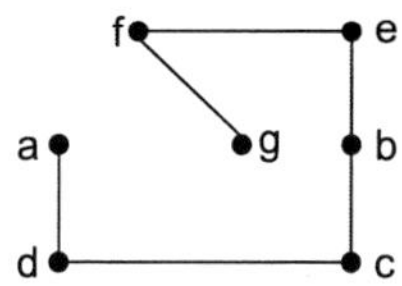

II. Let the vertex ordering be *gfedcba*

(*i*) We start with vertex g as root.

(*ii*) We add *f* to *g* and then *e* to *f*.

(*iii*) We can not add *d* and *c* to *e*. Add *b* to *e*.

(*iv*) We remain with vertices *d*, *c* and *a*. We can not add *d* to *b* but add *c* to *b*.

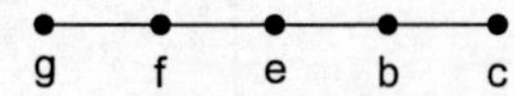

(*v*) The remaining vertices are *d* and *a*. Add *d* to *c* and then *a* to *d*.

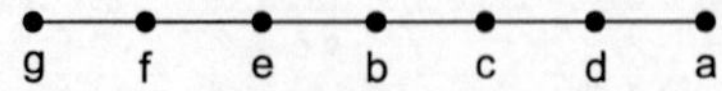

It contains all the vertices. Hence it is a spanning tree as follows :

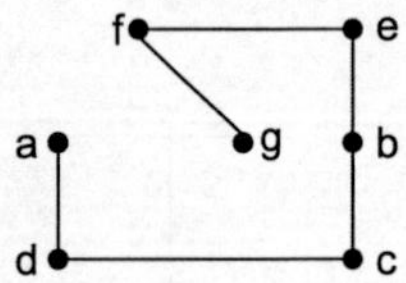

Depth-First Search Algorithm

Input : A connected graph G with vertices labelled v_1, v_2, ..., v_n.

Output : A spanning tree T for G.

Method :

1. (Visit a vertex). Let v_1 be the root of T and let L = v_1.
2. For all vertices adjacent to L, choose the edge {L, v_k}, where *k* is the minimum index such that adding {L, v_k} to T does not create a cycle. If no such edge exists, go to step 3 ; otherwise, add edge {L, v_k} to T and set L = v_k ; repeat step 2 at the new value for L.
3. (Back track or terminate). If *x* is the parent of L in T, set L = *x* and apply step 2 at the new value of L. If, on the other hand, L has no parent in T (so that L = v_1) then the depth-first search terminates and T is a spanning tree for G.

Example 3. Find a spanning tree for the graph using BFS and DFS.

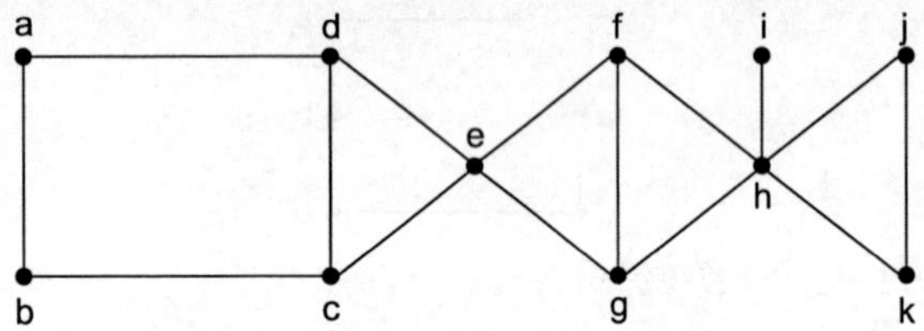

Using BFS

Let the vertex ordering be *abcdefghijk*

(*i*) We start with root as '*a*'. Add the edges {*a*, *b*} and {*a*, *d*}.

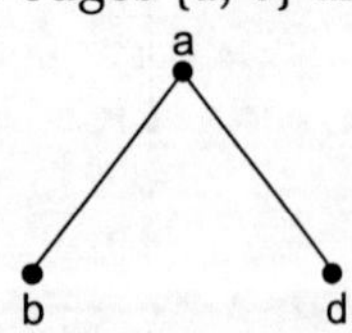

(*ii*) At *b* add the edge {*b*, *c*}.

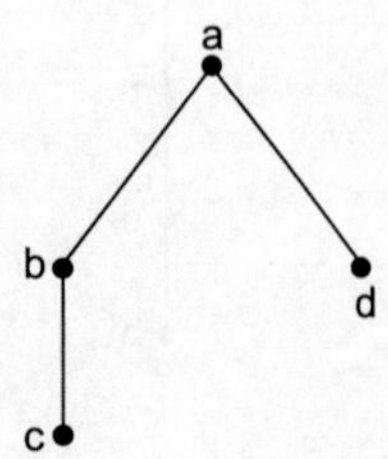

(*iii*) At d, we can not add $\{d, c\}$ as it forms a cycle. Add $\{d, e\}$.

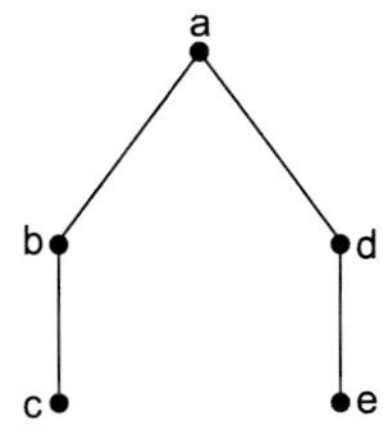

(*iv*) At c we can not add any more edges. At e add $\{e, f\}$ and $\{e, g\}$.

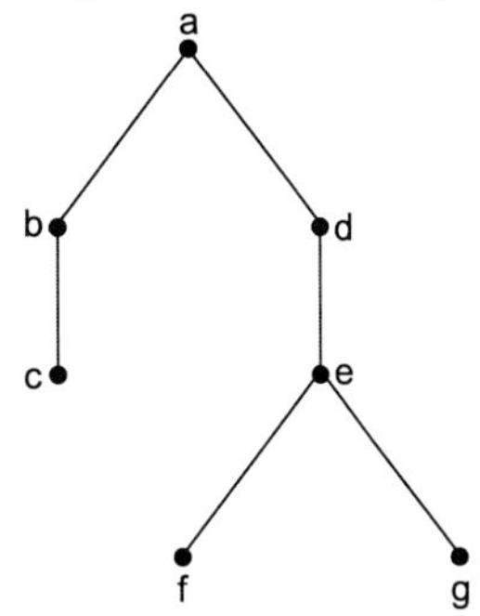

(*v*) At f add $\{f, h\}$.

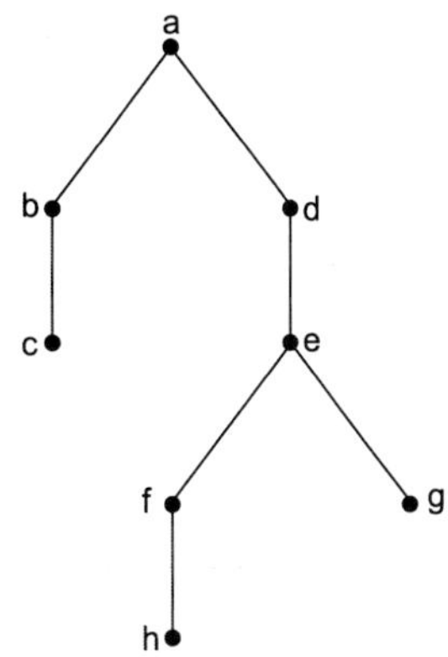

(*vi*) At g we can not add $\{g, h\}$ as it forms a cycle. At h add $\{h, i\}$, $\{h, j\}$ and $\{h, k\}$.

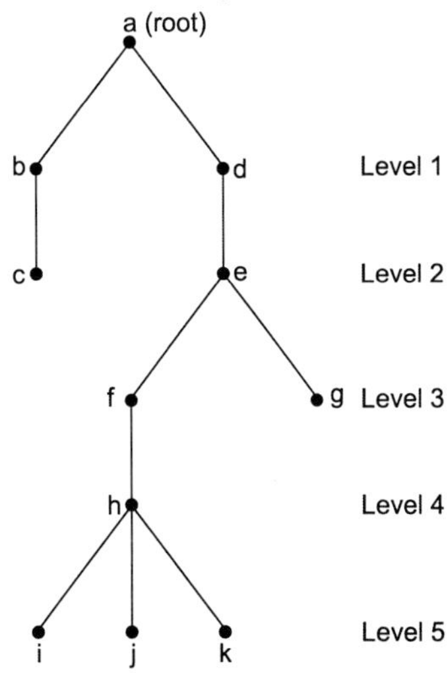

Thus we visited all the vertices. It is a spanning tree. We can write it as

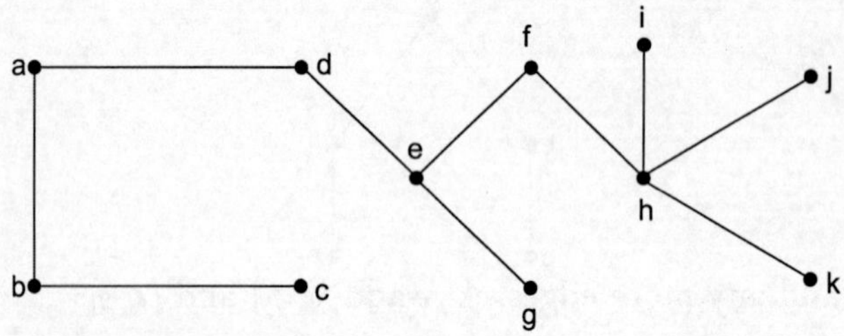

Using DFS

Let the vertex ordering be *abcdefghijk*

(*i*) We start with '*a*' as a root. We add edges successively {*a*, *b*}, {*b*, *c*}, {*c*, *d*}, {*d*, *e*}, {*e*, *f*}, {*f*, *g*}, {*g*, *h*}, {*h*, *i*}.

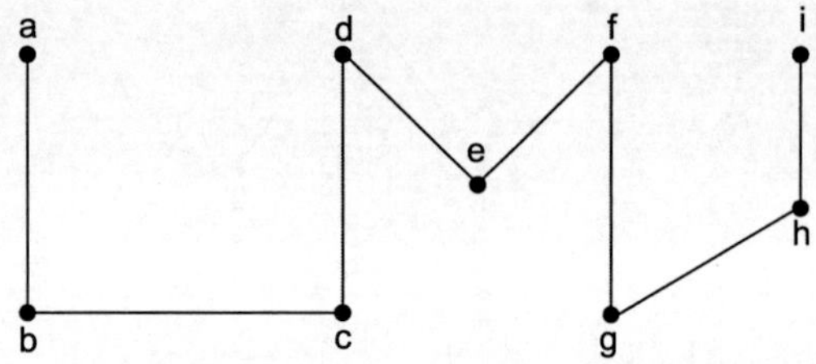

(*ii*) At '*i*' we can not add any more edges. Visit previous vertex *h*. We can add {*h*, *j*}. At '*j*' we can not add any edges. Again visit previous vertex *h* and add {*h*, *k*}.

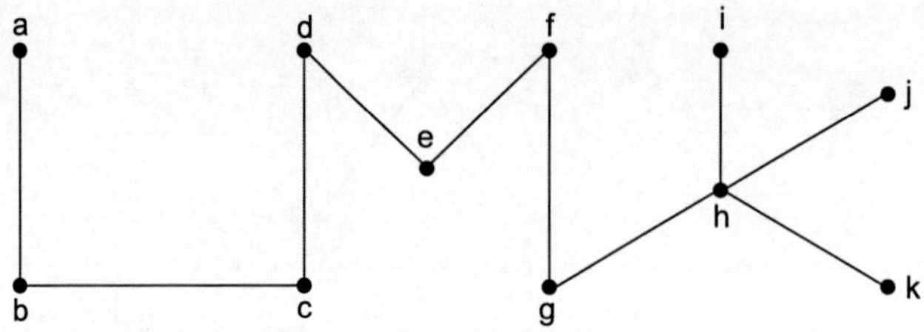

We visited all the vertices hence it is a spanning tree.

Example 4. Using the BFS and DFS method, determine the spanning trees T for the graph G with *e* as the root of T.

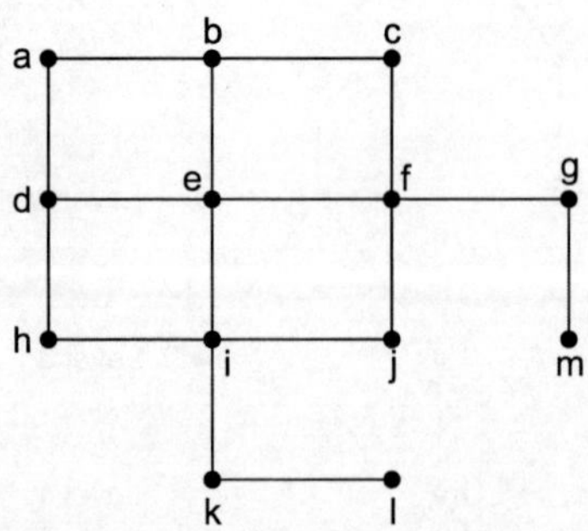

Using BFS method

Let the vertex ordering be *eabcdfghijklm*

(*i*) We take *e* as root of T, We find *b*, *d*, *f* and *i* at level 1.

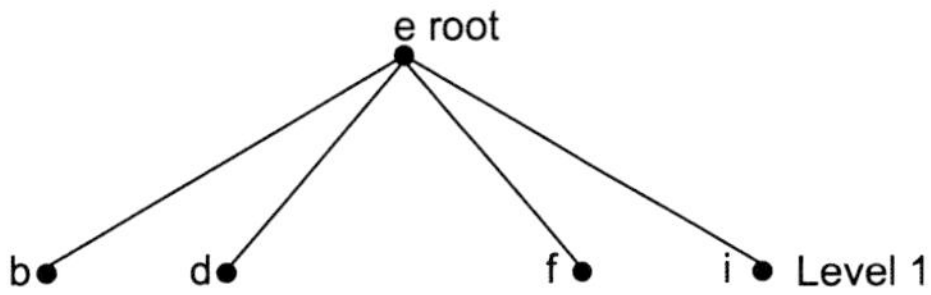

(*ii*) The remaining vertices are *acghjklm*. The level 2 vertices for '*b*' are '*a*' and '*c*'; for *d* is *h*; for *f* are *g* and *j* and for *i* only one vertex *k*.

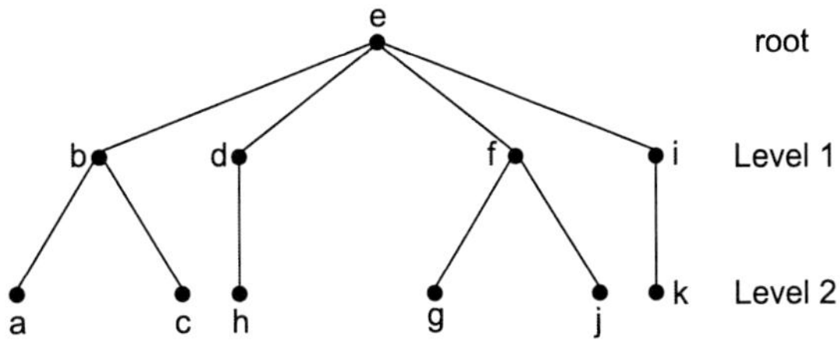

(*iii*) The remaining vertices are *l*, *m*.
At level 3 we add *m* to *g* and *l* to *k*.

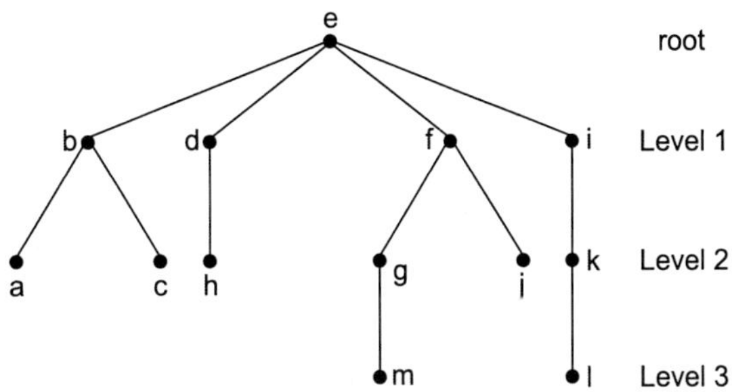

We visited all the vertices. Hence it is a spanning tree. It can also be written in original form as

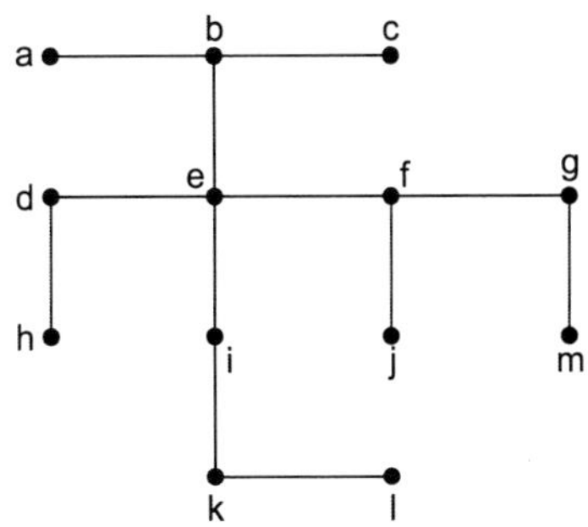

Using DFS method

Let the vertex ordering be *eabcdfghijklm*

(*i*) We take '*e*' as root of T. We start at '*e*' and successively add edges {*e*, *b*}, {*b*, *a*}, {*a*, *d*}, {*d*, *h*}, {*h*, *i*}, {*i*, *j*}, {*j*, *f*} and {*f*, *c*}. We stop at *c* since no more edges could be added.

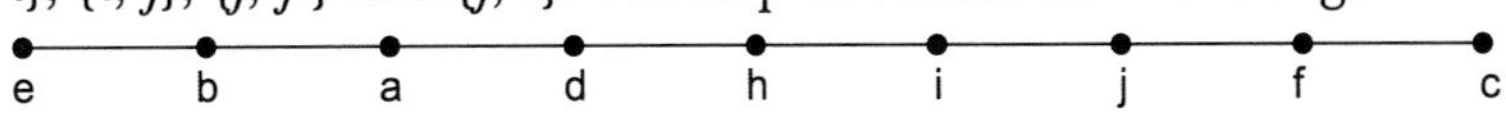

(*ii*) We back track to vertex *f* and find that edges {*f*, *g*} and {*g*, *m*} could be successively added. At *m* we cannot add any more edges.

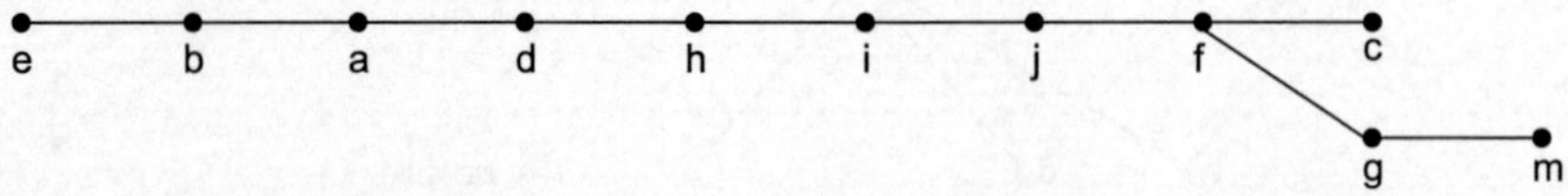

(*iii*) We back track to *g*, we cannot add any edges at *g*. Next back to *f*, no more edges added at *f*. Again back to *j*, no edges can be added at *j*. Now back to *i*, we can add successively the edges {*i*, *k*} and {*k*, l}.

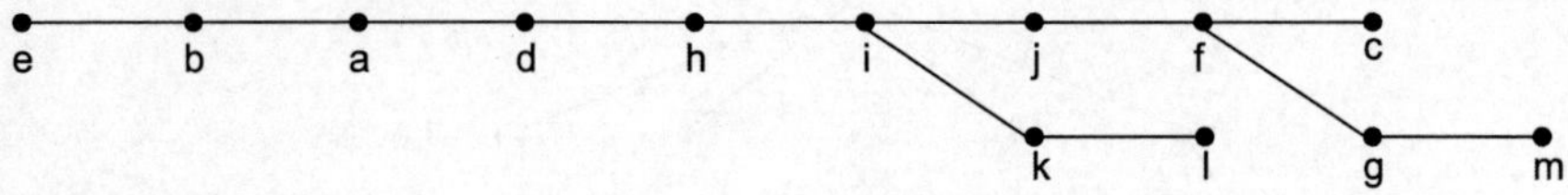

We find that all the vertices are included. Hence it is the spanning tree. It can be written in original form as

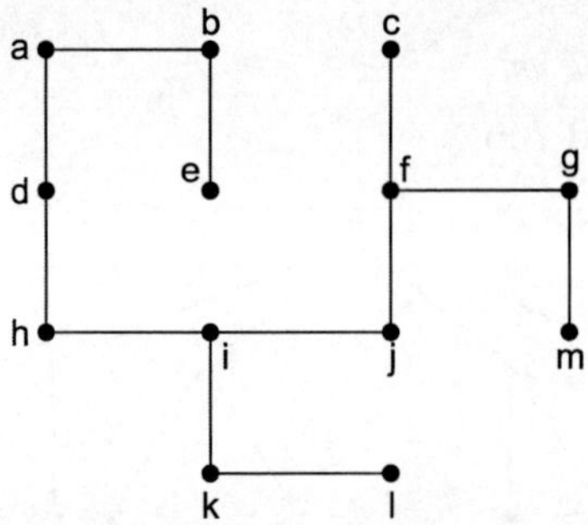

7.5 PLANAR GRAPHS

Definition. *A graph G is said to* ***be planar*** *if it can be drawn in a plane so that its edges do not cross over.*

Definition : *A planar graph is a* ***plane graph*** *if it is already drawn in the plane so that no two edges cross over.*

Example 1. The graph G shown below is planar.

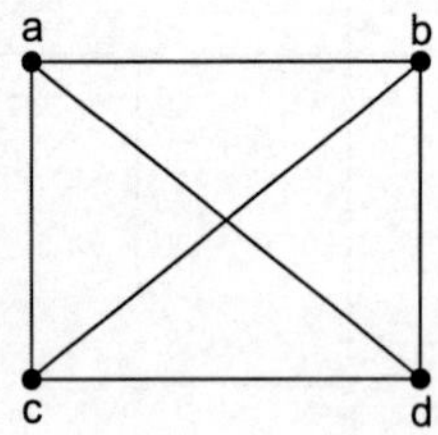

Since it can be written as

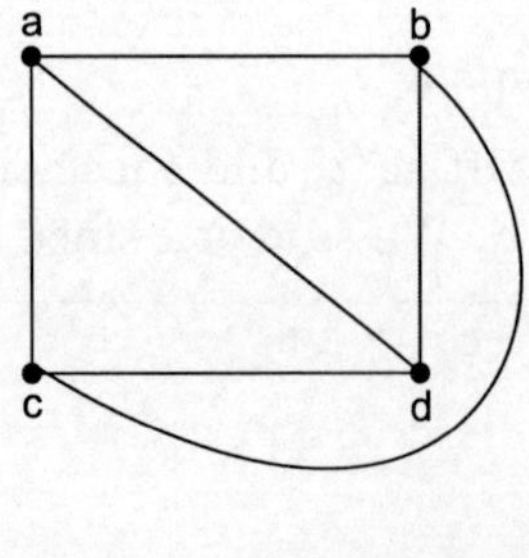

Example 2. Show that the following graph G is planar and draw plane graph of it.

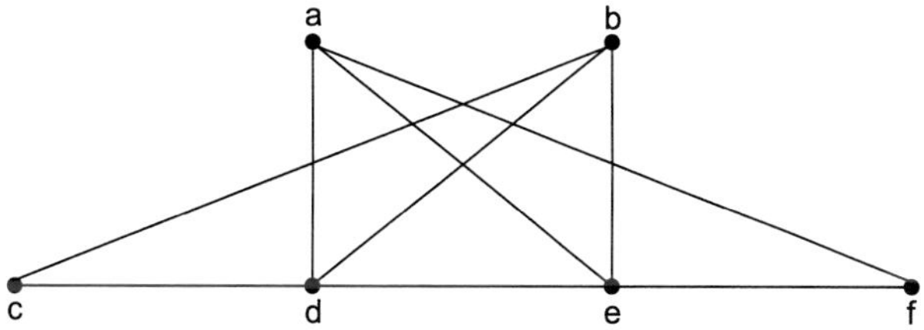

It is planar. It can be written as the following plane graph.

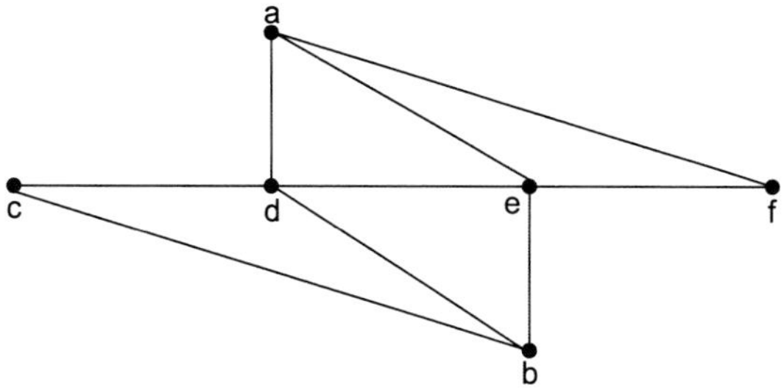

Definition : *A graph with 'n' vertices so that each of the 'n' vertices are adjacent to each of the other n-1vertices is called a* ***complete graph*** *and is denoted by* k_n.

Example 3. Show that k_n is planar for $n = 1, 2, 3, 4$.

k_1 is planar. 

k_2 is planar 

k_3 is planar 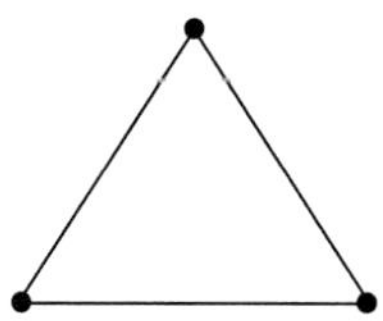

k_4 : 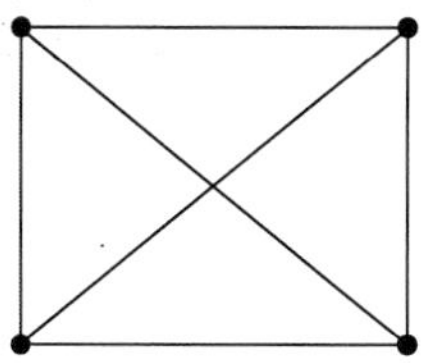

k_4 is planar since it can be written as

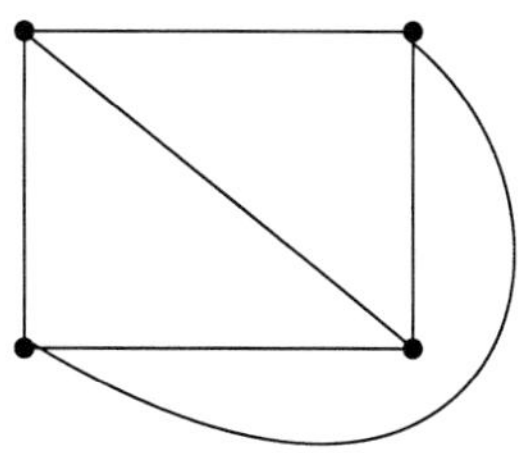

Example 4. Show that k_5 is not planar.

The graph k_5 is

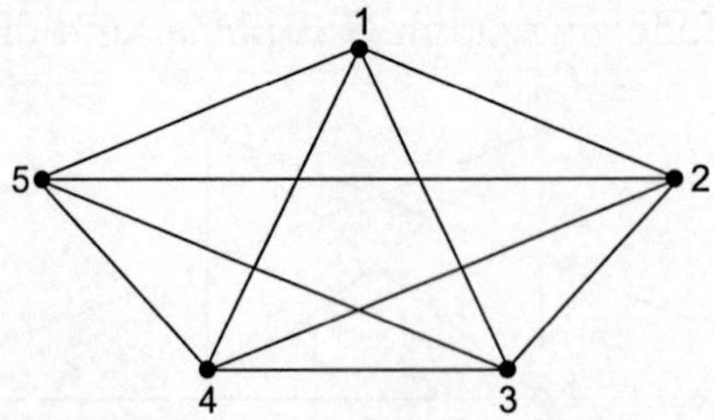

Now we attempt to write this k_5 without cross overs. Observe the vertices are symmetric.

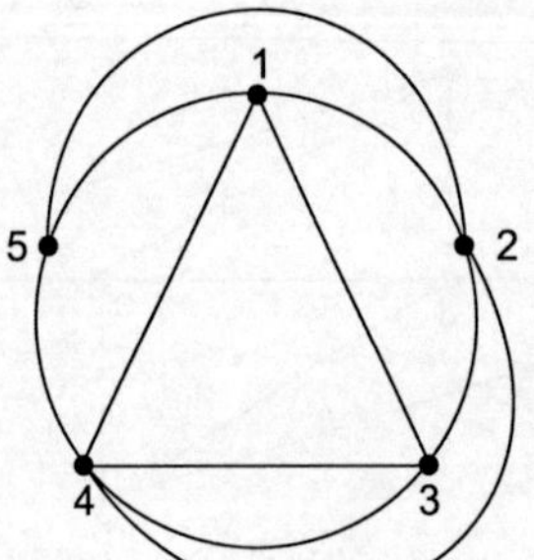

We remain to draw the edge {3, 5}. Since all the vertices forms a circle, the edge {3, 5} should be drawn entirely inside the circle or outside the circle, otherwise it crosses the circle. It is not possible to draw {3, 5} inside the circle since it crosses {1, 4}. Also it is not possible to draw {3, 5} outside the circle. Since it crosses the edge {2, 4}. Thus, it is impossible to draw the edge {3, 5} without cross over. Thus k_5 is non-planar.

Definition : *A graph with m + n vertices so that each of the first m vertices are adjacent to each of the second n vertices and there are no edges between the first m vertices also there are no edges between the second n vertices is called a* ***complete bipartite graph*** *and is denoted by* $k_{m,\,n}$.

Example : Show that $k_{m,\,n}$ is planar if either $m \le 2$ or $n \le 2$. (JNTU, MCA, Feb 2007)

Name	*Graph*	*Plane Graph*
$k_{2,\,3}$	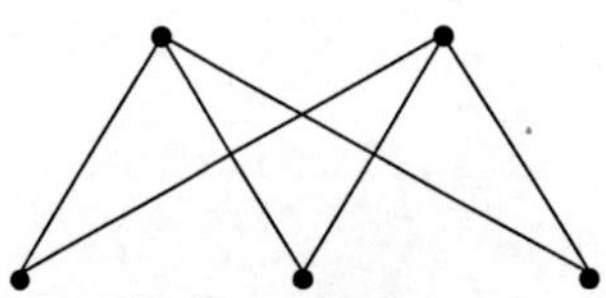	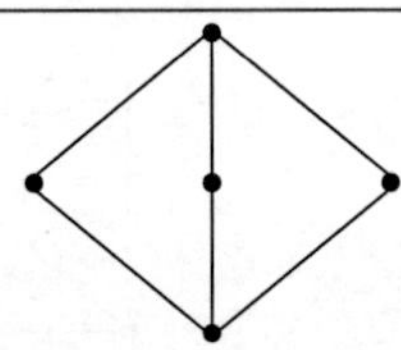
$k_{2,\,4}$	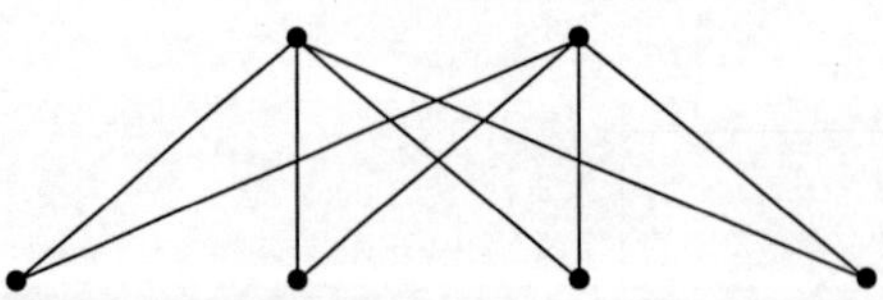	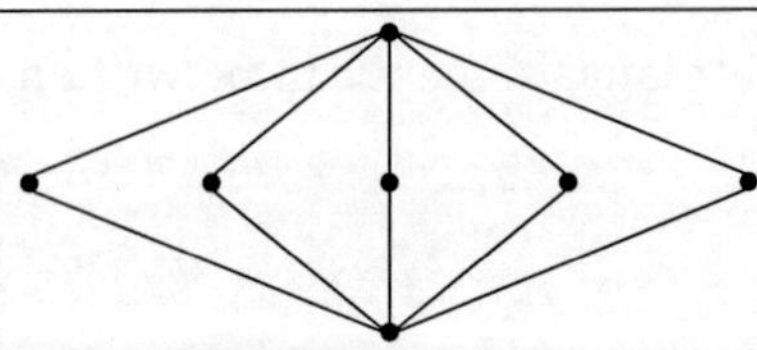
$k_{2,\,5}$	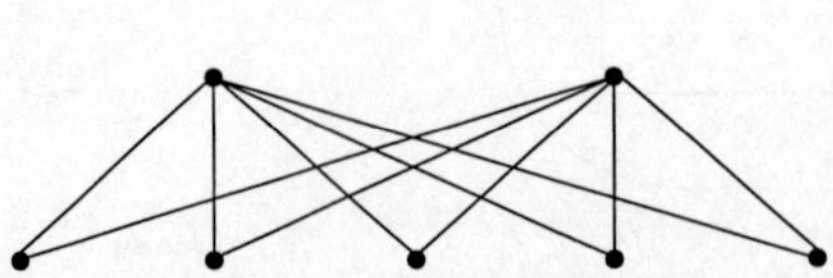	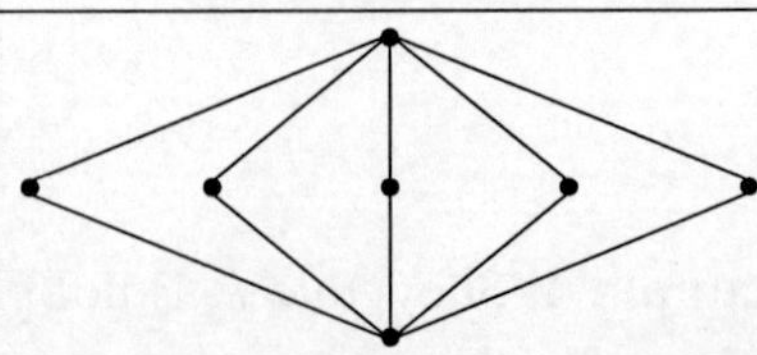

We can write $k_{2,n}$ as n vertices in a row and the 2 vertices, one vertex above the row and the other below the row. Thus, $k_{m,n}$ is planar if either $m \leq 2$ or $n \leq 2$.

Example 5. Show that $k_{3,3}$ is non planar.

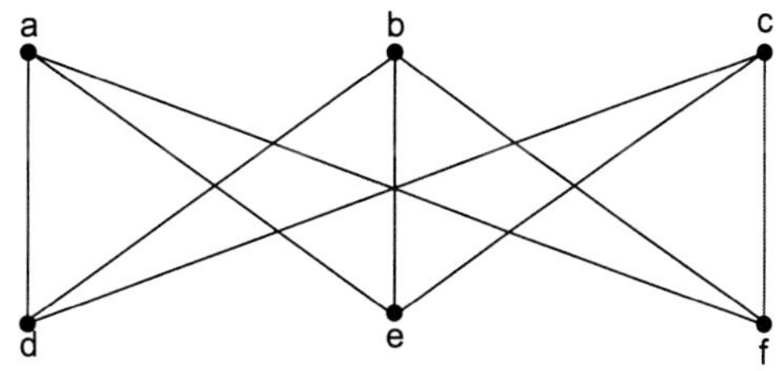

We form a circle containing all the vertices and then we try to draw remaining edges without cross over. Observe the vertices are symmetric :

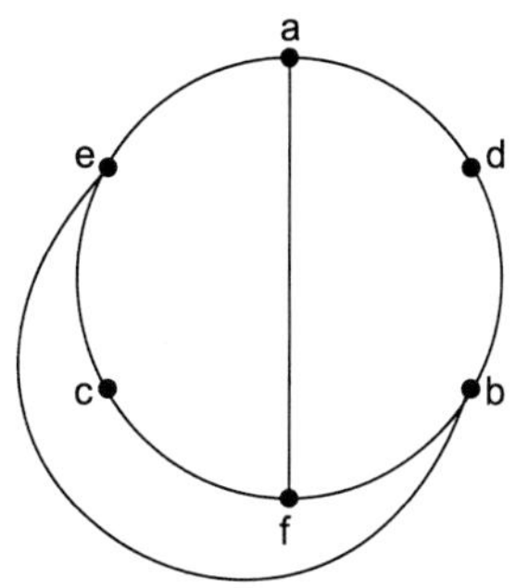

We remain to draw the edge $\{c, d\}$. It should be drawn either entirely inside the circle or outside the circle. When we attempt to draw $\{c, d\}$ inside the circle it crosses $\{a, f\}$. To draw $\{c, d\}$ outside the circle it crosses $\{b, e\}$. Thus, it is not possible to draw the edge $\{c, d\}$ without cross over. Therefore, $k_{3,3}$ is non-planar.

Example 6. Suppose we have 3 houses A, B and C and 3 utility outlets Electricity, Gas and Water. Is it possible to connect each utility to each of the 3 houses without the lines crossing.

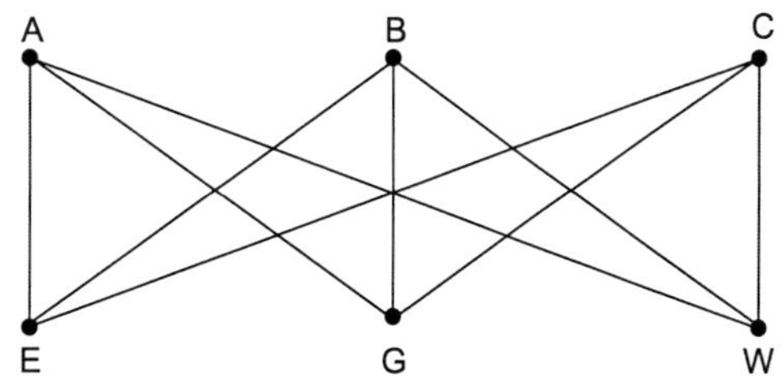

It is *a* $k_{3,3}$ graph. We know that $k_{3,3}$ is not planar, hence it is not possible to connect each of the 3 utilities to each of the 3 houses without the lines crossing.

Definition : *A plane graph divides the plane into connected areas called* ***regions*** *or* ***faces****. Each plane graph G determines a region of infinite area called the exterior region of G.*

Example 7. The following plane graph G has 4 regions r_1, r_2, r_3, r_4 in which r_4 is the exterior region.

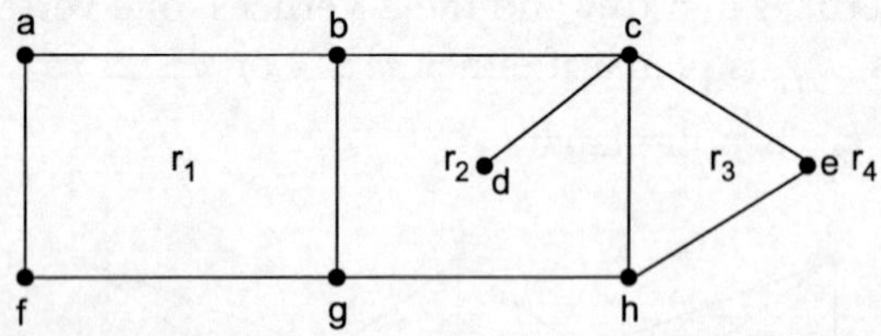

Definition : *The **degree of a region** is the length of the closed path bordering the region.*

Example 8. Find the degrees of the regions of the graph shown in example 4.8.

Region	border	degree
r_1	$a-b-g-f-a$	4
r_2	$b-c-d-c-h-g-b$	6
r_3	$c-e-h-c$	3
r_4	$a-b-c-e-h-g-f-a$	7

Note. V (G) = The set of all vertices of G

E (G) = The set of all edges of G

R (G) = The set of all regions of plane graph G.

Theorem : If G is a plane graph, then the sum of the degrees of the regions determined by G is twice the number of edges of G.

i.e., $$\sum_{r \in R(G)} \deg(r) = 2\,|E|$$

where | E | is the number of edges of G.

Proof : Each edge of the plane graph is a border of two regions or is contained in a region and will therefore occur twice in any path along the border of that region. Thus, every edge in the plane graph contribute two in determining the degrees of the regions of G.

Example 9. Verify sum of degrees theorem for the plane graph shown in example 4.8.

Number of edges in G is 10.

i.e., $$|E| = 10.$$

$$\sum_{r \in R(G)} \deg(r) = \deg(r_1) + \deg(r_2) + \deg(r_3) + \deg(r_4)$$

$$= 4 + 6 + 3 + 7$$

$$= 20 = 2 \times 10$$

$$= 2\,|E|$$

Example 10. Find the number of edges in a plane graph G containing 4 regions each of degree 3.

By sum of degrees theorem

$$\sum \deg(r) = 2\,(E)$$

$$\therefore \quad 4 \times 3 = 2\,|E| \Rightarrow |E| = 6.$$

Theorem. (Euler's Formula) : If G is a connected plane graph, then $|V| - |E| + |R| = 2$, where $|V|$, $|E|$ and $|R|$ are respectively the number of vertices, edges and regions of G.

(JNTU, November 2006, Set 3)

Proof : We prove the theorem by mathematical induction on number of regions determined by G.

Let $|R| = 1$, then G is a tree, hence we know that $|E| = |V| - 1$

$\therefore \quad |V| - |E| + |R| = |V| - (|V| - 1) + 1 = 2$

$\therefore$ The result is true for $|R| = 1$.

Let us assume that the result is true for $|R| = k$ for $k \geq 1$.

Now suppose G is a connected plane graph that determines $k + 1$ regions. Delete an edge common to the boundary of two separate regions. The resulting graph G′ has the same number of vertices, one fewer edge, but also one fewer region, since two previous regions have been merged by the removal of the edge. Thus, we have $|V'| = |V|$, $|E'| = |E| - 1$ and $|R'| = |R| - 1$

$$\begin{aligned} \therefore \quad |V| - |E| + |R| &= |V'| - (|E'| + 1) + (|R'| + 1) \\ &= |V'| - |E'| + |R'| \\ &= 2 \qquad \because \text{ The result is true for G}' \text{ which contains } k \text{ regions.} \end{aligned}$$

Thus, the theorem is proved by the mathematical induction.

Corollary : If G is a simple connected plane graph with $|E| > 1$ then $|E| \leq 3|V| - 6$

Proof : Since G is simple, we have

$$\deg(r) \geq 3, \forall\, r \in R(G)$$

$$\therefore \quad \sum_{r \in R(G)} \deg(r) \geq \sum_{r \in R(G)} 3$$

$$\therefore \quad 2|E| \geq 3|R| \qquad \text{...(1)}$$

$\because$ L.H.S. is obtained by the sum of degrees theorem. R.H.S. is addition of 3, the number of regions times.

By Euler's formula

$$|V| - |E| + |R| = 2 \qquad \text{...(2)}$$

To get the inequality we have to eliminate $|R|$ from (1) and (2). From (2), $|R| = 2 - |V| + |E|$.

Substituting in (1), we get

$$2|E| \geq 3(2 - |V| + |E|)$$

$$\therefore \quad 2|E| \geq 6 - 3|V| + 3|E|$$

$$\therefore \quad |E| \leq 3|V| - 6.$$

Theorem : A complete graph k_n is planar if and only if $n \leq 4$.

Proof : Since k_1, k_2, k_3 are subgraphs of k_4 and k_5 is subgraph of k_n for each $n \geq 6$, it suffices to show that k_4 is planar and k_5 is non-planar.

The graph k_4 is

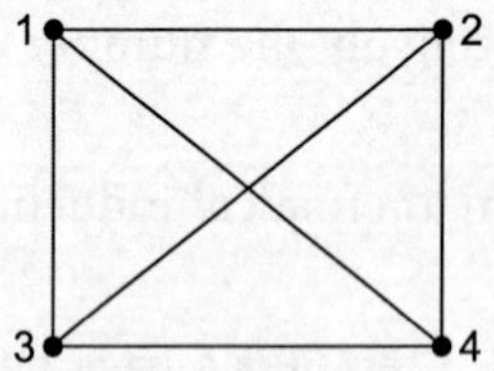

It can be written as without cross over of edges as

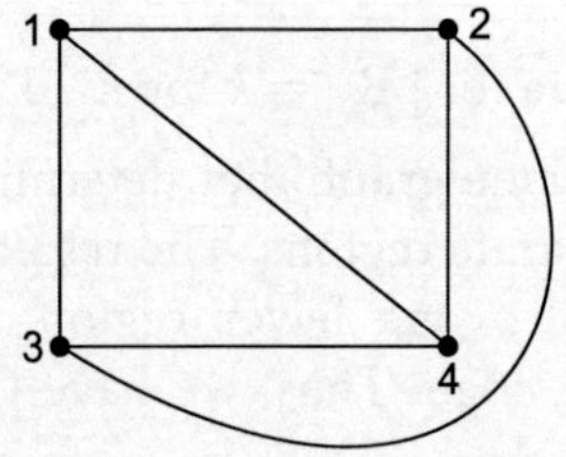

$\therefore$ k_4 is planar.

Now we prove k_5 is non-planar by an indirect argument. Suppose k_5 is planar

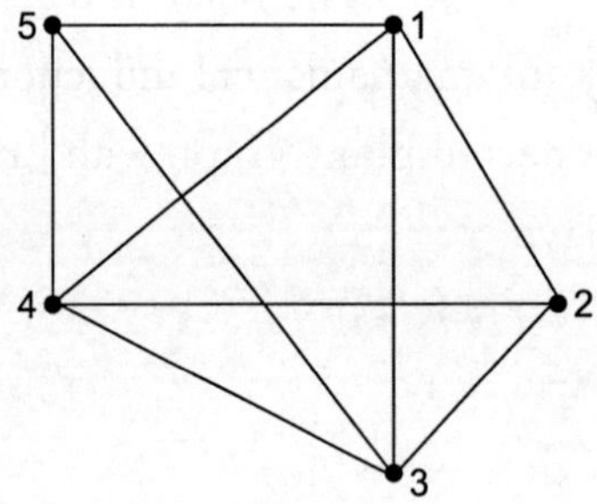

Also k_5 is simple, we would have

$$|E| \leq 3|V| - 6$$

$\therefore$ $$10 \leq 3 . 5 - 6$$

$\therefore$ $$10 \leq 9$$

which is a contradiction. Hence the proof.

Theorem : A complete bipartite graph $k_{m, n}$ is planar if and only if $m \leq 2$ or $n \leq 2$.

Proof : It is sufficient to show that $k_{2, n}$ is planar and $k_{3, 3}$ is non-planar.

The graph $k_{2, n}$ is

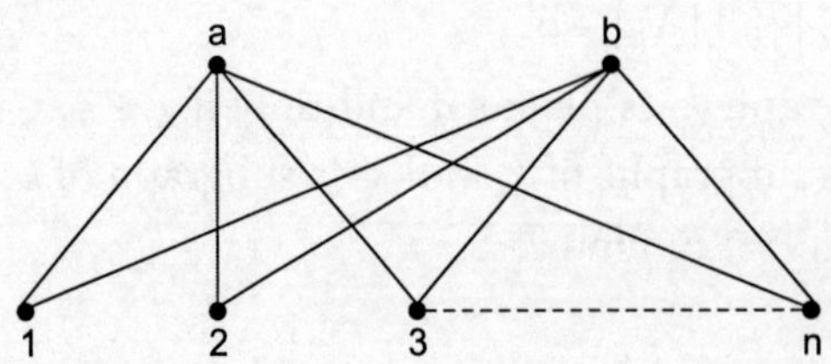

It can be written as

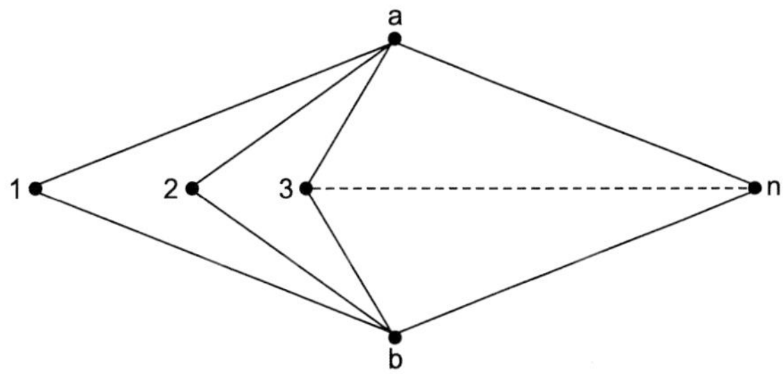

$\therefore$ $k_{2,\,n}$ is planar.

Now we show $k_{3,\,3}$ is nonplanar by indirect argument.

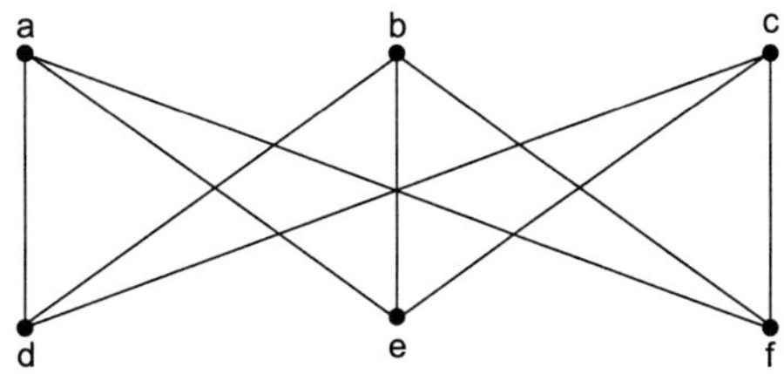

Suppose $k_{3,\,3}$ is planar. Observe each cycle of $k_{3,\,3}$ has length ≥ 4. $k_{3,\,3}$ has 6 vertices and 9 edges.

Hence deg $(r) \geq 4$ for each region r of $k_{3,\,3}$

$$\therefore \quad \sum \deg(r) \geq \sum 4$$

$$\therefore \quad 2\,|\,E\,| \geq 4\,|\,R\,|$$

$$\therefore \quad 2\,|\,E\,| \geq 4\,(2 - |\,V\,| + |\,E\,|) \qquad \because \text{ sum of degrees theorem}$$

$$\therefore \quad |\,E\,| \geq 2\,(2 - |\,V\,| + |\,E\,|) \qquad \because \ |\,V\,| - |\,E\,| + |\,R\,| = 2. \text{ (Euler's formula)}$$

$$\therefore \quad |\,E\,| \leq 2\,|\,V\,| - 4$$

$$\therefore \quad 9 \leq 2\,.\,6 - 4$$

$$\therefore \quad 9 \leq 8$$

which is a contradiction.

Example 11. Suppose G is a connected planar graph with 14 regions each of degree 4, find the number of vertices in G.

By sum of degrees theorem

$$\sum_{r \in R(G)} \deg(r) = 2\,|\,E\,|$$

$$\therefore \quad 14 \times 4 = 2\,|\,E\,|$$

$$\therefore \quad |\,E\,| = 28$$

By Eulers formula,

$$|\,V\,| - |\,E\,| + |\,R\,| = 2$$

$$\therefore \quad |\,V\,| = 2 + |\,E\,| - |\,R\,|$$

$$= 2 + 28 - 14$$

$$= 16.$$

Example 12. Is the following graph planar?

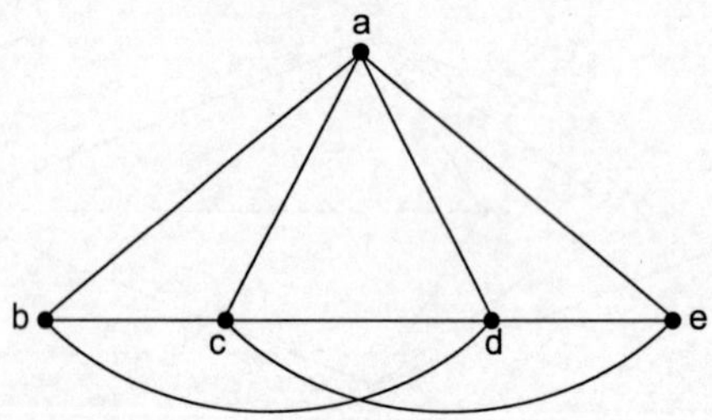

Form a circle which contains all the vertices of G.

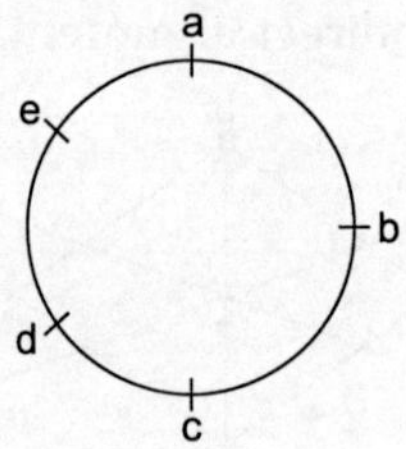

We try to draw remaining edges without cross overs.

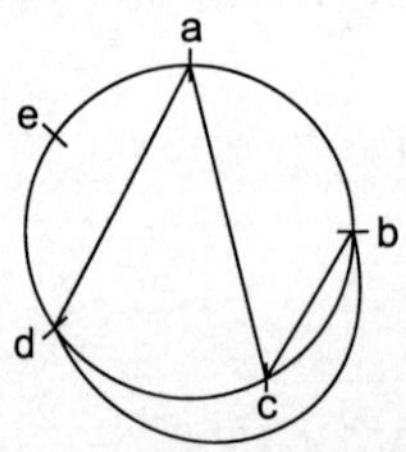

We still remain to draw the edge $\{c, e\}$ which is not possible to draw without cross other edges. But don't come to the decision G is non-planar, since the vertices are not symmetric in G.

Now we form a circle containing all the vertices in another way.

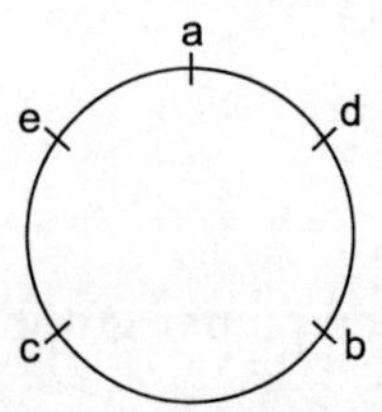

We try to draw remaining edges of G without cross overs.

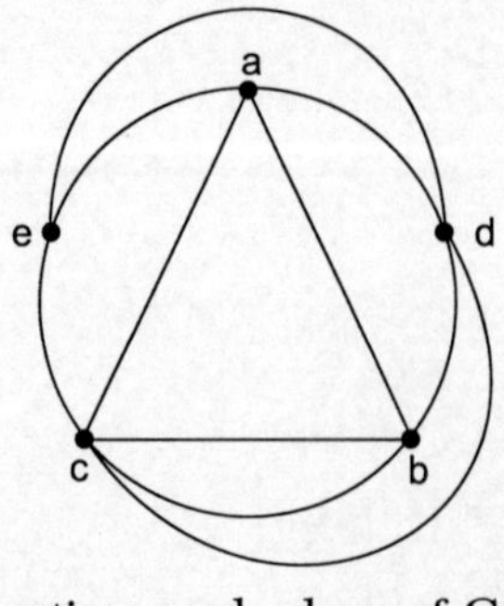

The graph now contains all the vertices and edges of G. Thus G is planar.

Example 13. Is the following graph planar?

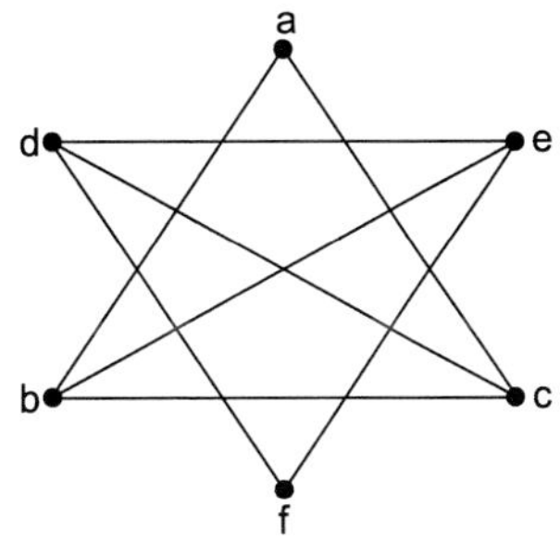

We form a circle containing all the vertices.

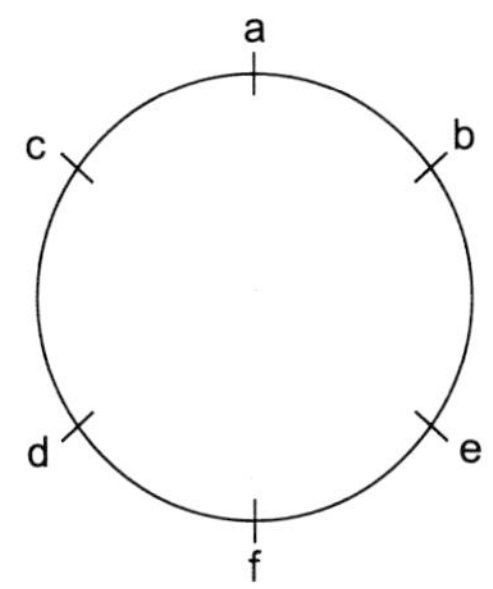

Now we attempt to draw remaining edges without cross overs.

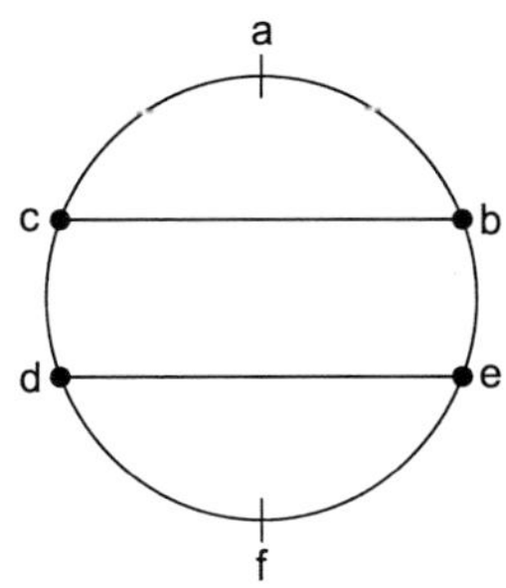

It contains all the edges and vertices of G. Hence G is planar.

EXERCISE 7.1

1. Find the adjacency matrix of the following graphs :

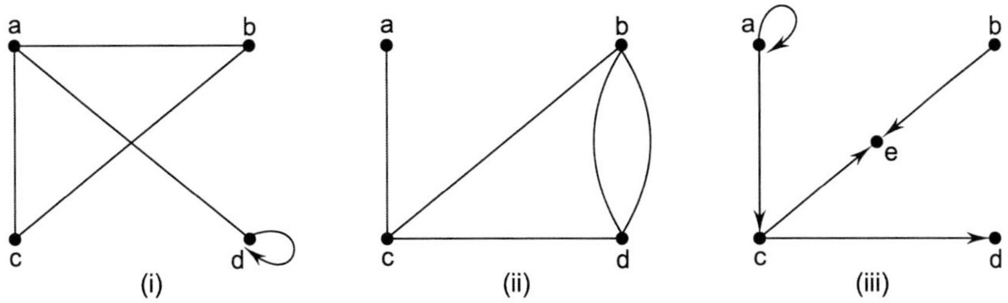

2. Draw the graphs for the following adjacency matrices.

$$\begin{bmatrix} 1 & 0 & 0 & 1 \\ 0 & 0 & 1 & 1 \\ 0 & 1 & 0 & 0 \\ 1 & 1 & 0 & 0 \end{bmatrix} \qquad \begin{bmatrix} 1 & 2 & 1 \\ 2 & 0 & 3 \\ 1 & 3 & 0 \end{bmatrix} \qquad \begin{bmatrix} 0 & 1 & 0 \\ 0 & 1 & 1 \\ 1 & 0 & 0 \end{bmatrix}$$

(*i*) (*ii*) (*iii*)

3. Explain about the adjacency matrix representation of graphs. Illustrate with an example.
(JNTU, Nov. 2006, Sets 1, 2, 4, Feb. 2007, Sets 2, 3).
4. What are the advantages of adjacency matrix representation?
(JNTU, Nov. 2006, Sets 1, 2, 4, Feb. 2007, Sets 2, 3).
5. What are the steps involving in graph traversal using Breadth-First Search (BFS) algorithm? Illustrate with a graph example.
6. Explain the algorithm for Depth-First Search (DFS) traversal of a graph.
7. Define a spanning tree. Find all spanning trees of the following graph.
(JNTU, Nov. 2006, Sets 3)

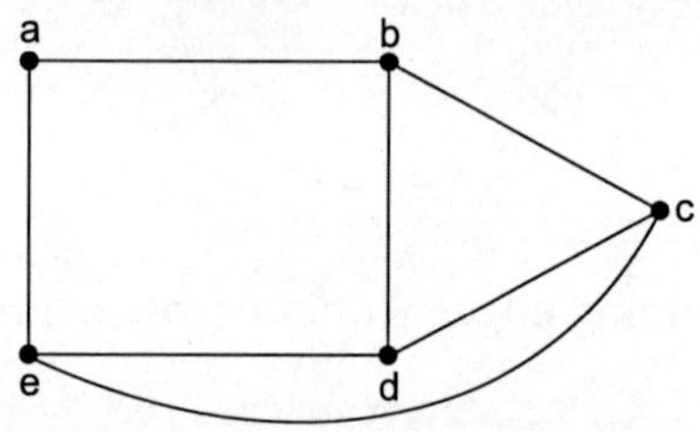

8. Describe how to obtain all possible spanning trees for the following graphs and list the total number of different spanning trees.

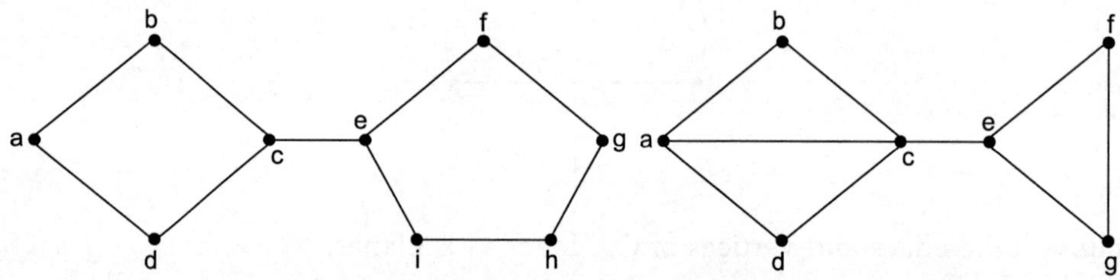

(i) (ii)

9. Find spanning trees by using algorithms BFS and DFS with the vertex orderings abcdefg and gfedcba for the following graph.

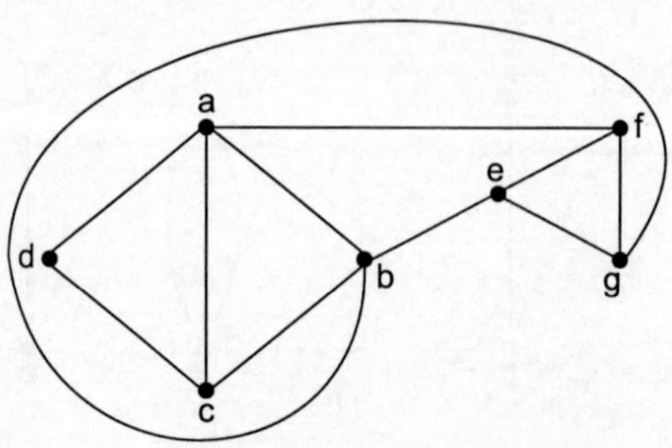

10. Define planar graph. Identify which of the following are planar.

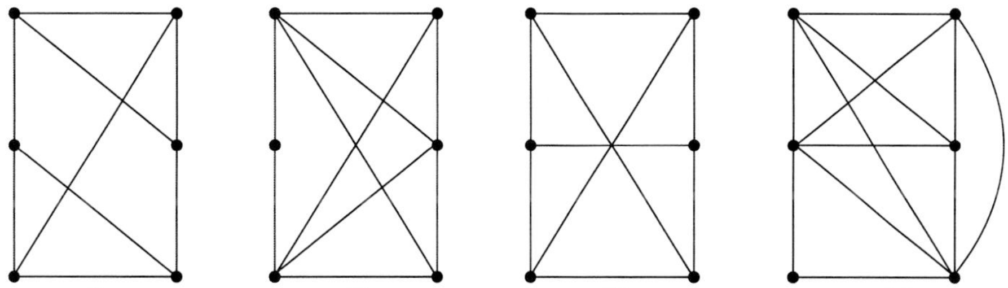

(i) (ii) (iii) (iv)

11. Define the degree of a region. Find the degree of each region of the plane graph G.

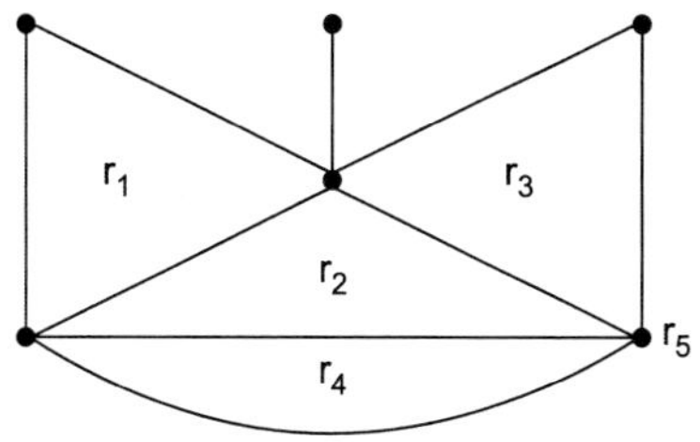

12. Show that the sum of the degrees of the regions of a plane graph G is equal to twice the number of edges in G.

13. Find the number of edges in a planar graph containing 4 regions of degree 3 and one region of degree 4.

14. State and prove Euler's formula for a connected planar graph.

15. Verify Eulers formula for the graph in example 11.

16. Prove that a complete bipartite graph $k_{m,\,n}$ is planar if $m \leq 2$ or $n \leq 2$.

17. Show that a complete graph k_n is planar iff $n \leq 4$.

18. Suppose that G is a connected planar graph. Determine | V | if G has 35 regions each of degree 6.

19. If G is a connected plane graph with 6 vertices and 12 edges prove that the degree of each region is 3.

20. Give an example of a connected plane graph such that

(*a*) $|E| = 3|V| - 6$

(*b*) $|E| < 3|V| - 6$

ANSWERS

1. (*i*)

	a	*b*	*c*	*d*
a	0	1	1	1
b	1	0	1	0
c	1	1	0	0
d	1	0	0	1

(*ii*)

	a	*b*	*c*	*d*
a	0	0	1	0
b	0	0	1	2
c	1	1	0	1
d	0	2	1	0

(*iii*)

	a	*b*	*c*	*d*	*e*
a	1	0	1	0	0
b	0	0	0	0	1
c	0	0	0	1	1
d	0	0	0	0	0
e	0	0	0	0	0

9. BFS (*abcdefg*)

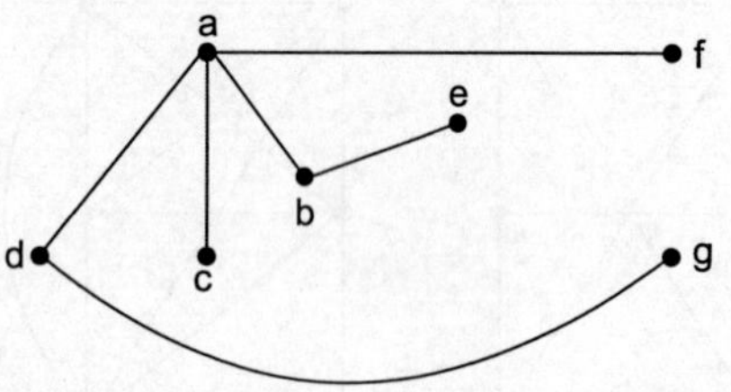

BFS (gfedcba)

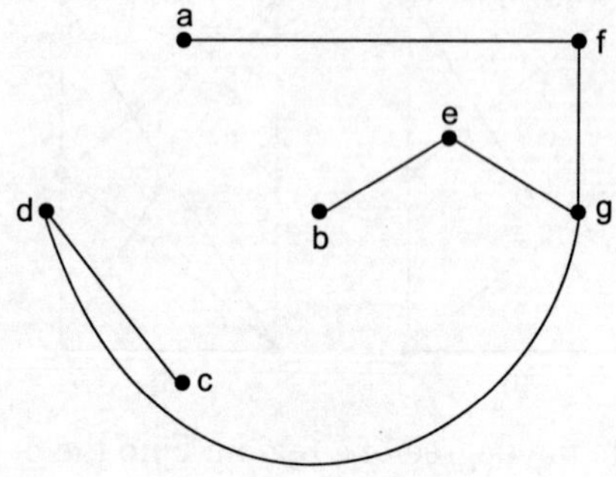

DFC (*abcdefg*)

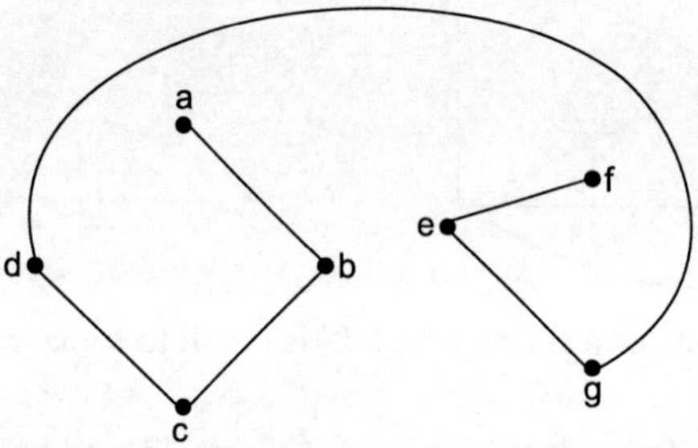

DFS (gfedcba)

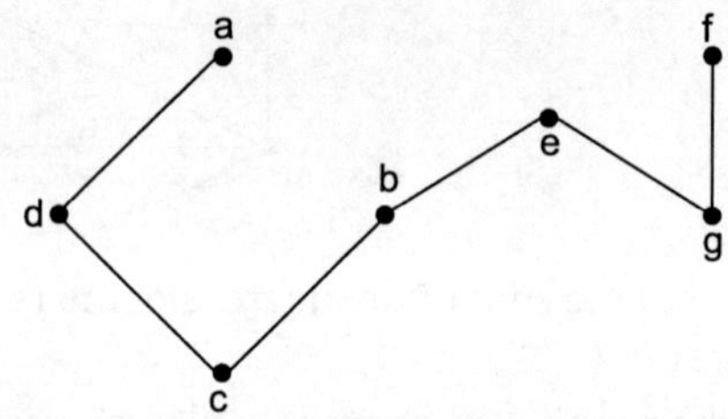

10. (*i*), (*ii*) are planar

(*iii*), (*iv*) are not planar.

11.

Region	r_1	r_2	r_3	r_4	r_5
degree	3	3	3	2	7

13. 8

18. 72.

CHAPTER

8

Graph Theory and Applications

INTRODUCTION

Graph theory begins with very simple geometric ideas which are used to conceptualize various complex systems in a much simpler way. It is applied in many areas such as Engineering, Computer Science, Economics, Sociology, Psychology, Linguistics etc. To a computer professional graph theory provides various techniques to study and analyze different problems associated with Computer design, Logic design, Operating systems, Formal languages, Artificial Intelligence, Code optimization. Information gathering and retrieval etc.,

8.1 BASIC CONCEPTS

Definition : *A **graph** $G = (V, E)$ consists of a set of objects $V = \{v_1, v_2, ...\}$ called vertices (nodes, points) and another set $E = \{e_1, e_2, ...\}$ whose elements are called edges such that each edge e_k is identified with an un-ordered pair $\{v_i, v_j\}$ of vertices.*

Example 1. Give an example of a graph :

Let $G = (V, E)$ where $V = \{v_1, v_2, v_3, v_4\}$, $E = \{e_1, e_2, e_3, e_4, e_5\}$ and the edges are identified as $e_1 = \{v_1, v_2\}$, $e_2 = \{v_2, v_3\}$, $e_3 = \{v_2, v_4\}$, $e_4 = \{v_3, v_4\}$ and $e_5 = \{v_4, v_1\}$.

Diagramatically G is shown as follows :

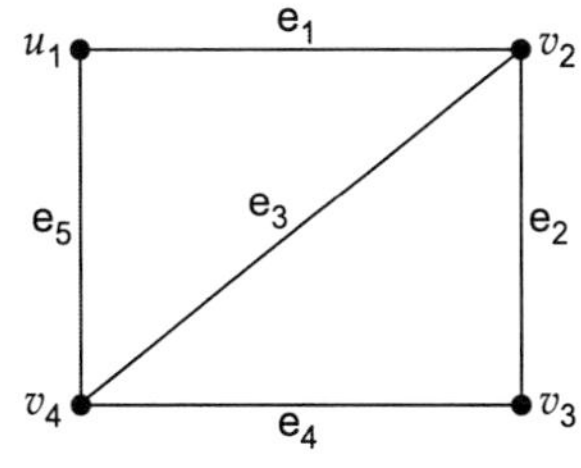

Figure. 8.1.

Definition : *An edge drawn from a vertex to itself is called a loop.*

Example 2. Give an example of a graph containing a loop.

Let G = (V, E), where V = $\{v_1, v_2, v_3\}$, E = $\{e_1, e_2, e_3, e_4\}$ and $e_1 = \{v_1, v_1\}$, $e_2 = \{v_1, v_2\}$, $e_3 = \{v_2, v_3\}$, $e_4 = \{v_3, v_1\}$. Here the edge e_1 is a loop.

Diagramatically the graph G is

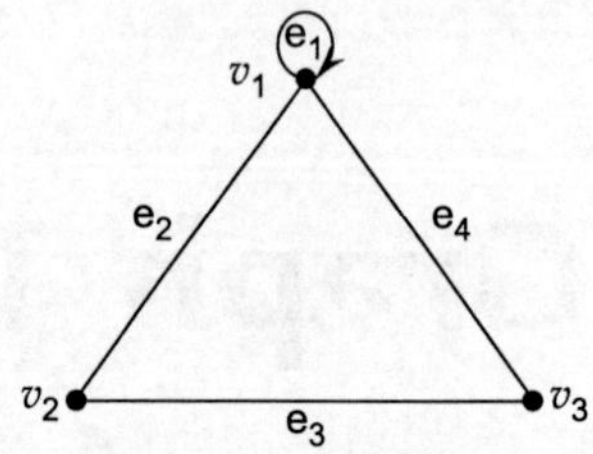

Figure. 8.2.

Definition : *In a graph if some pair of vertices are joined by more than one edge, such edges are called* ***parallel edges.***

Example 3. Give an example of a graph containing parallel edges.

Let G = (V, E), where V = $\{v_1, v_2, v_3\}$, E = $\{e_1, e_2, e_3, e_4\}$ and $e_1 = \{v_1, v_2\}$, $e_2 = \{v_2, v_3\}$, $e_3 = \{v_2, v_3\}$, $e_4 = \{v_1, v_3\}$. Here e_2, e_3 are parallel edges.

Diagramatically G is

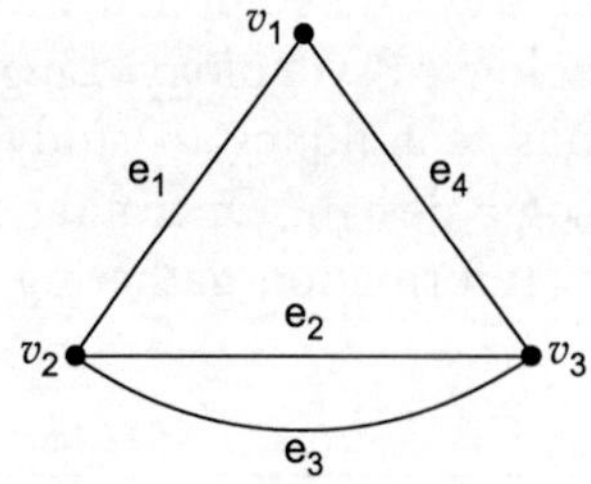

Figure. 8.3.

Definition. *A graph without loops and parallel edges is called a* ***simple graph.***

Example 4. Graphs shown in Figure. 1.1 is simple but Figures 1.2, 1.3 are not simple.

Definition : *If G = (V, E) is a finite graph then the number of vertices in G is called the* ***order*** *of G and the number of edges in G is called the* ***size*** *of G.*

Example 5. Find the order and size for Figures 1.1, 1.2, 1.3.

Figure	Order	Size
1.1	4	5
1.2	3	4
1.3	3	4

Definition : *The* ***degree of a vertex*** *is determined by counting each loop incident on v twice and each other edge incident on v once. It is denoted by deg (v).*

Example 6. Find the degrees of all the vertices shown in the following graph.

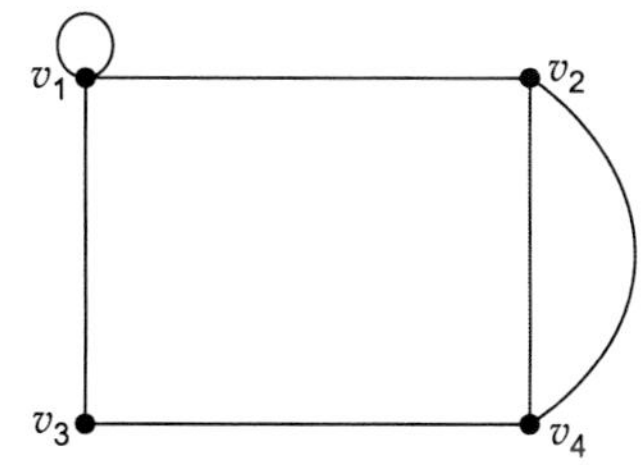

Figure. 8.4.

Vertex	Degree
v_1	4
v_2	3
v_3	2
v_4	3

Note : δ (G) is denoted as the minimum of all the degrees of the vertices of a graph G and Δ (G) is denoted as the maximum of all the degrees of the vertices of G.

Example 7. Find δ (G), Δ (G) for the graph shown in Figure. 1.4.

$$\delta (G) = \text{minimum of } \{\deg (v_i) / v_i \in V\}$$
$$= \text{min. of } \{4, 3, 2, 3\} = 2$$
$$\Delta (G) = \text{Maximum of } \{\deg (v_i) / v_i \in V\}$$
$$= \text{Max. of } \{4, 3, 2, 3\} = 4.$$

Note. For any graph G we have $\delta (G) \le \deg (v_i) \le \Delta (C)$ for all $v_i \in V$.

Definition : *Let G be a graph with vertex set $V = \{v_1, v_2, \ldots\ldots, v_n\}$ and let $d_i = \deg (v_i)$, then the sequence $(d_1, d_2, \ldots\ldots, d_n)$ is called the* ***degree sequence*** *of G.*

The vertices of a graph G can be ordered so that the degree sequence is monotonically increasing or monotonically decreasing.

Example 8. Find the degree sequence of the graph shown in Figure. 1.4.

The degree sequence for the vertex ordering v_1, v_2, v_3, v_4 is (4, 3, 2, 3).

We can write the degree sequence in increasing order as (2, 3, 3, 4). The vertex ordering for this is v_3, v_2, v_4, v_1.

Definition : *A graph G is said to be* ***k-regular*** *if each vertex in it is of degree k. That is $\delta (G) = \Delta (G) = k$.*

Example 9. Draw all 2-regular graphs of order 4.

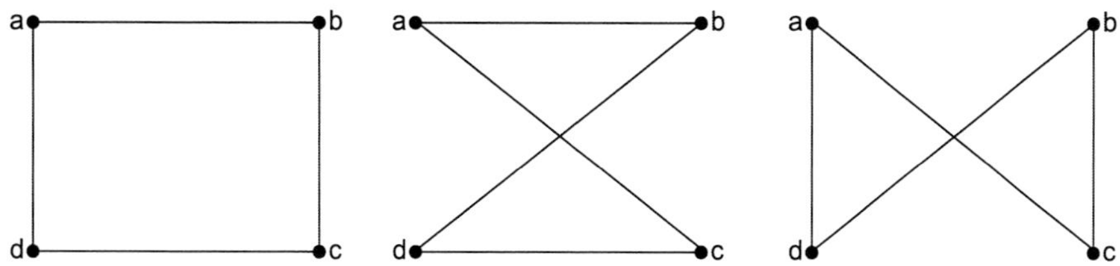

Figure. 8.5.

Theorem : (The sum of degrees theorem)

Let G be a graph with $|V| = n$ vertices and $|E|$ edges then $\sum_{i=1}^{n} \deg(v_i) = 2|E|$

Proof : Let $v_1, v_2, \ldots\ldots, v_n$ be the vertices of a graph G. Since each edge has two end points and each end contributes one to the sum of degrees then the sum of degrees of all the vertices in G is twice the number of edges in G.

i.e., $$\sum_{i=1}^{n} \deg(v_i) = 2|E|.$$

Example 10. Verify the sum of degrees theorem to the following graph.

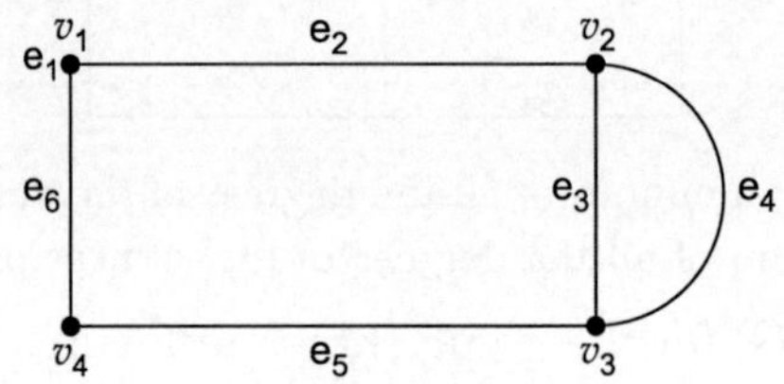

Figure. 8.6.

Vertex	v_1	v_2	v_3	v_4
Degree	4	3	3	2

$$\sum \deg(v_i) = 4 + 3 + 3 + 2 = 12$$

$$2|E| = 2 \times \text{number of edges} = 2 \times 6 = 12$$

$$\therefore \quad \sum \deg(v_i) = 2|E|$$

Example 11. Find the number of edges in a graph containing 2 vertices of degree 3, 3 vertices of degree 4 and a vertex of degree six.

By sum of degrees theorem,

$$\sum \deg(v_i) = 2|E|$$

$$\therefore \quad 2 \times 3 + 3 \times 4 + 1 \times 6 = 2|E|$$

$$\therefore \quad 6 + 12 + 6 = 2|E|$$

$$\therefore \quad |E| = 12$$

∴ The number of edges in G is 12.

Example 12. Find the number of vertices in a graph containing 3 vertices of degree 4, 2 vertices of degree 3 and remaining vertices of degree 2. Given that number of edges in G is 11.

By sum of degree's theorem.

$$\sum \deg(v_i) = 2|E|$$

$$\therefore \quad 3 \times 4 + 2 \times 3 + x \times 2 = 2 \times 11$$

where x is the number of remaining vertices

$$\therefore \quad 12 + 6 + 2x = 22$$

$$\therefore \quad x = 2.$$

$\therefore$ The number of vertices in G is $3 + 2 + x = 3 + 2 + 2 = 7$.

Theorem : In any graph, the number of odd degree vertices is an even number.

Proof : Let $V = \{v_1, v_2, ..., v_m, v_1', v_2', ... v_n'\}$ be the set of vertices in which each v_i is of odd degree and each v_i' is of even degree. Let $\deg(v_i) = 2p_i + 1$ and $\deg(v_i') = 2q_i$.

By sum of degrees theorem,

$$\therefore \quad \sum_{i=1}^{m} \deg(v_i) + \sum_{i=1}^{n} \deg(v_i') = 2\,|\,E\,|$$

$$\sum_{i=1}^{m} (2p_i + 1) + \sum_{i=1}^{n} 2q_i = 2\,|\,E\,|$$

$$\therefore \quad 2\sum_{i=1}^{m} p_i + m + 2\sum_{i=1}^{n} q_i = 2\,|\,E\,|$$

$$\therefore \quad m = 2\left[|\,E\,| - \sum_{i=1}^{m} p_i - \sum_{i=1}^{n} q_i\right]$$

$\therefore$ m is an even number.

Theorem : If G is a graph with $|\,V\,|$ vertices and $|\,E\,|$ edges then $\delta(G) \leq \dfrac{2|\,E\,|}{|\,V\,|} \leq \Delta(G)$

Proof : In any graph G we know that

$$\delta(G) \leq \deg(v) \leq \Delta(G)\ ;\ \forall\, v \in V$$

$$\therefore \quad \sum_{v \in V} \delta(G) \leq \sum_{v \in V} \deg(v) \leq \sum_{v \in V} \Delta(G)$$

$$\therefore \quad \delta(G)\,.\,|\,V\,| \leq 2\,|\,E\,| \leq \Delta(G)\,.\,|\,V\,| \qquad \because \text{ By sum of degrees theorem.}$$

$$\therefore \quad \delta(G) \leq \frac{2|\,E\,|}{|\,V\,|} \leq \Delta(G).$$

Theorem : If G is a k-regular graph then,

$$k\,.\,|\,V| = 2\,|\,E\,|$$

Proof : Since G is k-regular, we have

$$\deg(v) = k,\ \forall\, v \in V$$

$$\therefore \quad \sum_{v \in V} \deg(v) = \sum_{v \in V} k$$

$$\therefore \quad 2\,|\,E\,| = k\,|\,V\,| \qquad \because \text{ By sum of degrees theorem.}$$

Example 13. Find the number of edges in a 4-regular graph of order 3.

Given G is 4-regular

Here $k = 4$, and $|V| = 3$

We know that $k|V| = 2|E|$

$\therefore$ $4 \times 3 = 2 \times |E|$

$\therefore$ $|E| = 6$

$\therefore$ The number of edges in G is 6.

Example 14. Is there a graph with degree sequence (1, 3, 3, 3, 4, 4, 5, 6) ?

No, since the number of vertices with odd degree is not even number as we have 5 odd degrees 1, 3, 3, 3, 5.

Definition : *A sequence $d = (d_1, d_2, \ldots\ldots, d_n)$ is* ***graphic*** *if there exists a simple graph with degree sequence d.*

Example 15. Is the degree sequence (2, 2, 3, 3, 4) a graphic ?

We should have 5 vertices. Let these be *a, b, c, d, e* with degrees respectively 4, 3, 3, 2, 2.

We first draw all 4 edges from *a* to other vertices *b, c, d, e.*

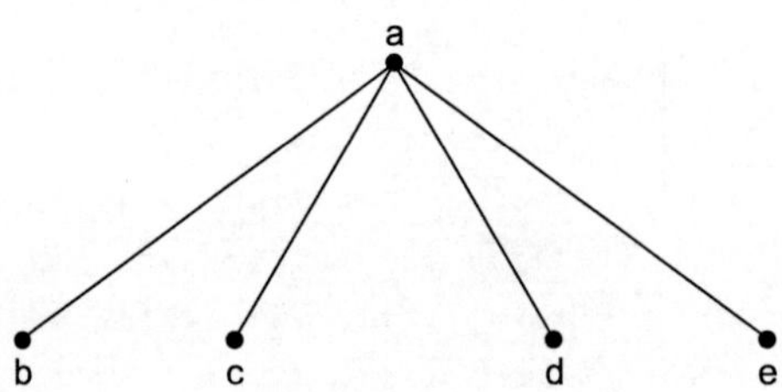

$\therefore$ The number of edges adjacent to the vertices *b, c, d, e* in the remaining portion of the graph are respectively 2, 2, 1, 1. Now draw 2 edges from the vertex *b.*

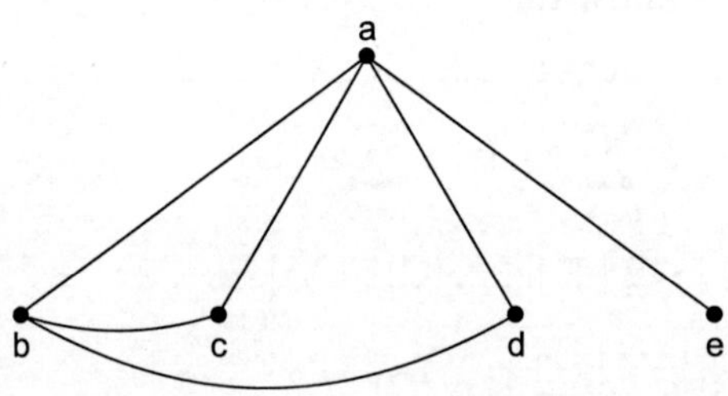

$\therefore$ In the remaining portion of the graph the number of edges adjacent to *c, d, e* are respectively 1, 0, 1. Draw an edge from *c* to *e.*

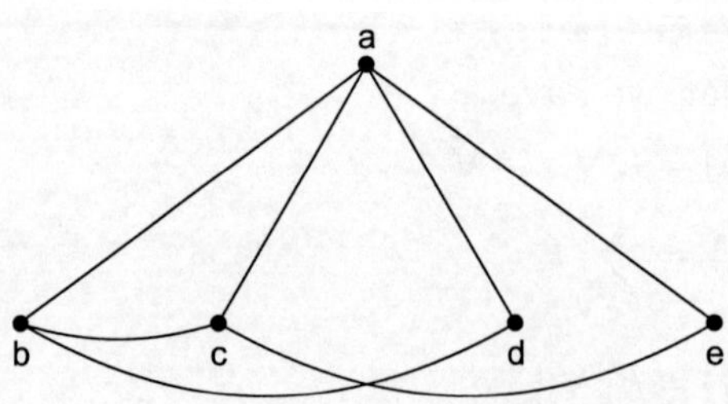

Thus, the degree sequence (2, 2, 3, 3, 4) is graphic.

Note : (Havel - Hakimi). The degree sequence $(d_1, d_2, ..., d_n)$ is graphic if and only if $(d_2 - 1, d_3 - 1, ..., d_{d_1} - 1, d_{d_1+1} - 1, d_{d_1+2}, ..., d_n)$ is graphic. (Assume $d_1 \geq d_2 \geq ...... \geq d_n$).

Example 16. *Is there a simple graph with degree sequence (1, 3, 3, 3, 4, 5, 5) ?*

Solution. We first write the sequence in descending order 5, 5, 4, 3, 3, 3, 1.

Delete the first 5 and subtract one from each of the next 5 terms. We get the sequence 4, 3, 2, 2, 2, 1.

Now delete the first 4 and subtract one from each of the next 4 terms gives the sequence 2, 1, 1, 1, 1

Now delete the first 2 and subtract one from each of the next 2 terms gives 0, 0, 1, 1 which is graph.

Thus, we can draw a simple graph with degree sequence (1, 3, 3, 3, 4, 5, 5).

Example 17. *Is the degree sequence (1, 2, 3, 3, 3, 5, 5) a graphic ?*

Solution. We first write the sequence in descending order

a	*b*	*c*	*d*	*e*	*f*	*g*
5,	5,	3,	3,	3,	2,	1
	4,	2,	2,	2,	1,	1
		1,	1,	1,	0,	1
			0,	1,	0,	1

which is graphic.

∴ The given degree sequence is graphic.

Example 18. *Is there a simple graph with degree sequence (1, 1, 3, 3, 3, 4, 6, 7) ?*

Solution.

a	*b*	*c*	*d*	*e*	*f*	*g*	*h*
7	6	4	3	3	3	1	1
	5	3	2	2	2	0	0
		4					

which is not graphic since we can not find a vertex with degree 5.

∴ The given sequence is not graphic.

Example 19. *Is there a graph with degree sequence (1, 1, 3, 3, 3, 4, 6, 7) ?*

Solution.

a	*b*	*c*	*d*	*e*	*f*	*g*	*h*	
7	6	4	3	3	3	1	1	
	5	3	2	2	2	0	0	
	3	3	2	2	2	0	0	(draw a loop at *b*)
		2	1	1	2	0	0	
			0	0	2	0	0	(draw a loop at *f*)

We draw the graph in the following manner.

(*i*) Draw seven edges from *a* to all other edges.

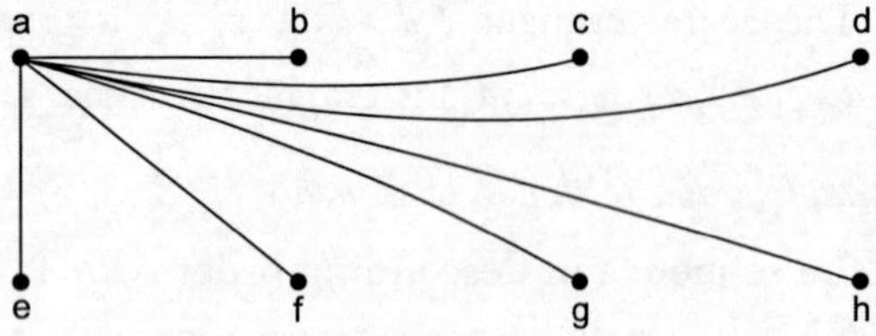

(*ii*) Now draw a loop at *b* and then 3 edges from *b* to the vertices *c*, *d*, *e*.

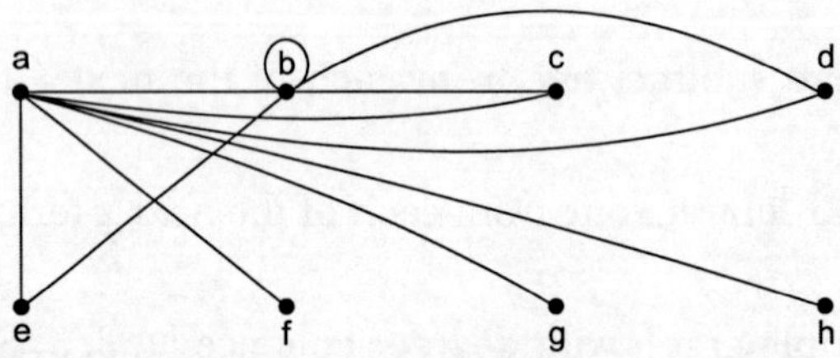

(*iii*) Draw 2 edges from *c* to *d* and *e*.

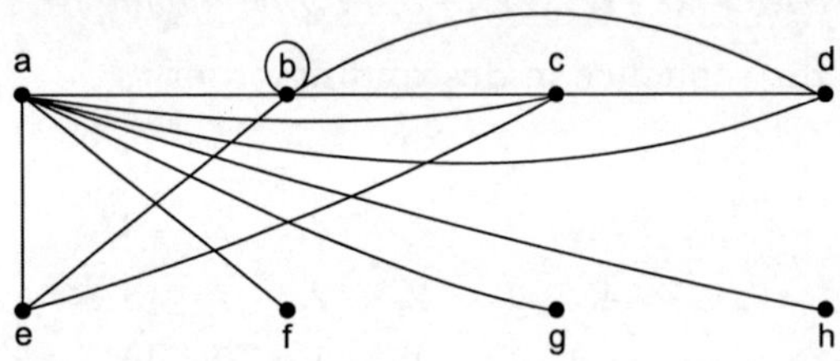

(*iv*) Draw a loop at *f*

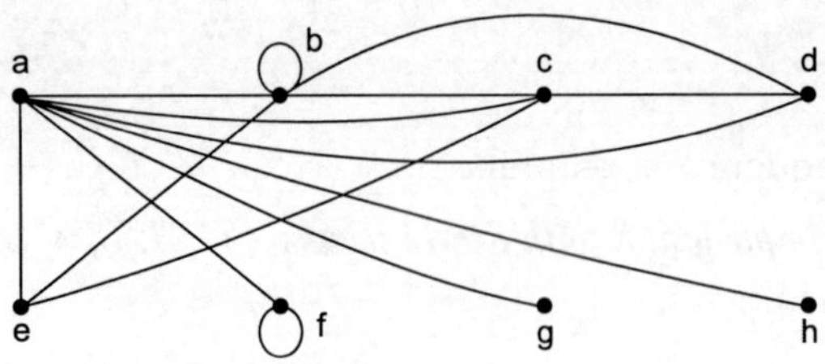

∴ The graph shown above is a graph with degree sequence (1, 1, 3, 3, 3, 3, 4, 6, 7).

Definition : ***A walk*** *in a graph is a finite alternating sequence of vertices and edges* $v_o\ e_1\ v_1\ e_2\ v_2$, ..., $e_n\ v_n$. *Such that* $e_i = (v_{i-1}, v_i)$, $1 \le i \le n$.

If $v_o = v_n$ it is called a **closed walk**, otherwise it is an **open walk**.

Example 20.

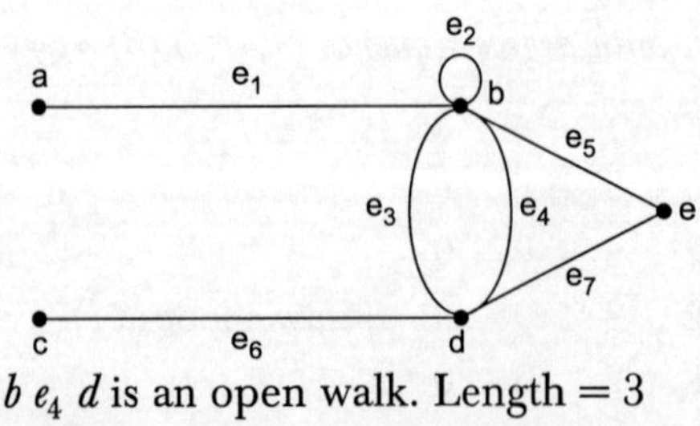

$a\ e_1\ b\ e_2\ b\ e_4\ d$ is an open walk. Length = 3

$a\ e_1\ b\ e_2\ b\ e_1\ a$ is a closed walk. Length = 3.

$a\ e_1\ b\ e_6\ c$ is not a walk.

Definition : *A walk is called a **trial** if all its edges are distinct.*

An open trial is called a **path**

A closed trial is called a **circuit.**

Definition : *A circuit in which all the vertices are distinct except the first and last vertices is called a **cycle**.*

Example 21. In the graph shown in example 1.20,

$b\ e_5\ e\ e_7\ d$ is a path

$b\ e_5\ e\ e_7\ d\ e_4\ b$ is a circuit

$b\ e_4\ d\ e_3\ b\ e_4\ d$ is a walk but not a trial $\because$ e_4 is repeated.

$b\ e_5\ e\ e_7\ d\ e_4\ b$ is a cycle.

$b\ e_5\ e\ e_7\ d\ e_4\ b\ e_2\ b$ is not a cycle but *a* circuit since b is repeated.

Example 22. Give an example of a circuit which is not a cycle in a simple graph.

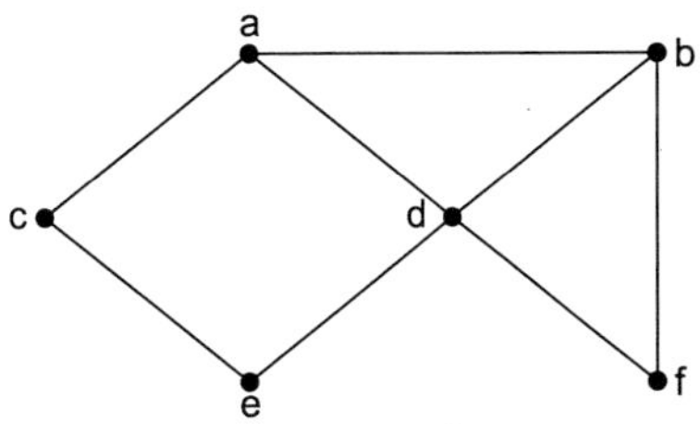

$a - d - f - b - d - e - c - a$ is a circuit but not a cycle since an internal vertex d is repeated.

Definition : *A graph is called **connected** if there exists a path from a vertex to any other vertex.*

Example 23.

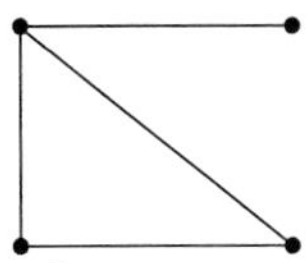

is a connected graph.

Example 24. The graph 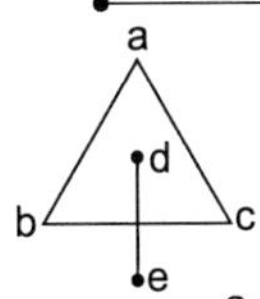

is not connected since there is no path from '*a*' to '*d*'. It has two connected components 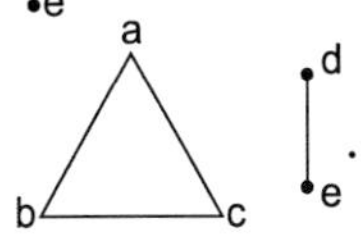

.

Example 25. Write the connected components of the following graph :

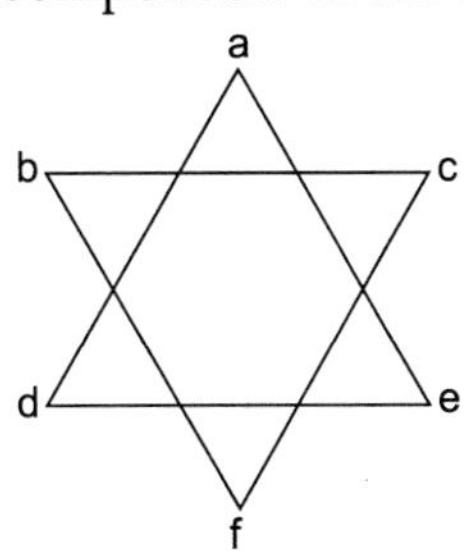

The graph is not connected since there is no path from '*a*' to '*b*'.

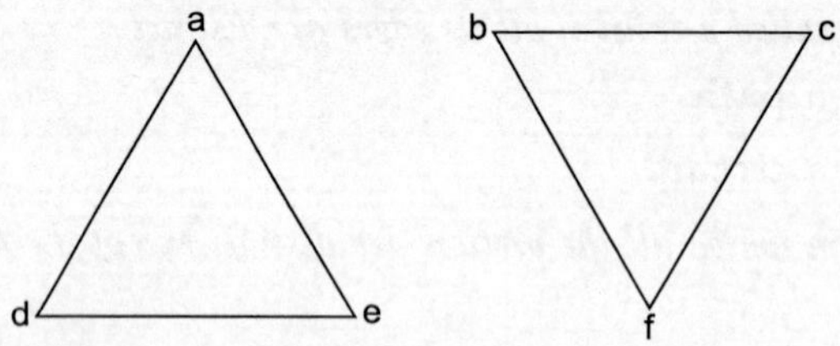

are the two connected components.

Definition : *The **adjacency matrix** of a simple graph* $G = (V, E)$ *where* $V = \{v_1, v_2, \ldots\ldots, v_n\}$ *are the ordered vertices is an* $n \times n$ *matrix* $A_G = (a_{ij})$

where $$a_{ij} = \begin{cases} 1 & \text{if} \quad \text{an edge connects } v_i \text{ to } v_j \\ 0 & \text{if} \quad \text{no edge connects } v_i \text{ to } v_j \end{cases}$$

Example 26. Determine the adjacency matrix for the simple graph shown in the following figure.

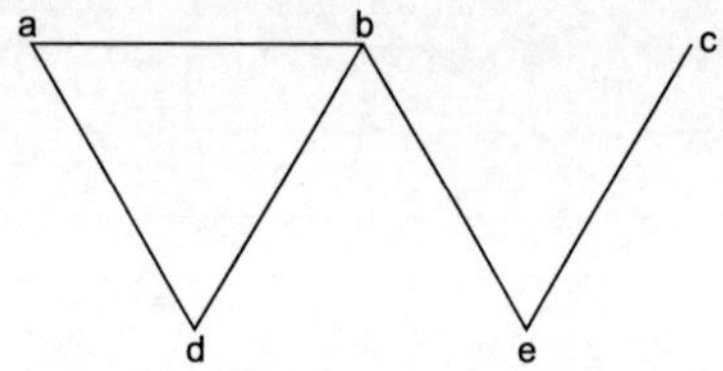

We first order the vertices as *a, b, c, d, e.* The adjacency matrix of the graph is

$A_G =$

	a	*b*	*c*	*d*	*e*
a	0	1	0	1	0
b	1	0	0	1	1
c	0	0	0	0	1
d	1	1	0	0	0
e	0	1	1	0	0

Note 1. If there is a loop at the vertex v_i we write $a_{ii} = 1$.

2. If there is more than one edge between the vertices v_i and v_j we write

$$a_{ij} = \text{Number of edges between } v_i \text{ and } v_j.$$

Example 27. Write the adjacency matrix of the following multigraph.

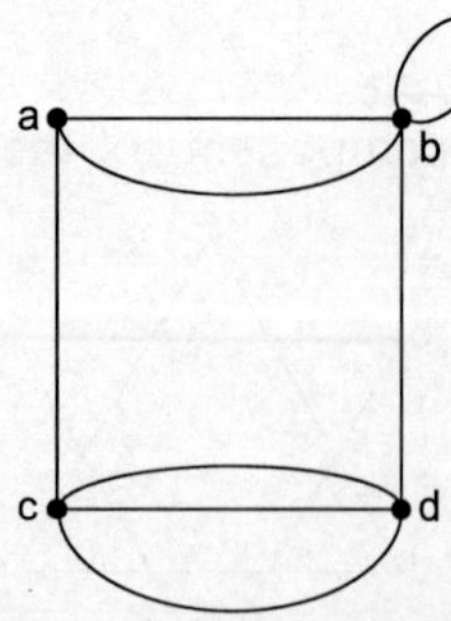

Let *a, b, c, d* be the ordered vertices of the multigraph G.

The adjacency matrix of G is

$A_G =$		a	b	c	d
	a	0	2	1	0
	b	2	1	0	1
	c	1	0	0	3
	d	0	1	3	0

8.2 ISOMORPHISM AND SUBGRAPHS

Look at the following graphs. They appear in as different graphs.

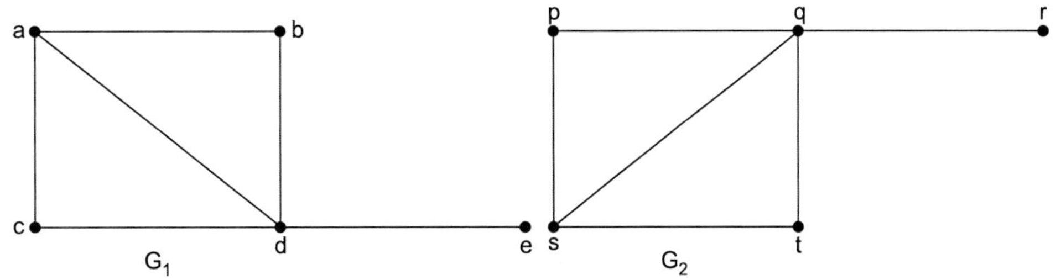

Figure. 2.1.

A careful look at the vertices and edges of G_1 and G_2 by ignoring the vertex labelling, they are the same graphs. Such graphs are called isomorphic to each other.

Definition : *Two graphs G_1 and G_2 are **isomorphic** if there is a function $f: V(G_1) \to V(G_2)$ such that*

(*i*) f is one-to-one

(*ii*) f is onto

and (*iii*) f-preserves adjacency

i.e., for each pair of vertices u and v of G,

$$\{u, v\} \in E(G_1) \text{ if and only if } \{f(u), f(v)\} \in E(G_2).$$

Discovering Isomorphism :

The problem of determining whether or not two graphs are isomorphic is known as the isomorphic problem. To solve it we verify the following in the two graphs :

(*i*) The same number of vertices

(*ii*) The same number of edges

(*iii*) The same number of connected components

(*iv*) The same degree sequences

(*v*) The same number of cycles of different lengths.

If any of the above criterion fails the graphs are not isomorphic. The above rules are only necessary but not sufficient for providing two graphs are isomorphic.

To show two graphs G_1 and G_2 are isomorphic we construct a one-to-one and onto function $f: V(G_1) \to V(G_2)$ such that $A_{G_1} = A_{G_2}$ where A_{G_1} = the adjacency matrix of G_1 for the vertex ordering $v_1, v_2, \ldots\ldots, v_n$ and A_{G_2} = the adjacency matrix of G_2 for the vertex ordering $f(v_1), f(v_2), \ldots\ldots f(v_n)$.

Example 1. Show that the graphs G_1 and G_2 shown in Figure. 2.1 are isomorphic.

G_1	Vertex	a	b	c	d	e
	degree	3	2	2	4	1

G_2	Vertex	p	q	r	s	t
	degree	2	4	1	3	2

We should have $a \to s$, $d \to q$ and $e \to r$

We take $b \to p$ and $c \to t$

Thus,

v	a	b	c	d	e
$f(v)$	s	p	t	q	r

The adjacency matrix of G_1 for the vertex ordering a, b, c, d, e is

$$A_{G_1} = \begin{array}{c|ccccc} & a & b & c & d & e \\ \hline a & 0 & 1 & 1 & 1 & 0 \\ b & 1 & 0 & 0 & 1 & 0 \\ c & 1 & 0 & 0 & 1 & 0 \\ d & 1 & 1 & 1 & 0 & 1 \\ e & 0 & 0 & 0 & 1 & 0 \end{array}$$

The adjacency matrix of G_2 for the vertex ordering s, p, t, q, r is

$$A_{G_2} = \begin{array}{c|ccccc} & s & p & t & q & r \\ \hline s & 0 & 1 & 1 & 1 & 0 \\ p & 1 & 0 & 0 & 1 & 0 \\ t & 1 & 0 & 0 & 1 & 0 \\ q & 1 & 1 & 1 & 0 & 1 \\ r & 0 & 0 & 0 & 1 & 0 \end{array}$$

$\therefore \quad A_{G_1} = A_{G_2}$

Thus f is one-to-one, onto and also preserves adjacency. We conclude the graphs G_1 and G_2 are isomorphic.

Example 2. Show that the following two graphs are isomorphic.

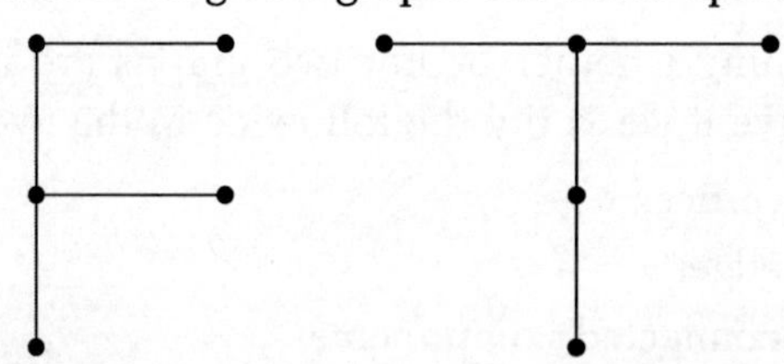

We first give the vertex labelling.

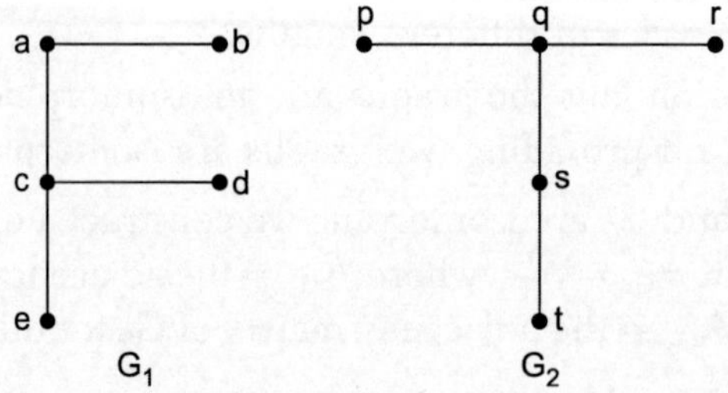

G_1	Vertex	a	b	c	d	e
	Degree	2	1	3	1	1

G_2	Vertex	p	q	r	s	t
	Degree	1	3	1	2	1

We should have $a \to s$ and $c \to q$.

$\{a, b\}$ is an edge in G_1 further deg $(a) = 2$ and deg $(b) = 1$. In G_2 we have deg (S) = 2 and {S, t} is an edge where deg $(t) = 1$

∴ We should take $b \to t$

Now take $d \to p$ and $e \to r$

v	a	b	c	d	e
$f(v)$	s	t	q	p	r

$$A_{G_1} = \begin{array}{c|ccccc} & a & b & c & d & e \\ \hline a & 0 & 1 & 1 & 0 & 0 \\ b & 1 & 0 & 0 & 0 & 0 \\ c & 1 & 0 & 0 & 1 & 1 \\ d & 0 & 0 & 1 & 0 & 0 \\ e & 0 & 0 & 1 & 0 & 0 \end{array} \qquad A_{G_2} = \begin{array}{c|ccccc} & s & t & q & p & r \\ \hline s & 0 & 1 & 1 & 0 & 0 \\ t & 1 & 0 & 0 & 0 & 0 \\ q & 1 & 0 & 0 & 1 & 1 \\ p & 0 & 0 & 1 & 0 & 0 \\ r & 0 & 0 & 1 & 0 & 0 \end{array}$$

$$\therefore \qquad A_{G_1} = A_{G_2}$$

Thus f is one-one, onto and preserves adjacency. The graph G_1 and G_2 are isomorphic.

Example 3. Determine whether the following two graphs are isomorphic.

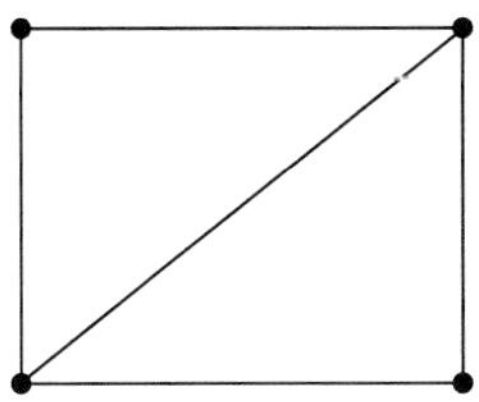

G_1

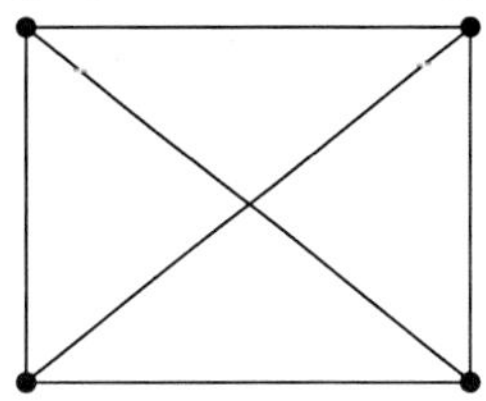

G_2

(*i*) Number of vertices

$$|V_1| = 4, |V_2| = 4$$

$$\therefore \qquad |V_1| = |V_2|$$

(*ii*) Number of edges

$$|E_1| = 5, |E_2| = 6$$

∴ $|E_1| \neq |E_2|$ hence G_1, G_2 are not isomorphic.

Example 4. Test whether the following graphs are isomorphic

G_1

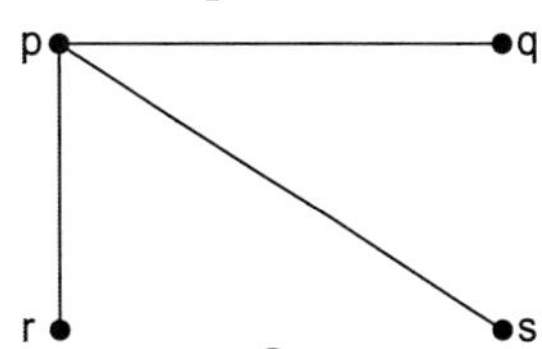

G_2

(*i*) Number of vertices

$$|V_1| = 4, |V_2| = 4$$

$\therefore \qquad |V_1| = |V_2|$

(*ii*) Number of edges

$$|E_1| = 3, |E_2| = 3$$

$\therefore \qquad |E_1| = |E_2|$

(*iii*) degree sequences

G_1	Vertex	*a*	*b*	*c*	*d*
	Degree	1	1	2	2

G_2	Vertex	*p*	*q*	*r*	*s*
	Degree	3	1	1	1

$\therefore \qquad d_1 = (1, 1, 2, 2) \quad \therefore \quad d_2 = (1, 1, 1, 3)$

$$d_1 \neq d_2.$$

Hence G_1, G_2 are not isomorphic.

Example 5. Prove or disprove that the following two graphs are isomorphic.

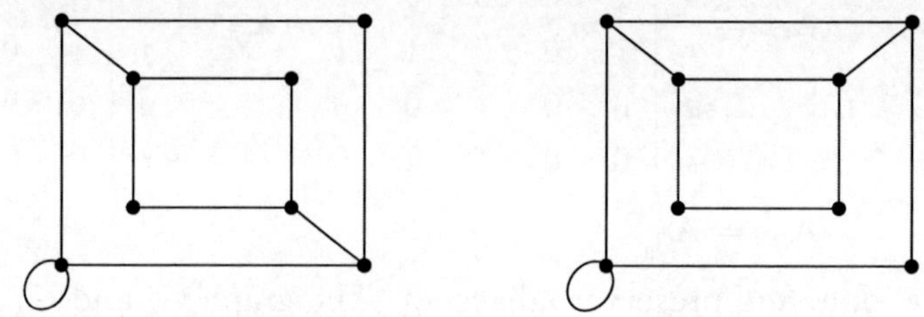

We first give vertex labelling for the graphs :

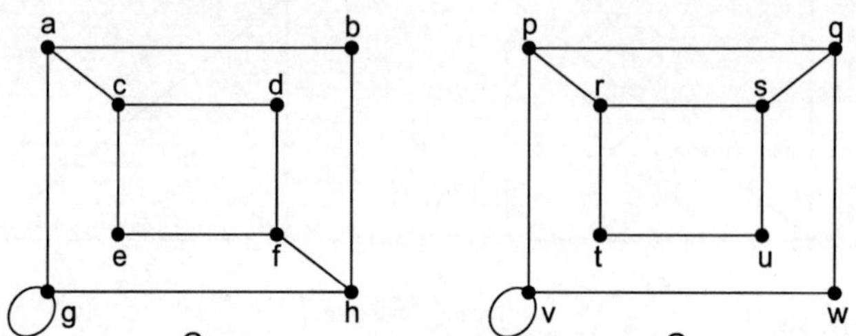

G_1 $\qquad$ G_2

(*i*) Number of vertices

$$|V_1| = 8, |V_2| = 8$$

$\therefore \qquad |V_1| = |V_2|$

(*ii*) Number of edges

$$|E_1| = 11, |E_2| = 11$$

$\therefore \qquad |E_1| = |E_2|$

(*iii*) degree sequences

G_1	Vertex	*a*	*b*	*c*	*d*	*e*	*f*	*g*	*h*
	Degree	3	2	3	2	2	3	4	3

$\therefore \qquad d_1 = (2, 2, 2, 3, 3, 3, 3, 4)$

G_2	Vertex	p	q	r	s	t	u	v	w
	Degree	3	3	3	3	2	2	4	2

$\therefore \qquad d_2 = (2, 2, 2, 3, 3, 3, 3, 4)$

$\therefore \qquad d_1 = d_2.$

(*iv*) Number of cycles of different lengths.

Each of G_1, G_2 has one loop (a cycle of length one).

G_1 has two cycles of length 4 (a-b-h-g-a ; c-d-f-e-c) But G_2 has three cycles of length 4 (p-q-w-v-p ; r-s-u-t-r ; p-q-s-r-p)

Thus, G_1, G_2 are not isomorphic.

Example 6. Determine whether the following graphs are isomorphic or not ?

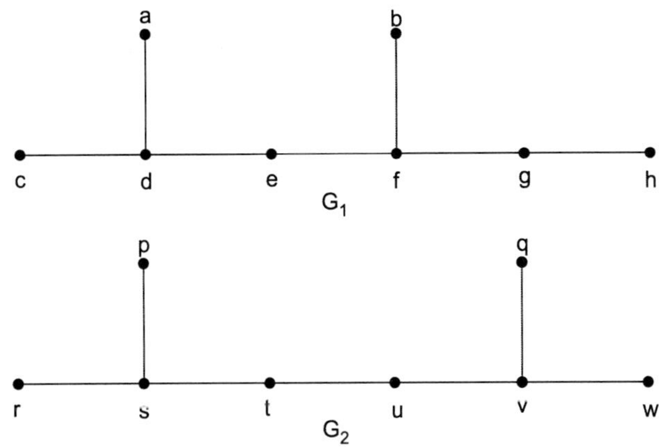

(*i*) Number of vertices

$$|V_1| = 8, |V_2| = 8$$

$$\therefore \qquad |V_1| = |V_2|$$

(*ii*) Number of edges

$$|E_1| = 7, |E_2| = 7$$

$$\therefore \qquad |E_1| = |E_2|$$

(*iii*) degree sequences

G_1	Vertex	a	b	c	d	e	f	g	h
	Degree	1	1	1	3	2	3	2	1

$$d_1 = (1, 1, 1, 1, 2, 2, 3, 3)$$

G_2	Vertex	p	q	r	s	t	u	v	w
	Degree	1	1	1	3	2	2	3	1

$$d_2 = (1, 1, 1, 1, 2, 2, 3, 3)$$

$$\therefore \qquad d_1 = d_2$$

(*iv*) Each of the graphs G_1, G_2 has no cycles.

But still the graphs are not isomorphic since the length of the path between 3 degree vertices in G_1 is 2 (*d-e-f*) while in G_2 is 3 (*s-t-u-v*).

Example 7. Determine whether the following two graphs are isomorphic or not ? If isomorphic give the one-to-one mapping.

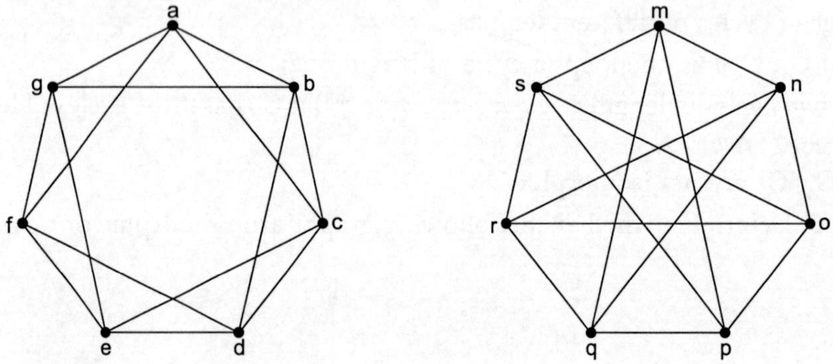

G_1 G_2

Both of the graphs G_1, G_2 have 7 vertices and 14 edges. Every vertex in each graph has degree 4. We construct an isomorphism by using the observation that on each edge of the perimeter in G_1 we have two triangles and on each internal edge we have one triangle where as in G_2 this criterion is reversed.

We start with vertex '*a*' in G_1, we can match '*a*' to any vertex in G_2, since each graph is symmetric. Let us take $a \to m$. Now the set of neighbours of '*a*' must be matched to the set of neighbours of '*m*'. We can match '*b*' to either '*p*' or '*q*', since $\{a, b\}$ is a perimeter edge of G_1, it should match to an internal edge in G_2. Let us take $b \to p$.

Now $\{a, c\}$ is an internal edge and *a*, *b*, *c* forms *a* triangle in G_1. It should match to the perimeter edge $\{m, s\}$ but not match with $\{m, n\}$ as *m*, *p*, *n* does not form a triangle. We should take $c \to s$.

The edge $\{a, f\}$ should match to the remaining perimeter edge $\{m, n\}$ and $\{a, g\}$ should match to $\{m, q\}$. Thus, we have $f \to n$ and $g \to q$.

Clearly $\{d, e\}$ should match to $\{r, o\}$. We should have $d \to 0$. Since '*d*' is adjacent to *b*, *c*, *f* and the vertex in G_2 which is adjacent to *p*, *s*, *n* is '*o*' but not '*r*'. At last we have $e \to r$.

∴

v	a	b	c	d	e	f	g
$f(v)$	m	p	s	o	r	n	q

$A_{G_1} =$

	a	b	c	d	e	f	g
a	0	1	1	0	0	1	1
b	1	0	1	1	0	0	1
c	1	1	0	1	1	0	0
d	0	1	1	0	1	1	0
e	0	0	1	1	0	1	1
f	1	0	0	1	1	0	1
g	1	1	0	0	1	1	0

$A_{G_2} =$

	m	*p*	*s*	*o*	*r*	*n*	*q*
m	0	1	1	0	0	1	1
p	1	0	1	1	0	0	1
s	1	1	0	1	1	0	0
o	0	1	1	0	1	1	0
r	0	0	1	1	0	1	1
n	1	0	0	1	1	0	1
q	1	1	0	0	1	1	0

$\therefore \quad A_{G_1} = A_{G_2}$.

Thus, f is one-one, onto and preserves adjacency. Hence G_1, G_2 are isomorphic.

Definition : *A graph $H = (V_1, E_1)$ is called a* ***subgraph*** *of $G = (V, E)$ if and only if $V_1 \subseteq V$ and $E_1 \subseteq E$.*

Definition : *A graph $H = (V_1, E_1)$ is called a* ***spanning subgraph*** *of $G = (V, E)$ if and only if $V_1 = V$ and $E_1 \subseteq E$.*

Definition : *Let $G = (V, E)$ be a graph and $W \subseteq V$. The* ***subgraph induced*** *by W is the subgraph H obtained by taking the vertex set of H is W and the edges of H are to be those edges of G that join pairs of vertices in W.*

Example 8. Let the graph G be as follows :

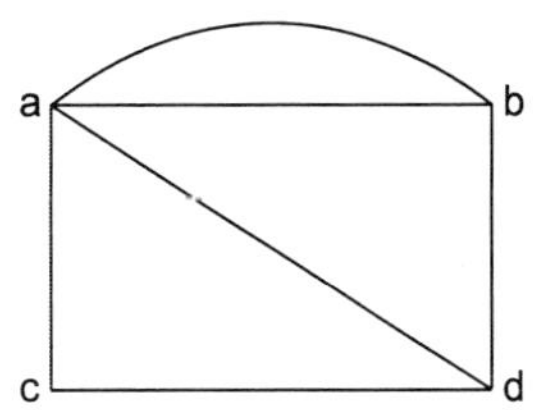

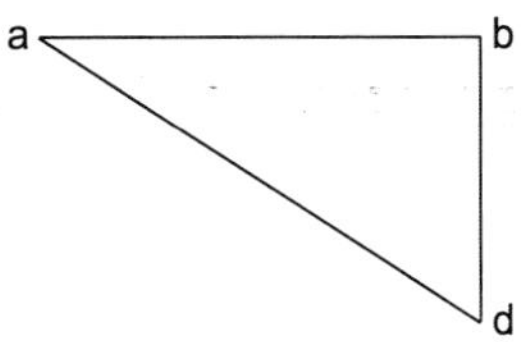

is a subgraph of G

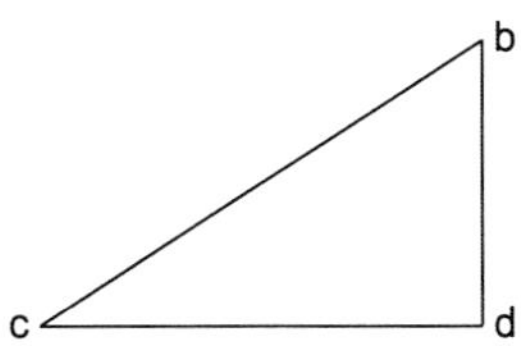

is not a subgraph of G

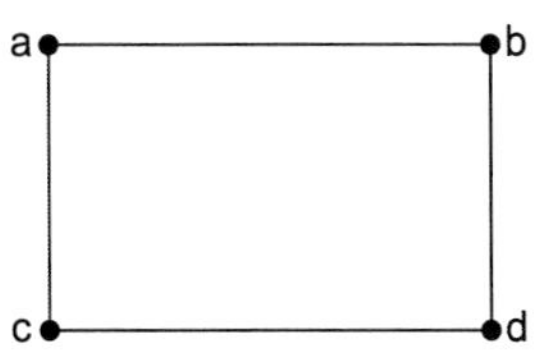

,

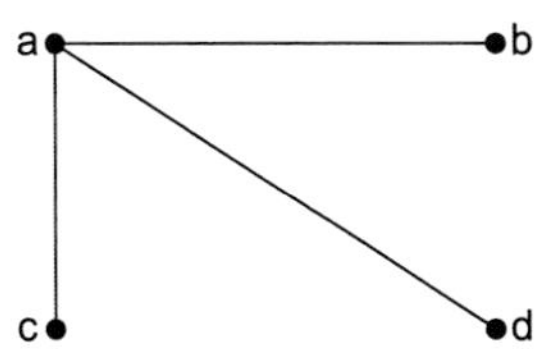

are spanning subgraphs of G

Let $W = \{a, b, c\}$. The subgraph induced by W of G is

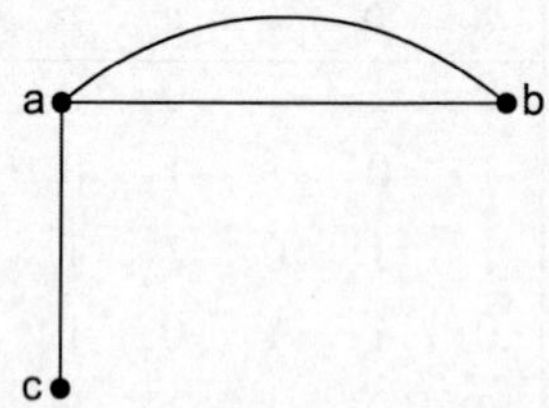

Special Simple Graphs :

There are some special simple graphs which could be used as models in many real applications.

Definition : *A graph in which every vertex is connected to every other vertex is called a* ***complete graph***.

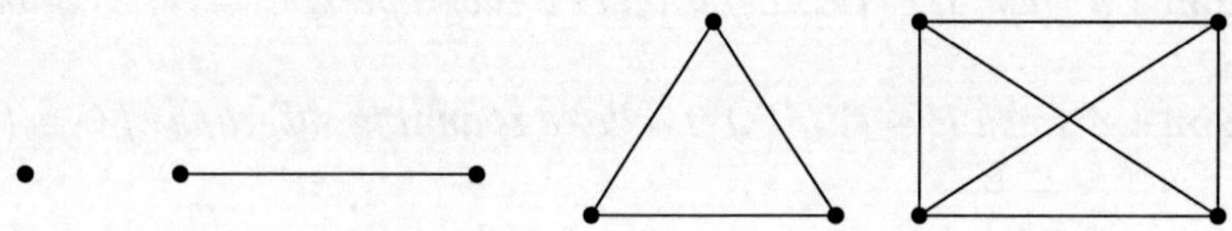

K_1 K_2 K_3 K_4

The number of edges in k_n is $\frac{n(n-1)}{2}$.

Definition : *A graph in which all the edges forms a cycle is called a* ***cycle graph***.

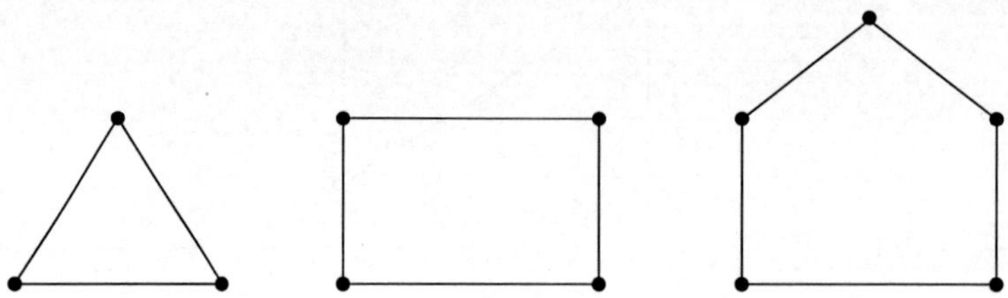

C_3 C_4 C_5

Definition : *A graph whose edges forms a* ***path*** *is called a path graph.*

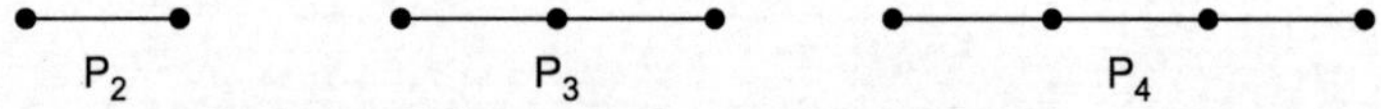

P_2 P_3 P_4

Definition : *A graph obtained from a cycle graph by joining a single new vertex (the hub) to each vertex of the cycle is called a* ***wheel graph***.

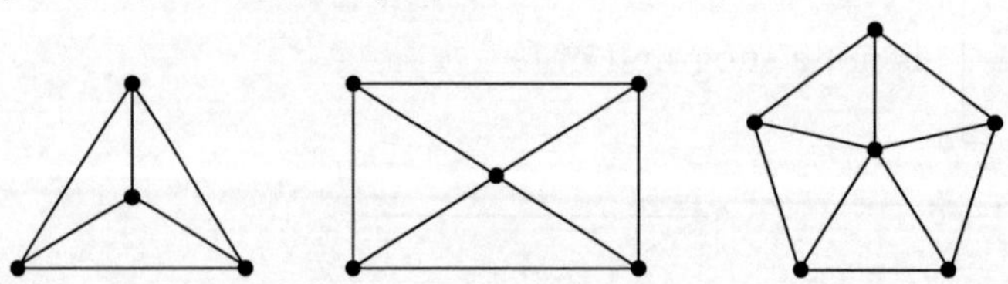

W_4 W_5 W_6

Definition : *A graph with no edges is called a* ***null graph***.

N_1 N_2 N_3

Definition : *The graph Q_n is called a **cube**. The vertices of Q_n are all binary strings of length n. Two vertices are adjacent if they differ in exactly one bit position.*

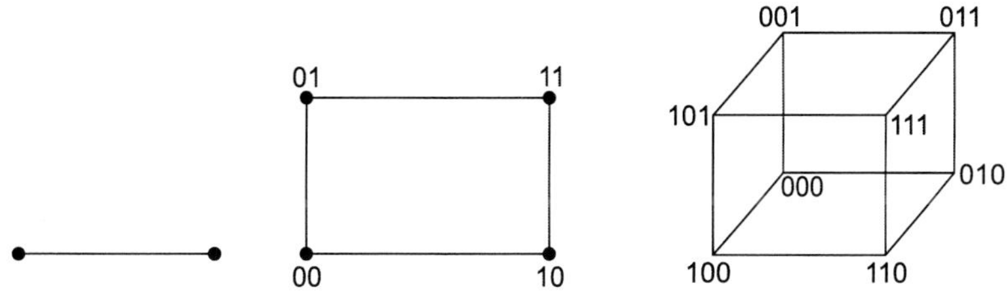

Q_1 Q_2 Q_3

Definition : *A graph in which the set of vertices can be partitioned into two sets M and N in such a way that each edge joins a vertex in M to a vertex in N is called a **bipartite graph**.*

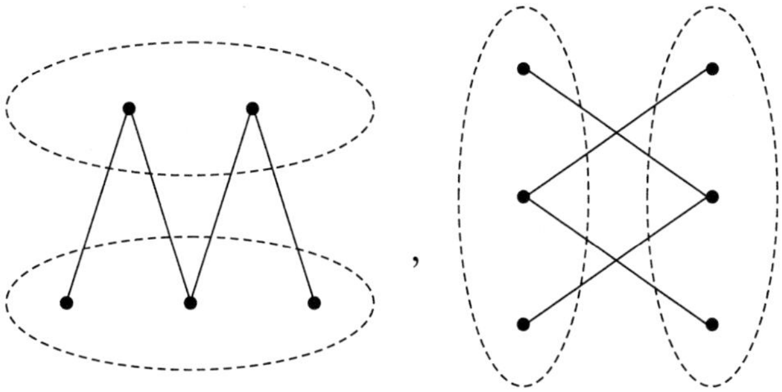

are bipartite.

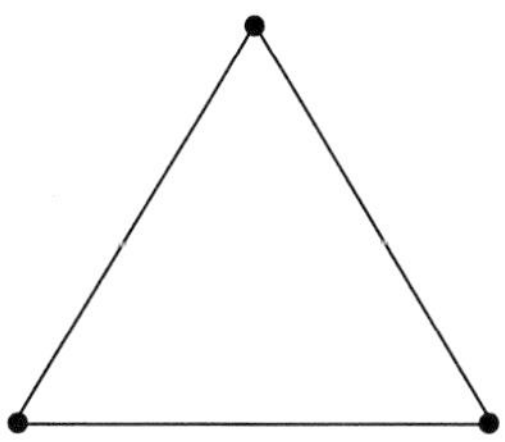

is not a bipartite graph.

Definition : *A bipartite graph in which every vertex of M is adjacent to every vertex of N is called a **complete bipartite graph.***

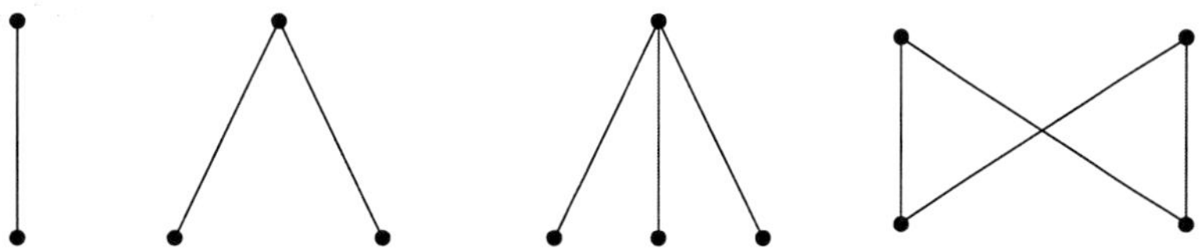

$K_{1,1}$ $K_{1,2}$ $K_{1,3}$ $K_{2,2}$

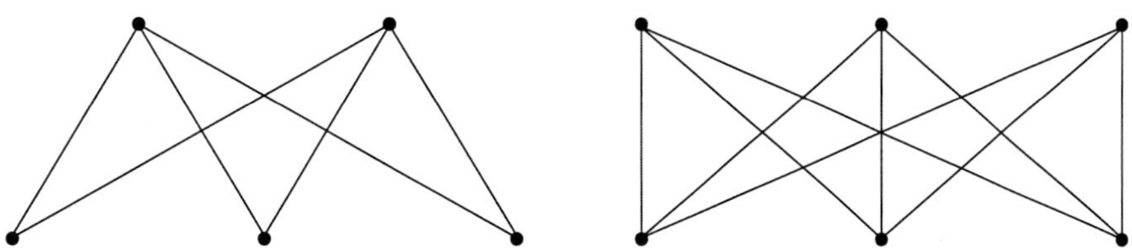

$K_{2,3}$ $K_{3,3}$

$k_{m,\,n}$ has $m + n$ vertices and mn edges.

Definition : *A complete bipartite graph $k_{1,n}$ is called a **star graph**.*

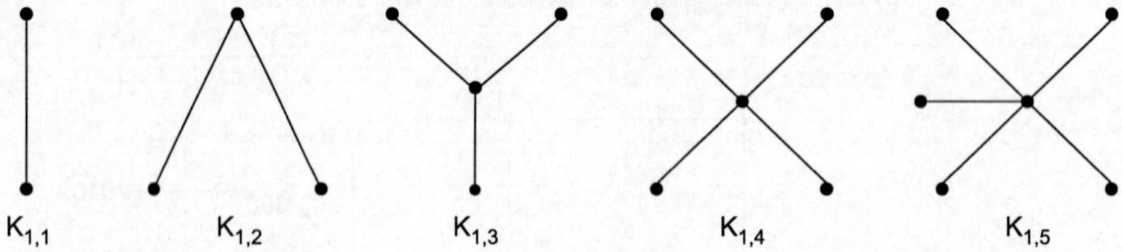

Operations on Graphs

Complement of a graph : The complement of a simple graph $G = (V, E)$ is the graph $\overline{G} = (V, \overline{E})$ where $\overline{E} = \{v_i, v_j\}/\{v_i, v_j\} \notin E\}$.

Example 9. Find the complement of the graph.

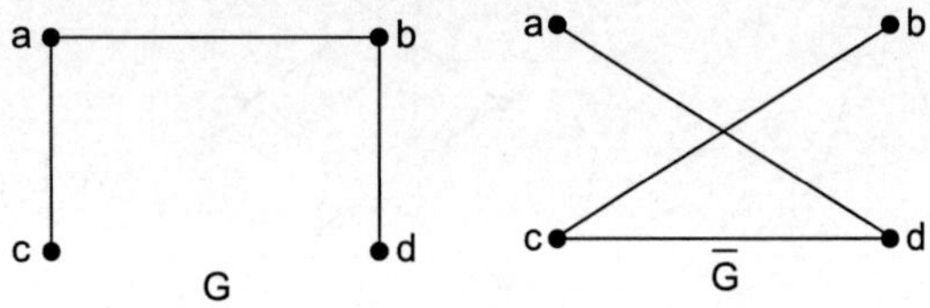

Note:1. If G is a graph with 'n' vertices then $\overline{G}$ is obtained from k_n by simply deleting the edges of G.

2. If G has 'n' vertices and 'p' edges then $\overline{G}$ has 'n' vertices and $\frac{n(n-1)}{2} - p$ edges.

3. Let G_1 and G_2 be simple graphs. Then G_1 and G_2 are isomorphic if and only if $\overline{G}_1$ and $\overline{G}_2$ are isomorphic.

Union of Graphs : Let $G_1 = (V_1, E_1)$ and $G_2 = (V_2, E_2)$ be two graphs. Then the union of G_1 and G_2 is defined as $G_1 \cup G_2 = (V, E)$ where $V = V_1 \cup V_2$ and $E = E_1 \cup E_2$.

Intersection of Graphs : Let $G_1 = (V_1, E_2)$ and $G_2 = (V_2, E_2)$ be two graphs. Then the intersection of G_1 and G_2 is defined as $G_1 \cap G_2 = (V, E)$ where $V = V_1 \cap V_2$ and $E = E_1 \cap E_2$.

Example 10. Determine the union and intersection of the following graphs :

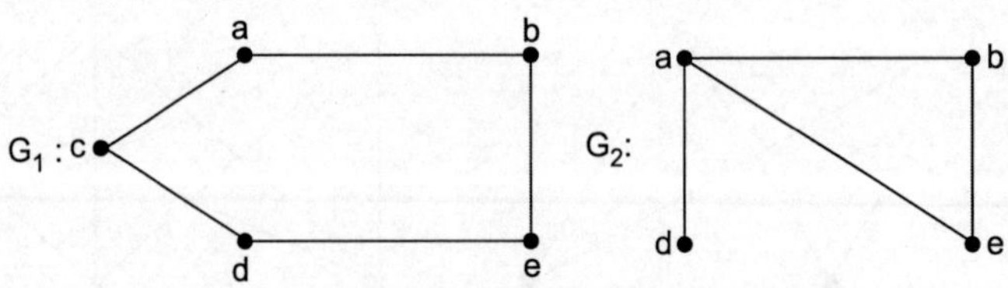

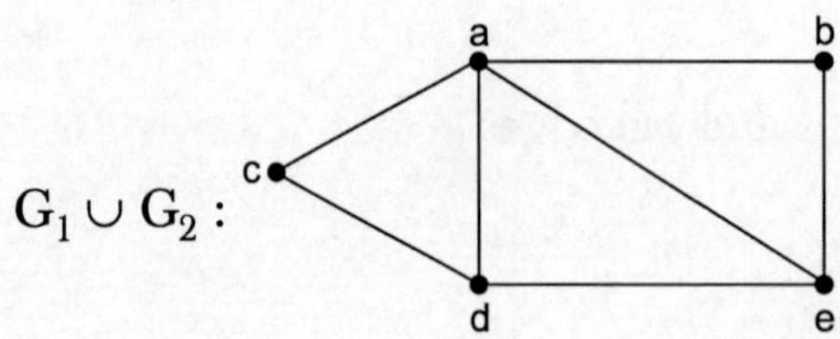

$G_1 \cap G_2$:

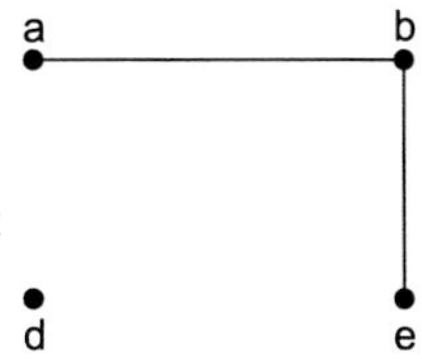

Subgraph G-v, Cut Points : G-v is the subgraph of G obtained by deleting the vertex v from the vertex set V (G) and deleting all edges in E (G) which are incident on v.

A vertex v is called a **cut vertex** or a **cut point** for a connected graph G if G-v is disconnected (More generally, v is a cut point for any graph G, if G-v has more connected components than G).

Example 11. Find G-a, G-c and G-e of the following graph G.

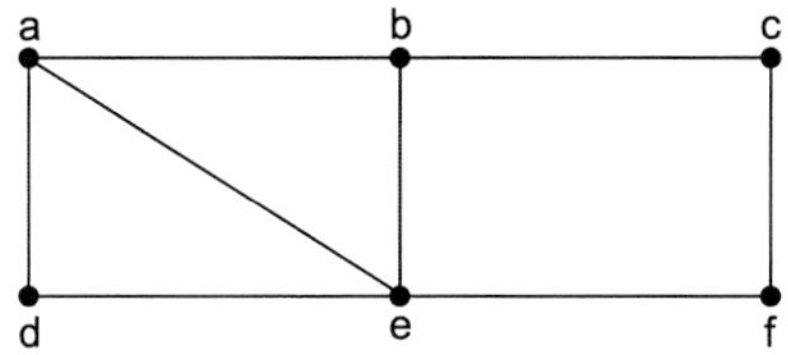

G

G-a :

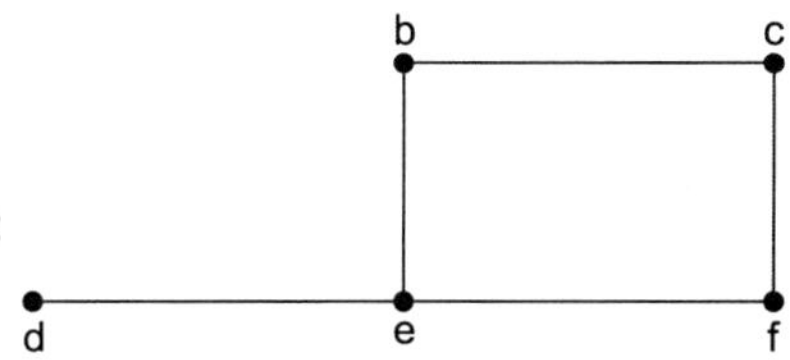

G-c :

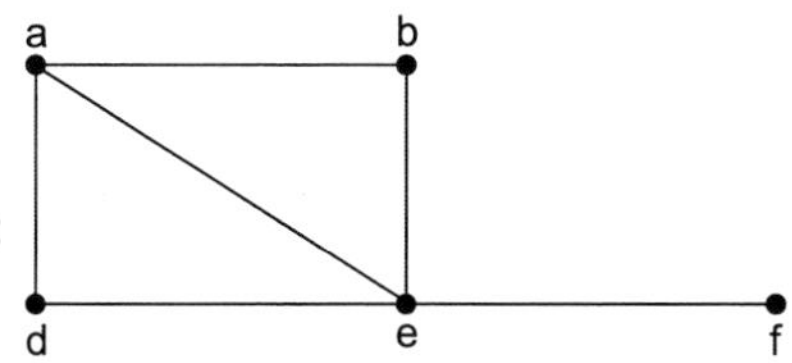

G-e :

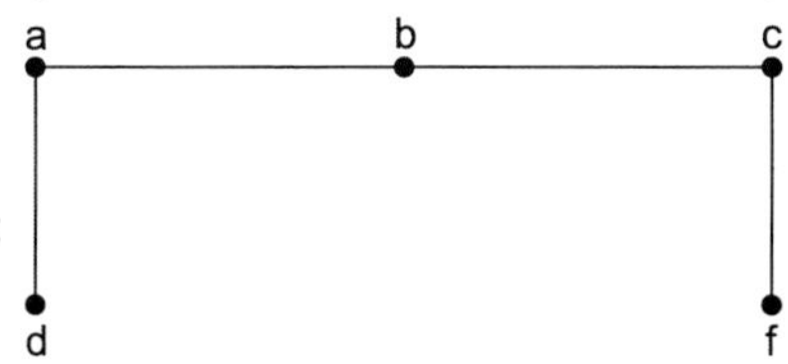

Example 12. Find all cut points of the following graph G.

G :

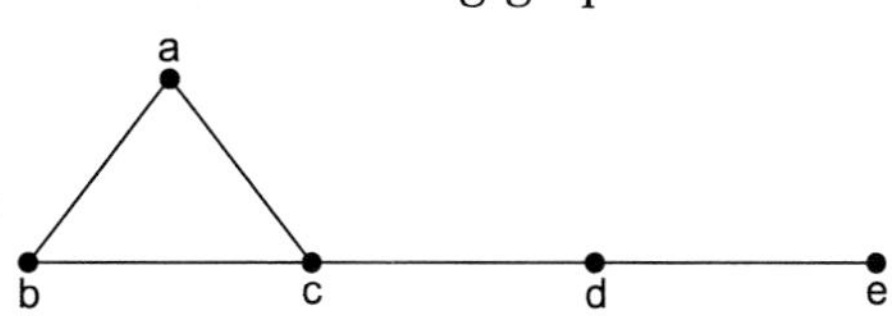

G-a :

G-*b* :

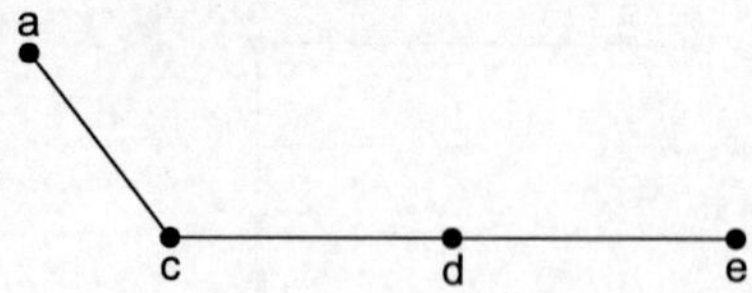

G-*c* :

G-*d* :

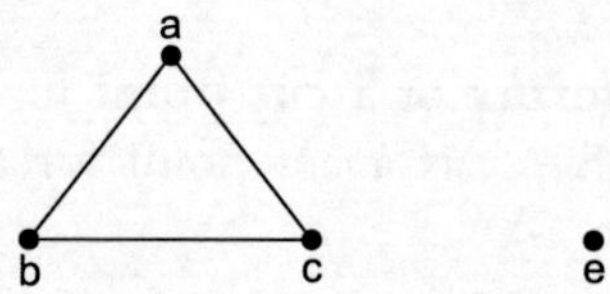

G-*e* :

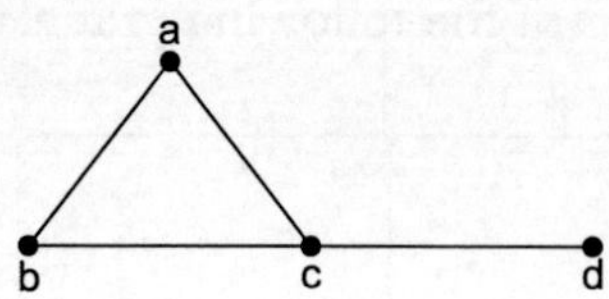

Here G-*c* and G-*d* are disconnected graphs. Hence the vertices '*c*' and '*d*' are the cut points of G.

Subgraph G-*e*, Bridges : G-*e* is the subgraph of G obtained by simply deleting the edge '*e*' from the edge set E (G).

An edge '*e*' is called a **cut edge** or a **bridge** for a connected graph G if G-*e* is disconnected. (In general '*e*' is a bridge for any graph G if G-*e* has more connected components than G has).

Example 13. Find G-{*a*, *b*} and G-{*c*, *d*} of the following graph G.

G :

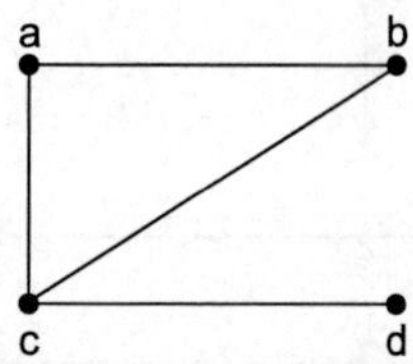

G-{*a*, *b*} :

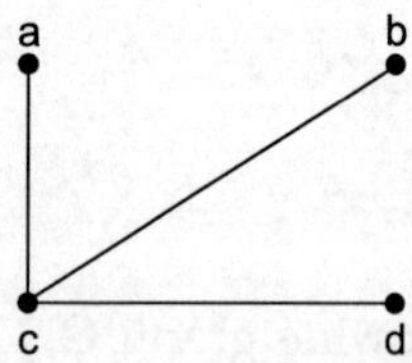

G-{*c*, *d*} :

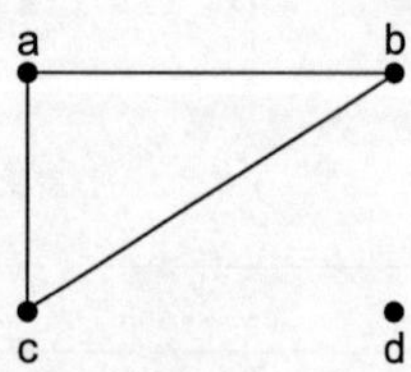

Here {*c*, *d*} is a bridge.

Example 14. Find the bridges of the following graph G.

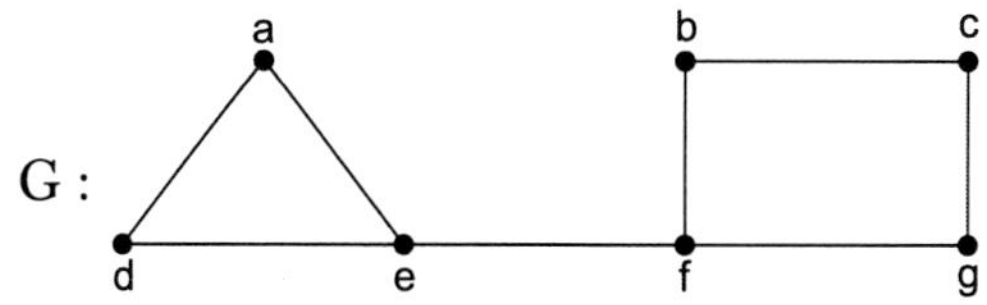

It has only one bridge $\{e, f\}$ since $G = \{e, f\}$ is disconnected graph :

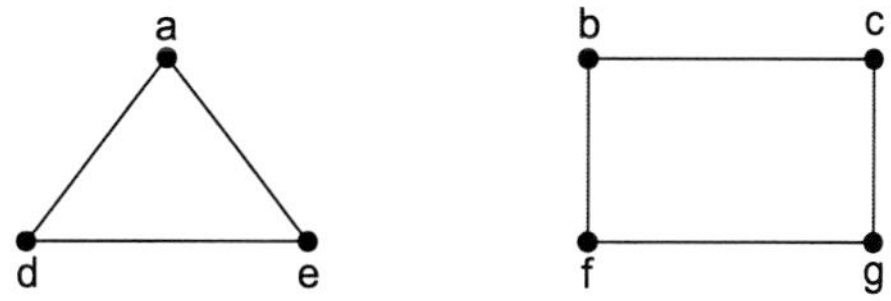

8.3 MULTIGRAPHS AND EULER CIRCUITS

Definition : A graph containing parallel edges is called a **multigraph**.

Example 1. G : 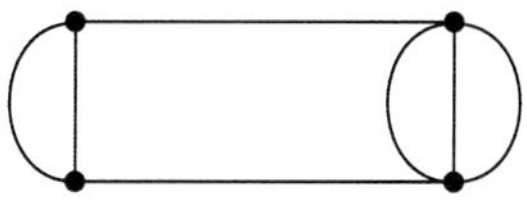

G is a multigraph.

Example 2. State Konigsberg bridge problem and write the multigraph of it given by Euler.

Statement : The 18th Century East Prussian city of Konigsberg included two islands and seven bridges as shown in the following figure.

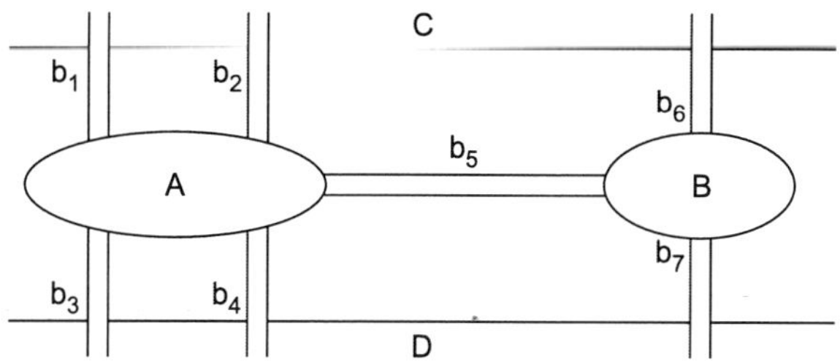

A and B are islands. C and D are banks of pregal river. b_1, b_2, ..., b_6 and b_7 are bridges.

Problem : Beginning at any of the 4 land areas and ending anywhere, can a person walk through city crossing all 7 bridges but not crossing any bridge more than once.

Euler's graphical representation :

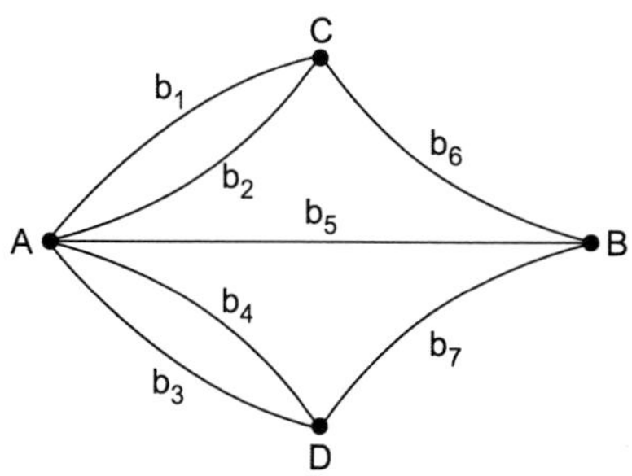

Definition : *An **Euler Path** in a multigraph is a path that includes each edge of the multigraph exactly once and intersects each vertex of the multigraph at least once.*

Definition : *A multigraph is said to be **traversable** if it has an Euler Path.*

Definition : *An **Euler Circuit** is an Euler Path whose end points are identical.*

Definition : *A multigraph is said to be an **Eulerian multigraph** if it has an Euler circuit.*

Example 3. Give an example of a traversable multigraph.

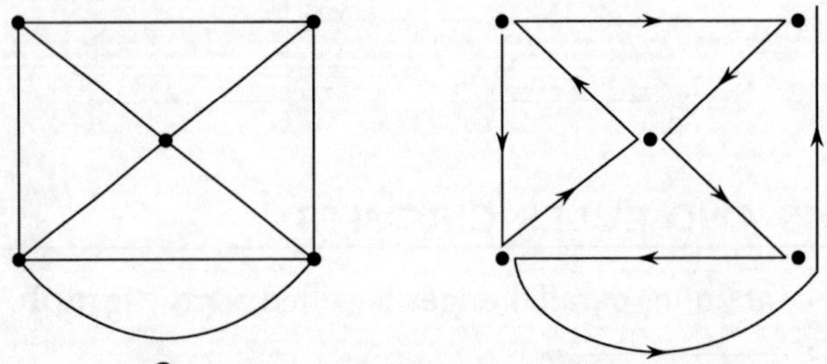

G_1 Euler Path in G_1

Example 4. Give an example of an Eulerian graph.

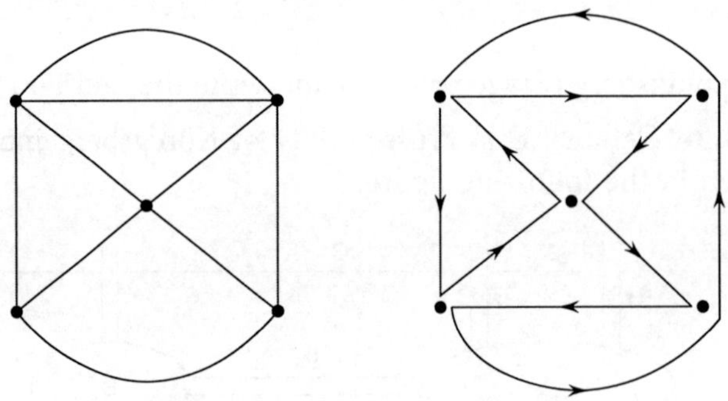

G_2 Euler circuit in G_2

Note. Both G_1 and G_2 of above are traversable graphs but G_1 is not an Eulerian graph.

Example 5. Give an example which is not traversable

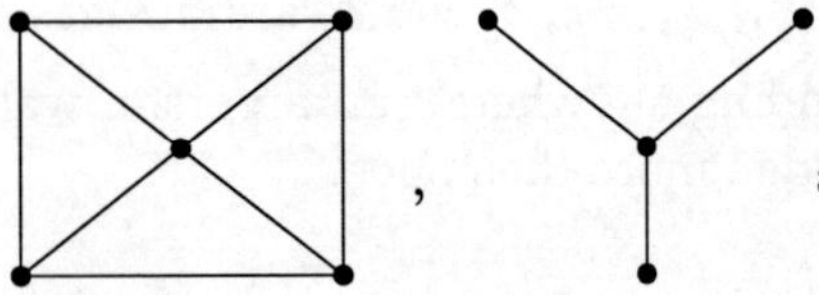

are not traversable.

Theorem : *A graph with more than two odd degree vertices is not traversable.*

Proof : Suppose G is traversable and let π be an Euler path in G. Let *v* be any internal vertex of the path π. Whenever the traversable trail enters *v* by an edge, there must always be an edge not previously used by which the trail can leave *v*. Hence *v* has an even degree. Therefore, the vertices with odd degree are at most the end two points. Thus the proof.

Theorem : *A connected graph G is an Eulerian if and only if each vertex of G has even degree.*

Proof (Necessary) : Suppose G is an Eulerian and π is an Euler circuit in G. Let *v* be any vertex of G, the trail enters and leaves '*v*' the same number of times without repeating any edge. Hence the degree of *v* is even.

(**Sufficient**) Suppose each vertex of G has even degree. We construct an Euler circuit in G. We begin a trail π_1 at any vertex 'u' of G. We pass through the edges one after another. Let π_1 ends at $v \neq u$. Then the number of edges incident on 'v' in π_1 must be odd. Since the degree of 'v' is even in G, there must be another edge incident on 'v' which is not in π_1. Thus, we can continue to extend π_1 until $v = u$ *i.e.,* until $\pi_1$is closed. If π_1 includes all the edges of G, then π_1 is an Euler circuit.

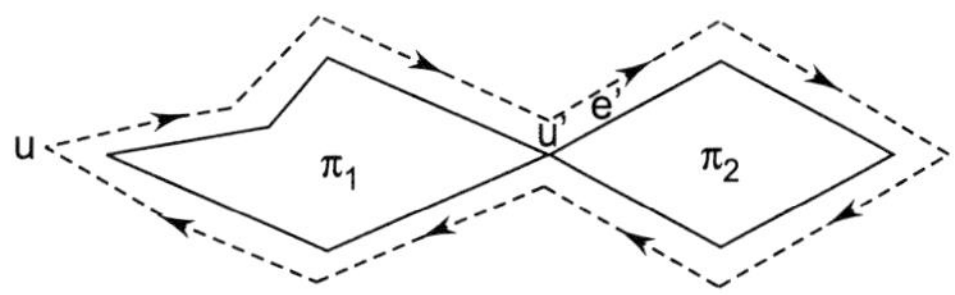

Suppose π_1 does not include all edges of G. Let H be a subgraph of G by deleting all edges of π_1 from G. H may not be connected but each vertex of H has even degree, since π_1 contains an even number of edges incident on any vertex. Since G is connected there is an edge e' of H which has an end point u' in π_1. We construct a trail π_2 in H beginning at u' and using e'. Since all vertices in H have even degree, we can continue to extend π_2 in H until π_2 returns to u' as shown in above figure. We can clearly put π_1 and π_2 to form a larger closed trail in G. We continue this process until all the edges of G are used. We finally obtain an Euler circuit and so G is an Eulerian.

Corollary : A connected graph with exactly two odd degree vertices is traversable. A traversable trail may begin at either odd degree vertex and will end at the other odd degree vertex.

Proof : Let G′ = (V, E) be a connected graph with exactly two odd degree vertices, say 'u' and 'v'. Add an edge 'e' between u and v to the graph G to form the new graph G′ = (V, E′) where $E' = E \cup \{e\}$. Then all the vertices of G′ are even degree. By the above theorem, there is an Euler circuit π in G′. Hence $\pi - \{e\}$ is an Euler path in G with 'u' and 'v' are end points and so G is traversable.

Theorem : A connected graph G is traversable if and only if G has zero or exactly two vertices of odd degree.

Proof is obtained by the theorem 3.1 and the corollary of theroem 3.2.

Example 6. Which of the following graphs are traversable, Eulerian or neither.

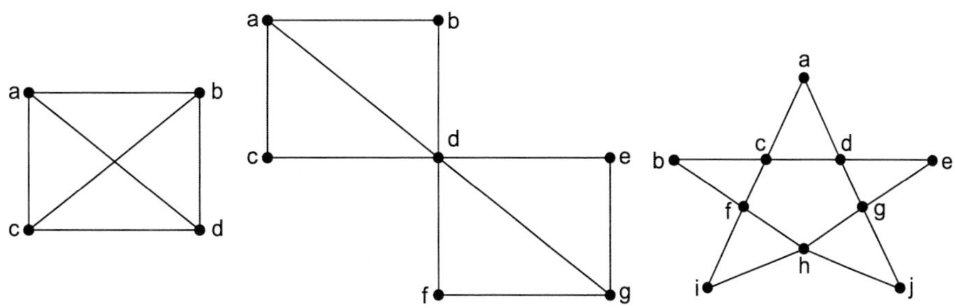

G_1 G_2 G_3

G_1 : Each vertex is of degree 3 *i.e.,* it has more than 2 vertices of odd degree. Hence it is neither traversable nor Eulerian.

G_2 : It has exactly 2 vertices 'a' and 'g' of odd degree hence it is traversable but not Eulerian.

A traversable trial in G_2 is *a-b-d-c-a-d-e-g-f-d-g.*

G_3 : Each vertex is of degree even. Hence it is both traversable and Eulerian.

An Euler circuit (or Euler path) in G_3 is *a-c-f-i-h-g-e-d-c-b-f-h-j-g-d-a.*

Example 7. Show that the problem of Konigsberg Bridges is unsolvable (Euler).

The multigraph of the Konigsberg bridges is as follows :

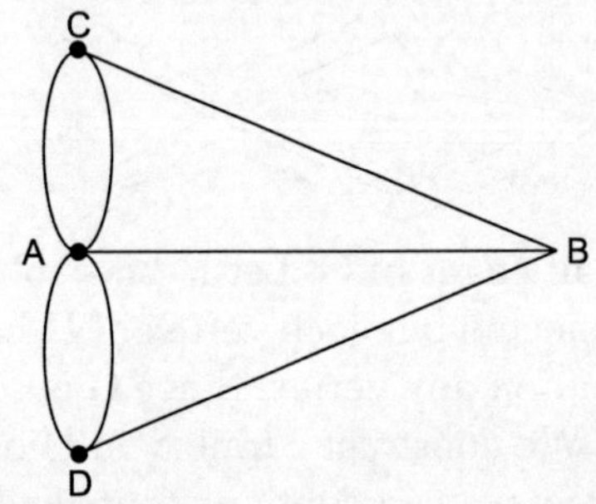

Here deg (A) = 5, deg (B) = 3, deg (C) = 3, deg (D) = 3.

It has more than two vertices of odd degree. Hence it is not traversable.

Example 8. Find which of the following multigraphs have Euler paths, circuits or neither.

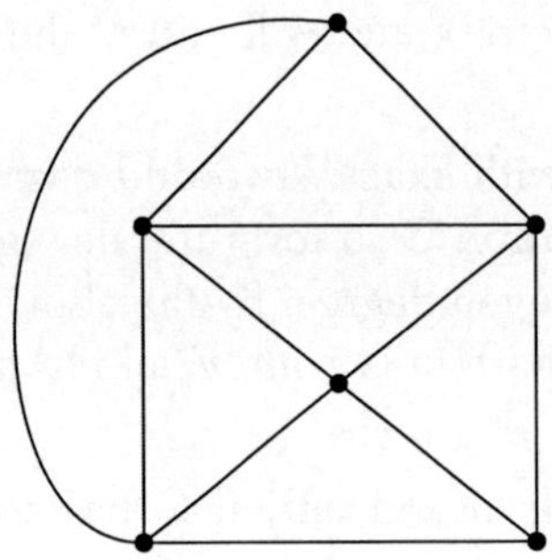

(i)

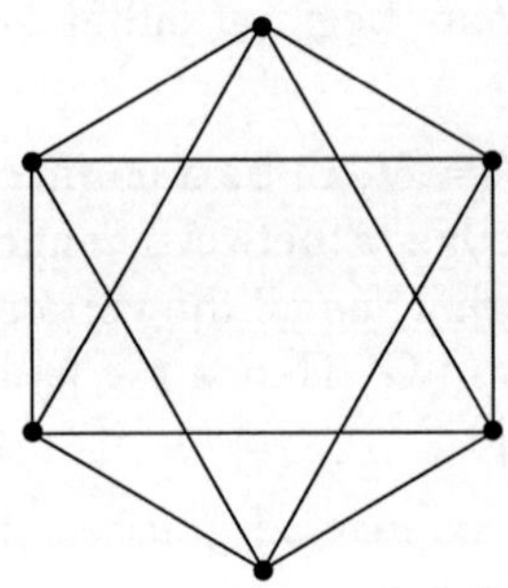

(ii)

Let us give vertex labelling.

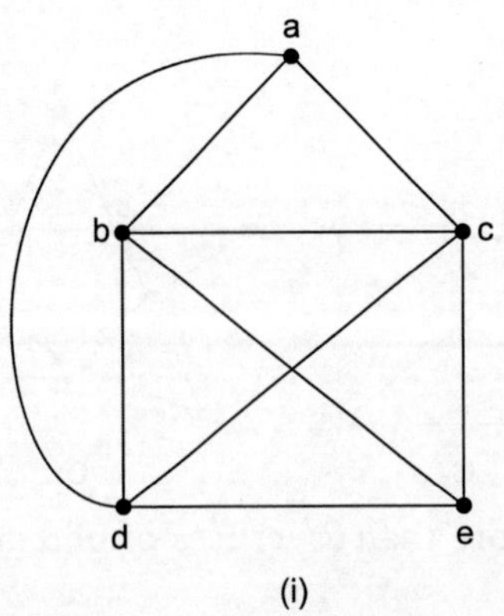

(i)

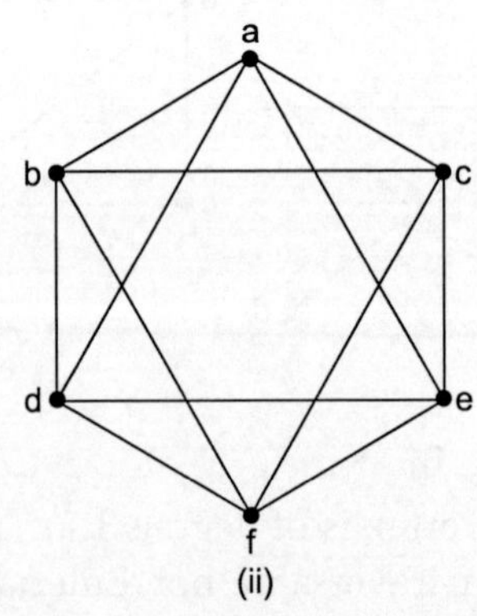

(ii)

(*i*)	Vertex	*a*	*b*	*c*	*d*	*e*
	Degree	3	4	4	4	3

It has two vertices '*a*' and '*e*' of odd degree. Hence it is traversable but not Eulerian. *i.e.,* we can find Euler path (between '*a*' and '*e*') but there is no Euler circuit.

An Euler path in (*i*) is

a-b-c-a-d-b-e-c-d-e.

(*ii*)	Vertex	*a*	*b*	*c*	*d*	*e*	*f*
	Degree	4	4	4	4	4	4

Here each vertex is of even degree.

Hence it is both traversable and Eulerian.

An Euler path or an Euler circuit in (*ii*) is *a-b-d-f-e-c-b-f-c-a-d-e-a.*

8.4 HAMILTONIAN GRAPHS

Definition : *A path in a graph G is called a* ***Hamiltonian path*** *if it contains every vertex of G.*

Definition : *A cycle in a graph G is called a* ***Hamiltonian cycle*** *if it contains every vertex of G.*

Definition : *A graph G is said to be a* ***Hamiltonian graph*** *if it contains a Hamiltonian cycle.*

Note : An Eulerian cycle uses every edge exactly once but may repeat vertices, while a Hamiltonian cycle uses each vertex exactly once (except for the first and last) but may skip edges.

Example 1. Determine which of the following are Hamiltonian graphs :

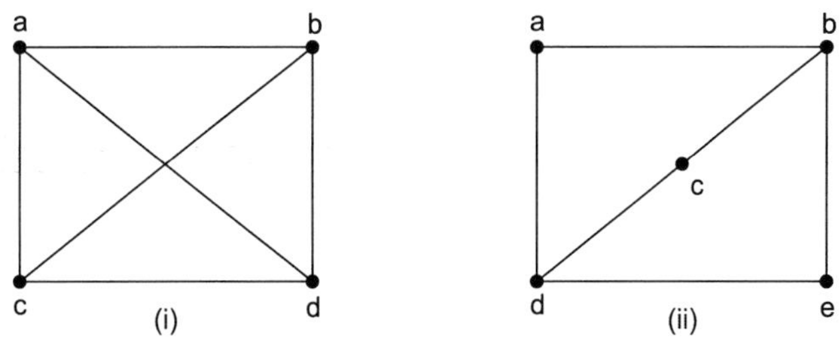

(*i*) is Hamiltonian graph since it has an Hamiltonian cycle *a-b-c-d-a.*

(*ii*) is not Hamiltonian graph since it has no Hamiltonian cycle, but it has a Hamiltonian path *a-d-e-b-c.*

Rules for Constructing Hamiltonian Paths and Cycles in a graph G :

Rule 1. If G is a graph with '*n*' vertices then a Hamiltonian cycle must contain exactly '*n*' edges, and a Hamiltonian path must contain exactly '*n* – 1' edges.

Rule 2. If '*v*' is a vertex of degree 2 then both edges incident on '*v*' should contain in every Hamiltonian cycle and every Hamiltonian path should contain at least one edge incident on *v*.

In general if degree of '*v*' is '*k*' then a Hamiltonian cycle contain exactly two edges incident on *v* and a Hamiltonian path must contain at least one edge incident on '*v*' and at most two edges incident on '*v*'.

Rule 3. No cycle that does not contain all the vertices of G can be formed when building a Hamiltonian path or cycle.

Rule 4. Once a Hamiltonian cycle we are constructing has passed through a vertex '*v*', then all the unused edges incident on *v* can be deleted.

Example 2. Find the Hamiltonian cycle in the following graph.

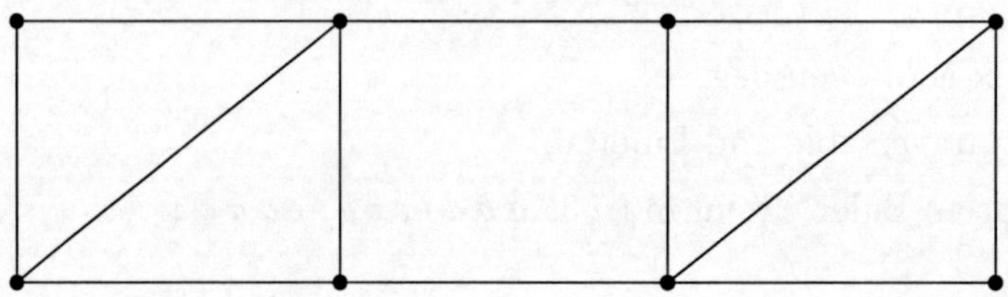

The Hamiltonian cycle in the graph is

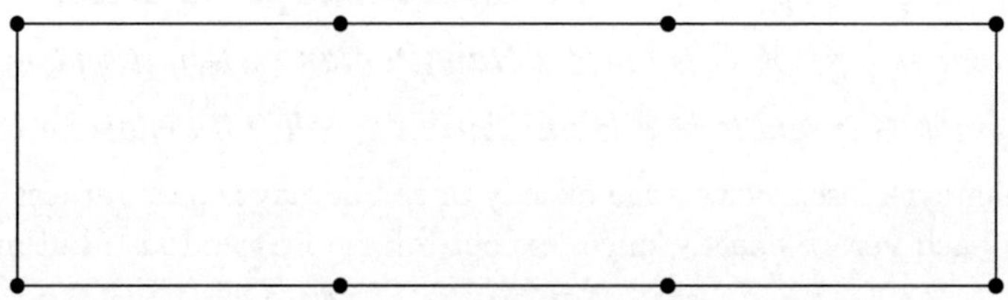

Example 3. Show that the following graph is not Hamiltonian.

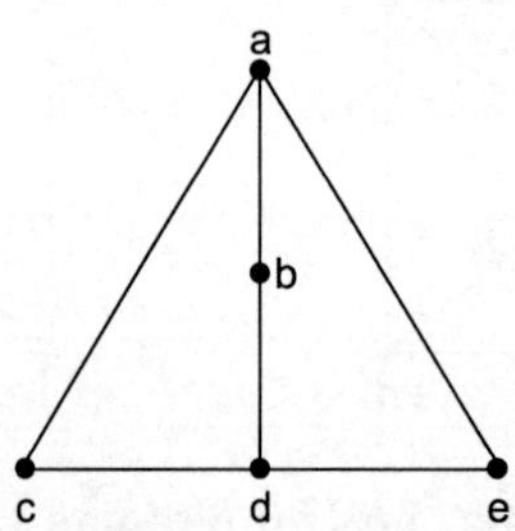

Here deg (b) = 2, hence {b, a}, {b, d} should be included in every Hamiltonian cycle. Also deg (c) = 2, therefore {c, a} and {c, d} should be in every Hamiltonian cycle. Thus *a-b-d-c-a* forms a cycle without containing the vertex '*e*'. Hence it is not an Hamiltonian graph.

Example 4. Show that the following graph has no Hamiltonian cycle. But the graph has a Hamiltonian path.

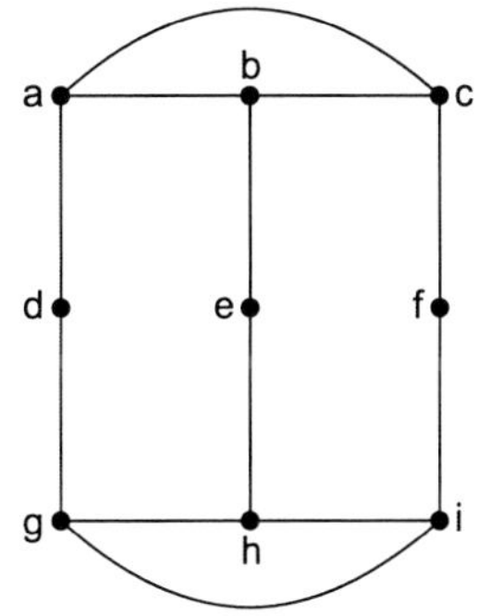

G

Since the degree of each of the vertices *d*, *e* and *f* is 2, any Hamiltonian cycle must include the 6 edges {*d*, *a*}, {*d*, *g*}, {*e*, *b*}, {*e*, *h*}, {*f*, *c*} and {*f*, *i*}. Now consider the vertex *b*. We can include only one edge either {*b*, *a*} or {*b*, *c*} in the Hamiltonian cycle, since {*b*, *e*} is already included. Let us include the edge {*b*, *a*} in Hamiltonian cycle and delete the edge {*b*, *c*}. Now the degree of *c* reduces to 2, so then a Hamiltonian cycle must include the edge {*c*, *a*} also. But then there would be 3 edges incident at a. Hence no Hamiltonian cycle exists in G. But a Hamiltonian path *a-d-g-h-e-b-c-f-i* exists in G.

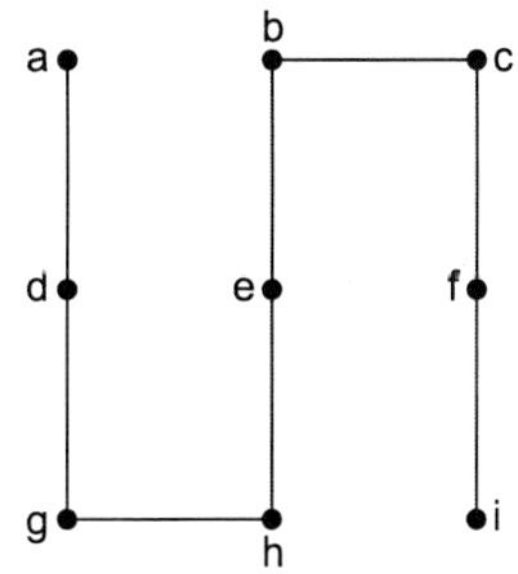

Example 5. Give an example of a graph which is Hamiltonian but not Eulerian.

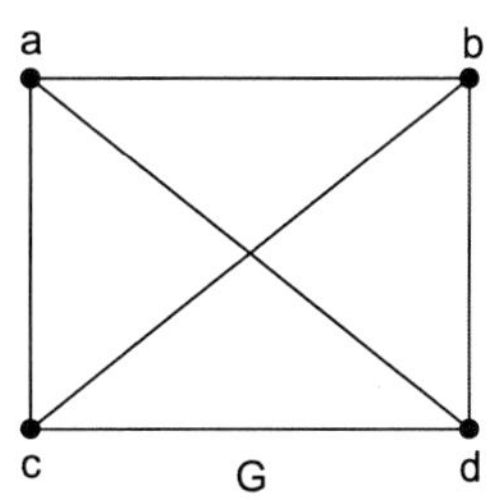

G is Hamiltonian, since it has an Hamiltonian cycle *a-b-d-c-a*.

But G is not Eulerian, since it has more than 2 vertices of odd degree (each vertex is of degree 3).

Example 6. Give an example of a graph which is Eulerian but not Hamiltonian.

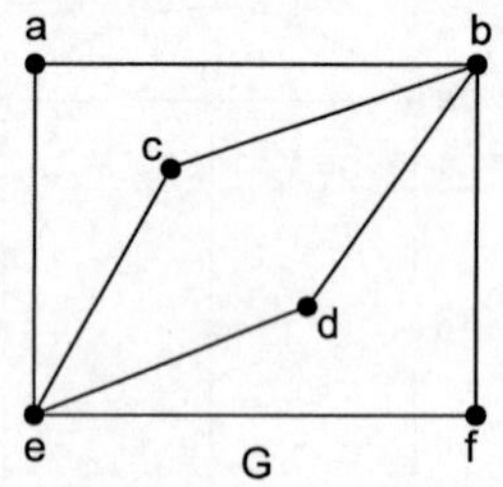

G is Eulerian, since each vertex is of even degree. An Eulerian cycle in G is *a-b-c-e-d-b-f-e-a.*

But G is not Hamiltonian. Since we have 6 vertices in which 4 are two degree vertices. Hence the 8 edges must be included in every Hamiltonian cycle but a Hamiltonian cycle of G should contain only 6 edges.

Grinberg's Theorem :

Statement : Let G be a simple plane graph with n vertices. suppose C is a Hamiltonian cycle in G. Then with respect to this cycle C,

$$\sum_{i=3}^{n} (i-2)(r_i - r_i') = 0.$$

where r_i is the number of regions of degree i in the interior of C and r_i' is the number of regions of degree i in the exterior of C.

Proof : G is a simple plane graph, therefore none of the edges of G intersect. We first consider the interior of C. A diagonal splits the region through which it passes into two parts. Each time a diagonal is inserted in the interior of C, the number of regions in C is increased by 1. Suppose that exactly '*d*' diagonals occur in the interior of C. Hence,

$$\sum_{i=3}^{n} r_i = d + 1 \;\Rightarrow\; d = \sum_{i=3}^{n} r_i - 1 \qquad ...(1)$$

If N denotes the sum of the degrees of all regions interior to C, then

$$N = \sum_{i=3}^{n} i\, r_i$$

N counts each diagonal twice and each edge of C exactly once, therefore,

$$N = \sum_{i=3}^{n} i\, r_i = 2d + n \qquad ...(2)$$

Substituting for *d*, we have

$$\sum_{i=3}^{n} i\, r_i = 2\left(\sum_{i=3}^{n} r_i - 1\right) + n$$

$$= 2\sum_{i=3}^{n} r_i - 2 + n$$

$$\therefore \qquad \sum_{i=3}^{n}(i-2)\, r_i = n-2 \qquad \text{...(3)}$$

Similarly, by considering the exterior of C, we get

$$\sum_{i=3}^{n}(i-2)\, r_i' = n-2 \qquad \text{...(4)}$$

∴ From (3) and (4), we get

$$\sum_{i=3}^{n}(i-2)\, r_i = \sum_{i=3}^{n}(i-2)\, r_i'$$

$$\therefore \qquad \sum_{i=3}^{n}(i-2)\,(r_i - r_i') = 0$$

Example 7. Show that the following graph has no Hamiltonian cycle.

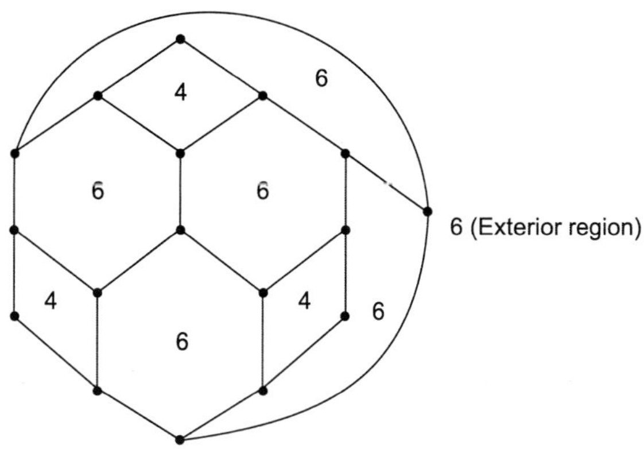

Given one is a plane graph with 3 regions of degree 4 and 6 regions of degree 6.

$$\therefore \qquad r_4 + r_4' = 3 \quad \text{and} \quad r_6 + r_6' = 6 \qquad \text{...(1)}$$

Applying Grinberg's theorem, we get

$$(4-2)\,(r_4 - r_4') + (6-2)\,(r_6 - r_6') = 0$$

$$\Rightarrow \qquad 2\,(r_4 - r_4') + 4\,(r_6 - r_6') = 0$$

$$\Rightarrow \qquad r_4 - r_4' + 2\,(r_6 - r_6') = 0$$

$$\Rightarrow \qquad r_4 - (3 - r_4) + 2r_6 - 2\,(6 - r_6) = 0 \qquad \because \text{ from (1)}$$

$$\Rightarrow \qquad 2\,(r_4 + 2r_6) = 15$$

which is a contradiction since L.H.S. is even where as R.H.S. is odd.

Therefore, there cannot be a Hamiltonian cycle in the given graph.

Example 8. Show that there are no planar Hamiltonian graphs with regions of degree 4, 5 and 8 and only one region of degree 4.

From Grinberg's theorem, we have

$$(4-2)(r_4 - r_4') + (5-2)(r_5 - r_5') + (8-2)(r_8 - r_8') = 0$$

$$\Rightarrow \quad 2(r_4 - r_4') + 3(r_5 - r_5') + 6(r_8 - r_8') = 0$$

There is exactly one region of degree 4, therefore we get

$$2(\pm 1) + 3(r_5 - r_5') + 6(r_8 - r_8') = 0$$

Which is of the form

$$3x + 6y = \pm 2$$

$$\Rightarrow \quad x = \frac{-6y \pm 2}{3}$$

$$\Rightarrow \quad x = -2y \pm \frac{2}{3}$$

whenever y is an integer x is not an integer which is a contradiction.

Hence there cannot be Hamiltonian plane graph with the given regions.

8.5 CHROMATIC NUMBERS

Definition : *A **vertex colouring** of a graph G is an assignment of colours to the vertices of G, one colour to each vertex, so that adjacent vertices are assigned different colours.*

An *n*-colouring of G is a colouring of G using *n* colours. If G has an *n*-colouring, then G is said to be *n*-colourable.

5.1. Example 1.

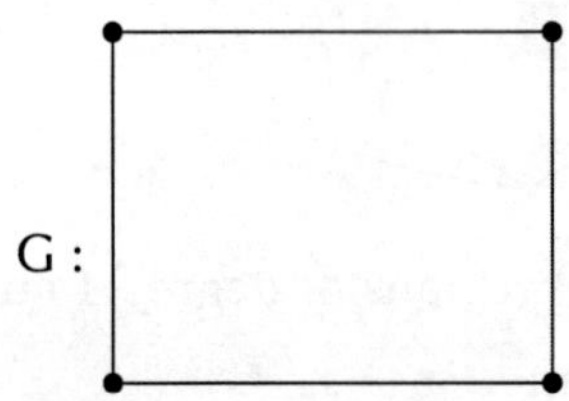

Each vertex of G can be assigned different colour.

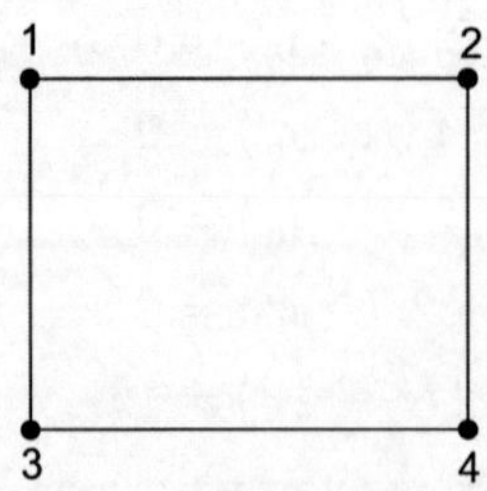

4-colourable

We can use colours 1, 2 and 3 for G.

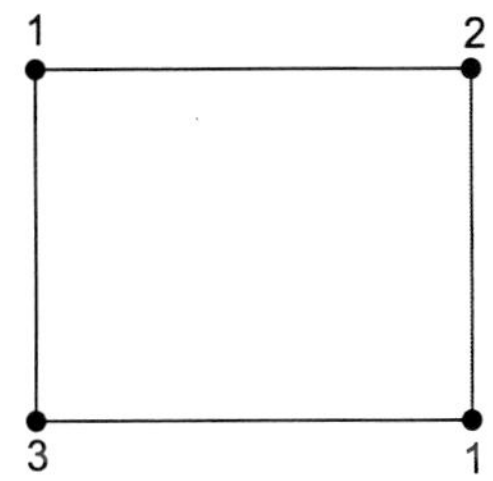

3-colourable

We can also use only two colours 1 and 2 for G.

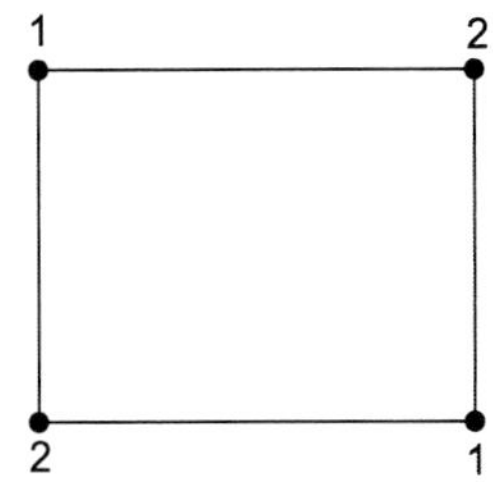

2-colourable.

But G is not 1-colourable, since one colour is not sufficient to assign colours to G so that adjacent colours are different.

Definition : *The **chromatic number** of a graph G, is the minimum number of colours required to colour the vertices of G, such that no two adjacent vertices receive the same colour. We denote the chromatic number of G by χ (G), and if χ (G) = k we say that G is k-**chromatic**.*

Example 2. The chromatic number of isolated vertex is one.

Example 3. The chromatic number of a path graph P_n (for $n \geq 2$) is 2.

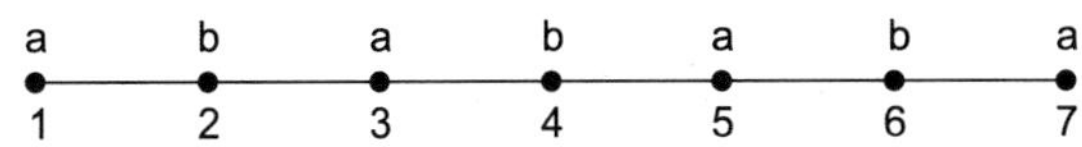

$\chi(P_7) = 2.$

Example 4. The chromatic number of a cycle graph C_n (for $n \geq 3$) is 2 when n is even and is 3 while n is odd

i.e.,
$$\chi(C_n) = \begin{cases} 2 & \text{if} \quad n \text{ is even} \\ 3 & \text{if} \quad n \text{ is odd} \end{cases}$$

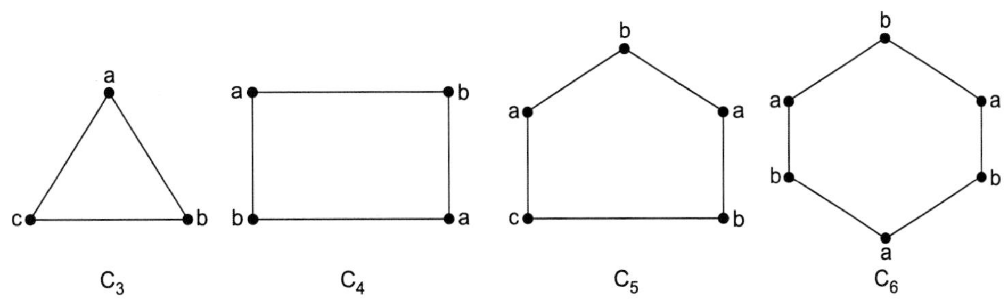

$$\chi(C_3) = 3, \chi(C_4) = 2, \chi(C_5) = 3, \chi(C_6) = 2.$$

Example 5. The chromatic number of a complete bipartite graph $k_{m,\,n}$ is 2.

(Every bipartite graph is 2-colourable)

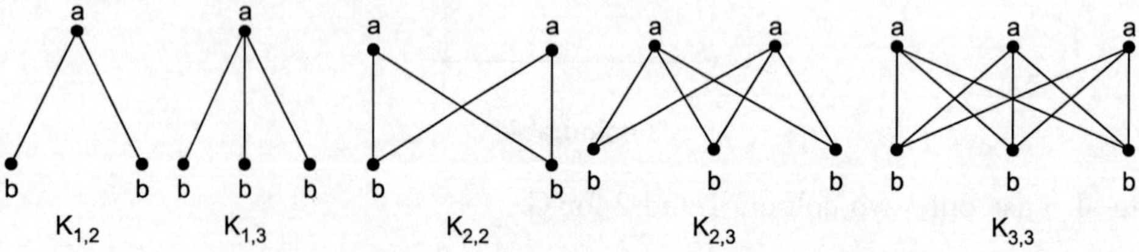

$$\chi(k_{m,\,n}) = 2.$$

Example 6. The chromatic number of a complete graph on '*n*' vertices k_n is *n*.

k_n has '*n*' vertices and each vertex is adjacent to all the other vertices. Hence each vertex should be assigned with different colours. Thus, $\chi(k_n) = n$.

Some Basic Rules

Rule 1. $\chi(G) \le |V|$, where $|V|$ is the number of vertices in G.

Rule 2. If H is a subgraph of G, then $\chi(H) \le \chi(G)$.

Rule 3. If deg $(v) = d$, then at most d colours are required to colour the vertices adjacent to v.

Rule 4. $\chi(G) = \text{Maximum}\{\chi(C) / C \text{ is a connected component of } G\}$.

Rule 5. $\chi(G) \le 1 + \Delta(G)$ for any graph G, where $\Delta(G) = \text{Maximum}\{\deg(v)/v \in V(G)\}$

Welch-Powell algorithm to colour a graph G

First order the vertices according to decreasing degrees. Then use the first colour to the first vertex and to colour, in sequential order, each vertex which is not adjacent to a previously coloured vertex. Repeat the process using the second colour and the remaining unpainted vertices. Continue repeating with additional colours until all the vertices have been painted. This algorithm gives an upper bound to $\chi(G)$.

Example 7. Find the chromatic number of the following wheel graph G.

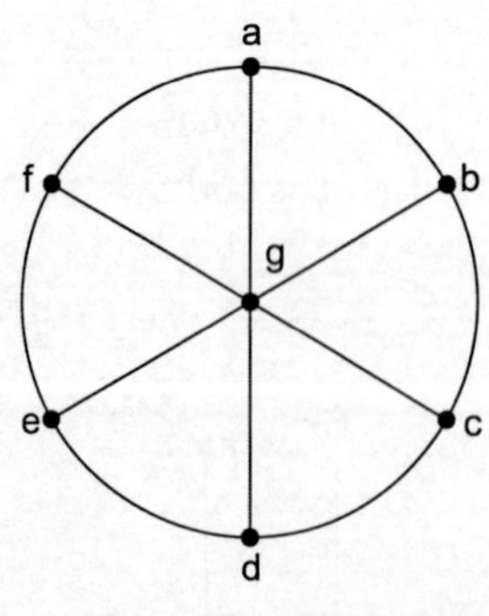

Degree	6	3	3	3	3	3	3
Vertex	*g*	*a*	*b*	*c*	*d*	*e*	*f*
Colour	1	2	3	2	3	2	3

$\therefore \qquad \chi(G) \leq 3$

The subgraph with vertices a, b, g is k_3.

$\therefore \qquad \chi(G) \geq 3$

Thus $\qquad \chi(G) = 3.$

Example 8. Find the chromatic number of the following graph G.

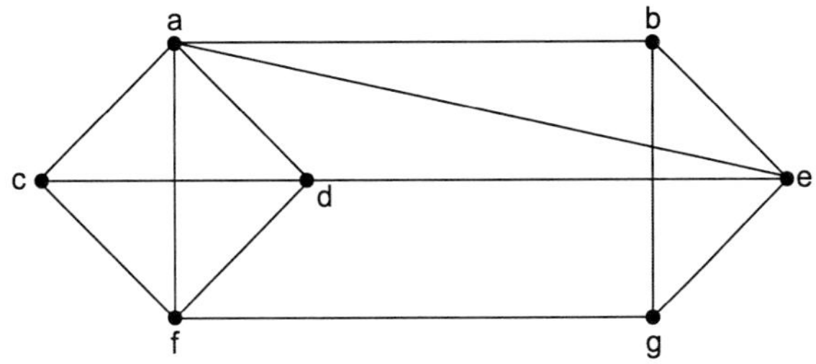

Degree	5	4	4	4	3	3	3
Vertex	*a*	*d*	*e*	*f*	*b*	*c*	*g*
Colour	1	2	3	3	2	4	1

$\therefore \qquad \chi(G) \leq 4$

The subgraph with vertices a, c, d, f is k_4

$\therefore \qquad \chi(G) \geq 4$

Thus $\qquad \chi(G) = 4.$

Example 9. Find the chromatic number of the following :

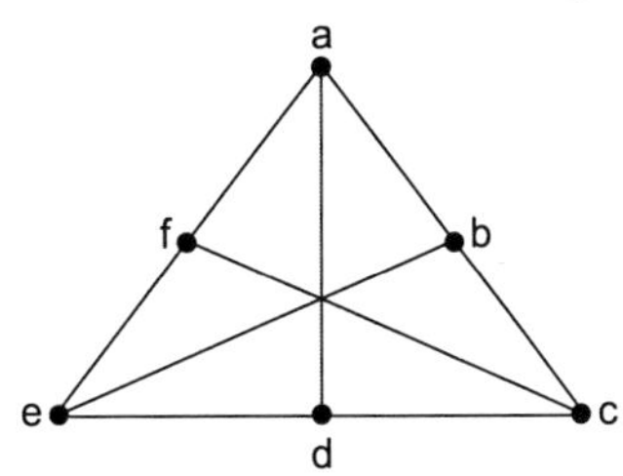

Degree	3	3	3	3	3	3
Vertex	*a*	*b*	*c*	*d*	*e*	*f*
Colour	1	2	1	2	1	2

$\therefore \qquad \chi(G) \leq 2$

The $\{a, b\}$ requires two colours

$\therefore \qquad \chi(G) = 2.$

EXERCISE 8.1

1. Define the following graphs with one suitable examples for each graphs :
   (*i*) Complement graph

(*ii*) Sub graph
(*iii*) Induced sub graph
(*iv*) Spanning sub graph

2. Determine the number of edges in
   (*i*) Complete graph K_n
   (*ii*) Complete bipartite graph $K_{m, n}$
   (*iii*) Cycle graph C_n and
   (*iv*) Path graph P_n.
3. State and prove the first theorem of graph theory.
4. Prove that in any non-directed graph there is an even number of vertices of odd degree.
5. How many vertices will the following graphs have if they contain
   (*i*) 16 edges and all vertices of degree 2.
   (*ii*) 21 edges, 3 vertices of degree 4, and the other vertices of degree 3.
   (*iii*) 24 edges and all vertices of the same degree.
6. Is the following degree sequences are graphic.
   (*i*) (0, 1, 2, 3, 4, 4)
   (*ii*) (1, 2, 3, 4, 4, 5, 6, 7).
7. Distinguish between cycle and circuit. Give suitable examples for each.
8. Define isomorphism. What are the steps followed in discovering the isomorphism.
9. Prove or disprove that the following two graphs are isomorphic.

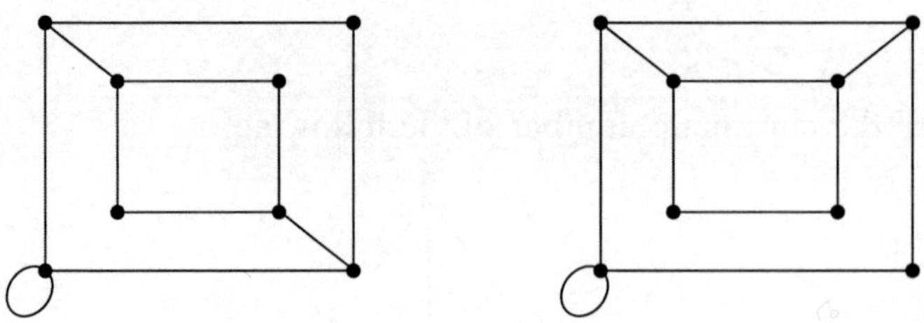

10. Determine whether the following two graphs are isomorphic or not. If isomorphic, give the one to one mapping.

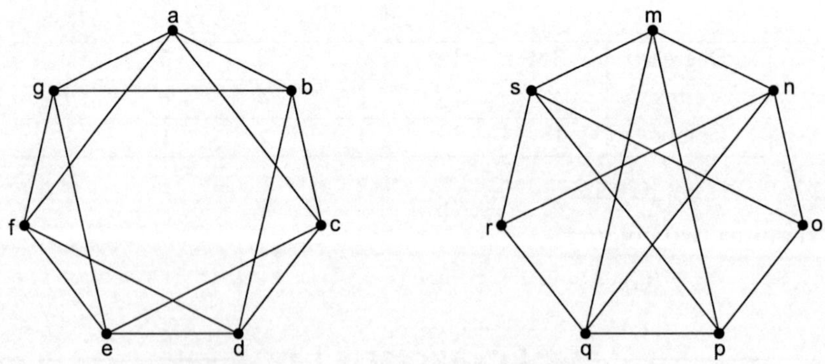

11. Show that the following two graphs are isomorphic.

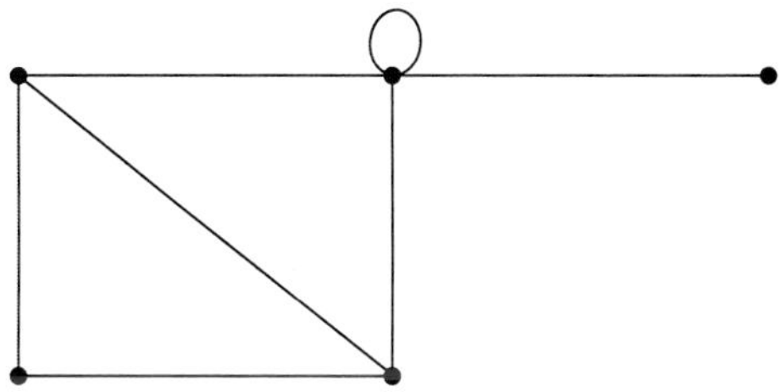

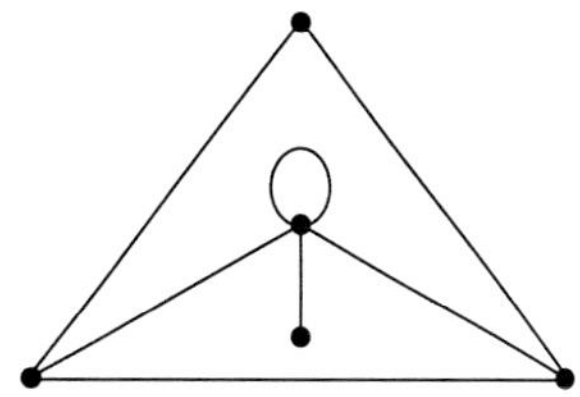

12. Define the following terms. Give one suitable example for each.

(*i*) Euler path

(*ii*) Euler circuit

(*iii*) Multi graph

(*iv*) Hamiltonian graph

13. Find which of the following multigraphs have Euler paths, circuits or neither.

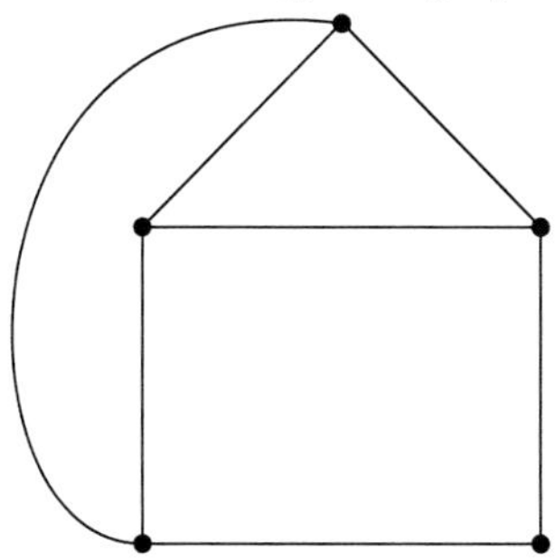

G_1

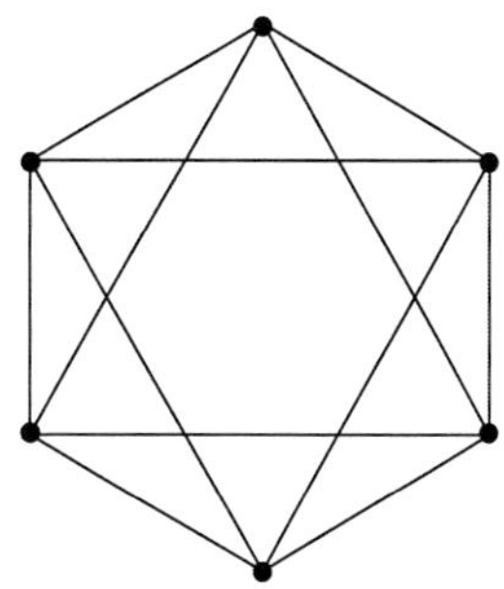

G_2

14. Prove that a non-directed multigraph has an Euler circuit iff it is connected and all of its vertices are of even degree.

15. Find a Hamiltonian cycle in each of the following graphs :

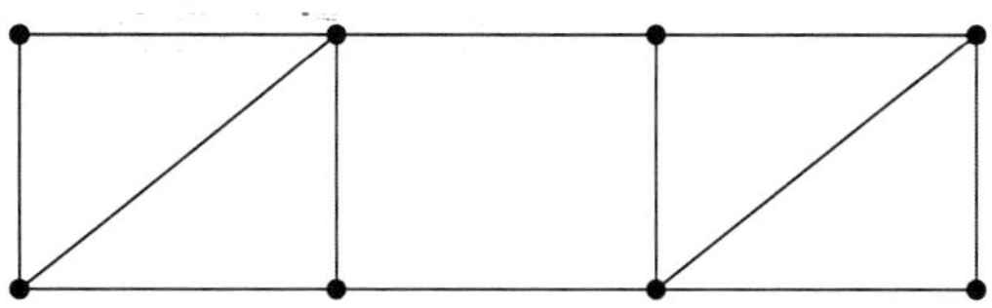

(i)

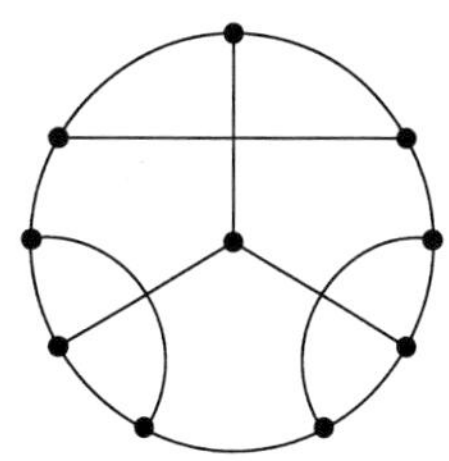

(ii)

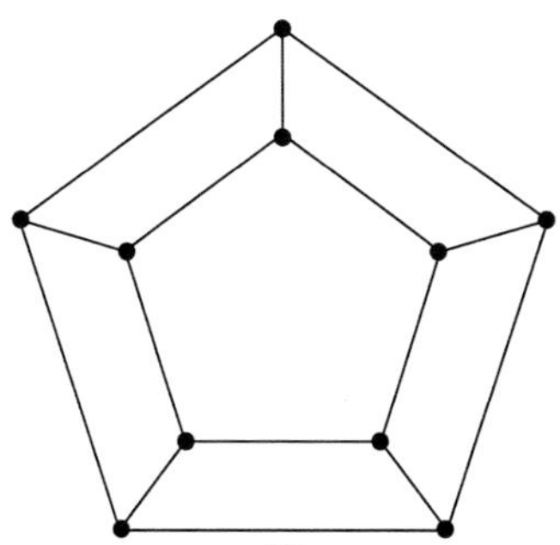

(iii)

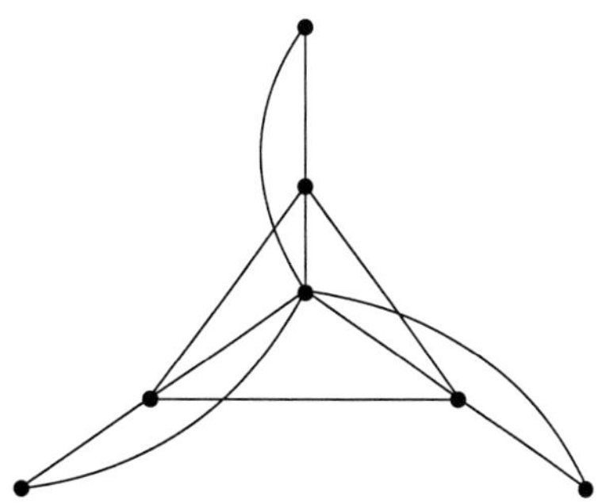

(iv)

16. State and prove Grinberg's theorem to determine whether Hamiltonian cycle exists in a given graph or not.

17. Use Grinberg's theorem to show that there are no planar Hamiltonian graphs with
    (*i*) regions of degree 5, 8 and 9 with exactly one region of degree 9.
    (*ii*) regions of degree 5, 8, 9 and 11 with exactly one region with degree 9.
18. Prove that there is no Hamiltonian cycle in each of the following graphs.

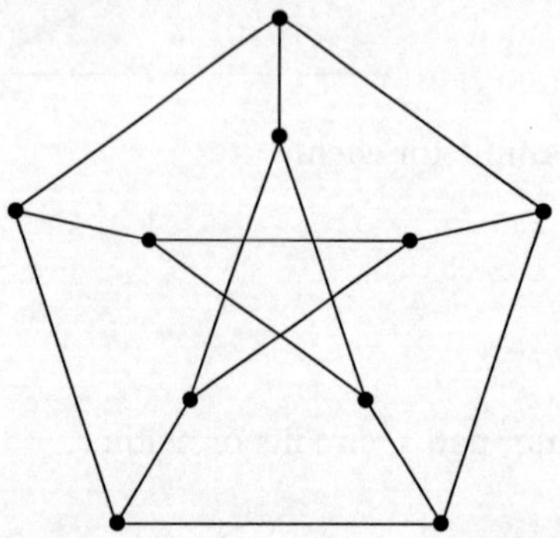

G_1

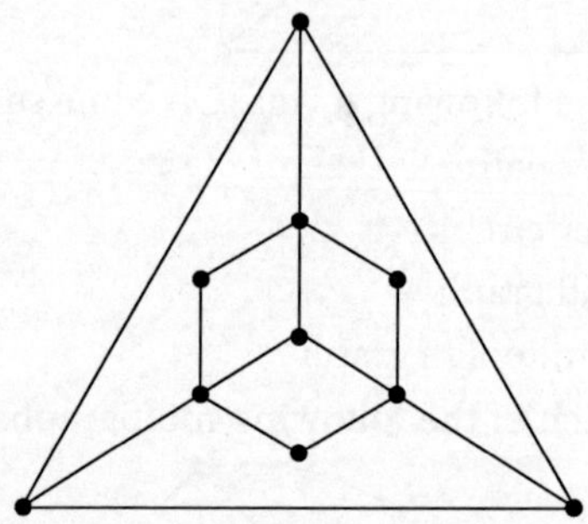

G_2

19. Define chromatic number of a graph. Find the chromatic number of each of the graphs shown in example 18.
20. Find the chromatic number of (*i*) $k_{3,3}$ and (*ii*) a cycle with even number of vertices.

ANSWERS

2. (*i*) $\dfrac{n(n-1)}{2}$ (*ii*) mn
   (*iii*) n (*iv*) $n - 1$
5. (*i*) 16 (*ii*) 13 (*iii*) factors of 48.
6. (*i*) No (*ii*) Yes
9. not isomorphic
10. isomorphic
13. G_1 has neither Euler path nor Euler circuit. G_2 has both Euler path and Euler circuit.
19. $\chi(G_1) = 3$, $\chi(G_2) = 3$
20. (*i*) 2 (*ii*) 2.

INDEX